# ELEMENTS OF VIBRATION ANALYSIS

# ELEMENTS OF VIBRATION ANALYSIS

## Second Edition

**Leonard Meirovitch**

*College of Engineering*
*Virginia Polytechnic Institute and State University*

**McGraw-Hill Book Company**

New York   St. Louis   San Francisco   Auckland   Bogotá   Hamburg
Johannesburg   London   Madrid   Mexico   Montreal   New Delhi
Panama   Paris   São Paulo   Singapore   Sydney   Tokyo   Toronto

To My Wife and to the
Memory of My Parents

This book was set in Times Roman by Eta Services Ltd.
The editor was Anne Murphy;
the cover was designed by Karl H. Steinbrenner;
the production supervisor was Charles Hess.
Project supervision was done by Albert Harrison, Harley Editorial Services.
R. R. Donnelley & Sons Company was printer and binder.

**ELEMENTS OF VIBRATION ANALYSIS**

234567890 DOCDOC 8987

ISBN 0-07-041342-8

**Library of Congress Cataloging in Publication Data**
Meirovitch, Leonard.
  Elements of vibration analysis.

  Bibliography: p.
  Includes index.
  1. Vibration.   I. Title.
QA935.M53   1986      531.32      85-14904
ISBN 0-07-041342-8

# PREFACE

In the last several decades, impressive progress has been made in vibration analysis, prompted by advances in technology. On the one hand, the requirement for the analysis of increasingly complex systems has been instrumental in the development of powerful computational techniques. On the other hand, the development of fast digital computers has provided the means for the numerical implementation of these techniques. Indeed, one of the most significant advances in recent years is the finite element method, a method developed originally for the analysis of complex structures. The method has proved to be much more versatile than conceived originally, finding applications in other areas, such as fluid mechnics and heat transfer. At the same time, significant progress was being made in linear system theory, permitting efficient derivation of the response of large-order systems. *Elements of Vibration Analysis* was written in recognition of these advances.

The second edition of *Elements of Vibration Analysis* differs from the first edition in several respects. In the first place, the appeal of the first few chapters has been broadened by the inclusion of more applied topics, as well as additional explanations, examples, and homework problems. Advanced material has been transferred to later chapters. The chapter on the finite element method, Chap. 8, has been rewritten almost entirely so as to reflect the more current thinking on the subject, as well as to include more recent developments. The section on the Routh-Hurwitz criterion in Chap. 9 has been expanded. On the other hand, some advanced material in Chap. 10 has been deleted. The chapter on random vibrations, Chap. 11, has been enlarged by absorbing material on Fourier transforms from Chap. 2 and by expanding the discussion of narrowband processes. Chapter 12 represents an entirely new chapter, devoted to techniques for the computation of the response on digital computers. The chapter includes material indispensable in a modern course in vibrations. Finally, some of the material in App. C has been rewritten by placing the emphasis on physical implications. As a result of these revisions, the first part of the second edition is more accessible to juniors and should have broader appeal than the first edition. Moreover, the material in later chapters makes this second edition an up-to-date book on vibration analysis.

The book contains material for several courses on vibrations. The material covers a broad spectrum of subjects, from the very elementary to the more advanced, and is arranged in increasing order of difficulty. The first five chapters of the book are suitable for a beginning course on vibrations, offered at the junior or senior level. The material in Chaps. 6–12 can be used selectively for courses on dynamics of structures, nonlinear oscillations, random vibrations, and advanced vibrations, either at the senior, or first-year graduate level. To help the instructor in tailoring the material to his or her needs, the book is reviewed briefly:

Chapter 1 is devoted to the free vibration of single-degree-of-freedom linear systems. This is standard material for a beginning course on vibrations.

Chapter 2 discusses the response of single-degree-of-freedom linear systems to external excitation in the form of harmonic, periodic, and nonperiodic forcing functions. The response is obtained by the classical and Laplace transformation methods. A large number of applications is presented. If the response by Laplace transformation is not to be included in a first course on vibrations, then Secs. 2.17 and 2.18 can be omitted.

Chapter 3 is concerned with the vibration of two-degree-of-freedom systems. The material is presented in a way that makes the transition to multi-degree-of-freedom systems relatively easy. The subjects of beat phenomenon and vibration absorbers are discussed. The material is standard for a first course on vibrations.

Chapter 4 presents a matrix approach to the vibration of multi-degree-of-freedom systems, placing heavy emphasis on modal analysis. The methods for obtaining the system response are ideally suited for automatic computation. The material is suitable for a junior level course. Sections 4.11 through 4.13 can be omitted on a first reading.

Chapter 5 is devoted to exact solutions to response problems associated with continuous systems, such as strings, rods, shafts, and bars. Again the emphasis is on modal analysis. The intimate connection between discrete and continuous mathematical models receives special attention. The material is suitable for juniors and seniors.

Chapter 6 provides an introduction to analytical dynamics. Its main purpose is to present Lagrange's equations of motion. The material is a prerequisite for later chapters, where efficient ways of deriving the equations of motion are necessary. The chapter is suitable for a senior-level course.

Chapter 7 discusses approximate methods for treating the vibration of continua for which exact solutions are not feasible. Discretization methods based on series solutions, such as the Rayleigh-Ritz method, and lumped methods are presented. The material is suitable for seniors.

Chapter 8 is concerned with the finite element method. The earlier material is presented in a manner that can be easily understood by seniors. Later material is more suitable for beginning graduate students.

Chapter 9 is the first of two chapters on nonlinear systems. It is devoted to such qualitative questions as stability of equilibrium. The emphasis is on geometric description of the motion by means of phase plane techniques. The material is suitable for seniors or first-year graduate students, but Secs. 9.6 and 9.7 can be omitted on a first reading.

Chapter 10 uses perturbation techniques to obtain quantitive solutions to response problems of nonlinear systems. Several methods are presented, and phenomena typical of nonlinear systems are discussed. The material can be taught in a senior or a first-year graduate course.

Chapter 11 is devoted to random vibrations. Various statistical tools are introduced, with no prior knowledge of statistics assumed. The material in Secs. 11.1 through 11.12 can be included in a senior-level course. In fact, its only prerequisites are Chaps. 1 and 2, as it considers only the response of single-degree-of-freedom linear systems to random excitation. On the other hand, Secs. 11.13 through 11.18 consider multi-degree-of-freedom and continuous systems and are recommended only for more advanced students.

Chapter 12 is concerned with techniques for the determination of the response on a digital computer. Sections 12.2 through 12.5 discuss the response of linear systems in continuous time by the transition matrix and Sec. 12.6 presents discrete-time techniques. Section 12.7 is concerned with the response of nonlinear systems. All this material is intended for a senior, or first-year graduate course. Sections 12.8 through 12.12 are concerned with frequency-domain techniques and in particular with aspects of implementation on a digital computer. The material is suitable for a graduate course.

Appendix A presents basic concepts involved in Fourier series expansions, App. B is devoted to elements of Laplace transformation, and App. C presents certain concepts of linear algebra, with emphasis on matrix algebra. The appendixes can be used for acquiring an elementary working knowledge of the subjects, or for review if the material was studied previously.

It is expected that the material in Chaps. 1 through 3 and some of that in Chap. 4 will be used for a one-quarter, elementary course, whether at the junior or senior level. For a course lasting one semester, additional material from Chap. 4 and most of Chap. 5 can be included. A second-level course on vibrations has many options. Independent of these options, however, Chap. 6 must be regarded as a prerequisite for further study. The choice among the remaining chapters depends on the nature of the intended course. In particular, Chaps. 7 and 8 are suitable for a course whose main emphasis is on deterministic structural dynamics. Chapters 9 and 10 can form the core for a course on nonlinear oscillations. Chapter 11 can be used for a course on random vibrations. Finally, Chap. 12 is intended for an advanced, modern course on vibration analysis, with emphasis on numerical results obtained on a digital computer.

The author wishes to thank Jeffrey K. Bennighof, Virginia Polytechnic Institute and State University; Andrew J. Edmondson, University of Tennessee; Henryk Flashner, University of Southern California; Carl H. Gerhold, Texas A & M University; Charles M. Krousgrill, Purdue University; Donald L. Margolis, University of California, Davis; Kenneth G. McConnell, Iowa State University; David A. Peters, Georgia Institute of Technology; Roger D. Quinn, Virginia Polytechnic Institute and State University; Robert F. Steidel, University of California, Berkeley; Benson Tongue, Georgia Institute of Technology; and Wayne W. Walter, Rochester Institute of Technology; who made valuable suggestions. Thanks are due also to Mark A. Norris, Virginia Polytechnic Institute and State University, for producing some of the plots. Last but not least, the author wishes to thank Norma B. Guynn for her excellent job in typing the manuscript.

*Leonard Meirovitch*

# CONTENTS

# INTRODUCTION

The study of the relation between the motion of physical systems and the forces causing the motion is a subject that has fascinated the human mind since ancient times. For example, philosophers such as Aristotle tried in vain to find the relation; the correct laws of motion eluded him. It was not until Galileo and Newton that the laws of motion were formulated correctly, within certain limitations. These limitations are of no concern unless the velocities of the bodies under consideration approach the speed of light. The study relating the forces to the motion is generally referred to as *dynamics*, and the laws governing the motion are the well-known Newton's laws.

An important part of modern engineering is the analysis and prediction of the dynamic behavior of physical systems. An omnipresent type of dynamic behavior is *vibratory motion*, or simply *vibration*, in which the system oscillates about a certain equilibrium position. This text is concerned with the oscillation of various types of systems, and in particular with the vibration of mechanical systems.

Physical systems are in general very complex and difficult to analyze. More often than not they consist of a large number of components acting as a single entity. To analyze such systems, the various components must be identified and their physical properties determined. These properties, which govern the dynamic behavior of the system, are generally determined by experimental means. As soon as the characteristics of every individual component are known, the analyst is in a position to construct a mathematical model, which represents an idealization of the actual physical system. For the same physical system it is possible to construct a number of mathematical models. The most desirable is the simplest model that retains the essential features of the actual physical system.

The physical properties, or characteristics, of a system are referred to as *parameters*. Generally real systems are continuous and their parameters distributed. However, in many cases it is possible to simplify the analysis by replacing the distributed characteristics of the system by discrete ones. This is accomplished by a suitable "lumping" of the continuous system. Hence, mathematical models can be divided into two major types: (1) *discrete-parameter systems*, or *lumped systems*, and (2) *distributed-parameter systems*, or *continuous systems*.

The type of mathematical model considered is of fundamental importance in analysis because it dictates the mathematical formulation. Specifically, the behavior of discrete-parameter systems is described by ordinary differential equations, whereas that of distributed-parameter systems is generally governed by partial differential equations. For the most part, discrete systems are considerably simpler to analyze than distributed ones. In this text, we discuss both discrete and distributed systems.

Although there is an appreciable difference in the treatment of discrete and distributed systems, there is an intimate relation between the two types of mathematical models when the models represent the same general physical system. Hence, the difference is more apparent than real. Throughout this text, special emphasis is placed on the intimate relation between discrete and distributed models by pointing out common physical features and parallel mathematical concepts.

Vibrating systems can also be classified according to their behavior. Again the systems can be divided into two major types, namely, *linear* and *nonlinear*. The classification can be made by merely inspecting the system differential equations. Indeed, if the *dependent variables* appear to the first power only, and there are no cross products thereof, then the system is linear. On the other hand, if there are powers higher than one, or fractional powers, then the system is nonlinear. Note that systems containing terms in which the *independent variables* appear to powers higher than one, or to fractional powers, are merely *systems with variable coefficients* and not necessarily nonlinear systems.

Quite frequently the distinction between linear and nonlinear systems depends on the range of operation, and is not an inherent property of the system. For example, the restoring torque in a simple pendulum is proportional to sin $\theta$, where $\theta$ denotes the amplitude. For large amplitudes sin $\theta$ is a nonlinear function of $\theta$, but for small amplitudes sin $\theta$ can be approximated by $\theta$. Hence, the same pendulum can be classified as a linear system for small amplitudes and as a nonlinear system for large amplitudes. Nonlinear systems require different mathematical techniques than linear systems, as we shall have the opportunity to find out.

At times the approach to the response problem is dictated not by the system itself but by the excitation. Indeed, for the most part, the excitations are known functions of time. In such cases, the excitation is said to be *deterministic*, and the response is also deterministic. On the other hand, the excitation produced by an earthquake on a building is random in nature, in the sense that its value at any given instant of time cannot be predicted. Such excitation is said to be *nondeterministic*. Perhaps if all the factors contributing to the excitation were known, the excitation could be regarded as deterministic. However, the complexity involved in handling irregular functions renders the deterministic approach impractical, and the excitation and response must be expressed in terms of *statistical averages*. This text is concerned with the response of systems to both deterministic and nondeterministic excitations.

Finally, one must distinguish between *continuous-time systems* and *discrete-time systems*. In practice, most systems are continuous in time. However, if the solution for the response is to be obtained on a digital computer, then continuous-time systems must be treated as discrete in time. A similar situation exists for *frequency-domain* solutions. This text discusses both discrete-time systems and *discrete-frequency* techniques.

# ABOUT THE AUTHOR

Leonard Meirovitch is a well-known researcher and educator. He is currently a University Distinguished Professor in the College of Engineering at Virginia Polytechnic Institute and State University (VPI&SU). He is the author of a very large number of journal publications in structural dynamics and control of structures and of the books *Analytical Methods in Vibrations* (Macmillan, 1967); *Methods of Analytical Dynamics* (McGraw-Hill, 1970); *Elements of Vibration Analysis*, first edition (McGraw-Hill, 1975); *Computational Methods in Structural Dynamics* (Sijthoff & Noordhoff, 1980); and *Introduction to Dynamics and Control* (Wiley, 1985). Dr. Meirovitch is a Fellow of the American Institute of Aeronautics and Astronautics (AIAA) and the recipient of the 1981 VPI&SU Alumni Award for Research Excellence, the 1983 AIAA Structures, Structural Dynamics, and Materials Award, and the 1984 AIAA Pendray Aerospace Literature Award. He is the editor for the series *Mechanics: Dynamical Systems*, Martinus Nijhoff Publishers, The Netherlands; a member of the international editorial board of *Journal de Mécanique Théorique et Appliquée*; and an associate editor for *Journal of Optimization Theory and Applications*.

# FREE RESPONSE OF SINGLE-DEGREE-OF-FREEDOM LINEAR SYSTEMS

## 1.1 GENERAL CONSIDERATIONS

As mentioned in the Introduction, systems can be classified according to two distinct types of mathematical models, namely, discrete and continuous. Discrete models possess a finite number of degrees of freedom, whereas continuous models possess an infinite number of degrees of freedom. The number of degrees of freedom of a system is defined as the number of independent coordinates required to describe its motion completely (see also Sec. 4.2). Of the discrete mathematical models, the simplest ones are those described by a first-order or a second-order ordinary differential equation with constant coefficients. A system described by a single second-order differential equation is commonly referred to as a *single-degree-of-freedom system*. Such a model is often used as a very crude approximation for a generally more complex system, so that one may be tempted to regard its importance as being only marginal. This would be a premature judgment, however, because in cases in which a technique known as modal analysis can be employed, the mathematical formulation associated with many linear multi-degree-of-freedom discrete systems and continuous systems can be reduced to sets of *independent* second-order differential equations, each similar to the equation of a single-degree-of-freedom system. Hence, a thorough study of single-degree-of-freedom linear systems is amply justified. Unfortunately, the same technique cannot be used for nonlinear multi-degree-of-freedom discrete and continuous systems. The reason is that the above reduction is based on the principle of superposition, which applies only to linear systems (see Sec. 2.11). Nonlinear systems are treated in Chaps. 9 and 10 of this text and require different methods of analysis than do linear systems.

The primary objective of this text is to study the behavior of systems subjected to given excitations. The behavior of a system is characterized by the motion caused by these excitations and is commonly referred to as the system *response*. The motion is generally described by displacements, and less frequently by velocities or accelerations. The excitations can be in the form of initial displacements and velocities, or in the form of externally applied forces. The response of systems to initial excitations is generally known as *free response*, whereas the response to externally applied forces is known as *forced response*.

In this chapter we discuss the free response of single-degree-of-freedom linear systems, whereas in Chap. 2 we present a relatively extensive treatment of forced response. No particular distinction is made in this text between damped and undamped systems, because the latter can be regarded merely as an idealized limiting case of the first. The response of both undamped and damped systems to initial excitations is presented.

## 1.2 CHARACTERISTICS OF DISCRETE SYSTEM COMPONENTS

The elements constituting a discrete mechanical system are of three types, namely, those relating forces to displacements, velocities, and accelerations, respectively.

The most common example of a component relating forces to displacements is the *spring* shown in Fig. 1.1a. Springs are generally assumed to be massless, so that a force $F_s$ acting at one end must be balanced by a force $F_s$ acting at the other end, where the latter force is equal in magnitude but opposite in direction. Due to the force $F_s$, the spring undergoes an elongation equal to the difference between the displacements $x_2$ and $x_1$ of the end points. A typical curve depicting $F_s$ as a function

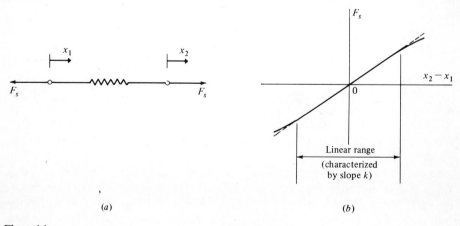

(a)

(b)

**Figure 1.1**

of the elongation $x_2 - x_1$ is shown in Fig. 1.1b; it corresponds to a so-called "softening spring," because for increasing elongations $x_2 - x_1$ the force $F_s$ tends to increase at a diminishing rate. If the force $F_s$ tends to increase at a growing rate for increasing elongations $x_2 - x_1$, the spring is referred to as a "stiffening spring." The force-elongation relation corresponding to Fig. 1.1b is clearly nonlinear. For small values of $x_2 - x_1$, however, the force can be regarded as being proportional to the elongation, where the proportionality constant is the slope $k$. Hence, in the range in which the force is proportional to the elongation the relation between the spring force and the elongation can be written in the form

$$F_s = k(x_2 - x_1) \tag{1.1}$$

A spring operating in that range is said to be *linear*, and the constant $k$ is referred to as the *spring constant*, or the *spring stiffness*. It is customary to label the spring, when it operates in the linear range, by its stiffness $k$. Note that the units of $k$ are pounds per inch (lb/in) or newtons per meter (N/m). The force $F_s$ is an *elastic force* known as the *restoring force* because, for a stretched spring, $F_s$ is the force that tends to return the spring to the unstretched configuration. In many cases the unstretched configuration coincides with the static equilibrium configuration (see Sec. 1.4).

The element relating forces to velocities is generally known as a damper; it consists of a piston fitting loosely in a cylinder filled with oil or water so that the viscous fluid can flow around the piston inside the cylinder. Such a damper is known as a *viscous damper* or a *dashpot* and is depicted in Fig. 1.2a. The damper is also assumed to be massless, so that a force $F_d$ at one end must be balanced by a corresponding force at the other end. If the forces $F_d$ cause smooth shear in viscous fluid, the curve $F_d$ versus $\dot{x}_2 - \dot{x}_1$ is likely to be linear, as shown in Fig. 1.2b, where dots designate time derivatives. Hence, the relation between the damper force and the velocity of one end of the damper relative to the other is

$$F_d = c(\dot{x}_2 - \dot{x}_1) \tag{1.2}$$

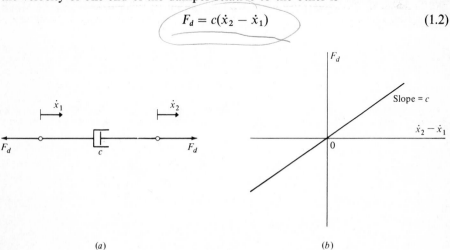

(a)  (b)

**Figure 1.2**

The constant of proportionality $c$, which is merely the slope of the curve $F_d$ versus $\dot{x}_2 - \dot{x}_1$, is called the *coefficient of viscous damping*. We shall refer to such dampers by their viscous damping coefficients $c$. The units of $c$ are pound·second per inch (lb·s/in) or newton·second per meter (N·s/m). The force $F_d$ is a damping force because it resists an increase in the relative velocity $\dot{x}_2 - \dot{x}_1$.

The element relating forces to accelerations is clearly the *discrete mass* (Fig. 1.3a). This relation has the form

$$F_m = m\ddot{x} \tag{1.3}$$

Equation (1.3) is a statement of Newton's second law of motion, according to which the force $F_m$ is proportional to the acceleration $\ddot{x}$, measured with respect to an inertial reference frame, where the proportionality constant is simply the mass $m$ (see Fig. 1.3b). The units of $m$ are pound·second$^2$ per inch (lb·s$^2$/in) or kilograms (kg). Note that in SI units the kilogram is a basic unit and the newton is a derived unit.

The physical properties of the components are recognized as being described in Figs. 1.1b, 1.2b, and 1.3b with the constants $k$, $c$, and $m$ playing the role of parameters. It should be reiterated that, unless otherwise stated, springs and dampers possess no mass. On the other hand, masses are assumed to behave like rigid bodies.

The preceding discussion is concerned exclusively with translational motion, although there are systems, such as those in torsional vibration, that undergo rotational motion. There is complete analogy between systems in axial and torsional vibration, with the counterparts of springs, viscous dampers, and masses being torsional springs, torsional viscous dampers, and disks possessing mass moments of inertia. Indeed, denoting the angular displacements at the two end points of a torsional spring by $\theta_1$ and $\theta_2$, and the restoring torque in the spring $k$ by $M_s$, the curve $M_s$ versus $\theta_2 - \theta_1$ is similar to that given in Fig. 1.1b. Moreover, denoting the damping torque by $M_d$ and the damping coefficient of the torsional viscous damper by $c$, the curve $M_d$ versus $\dot{\theta}_2 - \dot{\theta}_1$ is similar to that of Fig. 1.2b.

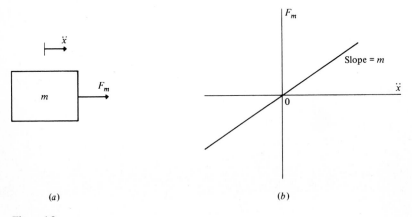

(a)                                             (b)

**Figure 1.3**

Finally, if the torsional system contains a disk of polar mass moment of inertia $I$, and the disk undergoes the angular displacement $\theta$, then the curve $M_I$ versus $\ddot{\theta}$ is similar to that of Fig. 1.3b, where $M_I$ is the inertia torque. Of course, the moment of inertia $I$ is simply the slope of that curve. Note that the units of the torsional spring $k$ are pound·inch per radian (lb·in/rad) or newton·meter per radian (N·m/rad), etc.

On occasions, certain dynamical systems consisting of distributed elastic members and lumped rigid masses can be approximated by strictly lumped systems. The approximation is based on the assumption that the mass of the distributed elastic member is sufficiently small, relative to the lumped masses, that it can be ignored. In this case, the fact that the elastic member is distributed loses all meaning, so that the elastic member can be replaced by an equivalent spring. The equivalent spring constant is determined by imagining a spring yielding the same displacement as the elastic member when subjected to the same force, or torque. The procedure is illustrated in Example 1.1 for a member in torsion and in Example 1.2 for a member in bending.

At times several springs are used in various combinations. Of particular interest are *springs connected in parallel* and *springs connected in series*, as shown in Figs. 1.4a and b, respectively. We shall be concerned here with linear springs. For the springs in parallel of Fig. 1.4a, the force $F_s$ divides itself into the forces $F_{s1}$ and $F_{s2}$ in the corresponding springs $k_1$ and $k_2$. Because the springs are linear, we have the relations

$$F_{s1} = k_1(x_2 - x_1) \qquad F_{s2} = k_2(x_2 - x_1) \tag{1.4}$$

But the forces $F_{s1}$ and $F_{s2}$ must add up to the total force $F_s$, or $F_s = F_{s1} + F_{s2}$, from which it follows that

$$F_s = k_{eq}(x_2 - x_1) \qquad k_{eq} = k_1 + k_2 \tag{1.5}$$

where $k_{eq}$ denotes the stiffness of an *equivalent spring* representing the combined effect of $k_1$ and $k_2$. If a number $n$ of springs of stiffnesses $k_i$ ($i = 1, 2, \ldots, n$) are arranged in parallel, then it is not difficult to show that

$$k_{eq} = \sum_{i=1}^{n} k_i \tag{1.6}$$

For springs in series, as shown in Fig. 1.4b, we can write the relations

$$F_s = k_1(x_0 - x_1) \qquad F_s = k_2(x_2 - x_0) \tag{1.7}$$

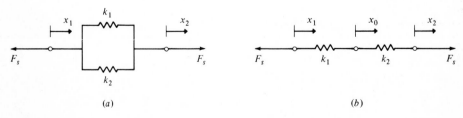

(a)                                    (b)

**Figure 1.4**

Eliminating $x_0$ from Eqs. (1.7), we arrive at

$$F_s = k_{eq}(x_2 - x_1) \qquad k_{eq} = \left(\frac{1}{k_1} + \frac{1}{k_2}\right)^{-1} \tag{1.8}$$

and, if there are $n$ springs connected in series, we conclude that

$$k_{eq} = \left(\sum_{i=1}^{n} \frac{1}{k_i}\right)^{-1} \tag{1.9}$$

In an analogous manner, it is possible to derive expressions for equivalent spring constants for torsional springs in parallel and in series, where the expressions are similar in structure to Eqs. (1.6) and (1.9).

**Example 1.1** The uniform circular shaft in torsion shown in Fig. 1.5 is fixed at the end $x = 0$ and has a rigid disk attached at the end $x = L$. Assume that the mass of the shaft is small relative to the mass of the disk and determine the equivalent spring constant of the system. The torsional stiffness of the shaft is $GJ(x) = GJ = \text{const}$, where $G$ is the shear modulus and $J$ is the polar moment of inertia of the cross-sectional area of the shaft.

It should be pointed out that, even though this is a dynamical system, the spring constant is a static concept and it simply expresses a load-deformation relation. Hence, in determining the equivalent spring constant, the mass moment of inertia of the disk plays no role, but the location of the disk does. The equivalent spring constant is defined as

$$k_{eq} = \frac{M}{\theta} \tag{a}$$

where $M$ is the torque applied on the disk and $\theta = \theta(L)$ is the angular displacement of the disk at $x = L$, as shown in Fig. 1.5. To calculate the torsional displacement $\theta(x)$ at any point $x$ along the shaft, we recall from mechanics of materials that $\theta(x)$ satisfies the differential equation

$$\frac{d}{dx}\left[GJ(x)\frac{d\theta(x)}{dx}\right] = m(x) = 0 \qquad 0 < x < L \tag{b}$$

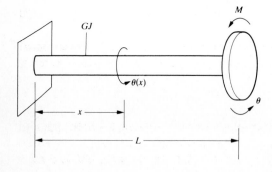

**Figure 1.5**

where $m(x)$ is the torque per unit length of shaft, which is zero in the case under consideration. This being a second-order differential equation, the rotation $\theta(x)$ must satisfy two boundary conditions, one at each end. The boundary conditions can be shown to be

$$\theta(0) = 0 \qquad GJ \frac{d\theta}{dx}\bigg|_{x=L} = M \qquad (c)$$

where the first boundary condition reflects the fact that the rotation must be zero at $x = 0$ and the second boundary condition states that the resultant torque of the internal stresses at the end $x = L$ is balanced by the external torque $M$. Because the torsional stiffness is constant, the differential equation, Eq. (b), reduces to

$$\frac{d^2\theta(x)}{dx^2} = 0 \qquad 0 < x < L \qquad (d)$$

The boundary conditions remain in the form (c).

The general solution of Eq. (d) is simply

$$\theta(x) = c_1 + c_2 x \qquad (e)$$

where $c_1$ and $c_2$ are constants of integration. The constants $c_1$ and $c_2$ are evaluated by invoking boundary conditions (c), with the result

$$c_1 = 0 \qquad c_2 = \frac{M}{GJ} \qquad (f)$$

so that the solution becomes

$$\theta(x) = \frac{Mx}{GJ} \qquad (g)$$

The rotation at $x = L$ is simply

$$\theta = \theta(L) = \frac{ML}{GJ} \qquad (h)$$

so that, recalling Eq. (a), we obtain the equivalent spring constant

$$k_{eq} = \frac{M}{\theta} = \frac{GJ}{L} \qquad (i)$$

The spring constant given by Eq. (i) is for the case in which the torsional stiffness is constant. When the torsional stiffness varies with $x$, one must work with Eq. (b).

**Example 1.2** A uniform beam in bending is simply supported at both ends and has a lumped mass at a distance $x = a$ from the left end (Fig. 1.6a). Assume that the mass of the beam is small relative to the lumped mass and determine

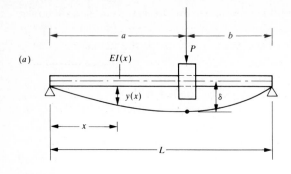

(a)

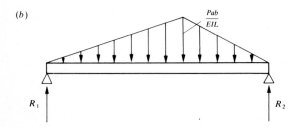

(b)

$R_1$          $R_2$

**Figure 1.6**

the equivalent spring constant. The bending stiffness of the beam is $EI(x) = EI = $ const, where $E$ is Young's modulus and $I$ is the cross-sectional area moment of inertia.

Following the pattern of Example 1.1, the equivalent spring constant is

$$k_{eq} = \frac{P}{\delta} \qquad (a)$$

where $P$ is a transverse force applied on the lumped mass and $\delta$ is the deflection of the point of application of $P$ and in the same direction as $P$. To determine $\delta$, we first calculate the deflection $y(x)$ of a nominal point $x$ and then write $\delta = y(a)$. The deflection $y(x)$ can be obtained by first integrating the differential equation

$$\frac{d^2}{dx^2}\left[ EI(x)\frac{d^2y(x)}{dx^2} \right] = 0 \qquad (b)$$

over the segments $0 < x < a$ and $a < x < L$ and then applying appropriate boundary conditions to evaluate the constants of integration. Of course, in the case at hand $EI(x) = EI = $ const. We do not pursue this approach here, but leave it as an exercise to the reader. Instead, we propose to use the area-moment method, according to which the deflection $y(x)$ is obtained by considering a fictitious beam, sometimes called a conjugate beam, subjected to a distributed load equal to the actual bending moment at any point $x$ divided by $EI(x)$. Then, $y(x)$ is equal to the moment of the fictitious loading about point $x$. In using this approach care must be exercised, as the supports of the fictitious beam are not always the same as those of the actual beam. For example, whereas to an actual hinged end corresponds a fictitious hinged end, to an actual free end corresponds a fictitious fixed end, and vice versa.

Figure 1.6$b$ shows the fictitious beam loaded with the actual bending moment diagram divided by $EI$. From the figure, we calculate the left reaction

$$R_1 = \frac{1}{L}\left[\frac{Pab}{EIL}\frac{a}{2}\left(b + \frac{a}{3}\right) + \frac{Pab}{EIL}\frac{b}{2}\frac{2b}{3}\right] = \frac{Pb}{6EIL}(L^2 - b^2) \qquad (c)$$

Then, the displacement is simply

$$y(x) = R_1 x - \frac{Pxb}{EIL}\frac{x}{2}\frac{x}{3} = \frac{Pbx}{6EIL}(L^2 - b^2 - a^2) \qquad (d)$$

Letting $x = a$ in Eq. ($d$), we obtain

$$\delta = y(a) = \frac{Pab}{6EIL}(L^2 - b^2 - a^2) = \frac{Pa^2b^2}{3EIL} \qquad (e)$$

Hence, the equivalent spring is

$$k_{eq} = \frac{P}{\delta} = \frac{3EIL}{a^2b^2} \qquad (f)$$

**Example 1.3** The shaft in torsion shown in Fig. 1.7$a$, consisting of two segments of different length and torsional rigidity, is fixed at the left end and has a disk attached at the right end. The data concerning the two shaft segments are as follows:

$$GJ_1 = 10^7 \text{ lb} \cdot \text{in}^2 \qquad L_1 = 160 \text{ in}$$

$$GJ_2 = 5 \times 10^6 \text{ lb} \cdot \text{in}^2 \qquad L_2 = 120 \text{ in}$$

Assuming that the shaft acts like a linear torsional spring, calculate the angular displacement of the disk caused by a torque $M = 1000 \text{ lb} \cdot \text{in}$, as shown. Then use the result to calculate the equivalent spring constant.

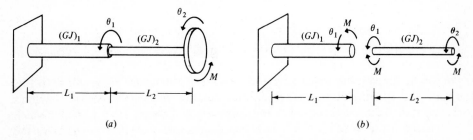

**Figure 1.7**

Let us denote by $\theta_1$ and $\theta_2$ the angular displacements of the right ends of the shaft segments 1 and 2, respectively (Fig. 1.7b). Because the torque $M$ acts everywhere along the shaft, we conclude from Example 1.1, that the rotation $\theta_1$ has the value

$$\theta_1 = \frac{ML_1}{GJ_1} = \frac{1000 \times 160}{10^7} = 1.6 \times 10^{-2} \text{ rad} \tag{a}$$

whereas the rotation of the right end of shaft segment 2 relative to the left end is

$$\theta_2 - \theta_1 = \frac{ML_2}{GL_2} = \frac{1000 \times 120}{5 \times 10^6} = 2.4 \times 10^{-2} \text{ rad} \tag{b}$$

Hence, the angular displacement of the disk is

$$\theta_2 = 1.6 \times 10^{-2} + 2.4 \times 10^{-2} = 4.0 \times 10^{-2} \text{ rad} \tag{c}$$

This enables us to calculate immediately the equivalent spring constant

$$k_{eq} = \frac{M}{\theta_2} = \frac{1000}{4.0 \times 10^{-2}} = 2.5 \times 10^4 \text{ lb·in/rad} \tag{d}$$

The same problem can be solved by regarding the two shaft segments as representing two torsional springs in series. Indeed, using Eqs. (a) and (b), the spring constants for the two shaft segments can be written in the form

$$k_1 = \frac{M}{\theta_1} = \frac{GJ_1}{L_1} = \frac{10^7}{160} \text{ lb·in/rad}$$

$$\tag{e}$$

$$k_2 = \frac{M}{\theta_2 - \theta_1} = \frac{GJ_2}{L_2} = \frac{5 \times 10^6}{120} \text{ lb·in/rad}$$

so that, using Eq. (1.9), we obtain

$$k_{eq} = \frac{1}{1/k_1 + 1/k_2} = \frac{1}{(160/10^7) + 120/(5 \times 10^6)} = 2.5 \times 10^4 \text{ lb·in/rad} \tag{f}$$

which agrees with Eq. (d).

## 1.3 DIFFERENTIAL EQUATIONS OF MOTION FOR FIRST-ORDER AND SECOND-ORDER LINEAR SYSTEMS

One of the simplest mechanical systems is the spring-damper system shown in Fig. 1.8a. We derive the differential equation of motion for the system by Newton's second law. To this end, we consider the free-body diagram of Fig. 1.8b, in which $F(t)$ is the external force and $x(t)$ is the displacement of the system from the equilibrium position, which coincides with the position in which the spring is

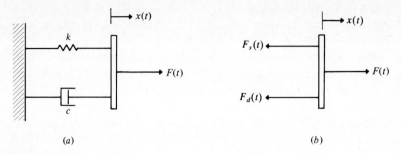

(a)                                                          (b)

**Figure 1.8**

unstretched. Invoking Newton's second law, and recognizing that this is a special case in which the system has no mass, we can write

$$F(t) - F_s(t) - F_d(t) = 0 \tag{1.10}$$

Because the left end is fixed and the displacement of the right end is $x(t)$, Eqs. (1.1) and (1.2) reduce to

$$F_s = kx(t) \qquad F_d = c\dot{x}(t) \tag{1.11}$$

Inserting Eqs. (1.11) into Eq. (1.10) and rearranging, we obtain the equation of motion

$$c\dot{x}(t) + kx(t) = F(t) \tag{1.12}$$

which is a first-order linear ordinary differential equation with constant coefficients. The constant coefficients $c$ and $k$ are known as the system parameters. We discuss the homogeneous solution of Eq. (1.12) later in this chapter and the particular solution in the next chapter.

A system of considerable interest in vibrations is the spring-damper-mass system of Fig. 1.9a. To derive the differential equation of motion, we use Newton's second law in conjunction with the free-body diagram shown in Fig. 1.9b and write

$$F(t) - F_s(t) - F_d(t) = m\ddot{x}(t) \tag{1.13}$$

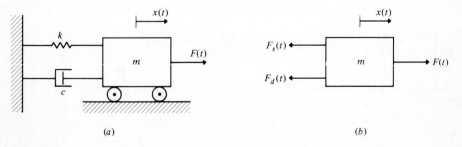

(a)                                                          (b)

**Figure 1.9**

Using Eqs. (1.11), Eq. (1.13) becomes

$$m\ddot{x}(t) + c\dot{x}(t) + kx(t) = F(t) \qquad (1.14)$$

which is a second-order linear ordinary differential equation with constant coefficients. The constant coefficients $m$, $c$, and $k$ represent the system parameters. A second-order system is commonly known as a *single-degree-of-freedom system*.

In the case of the system shown in Fig. 1.9a the *equilibrium position* coincides with the position in which the spring is unstretched. This is not always the case, however, and the question remains as to whether there exists a more convenient reference position. To answer this question, we consider the system of Fig. 1.10a and denote by $y(t)$ the displacement of $m$ from the unstretched spring position. Because of gravity, it differs from the equilibrium position by the static displacement $x_{st} = mg/k$. Using the unstretched spring position as a reference and referring to the free-body diagram of Fig. 1.10b, the differential equation of motion takes the form

$$m\ddot{y}(t) + c\dot{y}(t) + ky(t) + mg = F(t) \qquad (1.15)$$

Comparing Eqs. (1.14) and (1.15), we conclude that the latter contains the additional constant term $mg$. From Fig. 1.10a, however, we can write $y(t) = x(t) - x_{st}$. Inserting this value in Eq. (1.15), we obtain a differential equation in $x(t)$ that is identical in every respect to Eq. (1.14), because the terms $-kx_{st}$ and $mg$ cancel each other. The conclusion is that, in measuring displacements of a linear system from the static equilibrium position, we can omit the weight $mg$ because it is balanced at all times by an additional force $kx_{st}$ in the spring. It follows that the static equilibrium is a more convenient reference position than that corresponding to the unstretched configuration of the spring for the purpose of formulating the equation of motion.

As a matter of interest, it should be pointed out that the spring constant of a

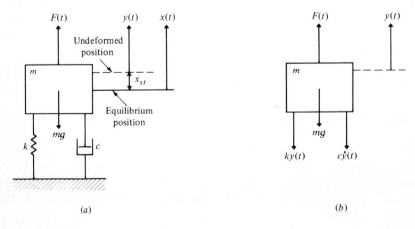

(a)

(b)

**Figure 1.10**

linear spring can be determined by simply attaching a mass of known weight to a hanging spring and measuring the static displacement.

Equation (1.14) is typical of a large class of systems known as single-degree-of-freedom damped systems. The structure of the differential equation is the same for all the systems in the class and the only difference lies in the definition of the system parameters. Because the structure of the differential equation is the same, the solution for one system in the class can be used for any other system in the same class by mere insertion of the proper parameters. Before a solution of Eq. (1.14) is attempted, however, it appears advisable to consider several special cases. These are discussed in the remainder of this chapter and in parts of Chap. 2.

## 1.4 SMALL MOTIONS ABOUT EQUILIBRIUM POSITIONS

It was demonstrated in Sec. 1.3 that the equilibrium position has the advantage that, when used as a reference position, it simplifies the equation of motion. This case, however, is a mere example of a more general theory concerning the motion in the neighborhood of equilibrium positions. In fact, the concept of equilibrium position is basic to linearization of nonlinear systems.

Let us consider a body of mass $m$ moving under the action of the force $F$ (Fig. 1.11), where $F$ is a given function of the displacement $y$ and velocity $\dot{y}$, $F = F(y, \dot{y})$. Using Newton's second law and dividing through by $m$, we obtain the equation of motion

$$\ddot{y} = \frac{1}{m} F(y, \dot{y}) = f(y, \dot{y}) \qquad (1.16)$$

where the notation is obvious. In general, $f$ is a nonlinear function of $y$ and $\dot{y}$, in which case a solution of Eq. (1.16) is likely to cause considerable difficulties.

Quite often, particularly in vibration problems, Eq. (1.16) admits special solutions characterized by the fact that both the velocity $\dot{y}$ and acceleration $\ddot{y}$ are zero. Hence, these special solutions are constant solutions and can be identified as describing *equilibrium positions*, in which the body is at rest. The interest lies in the motion characteristics in the neighborhood of equilibrium positions. In particular, the question arises as to how the system behaves if disturbed slightly from equilibrium. The various possibilities are as follows:

1. The system returns to equilibrium, in which case the equilibrium position is said to be *asymptotically stable*.

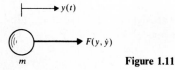

$\longrightarrow y(t)$

$\longrightarrow F(y, \dot{y})$

$m$         **Figure 1.11**

2. The system oscillates about the equilibrium without exhibiting any secular trend, i.e., neither tending to the equilibrium nor greatly departing from it, so that the motion remains bounded. In this case, the equilibrium is said to be merely *stable*.
3. The system tends away from equilibrium secularly, in which case the equilibrium is *unstable*.

Cases 1 and 2 are of particular interest in vibrations.

The questions remain as to how to identify equilibrium positions and how to determine the motion in the neighborhood of a given equilibrium position. To answer the first question, we recall that both $\dot{y}$ and $\ddot{y}$ are zero and that $y$ is constant at an equilibrium point. Denoting the equilibrium point by $y = y_e$ and considering Eq. (1.16), we conclude that the equilibrium positions must satisfy the equation

$$f(y_e, 0) = 0 \tag{1.17}$$

Equation (1.17) represents an algebraic equation, where the equation is in general nonlinear. Its solution yields the equilibrium positions $y_e$. In the special case in which $f$ is linear in $y$, there is only one equilibrium position, but in general for nonlinear systems there can be many equilibrium positions. When $f$ is a polynomial in $y$, there are as many equilibrium positions as the degree of the polynomial.

To determine the nature of the motion in the neighborhood of an equilibrium position, let us expand the function $f$ in a Taylor series about $y_e$, or

$$f(y, \dot{y}) = f(y_e, 0) + \left.\frac{\partial f(y, \dot{y})}{\partial y}\right|_e (y - y_e) + \left.\frac{\partial f(y, \dot{y})}{\partial \dot{y}}\right|_e \dot{y} + O(y, \dot{y}) \tag{1.18}$$

where $O(y, \dot{y})$ denotes nonlinear terms in $y$ and $\dot{y}$. The first term on the right side of Eq. (1.18) is zero by virtue of Eq. (1.17). Moreover, introducing the notation

$$\left.\frac{\partial f(y, \dot{y})}{\partial y}\right|_e = -b \qquad \left.\frac{\partial f(y, \dot{y})}{\partial \dot{y}}\right|_e = -a \tag{1.19}$$

ignoring the nonlinear terms in $y - y_e$ and $\dot{y}$ and letting $y - y_e = x$, Eq. (1.16) reduces to

$$\ddot{x} + a\dot{x} + bx = 0 \tag{1.20}$$

which represents the *linearized equation of motion about equilibrium.* The assumption leading to the linearized system is called the *small motions assumption.*

The solution of Eq. (1.20) has the general form

$$x(t) = Ae^{st} \tag{1.21}$$

Introducing Eq. (1.21) into Eq. (1.20) and dividing through by $Ae^{st}$, we conclude that the exponent $s$ must satisfy the algebraic equation

$$s^2 + as + b = 0 \tag{1.22}$$

which is known as the *characteristic equation*. Its solutions are

$$\begin{matrix} s_1 \\ s_2 \end{matrix} = -\frac{a}{2} \pm \sqrt{\left(\frac{a}{2}\right)^2 - b} \qquad (1.23)$$

so that the solution of Eq. (1.20) is

$$x(t) = A_1 e^{s_1 t} + A_2 e^{s_2 t} \qquad (1.24)$$

The nature of the solution, and hence the nature of the equilibrium, depends on the roots $s_1$ and $s_2$ of the characteristic equation. If $s_1$ and $s_2$ are real and negative, then exp $s_1 t$ and exp $s_2 t$ reduce to zero as $t \to \infty$, so that the solution dies out as time unfolds. If either $s_1$ or $s_2$ is real and positive, then the solution increases without bounds as $t \to \infty$. If the roots $s_1$ and $s_2$ are complex, then they are complex conjugates, and the nature of the solution depends on the real part of the roots. Indeed, the solution can be expressed as the product of two factors, one corresponding to the real part of the exponents and the other corresponding to the imaginary parts. The factor corresponding to the real part plays the role of a time-dependent amplitude and the factors corresponding to the imaginary parts vary harmonically with time. If the real part is negative, then the time-dependent amplitude approaches zero as $t \to \infty$, so that the solution represents a decaying oscillation. If the real part is positive, then the time-dependent amplitude increases without bounds as $t \to \infty$, so that the solution represents a divergent oscillation. If the real part is zero, in which case the roots are pure imaginary, the amplitude does not depend on time but is constant and the solution represents simple harmonic oscillation, which is bounded. Note that harmonic oscillation is a borderline case, separating decaying and divergent oscillations. Examining Eq. (1.23), we can list three cases according to the earlier classification:

1. $a > 0$, $b > 0$. In this case the roots are either real and negative or complex conjugates with negative real part, so that $x(t)$ approaches zero as $t \to \infty$. Hence, $y(t)$ approaches $y_e$, so that the equilibrium position is asymptotically stable.
2. $a = 0$, $b > 0$. The roots are pure imaginary, so that the solution $x(t)$ is oscillatory. Hence, the motion is bounded and the equilibrium position is stable.
3. $a < 0$, or $a = 0$, $b < 0$. The roots are either complex conjugates with positive real part or they are both real, with one root being positive and the other being negative. In either case the solution is divergent and the equilibrium position is unstable.

The subject of system stability is discussed in greater detail and in a more rigorous manner in Chap. 9.

In the special case in which $y_e = 0$ the equilibrium position is said to be *trivial*. This is a case encountered very frequently in practice.

**Example 1.4** The system shown in Fig. 1.12a represents a simple pendulum. It consists of a bob of mass $m$ attached to one end of an inextensible string of

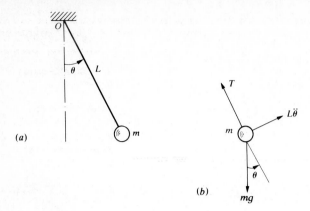

(a)

(b)

$mg$

**Figure 1.12**

length $L$, where the other end of the string is fixed at point $O$. Derive the equation for the angular motion $\theta(t)$ of the pendulum, identify the equilibrium positions and determine the nature of motion in the neighborhood of the equilibrium positions.

To derive the equation of motion, we consider the free-body diagram shown in Fig. 1.12$b$. The forces acting on the bob are the string tension $T$ and the gravity force $mg$. Note that the string is not capable of carrying transverse forces. Using Newton's second law and summing up forces in the transverse direction, we obtain

$$\sum F_t = -mg \sin \theta = ma_\theta = mL\ddot{\theta} \qquad (a)$$

where $a_\theta = L\ddot{\theta}$ is the acceleration in the transverse direction. Division of Eq. ($a$) through by $mL$ yields

$$\ddot{\theta} = f(\theta) \qquad (b)$$

where

$$f(\theta) = -\frac{g}{L} \sin \theta \qquad (c)$$

Equation ($b$) has the same form as Eq. (1.16), except that $f$ does not depend on the angular velocity $\dot{\theta}$.

To identify the equilibrium positions, we use Eq. (1.17) and write

$$f(\theta_e) = -\frac{g}{L} \sin \theta_e = 0 \qquad (d)$$

which has the solutions

$$\theta_e = 0, \pm \pi, \pm 2\pi, \ldots \qquad (e)$$

Although Eq. ($e$) indicates that mathematically there is an infinite number of equilibrium positions, physically there are only two positions

$$\theta_{e1} = 0 \qquad \theta_{e2} = \pi \qquad (f)$$

Of course, the first one is recognized as the trivial solution.

Next, let us use the notation $x = \theta - \theta_e$ and write the linearized equation of motion

$$\ddot{x} + bx = 0 \tag{g}$$

where, from the first of Eqs. (1.19),

$$b = -\left.\frac{\partial f}{\partial \theta}\right|_{\theta = \theta_e} = \frac{g}{L}\cos\theta_e \tag{h}$$

In the case of the equilibrium point $\theta_{e1} = 0$, we obtain

$$b = \frac{g}{L} > 0 \tag{i}$$

so that the equilibrium is stable. We will discuss Eq. (g) for this case extensively later in this chapter. In the case of the equilibrium point $\theta_{e2} = \pi$, we have

$$b = -\frac{g}{L} < 0 \tag{j}$$

so that the equilibrium is unstable.

The above results conform to expectation. Any small deviation from the equilibrium position in which the pendulum hangs down results in oscillation about the equilibrium. On the other hand, any small deviation from the upright equilibrium position tends to increase without bounds. The case in which the pendulum oscillates about the equilibrium position $\theta_{e1} = 0$ is by far the most important one, which explains why the equilibrium position $\theta_{e2} = \pi$ is discussed so seldom.

## 1.5 FORCE-FREE RESPONSE OF FIRST-ORDER SYSTEMS

Let us consider the spring-damper system of Sec. 1.3 and assume that the external excitation is zero. Setting $F(t) = 0$ in Eq. (1.12), we obtain the homogeneous equation

$$c\dot{x}(t) + kx(t) = 0 \tag{1.25}$$

Using the approach of Sec. 1.4, we let the solution of Eq. (1.25) have the exponential form

$$x(t) = Ae^{st} \tag{1.26}$$

Inserting Eq. (1.26) into Eq. (1.25) and dividing through by $Ae^{st}$, we obtain the characteristic equation

$$cs + k = 0 \tag{1.27}$$

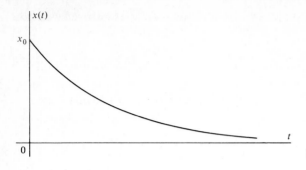

**Figure 1.13**

which has the single root

$$s = -\frac{k}{c} \tag{1.28}$$

so that the general solution of Eq. (1.25) is

$$x(t) = Ae^{-t/\tau} \tag{1.29}$$

where $A$ is a constant of integration and

$$\tau = \frac{c}{k} \tag{1.30}$$

is known as the *time constant*.

The constant of integration $A$ depends on the initial excitation. Letting $x(0) = x_0$ be the initial displacement, we can easily verify from Eq. (1.29) that $A = x_0$, so that the solution to the force-free problem is

$$x(t) = \begin{cases} x_0 e^{-t/\tau} & t > 0 \\ 0 & t < 0 \end{cases} \tag{1.31}$$

Equation (1.31) indicates that the response decays exponentially with time.

The homogeneous solution (1.31) is plotted in Fig. 1.13 as a function of time. We observe that, after being displaced initially by an amount $x_0$, the system returns to the zero equilibrium position without any oscillation. The time constant $\tau$ provides a measure of the speed of return of the system to equilibrium. Indeed, the rate of return is greater for small time constants and vice versa. Hence, for a stiff spring, or for light damping, the rate of return is fast, and vice versa.

## 1.6 HARMONIC OSCILLATOR

Let us consider now the second-order system described by Eq. (1.14). Before we discuss the general response, we wish to investigate the *force-free case*, namely, the case in which the force $F(t)$ is identically zero. Moreover, quite often damping is extremely small, so that for all practical purposes it can be ignored. Hence, we concern ourselves with the *undamped case* for which $c = 0$. Upon dividing Eq. (1.14) by $m$, the differential equation of motion reduces to

$$\ddot{x}(t) + \omega_n^2 x(t) = 0 \qquad \omega_n^2 = \frac{k}{m} \tag{1.32}$$

As shown in Sec. 1.4, the solution of Eq. (1.32) has the exponential form

$$x(t) = A e^{st} \tag{1.33}$$

Introducing Eq. (1.33) into Eq. (1.32) and dividing through by $A e^{st}$, we obtain the characteristic equation

$$s^2 + \omega_n^2 = 0 \tag{1.34}$$

which has the solutions

$$\begin{matrix} s_1 \\ s_2 \end{matrix} = \pm i\omega_n \tag{1.35}$$

where $i = \sqrt{-1}$. Inserting $s_1$ and $s_2$ into Eq. (1.33), the general solution of Eq. (1.32) can be written as

$$x(t) = A_1 e^{i\omega_n t} + A_2 e^{-i\omega_n t} \tag{1.36}$$

where $A_1$ and $A_2$ are constants of integration. Their values depend on the initial displacement $x(0)$ and initial velocity $\dot{x}(0)$.

Because the roots $s_1$ and $s_2$ are pure imaginary, we conclude from Sec. 1.4 that the solution, Eq. (1.36), must represent stable motion. This stable motion consists of pure oscillation and the quantity $\omega_n$ is known as the *natural frequency* of oscillation of the undamped system. The reason for the term natural frequency is that a force-free undamped second-order system, when set in motion by some initial conditions, will always oscillate with the same frequency $\omega_n$.

Solution (1.36) is in terms of complex quantities. Yet, on physical grounds, it can be argued that the solution must be real. Hence, the interest lies in reducing the solution to real form. To this end, consider the series

$$e^{i\omega_n t} = 1 + i\omega_n t + \frac{1}{2!}(i\omega_n t)^2 + \frac{1}{3!}(i\omega_n t)^3 + \frac{1}{4!}(i\omega_n t)^4 + \frac{1}{5!}(i\omega_n t)^5 + \cdots$$

$$= 1 - \frac{1}{2!}(\omega_n t)^2 + \frac{1}{4!}(\omega_n t)^4 - \cdots + i\left[\omega_n t - \frac{1}{3!}(\omega_n t)^3 + \frac{1}{5!}(\omega_n t)^5 - \cdots\right]$$

$$= \cos \omega_n t + i \sin \omega_n t \tag{1.37a}$$

In a similar manner, it is easy to verify that

$$e^{-i\omega_n t} = \cos \omega_n t - i \sin \omega_n t \tag{1.37b}$$

Inserting Eqs. (1.37) into Eq. (1.36), introducing the notation

$$A_1 + A_2 = A \cos \phi \qquad i(A_1 - A_2) = A \sin \phi \tag{1.38}$$

and recalling the trigonometric relation $\cos \alpha \cos \beta + \sin \alpha \sin \beta = \cos (\alpha - \beta)$, the solution becomes

$$x(t) = A \cos (\omega_n t - \phi) \tag{1.39}$$

where now the constants of integration are $A$ and $\phi$.

The constants $A$ and $\phi$ are referred to as the *amplitude* and *phase angle*, respectively. Because $A$ and $\phi$ depend on $A_1$ and $A_2$, they can also be regarded as constants of integration depending on the initial conditions $x(0)$ and $\dot{x}(0)$. Equation (1.39) indicates that the system executes *simple harmonic oscillation* with the natural frequency $\omega_n$, for which reason the system itself is called a *harmonic oscillator*. The motion described by Eq. (1.39) is the simplest type of vibration. The harmonic oscillator represents more of a mathematical concept than a physical reality. Nevertheless, the concept is valid for negligible damping, if the interest lies in the response for a time duration too short for extremely light damping to make its effect felt.

The discussion of the nature of harmonic oscillation is perhaps enhanced by the vector diagram shown in Fig. 1.14a. If **A** represents a vector of magnitude $A$ and the vector makes an angle $\omega_n t - \phi$ with respect to the vertical axis $x$, then the projection of the vector **A** on $x$ represents the solution $x(t) = A \cos (\omega_n t - \phi)$. The angle $\omega_n t - \phi$ increases linearly with time, with the implication that the vector **A** rotates counterclockwise with angular velocity $\omega_n$. As the vector rotates, the projection varies harmonically, so that the motion repeats itself every time the vector **A** sweeps a $2\pi$ angle. The projection $x(t)$ is plotted in Fig. 1.14b as a function of time.

The time necessary to complete one *cycle* of motion defines the *period T* given by

$$T = \frac{2\pi}{\omega_n} \tag{1.40}$$

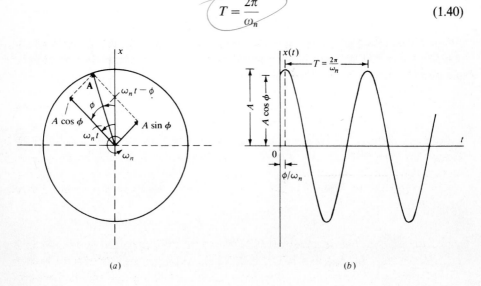

$(a)$                                                  $(b)$

**Figure 1.14**

where $\omega_n$ is measured in radians per second (rad/s) if $T$ is measured in seconds. Physically, $T$ represents the time necessary for one complete oscillation to take place; it is equal to the difference between two consecutive times at which the oscillator reaches the same state, where the state is to be interpreted as consisting of both position and velocity. As an illustration, the period is measured in Fig. 1.14$b$ between two consecutive peaks. It is also customary to measure the natural frequency in cycles per second (cps). In such a case the natural frequency is denoted by $f_n$, and because one cycle is equal to $2\pi$ radians we have

$$f_n = \frac{1}{2\pi} \omega_n = \frac{1}{T} \tag{1.41}$$

so that the natural frequency $f_n$ and the period $T$ are the reciprocals of each other. One cycle per second is a unit generally known as one hertz (Hz).

Finally, it will prove interesting to evaluate the constants of integration $A$ and $\phi$ in terms of the initial conditions. Introducing the notation $x(0) = x_0$, $\dot{x}(0) = v_0$, where $x_0$ is the initial displacement and $v_0$ the initial velocity, and using Eq. (1.39), it is easy to verify that the response of the harmonic oscillator to the initial conditions is

$$x(t) = x_0 \cos \omega_n t + \frac{v_0}{\omega_n} \sin \omega_n t \tag{1.42}$$

Moreover, we conclude that the amplitude $A$ and the phase angle $\phi$, when expressed in terms of the initial displacement and velocity, have the values

$$A = \sqrt{x_0^2 + \left(\frac{v_0}{\omega_n}\right)^2} \qquad \phi = \tan^{-1} \frac{v_0}{x_0 \omega_n} \tag{1.43}$$

A large variety of dynamical systems behave like harmonic oscillators, quite often when restricted to small motions. As an illustration, the simple pendulum of Example 1.4, when restricted to small angular motions about the trivial equilibrium $\theta = 0$, can be described by the differential equation

$$\ddot{\theta} + \omega_n^2 \theta = 0 \qquad \omega_n^2 = \frac{g}{L} \tag{1.44}$$

which represents a harmonic oscillator with the natural frequency $\omega_n = \sqrt{g/L}$. Note that Eq. (1.44) is valid as long as $\sin \theta \cong \theta$, which is approximately true for surprisingly large values of $\theta$. For example, $\theta = 30° = 0.5236$ rad and $\sin \theta = \sin 30° = 0.5000$ are close in value. In fact, there is less than 5 percent error in using $\theta$ instead of $\sin \theta$ for $\theta \leqslant 30°$.

**Example 1.5** The semicircular thin shell of radius $R$ shown in Fig. 1.15$a$ is allowed to rock on a rough horizontal surface. Derive the differential equation of motion for the case of no slip, show that for small motions the shell behaves like a harmonic oscillator, and calculate the natural frequency of the oscillator.

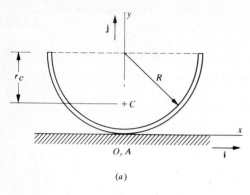

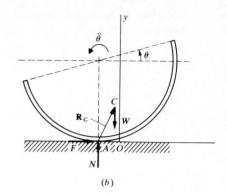

(a)                    (b)

**Figure 1.15**

This example provides us with the opportunity to derive the differential equation of motion for a relatively more complicated system than a spring-damper-mass system or a simple pendulum. To derive the equation of motion, we let the thin shell be tilted by an angle $\theta$ and denote by $C$ the center of mass of the shell and by $A$ the point of contact of the shell with the rough surface. The distance between the center of curvature of the shell and the mass center $C$ is denoted by $r_C$, where $r_C = 2R/\pi$.

Figure 1.15b shows a free-body diagram for the system. This being a case of planar motion, there are three equations of motion, namely, two force equations and one moment equation. Because the point of contact $A$ is in general a moving point, we will write the moment equation about the mass center $C$. This introduces the reactions $F$ and $N$ as unknowns, but these unknowns can be eliminated, thereby reducing the three equations to a single equation of motion in terms of $\theta$.

From planar rigid-body dynamics, the force and moment equations can be written in the general form

$$\sum F_x = ma_{Cx} \qquad \sum F_y = ma_{Cy} \qquad \sum M_C = I_C\alpha \qquad (a)$$

where $a_{Cx}$ and $a_{Cy}$ are the cartesian components of the acceleration vector $\mathbf{a}_C$ of the mass center of the shell and $\alpha$ is the angular acceleration of the shell. To calculate $\mathbf{a}_C$, we consider an inertial system of axes $x$, $y$ with the origin at point $O$, where $O$ and $A$ coincide when $\theta = 0$, and write the radius vector $\mathbf{R}_C$ from $O$ to $C$ in terms of cartesian components as follows:

$$\mathbf{R}_C = (-R\theta + r_C \sin \theta)\mathbf{i} + (R - r_C \cos \theta)\mathbf{j} \qquad (b)$$

in which $\mathbf{i}$ and $\mathbf{j}$ are unit vectors along $x$ and $y$, respectively. Taking the second derivative of $\mathbf{R}_C$ with respect to time, we obtain

$$\mathbf{a}_C = [(-R + r_C \cos \theta)\ddot{\theta} - \dot{\theta}^2 r_C \sin \theta]\mathbf{i} + r_C(\ddot{\theta} \sin \theta + \dot{\theta}^2 \cos \theta)\mathbf{j} \qquad (c)$$

The coefficients of $\mathbf{i}$ and $\mathbf{j}$ are recognized as $a_{Cx}$ and $a_{Cy}$, respectively. Hence, using the free-body diagram of Fig. 1.15$b$ and considering Eqs. ($a$) and ($c$), the equations of motion can be written in the explicit form

$$F = m[(-R + r_C \cos \theta)\ddot{\theta} - \dot{\theta}^2 r_C \sin \theta]$$

$$N - W = mr_C(\ddot{\theta} \sin \theta + \dot{\theta}^2 \cos \theta) \qquad (d)$$

$$F(R - r_C \cos \theta) - Nr_C \sin \theta = m(R^2 - r_C^2)\ddot{\theta}$$

Solving the first two of Eqs. ($d$) for $F$ and $N$ and inserting into the third of Eqs. ($d$), we obtain the equation of motion

$$2R(R - r_C \cos \theta)\ddot{\theta} + \dot{\theta}^2 Rr_C \sin \theta + gr_C \sin \theta = 0 \qquad (e)$$

Equation ($e$) represents a nonlinear second-order differential equation. Clearly, the trivial solution $\theta = 0$ is an equilibrium position. We propose to linearize the equation by considering small motions about equilibrium, which implies that the motion is restricted to small angles $\theta$. For small $\theta$, we have the approximate expressions $\sin \theta \cong \theta$, $\cos \theta \cong 1$. Moreover, ignoring any nonlinear terms in $\theta$ and $\dot{\theta}$ in Eq. ($e$), that is, terms of higher degree than the first, and recalling that $r_C = 2R/\pi$, we obtain the linearized equation of motion

$$\ddot{\theta} + \frac{g}{(\pi - 2)R} \theta = 0 \qquad (f)$$

Comparing Eq. ($f$) with Eq. (1.44), we conclude that for small angles $\theta$ the shell behaves like a harmonic oscillator with the natural frequency

$$\omega_n = \sqrt{\frac{g}{(\pi - 2)R}} \qquad (g)$$

**Example 1.6** Consider the harmonic oscillator described by Eq. (1.32), let $m = 2$ lb·s$^2$/in and $k = 600$ lb/in, and calculate the response $x(t)$ for the initial conditions $x_0 = x(0) = 1$ in, $v_0 = \dot{x}(0) = 10$ in/s. Plot $x(t)$ versus $t$.

The natural frequency of oscillation is simply

$$\omega_n = \sqrt{\frac{k}{m}} = \sqrt{\frac{600}{2}} = 10\sqrt{3} \text{ rad/s} \qquad (a)$$

Moreover, using Eqs. (1.43), we obtain the amplitude

$$A = \sqrt{x_0^2 + \left(\frac{v_0}{\omega_n}\right)^2} = \sqrt{1^2 + \left(\frac{10}{10\sqrt{3}}\right)^2} = \frac{2}{\sqrt{3}} \text{ in} \qquad (b)$$

and the phase angle

$$\phi = \tan^{-1} \frac{v_0}{x_0\omega_n} = \tan^{-1} \frac{10}{1 \times 10\sqrt{3}} = \tan^{-1} \frac{1}{\sqrt{3}} = \frac{\pi}{6} \text{ rad} \qquad (c)$$

Inserting Eqs. (*a*), (*b*), and (*c*) into Eq. (1.39), we obtain the response

$$x(t) = \frac{2}{\sqrt{3}} \cos \left( 10\sqrt{3}t - \frac{\pi}{6} \right) \text{ in} \qquad (d)$$

The plot $x(t)$ versus $t$ is shown in Fig. 1.14b.

## 1.7 FREE VIBRATION OF DAMPED SECOND-ORDER SYSTEMS

In the absence of external forces, $F(t) = 0$, the equation of motion of a second-order system, Eq. (1.14), reduces to a homogeneous differential equation. It is convenient to express the solution of this resulting homogeneous equation in terms of certain nondimensional parameters. To this end, let $F(t) = 0$ in Eq. (1.14), divide the result by $m$ and obtain the homogeneous differential equation

$$\ddot{x}(t) + 2\zeta\omega_n\dot{x}(t) + \omega_n^2 x(t) = 0 \qquad (1.45)$$

where $\omega_n$ is the natural frequency of the undamped oscillation, and is given by the second of Eqs. (1.32), and

$$\zeta = \frac{c}{2m\omega_n} \qquad (1.46)$$

is a nondimensional quantity known as the *viscous damping factor*. A general form of Eq. (1.45) was discussed in Sec. 1.4. Following the pattern established in Sec. 1.4, the solution of Eq. (1.45) can be assumed to have the form

$$x(t) = Ae^{st} \qquad (1.47)$$

Inserting Eq. (1.47) into Eq. (1.45) and dividing through by $Ae^{st}$, we obtain the characteristic equation

$$s^2 + 2\zeta\omega_n s + \omega_n^2 = 0 \qquad (1.48)$$

which has the roots

$$\begin{matrix} s_1 \\ s_2 \end{matrix} = (-\zeta \pm \sqrt{\zeta^2 - 1})\omega_n \qquad (1.49)$$

Clearly, the nature of the roots $s_1$ and $s_2$ depends on the value of $\zeta$. This dependence can be displayed in the $s$ plane, namely, the complex plane of Fig. 1.16, in the form of a diagram representing the locus of roots plotted as a function of the parameter $\zeta$. This allows for an instantaneous view of the effect of the parameter $\zeta$ on the natural behavior of the system, or, more specifically, on the system response. We see that for $\zeta = 0$ we obtain the imaginary roots $\pm i\omega_n$ leading to the harmonic solution discussed in Sec. 1.6. For $0 < \zeta < 1$ the roots $s_1$ and $s_2$ are complex conjugates, located symmetrically with respect to the real axis on a circle of radius $\omega_n$. As $\zeta$ approaches unity, the roots approach the point $-\omega_n$ on the real axis, and

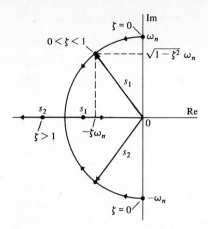

**Figure 1.16**

as $\zeta \to \infty$, $s_1 \to 0$ and $s_2 \to -\infty$. In the sequel, we relate the position of the roots $s_1$ and $s_2$ in the $s$ plane to the system behavior.

Inserting the roots given by (1.49) into (1.47), we can write the general solution

$$
\begin{aligned}
x(t) &= A_1 e^{s_1 t} + A_2 e^{s_2 t} \\
&= A_1 \exp\left[(-\zeta + \sqrt{\zeta^2 - 1})\omega_n t\right] + A_2 \exp\left[(-\zeta - \sqrt{\zeta^2 - 1})\omega_n t\right] \\
&= [A_1 \exp(\sqrt{\zeta^2 - 1}\,\omega_n t) + A_2 \exp(-\sqrt{\zeta^2 - 1}\,\omega_n t)]e^{-\zeta \omega_n t}
\end{aligned}
\tag{1.50}
$$

Solution (1.50) is in a form suitable for the cases in which $\zeta \geq 1$. For $\zeta > 1$ the motion is aperiodic and decaying exponentially with time. The exact shape of the curve depends on $A_1$ and $A_2$, which, in turn, can be evaluated in terms of the initial displacement $x_0$ and initial velocity $v_0$. The case $\zeta > 1$ is known as the *overdamped case*. Typical response curves are given in Fig. 1.17. In the special case in which $\zeta = 1$, Eq. (1.48) has a double root, $s_1 = s_2 = -\omega_n$. In this case the solution can be shown to have the form (see Prob. 1.29)

$$
x(t) = (A_1 + tA_2)e^{-\omega_n t}
\tag{1.51}
$$

which again represents an exponentially decaying response. The constants $A_1$ and $A_2$ depend on the initial conditions. The case $\zeta = 1$ is known as *critical damping*, and response curves for certain initial conditions are shown in Fig. 1.18. From the expression $\zeta = c/2m\omega_n$, we see that for $\zeta = 1$ the coefficient of viscous damping has the value $c_{cr} = 2m\omega_n = 2\sqrt{km}$. The importance of the concept should not be overstressed, because critical damping merely represents the borderline between the cases in which $\zeta > 1$ and $\zeta < 1$. It may be interesting to note, however, that for a given initial excitation a critically damped system tends to approach the equilibrium position the fastest (see Fig. 1.17c).

When $0 < \zeta < 1$, solution (1.50) is more conveniently written in the form

$$
\begin{aligned}
x(t) &= [A_1 \exp(i\sqrt{1 - \zeta^2}\,\omega_n t) + A_2 \exp(-i\sqrt{1 - \zeta^2}\,\omega_n t)]e^{-\zeta \omega_n t} \\
&= (A_1 e^{i\omega_d t} + A_2 e^{-i\omega_d t})e^{-\zeta \omega_n t}
\end{aligned}
\tag{1.52}
$$

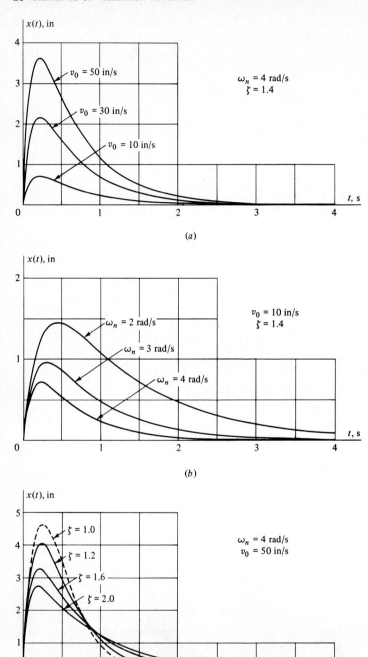

**Figure 1.17**

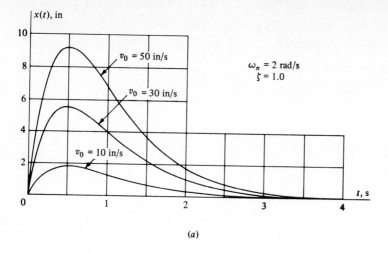

(a)

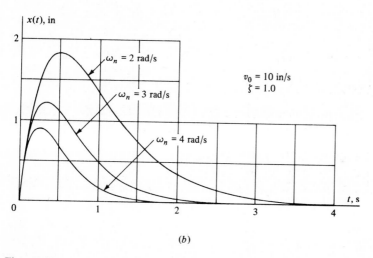

(b)

**Figure 1.18**

where

$$\omega_d = (1 - \zeta^2)^{1/2}\omega_n \tag{1.53}$$

is often called the *frequency of the damped free vibration*. From Eqs. (1.37), we can write $e^{\pm i\omega_d t} = \cos \omega_d t \pm i \sin \omega_d t$. Moreover, using the notation of Eqs. (1.38), Eq. (1.52) reduces to

$$x(t) = Ae^{-\zeta\omega_n t} \cos (\omega_d t - \phi) \tag{1.54}$$

which can be interpreted as an oscillatory motion with the constant frequency $\omega_d$ and phase angle $\phi$ but with the exponentially decaying amplitude $Ae^{-\zeta\omega_n t}$, where

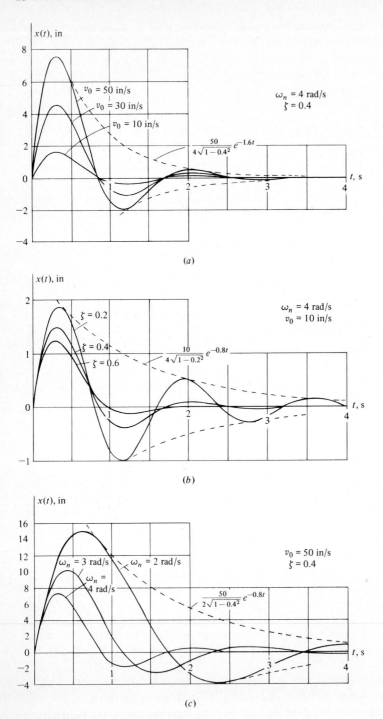

**Figure 1.19**

the constants $A$ and $\phi$ depend on the initial conditions. The case $0 < \zeta < 1$ is referred to as the *underdamped case*, and typical response curves are shown in Fig. 1.19. The curves $\pm Ae^{-\zeta\omega_n t}$ provide the envelope bounding the oscillatory response. Clearly, as $t \to \infty$, $x(t) \to 0$, so that the response eventually dies out, which represents the observed behavior of real systems.

**Example 1.7** Consider the system of Fig. 1.9 and calculate its response to the initial conditions $x(0) = 0$, $\dot{x}(0) = v_0$, for $\zeta > 1$, $\zeta = 1$, and $0 < \zeta < 1$.

For $\zeta > 1$ we make use of the formula (1.50), and write

$$x(0) = A_1 + A_2 = 0 \qquad A_2 = -A_1 \qquad (a)$$

so that now the solution has the form

$$x(t) = 2A_1 e^{-\zeta\omega_n t} \sinh \sqrt{\zeta^2 - 1}\, \omega_n t \qquad (b)$$

Differentiating Eq. (b) with respect to time, we obtain

$$\dot{x}(t) = 2A_1 \left( \sqrt{\zeta^2 - 1}\, \omega_n \cosh \sqrt{\zeta^2 - 1}\, \omega_n t - \zeta\omega_n \sinh \sqrt{\zeta^2 - 1}\, \omega_n t \right) e^{-\zeta\omega_n t} \qquad (c)$$

Letting $\dot{x}(0) = v_0$, Eq. (c) yields

$$2A_1 = \frac{v_0}{\sqrt{\zeta^2 - 1}\, \omega_n} \qquad (d)$$

It follows that for $\zeta > 1$ the general solution is

$$x(t) = \frac{v_0}{\sqrt{\zeta^2 - 1}\, \omega_n} e^{-\zeta\omega_n t} \sinh \sqrt{\zeta^2 - 1}\, \omega_n t \qquad (e)$$

For $\zeta = 1$, it is easy to show from Eq. (1.51) that $A_1 = 0$ and $A_2 = v_0$, so that the response is

$$x(t) = v_0 t e^{-\omega_n t} \qquad (f)$$

For $0 < \zeta < 1$, the initial displacement being equal to zero leads to $\phi = \pi/2$ in Eq. (1.54). Moreover, the amplitude is related to the initial velocity by $A = v_0/\omega_d$, so that Eq. (1.54) reduces to

$$x(t) = \frac{v_0}{\omega_d} e^{-\zeta\omega_n t} \sin \omega_d t \qquad \omega_d = \omega_n \sqrt{1 - \zeta^2} \qquad (g)$$

Expressions (e), (f), and (g), corresponding to overdamping, critical damping, and underdamping, are plotted in Figs. 1.17, 1.18, and 1.19, respectively, for the indicated values of the system parameters $\zeta$ and $\omega_n$ and the initial velocity $v_0$.

## 1.8 LOGARITHMIC DECREMENT

At times the amount of damping in a given system is not known and must be determined experimentally. We are concerned with the case in which damping is viscous and the system underdamped. As shown in Sec. 1.7, viscous damping causes the vibration to decay exponentially, where the exponent is a linear function of the damping factor $\zeta$. In this section, we wish to explore ways of determining $\zeta$ from the observation of this decay.

A convenient measure of the amount of damping in a single-degree-of-freedom system is provided by the extent to which the amplitude has fallen during one complete cycle of vibration. Let us denote by $t_1$ and $t_2$ the times corresponding to two consecutive displacements $x_1$ and $x_2$ measured one cycle apart (see Fig. 1.20), so that, using Eq. (1.54), we can form the ratio

$$\frac{x_1}{x_2} = \frac{Ae^{-\zeta\omega_n t_1} \cos(\omega_d t_1 - \phi)}{Ae^{-\zeta\omega_n t_2} \cos(\omega_d t_2 - \phi)} \tag{1.55}$$

Because $t_2 = t_1 + T$, where $T = 2\pi/\omega_d$ is the period of the damped oscillation, it follows that $\cos(\omega_d t_2 - \phi) = \cos[(\omega_d t - \phi) + \omega_d T] = \cos[(\omega_d t - \phi) + 2\pi] = \cos(\omega_d t_1 - \phi)$, so that Eq. (1.55) reduces to

$$\frac{x_1}{x_2} = \frac{e^{-\zeta\omega_n t_1}}{e^{-\zeta\omega_n(t_1 + T)}} = e^{\zeta\omega_n T} \tag{1.56}$$

In view of the exponential form of Eq. (1.56), it is customary to introduce the notation

$$\delta = \ln\frac{x_1}{x_2} = \zeta\omega_n T = \frac{2\pi\zeta}{\sqrt{1 - \zeta^2}} \tag{1.57}$$

where $\delta$ is known as the *logarithmic decrement*. Hence, to determine the amount of damping in the system, it suffices to measure any two consecutive displacements $x_1$

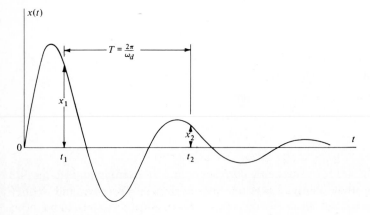

**Figure 1.20**

and $x_2$ one cycle apart, calculate the logarithmic decrement $\delta$ by taking the natural logarithm of the ratio $x_1/x_2$, and obtain $\zeta$ from

$$\zeta = \frac{\delta}{\sqrt{(2\pi)^2 + \delta^2}} \tag{1.58}$$

For small damping, $\delta$ is a small quantity, so that Eq. (1.58) can be approximated by

$$\zeta \cong \frac{\delta}{2\pi} \tag{1.59}$$

The damping factor $\zeta$ can also be determined by measuring two displacements separated by any number of complete cycles. Letting $x_1$ and $x_{j+1}$ be the amplitudes corresponding to the times $t_1$ and $t_{j+1} = t_1 + jT$, where $j$ is an integer, we conclude that

$$\frac{x_1}{x_{j+1}} = \frac{x_1}{x_2} \frac{x_2}{x_3} \frac{x_3}{x_4} \cdots \frac{x_j}{x_{j+1}} = (e^{\zeta \omega_n T})^j = e^{j\zeta \omega_n T} \tag{1.60}$$

because the ratio between any two consecutive displacements one cycle apart is equal to $e^{\zeta \omega_n T}$. Equation (1.60), in conjunction with Eq. (1.57), yields

$$\delta = \frac{1}{j} \ln \frac{x_1}{x_{j+1}} \quad = \frac{1}{n} \ln\left(\frac{x_o}{x_n}\right) \tag{1.61}$$

which can be introduced into Eq. (1.58), or Eq. (1.59), to obtain the viscous damping factor $\zeta$.

**Example 1.8** It was observed that the vibration amplitude of a damped single-degree-of-freedom system had fallen by 50 percent after five complete cycles. Assume that the system is viscously damped and calculate the damping factor $\zeta$.

Letting $j = 5$, Eq. (1.61) yields the logarithmic decrement

$$\delta = \frac{1}{5} \ln \frac{x_1}{x_6} = \frac{1}{5} \ln \frac{x_1}{0.5x_1} = \frac{1}{5} \ln 2 = \frac{1}{5}(0.6931) = 0.1386 \tag{a}$$

If the above value is inserted into Eq. (1.58), we must conclude that damping is relatively light. Hence, using Eq. (1.59), we obtain

$$\zeta \cong \frac{\delta}{2\pi} = \frac{0.1386}{2\pi} = 0.0221 \tag{b}$$

## 1.9 COULOMB DAMPING. DRY FRICTION

Coulomb damping arises when bodies slide on dry surfaces. For motion to begin, there must be a force acting upon the body that overcomes the resistance to motion caused by friction. The dry friction force is parallel to the surface and proportional to the force normal to the surface, where the latter is equal to the weight $W$ in the case of the mass-spring system shown in Fig. 1.21. The constant of proportionality is the static friction coefficient $\mu_s$, a number varying between 0 and 1 depending on the surface materials. Once motion is initiated, the force drops to $\mu_k W$, where $\mu_k$ is the kinetic friction coefficient, whose value is generally smaller than that of $\mu_s$. The friction force is opposite in direction to the velocity, and remains constant in magnitude as long as the forces acting on the mass $m$, namely, the inertia force and the restoring force due to the spring, are sufficient to overcome the dry friction. When these forces become insufficient, the motion simply stops.

Denoting by $F_d$ the magnitude of the damping force, where $F_d = \mu_k W$, the equation of motion can be written in the form

$$m\ddot{x} + F_d \operatorname{sgn}(\dot{x}) + kx = 0 \qquad (1.62)$$

where the symbol "sgn" denotes *sign of* and represents a function having the value $+1$ if its argument $\dot{x}$ is positive and the value $-1$ if its argument is negative. Mathematically, the function can be written as

$$\operatorname{sgn}(\dot{x}) = \frac{\dot{x}}{|\dot{x}|} \qquad (1.63)$$

Equation (1.62) is nonlinear, but it can be separated into two linear equations, one for positive and another one for negative $\dot{x}$, as follows:

$$m\ddot{x} + kx = -F_d \qquad \text{for } \dot{x} > 0 \qquad (1.64a)$$

$$m\ddot{x} + kx = F_d \qquad \text{for } \dot{x} < 0 \qquad (1.64b)$$

Although Eqs. (1.64) are nonhomogeneous, so that they can be regarded as representing forced vibration, the damping forces are passive in nature, so that discussion of these equations in this chapter is in order.

The solution of Eqs. (1.64) can be obtained for one time interval at a time, depending on the sign of $\dot{x}$. Without loss of generality, we assume that the motion starts from rest with the mass $m$ in the displaced position $x(0) = x_0$, where the initial displacement $x_0$ is sufficiently large that the restoring force in the spring exceeds the static friction force. Because in the ensuing motion the velocity is

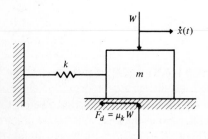

**Figure 1.21**

negative, we must solve Eq. (1.64b) first, where the equation can be written in the form

$$\ddot{x} + \omega_n^2 x = \omega_n^2 f_d \qquad \omega_n^2 = \frac{k}{m} \tag{1.65}$$

in which $f_d = F_d/k$ represents an equivalent displacement. Equation (1.65) is subject to the initial conditions $x(0) = x_0$, $\dot{x}(0) = 0$, so that its solution is simply

$$x(t) = (x_0 - f_d) \cos \omega_n t + f_d \tag{1.66}$$

which represents harmonic oscillation superposed on the average response $f_d$. Equation (1.66) is valid for $0 \leqslant t \leqslant t_1$, where $t_1$ is the time at which the velocity reduces to zero and the motion is about to reverse direction from left to right. Differentiating Eq. (1.66) with respect to time, we obtain

$$\dot{x}(t) = -\omega_n(x_0 - f_d) \sin \omega_n t \tag{1.67}$$

so that the lowest nontrivial value satisfying the condition $\dot{x}(t_1) = 0$ is $t_1 = \pi/\omega_n$, at which time the displacement is $x(t_1) = -(x_0 - 2f_d)$. If $x(t_1)$ is sufficiently large in magnitude to overcome the static friction, then the mass acquires a positive velocity, so that the motion must satisfy the equation

$$\ddot{x} + \omega_n^2 x = -\omega_n^2 f_d \tag{1.68}$$

where $x(t)$ is subject to the initial conditions $x(t_1) = -(x_0 - 2f_d)$, $\dot{x}(t_1) = 0$. The solution of Eq. (1.68) is

$$x(t) = (x_0 - 3f_d) \cos \omega_n t - f_d \tag{1.69}$$

Compared to (1.66), the harmonic component in solution (1.69) has an amplitude smaller by $2f_d$ and a negative constant component, namely $-f_d$. Solution (1.69) is valid in the time interval $t_1 \leqslant t \leqslant t_2$, where $t_2$ is the next value of time at which the velocity reduces to zero. This value is $t_2 = 2\pi/\omega_n$, at which time the velocity is ready to reverse direction once again, this time from right to left. The displacement at $t = t_2$ is $x(t_2) = x_0 - 4f_d$.

The above procedure can be repeated for $t > t_2$, every time switching back and forth between Eqs. (1.64a) and (1.64b). However, a pattern seems to emerge, rendering this task unnecessary. Over each half-cycle the motion consists of a constant component and a harmonic component with frequency equal to the natural frequency $\omega_n$ of the simple mass-spring system, where the duration of every half-cycle is equal to $\pi/\omega_n$. The average value of the solutions alternates between $f_d$ and $-f_d$, and at the end of each half-cycle the displacement magnitude is reduced by $2f_d = 2F_d/k$. It follows that for Coulomb damping the decay is linear with time, as opposed to the exponential decay for viscous damping. The motion stops abruptly when the displacement at the end of a given half-cycle is not sufficiently large for the restoring force in the spring to overcome the static friction. This occurs at the end of the half-cycle for which the amplitude of the harmonic component is

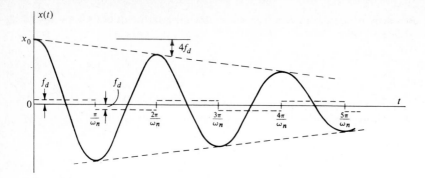

**Figure 1.22**

smaller than $2f_d$. Letting $n$ be the number of the half-cycle just prior to the cessation of motion, we conclude that $n$ is the smallest integer satisfying the inequality

$$x_0 - (2n - 1)f_d < 2f_d \tag{1.70}$$

The plot $x(t)$ versus $t$ can be obtained by combining solutions (1.66), (1.69), etc. Such a plot is shown in Fig. 1.22.

**Example 1.9** Let the parameters of the system of Fig. 1.21 have the values $m = 2 \ \text{lb} \cdot \text{s}^2/\text{in}$, $k = 800 \ \text{lb/in}$, and $\mu_k = 0.1$, and calculate the decay per cycle and the number of half-cycles until oscillation stops if the initial conditions are $x(0) = x_0 = 1.2 \ \text{in}$, $\dot{x}(0) = 0$.
  The decay per cycle is

$$4f_a = 4\frac{F_a}{k} = 4\frac{\mu_k mg}{k} = 4\frac{0.1 \times 2 \times 32.2 \times 12}{800}$$

$$= 4 \times 0.0966 = 0.3864 \ \text{in} \tag{a}$$

Moreover, $n$ must be the smallest integer satisfying the inequality

$$1.2 - (2n - 1) \times 0.0966 < 2 \times 0.0966 \tag{b}$$

from which we conclude that the oscillation stops after the half-cycle $n = 6$ with $m$ in the position $x(t_6) = x_0 - 12f_d = 1.2 - 12 \times 0.0966 = 0.0408 \ \text{in}$.

## PROBLEMS

**1.1** Consider two dashpots with viscous damping coefficients $c_1$ and $c_2$ and calculate the equivalent viscous damping coefficient for the cases in which the dashpots are arranged in parallel and in series, respectively.

**1.2** Consider the system of Fig. 1.23 and obtain an expression for the equivalent spring. Then, derive the differential equation of motion.

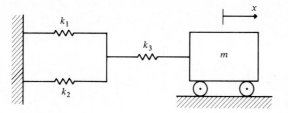

Figure 1.23

**1.3** Consider the system of Prob. 1.2, let $k_1 = k_2 = 500$ lb/in $(8.7563 \times 10^4$ N/m), $k_3 = 1500$ lb/in $(2.6269 \times 10^5$ N/m), and $m = 1.5$ lb·s²/in (262.69 kg) and calculate the system natural frequency.

**1.4** A buoy of uniform cross-sectional area $A$ and mass $m$ is depressed a distance $x$ from the equilibrium position, as shown in Fig. 1.24, and then released. Derive the differential equation of motion and obtain the natural frequency of oscillation. The mass density of the liquid in which the buoy floats is $\rho$.

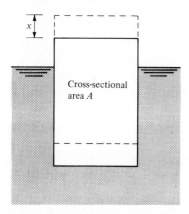

Cross-sectional area $A$

Figure 1.24

**1.5** The system shown in Fig. 1.25, consisting of an unknown mass $m$ and a spring with unknown spring constant $k$, has been observed to oscillate naturally with the frequency $\omega_n = 100$ rad/s. Determine the mass $m$ and spring constant $k$ knowing that when a mass $M = 0.9$ kg is added the modified natural frequency is $\Omega_n = 80$ rad/s.

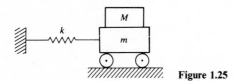

Figure 1.25

**1.6** Derive the differential equation of motion for the system shown in Fig. 1.26 and obtain the period of oscillation. Denote the mass density of the liquid by $\rho$ and the total length of the column of liquid by $L$.

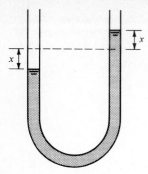

**Figure 1.26**

**1.7** The hinges of the rectangular door shown in Fig. 1.27 are mounted on a line making an angle $\alpha$ with respect to the vertical. Assume that the door has uniform mass distribution and determine the natural frequency of oscillation.

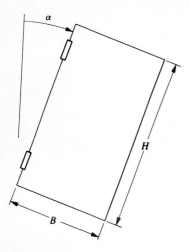

**Figure 1.27**

**1.8** To determine the centroidal mass moment of inertia $I_C$ of a tire mounted on a wheel, the system is suspended on a knife-edge, as shown in Fig. 1.28, and the natural period of oscillation $T$ is measured. Derive a formula for $I_C$ in terms of the mass $m$, the period $T$ of the system, and the radius $r$ from the center $C$ to the knife-edge.

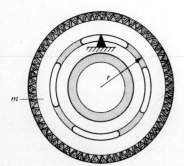

**Figure 1.28**

**1.9** A bead of mass $m$ is suspended on a massless string, as shown in Fig. 1.29. Assume that the string is subjected to the tension $T$, and that this tension does not change throughout the motion, and derive the differential equation for small motions from equilibrium, as well as the natural frequency of oscillation.

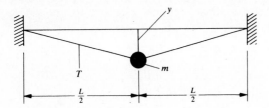

**Figure 1.29**

**1.10** A connecting rod of mass $m = 3 \times 10^{-3}$ kg and centroidal mass moment of inertia $I_C = 0.432 \times 10^{-4}$ kg·m² is suspended on a knife-edge about the upper inner surface of the wrist-pin bearing, as shown in Fig. 1.30. When disturbed slightly, the rod was observed to oscillate with the natural frequency $\omega_n = 6$ rad/s. Determine the distance $h$ between the support and the center of mass $C$.

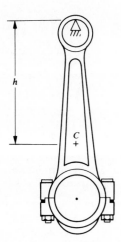

**Figure 1.30**

**1.11** A mass $m$ is attached to the end of a massless elastic blade of length $L$ and flexural stiffness $EI$ (see Fig. 1.31). Derive the equivalent spring constant of the blade and write the equation of motion for the transverse displacement of $m$. Calculate the period $T$.

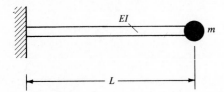

**Figure 1.31**

**1.12** A disk of mass moment of inertia $I$ is attached to the end of a massless uniform shaft of length $L$ and torsional rigidity $GJ$ (see Fig. 1.32). Derive the equation for the torsional vibration of the disk, and obtain the natural frequency of vibration.

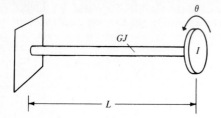

**Figure 1.32**

**1.13** A massless rigid bar is hinged at $O$, as shown in Fig. 1.33. Determine the natural frequency of oscillation of the system for the parameters $k_1 = 2500$ lb/in $(4.3782 \times 10^5$ N/m), $k_2 = 900$ lb/in $(1.5761 \times 10^5$ N/m), $m = 1$ lb·s²/in (175.13 kg), $a = 80$ in (2.03 m), and $b = 100$ in (2.54 m).

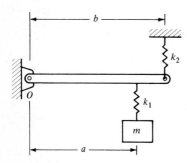

**Figure 1.33**

**1.14** A circular rigid disk of mass polar moment of inertia $I_p = 0.8$ kg·m² is mounted on a circular shaft made of two segments of different diameters and lengths, as shown in Fig. 1.34. The shaft is fixed at both ends. Let the shear modulus of the shaft material be $G = 80 \times 10^9$ N/m² and obtain the natural frequency of angular oscillation of the disk.

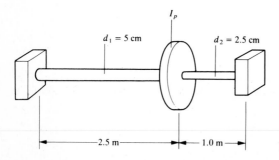

**Figure 1.34**

**1.15** The one-story building shown in Fig. 1.35 can be modeled in the first approximation as a single-degree-of-freedom system by regarding the columns as massless and the roof as rigid. Derive the differential equation of motion and determine the natural frequency. Assume that the mass $M$ can only translate horizontally, so that the columns undergo no rotation at the top.

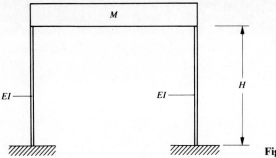

**Figure 1.35**

**1.16** Two gears $A$ and $B$ of mass moments of inertia $I_A$ and $I_B$, respectively, are attached to circular shafts of equal stiffness $GJ/L$ (Fig. 1.36). Derive the differential equation for the system and determine the natural frequency of the system for the case $R_A/R_B = n$. *Hint:* Draw one free-body diagram for each gear, recognizing that the reaction forces on the gears at the point of contact are equal in magnitude and opposite in direction, and that the angular motion of gear $B$ is $n$ times the angular motion of gear $A$.

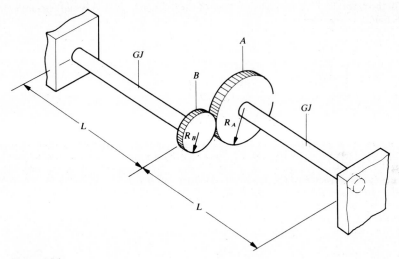

**Figure 1.36**

**1.17** A mass $m$ is suspended on a massless beam of bending stiffness $EI$ through a spring of stiffness $k$, as shown in Fig. 1.37. Derive the differential equation of motion and determine the natural frequency of oscillation.

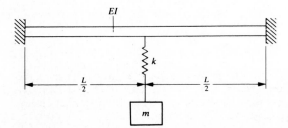

**Figure 1.37**

**1.18** The circular shaft shown in Fig. 1.38 has the torsional stiffness $GJ(x) = GJ[1 - \frac{1}{2}(x/L)^2]$. The shaft is fixed at $x = 0$ and has a rigid disk of polar mass moment of inertia equal to $I$ attached to the end $x = L$. Assume that the mass of the shaft is negligible, derive the differential equation of motion and obtain the natural frequency of oscillation.

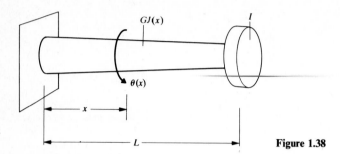

**Figure 1.38**

**1.19** A cantilever beam made of two sections has a lumped mass at $x = L$, as shown in Fig. 1.39. Assume that the mass of the beam can be ignored, derive the differential equation of motion and obtain the period of oscillation.

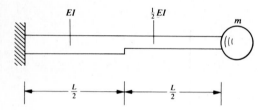

**Figure 1.39**

**1.20** A uniform rigid bar of mass $m$ is suspended by two inextensible massless strings of length $L$ (see Fig. 1.40). Such a system is referred to as a bifilar pendulum. Derive the differential equation for the oscillation $\theta$ about the vertical axis through the bar center. Note that the mass moment of inertia of the bar about its center is $I_C = \frac{1}{3}ma^2$.

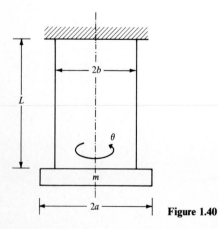

**Figure 1.40**

**1.21** Obtain the natural frequency of the system shown in Fig. 1.41. The spring is linear and the pulley has a mass moment of inertia $I$ about the center $O$. Let $k = 2500$ lb/in $(4.3782 \times 10^5$ N/m), $I = 600$ lb·in·s$^2$ (67.79 N·m·s$^2$), $m = 2.5$ lb·s$^2$/in (437.82 kg), and $R = 20$ in (0.51 m).

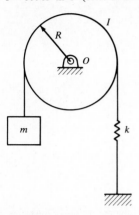

**Figure 1.41**

**1.22** A uniform disk of radius $r$ rolls without slipping inside a circular track of radius $R$, as shown in Fig. 1.42. Derive the equation of motion for arbitrarily large angles $\theta$. Then, show that in the neighborhood of the trivial equilibrium $\theta = 0$ the system behaves like a harmonic oscillator, and determine the natural frequency of oscillation.

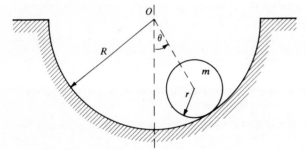

**Figure 1.42**

**1.23** The pendulum shown in Fig. 1.43 is attached to a linear spring of stiffness $k$. Derive the differential equation of motion of the system, then linearize the equation and determine the natural frequency of oscillation.

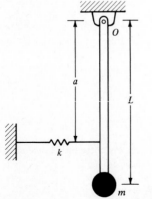

**Figure 1.43**

**1.24** A uniform bar of total mass $m$ and length $l$ rotates with the constant angular velocity $\Omega$ about a vertical axis, as shown in Fig. 1.44. Denote by $\theta$ the angle between the vertical axis and the bar, and:

(a) Determine the equilibrium positions as expressed by the constant angle $\theta_0$.
(b) Derive the differential equation for small motions $\theta_1$ about $\theta_0$.
(c) Determine a stability criterion for each equilibrium position based on the requirement that the motion $\theta_1$ be harmonic.
(d) Calculate the natural frequency of the oscillation $\theta_1$ for the stable cases.
(e) Determine the natural frequency for very large $\Omega$ and draw conclusions.

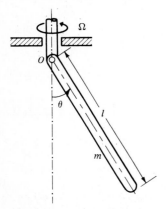

**Figure 1.44**

**1.25** The inverted pendulum of Fig. 1.45 is supported by a linear spring of stiffness $k$, as shown. Denote by $\theta$ the angle between the pendulum and the vertical through the hinge $O$ and:

(a) Determine the equilibrium positions, as expressed by the angle $\theta_0$.
(b) Derive the differential equation for small angular motions $\theta_1$ about $\theta_0$.
(c) Determine a stability criterion based on the requirement that the motion $\theta_1$ be harmonic.
(d) Calculate the natural frequency of the oscillation $\theta_1$.

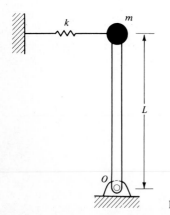

**Figure 1.45**

**1.26** An $L$-shaped massless rigid member is hinged at point $O$ and has a mass $m$ at the tip. The member is supported by a spring of stiffness $k$, as shown in Fig. 1.46. It is required to:

(a) Determine the equilibrium position, as expressed by the angle $\theta_0$ about $O$.
(b) Derive the differential equation for small angular motions $\theta_1$ about $\theta_0$.

(c) Calculate the natural frequency of oscillation $\theta_1$.
(d) Determine the height $H$ for which the system becomes unstable.

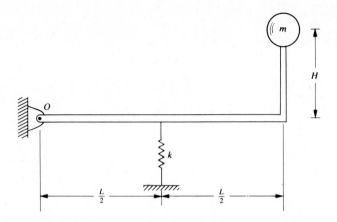

**Figure 1.46**

**1.27** The system of Prob. 1.12 is immersed in viscous liquid, so that there is a damping force $c\dot{\theta}$ resisting the motion. Calculate the period of the damped oscillation, where the period refers to the harmonic factor in the response.

**1.28** The simple pendulum of Fig. 1.47 is immersed in viscous liquid so that there is a force $c\dot{\theta}$ resisting the motion. Derive the equation of motion for arbitrary amplitudes $\theta$, then linearize the equation and obtain the frequency of the damped oscillation.

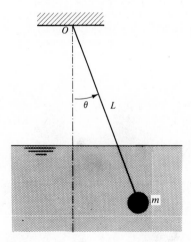

**Figure 1.47**

**1.29** Show that solution (1.50) can also be written in the form

$$x(t) = (C_1 \cosh \sqrt{\zeta^2 - 1}\, \omega_n t + C_2 \sinh \sqrt{\zeta^2 - 1}\, \omega_n t)e^{-\zeta\omega_n t}$$

Then let $\zeta \to 1$, set $C_1 = A_1$ and $C_2\sqrt{\zeta^2 - 1}\,\omega_n = A_2$, and prove Eq. (1.51).

**1.30** Calculate the frequency of the damped oscillation of the system shown in Fig. 1.48 for the values $k = 4000$ lb/in $(7.0051 \times 10^5$ N/m), $c = 20$ lb·s/in $(3502.54$ N·s/m), $m = 10$ lb·s$^2$/in $(1751.27$ kg), $a = 50$ in $(1.27$ m), and $L = 100$ in $(2.54$ m). Determine the value of the critical damping.

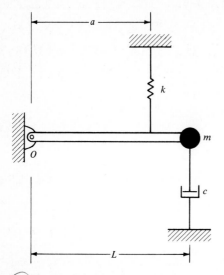

**Figure 1.48**

**1.31** Consider the system of Example 1.7, and determine the response $x(t)$ to the initial conditions $x(0) = x_0$, $\dot{x}(0) = 0$ for $\zeta > 1$, and $\zeta < 1$.

**1.32** Plot the response of the system of Prob. 1.31 to the initial displacement $x_0 = 10$ in $(0.254$ m) for the values of the damping factor $\zeta = 2, 1, 0.1$. Let $\omega_n = 5$ rad/s and consider the time interval $0 \leqslant t \leqslant 6$ s.

**1.33** Devise a vector construction representing Eq. (1.54).

**1.34** From the observation of the oscillation of a damped system it was determined that the maximum displacement amplitude during the second cycle is 75 percent of the first. Calculate the damping factor $\zeta$. Determine the maximum displacement amplitude after $4\frac{1}{2}$ cycles as a fraction of the first amplitude.

**1.35** Prove inequality (1.70).

**1.36** Plot $x(t)$ versus $t$ for the system of Example 1.9.

# FORCED RESPONSE OF
# SINGLE-DEGREE-OF-FREEDOM
# LINEAR SYSTEMS

## 2.1 GENERAL CONSIDERATIONS

A very important subject in vibrations is the response of systems to external excitations. The excitations, for example, can be in the form of initial displacements, initial velocities, or both. However, the excitation can also be in the form of forces which persist for an extended period of time. The response to such forces is called *forced response* and is the subject of this chapter. For linear systems, it is possible to obtain the response to initial conditions and external forces separately, and then combine them to obtain the total response of the system. This is based on the so-called principle of superposition.

The procedure for obtaining the response of a system to external forces depends to a large extent on the type of excitation. In this chapter, we follow a pattern of increasing complexity, beginning the discussion with simple harmonic excitation, extending it to periodic excitation and culminating with nonperiodic excitation. Because of its fundamental nature and because it has a multitude of practical applications, the case of harmonic excitation is discussed in great detail. The principle of superposition receives special attention, as it forms the basis for the analysis of linear systems. A rigorous discussion of the principle is provided. The case of periodic excitation can be reduced to that of harmonic excitation by regarding the periodic forcing function as a superposition of harmonic functions through the use of standard Fourier series. To discuss the response to nonperiodic excitation, the impulse response and the convolution integral are introduced. Finally, the system response by the Laplace transformation method is introduced, and the many advantages of this last method are pointed out.

## 2.2 RESPONSE OF FIRST-ORDER SYSTEMS TO HARMONIC EXCITATION. FREQUENCY RESPONSE

The differential equation of motion for a first-order system in the form of a damper-spring system was shown in Sec. 1.3 to be

$$c\dot{x}(t) + kx(t) = F(t) \tag{2.1}$$

where all the quantities are as defined in Sec. 1.3. The homogeneous solution of Eq. (2.1), obtained by letting $F(t) = 0$, was discussed in Sec. 1.3 and will not be repeated here. In this section we focus our attention on the particular solution, which represents the response to external forces. First, we consider the simplest case, namely, the response to harmonic excitation. To this end, it is convenient to let the force $F(t)$ have the form

$$F(t) = kf(t) = kA \cos \omega t \tag{2.2}$$

where $\omega$ is the *excitation frequency*, sometimes referred to as the *driving frequency*. Note that $f(t)$ and $A$ have units of displacement. The reason for writing the excitation in the form (2.2) is so as to permit expressing the response in terms of a nondimensional ratio, as we shall see shortly. Nondimensional ratios often enhance the usefulness of a solution by extending its applicability to a large variety of cases. Inserting Eq. (2.2) into Eq. (2.1) and dividing through by $c$, we obtain

$$\dot{x}(t) + ax(t) = Aa \cos \omega t \tag{2.3}$$

where

$$a = \frac{k}{c} = \frac{1}{\tau} \tag{2.4}$$

in which $\tau$ is the time constant, first encountered in Sec. 1.5.

The solution of the homogeneous differential equation, obtained by letting $A = 0$ in Eq. (2.3), decays exponentially with time (see Sec. 1.5), for which reason it is called the *transient solution*. On the other hand, the particular solution does not vanish as time unfolds and is known as the *steady-state solution* to the harmonic excitation in question. By virtue of the fact that the system is linear, the principle of superposition (see Sec. 2.11) holds, so that the homogeneous solution and the particular solution can be obtained separately and then combined linearly to obtain the complete solution.

Because the excitation force is harmonic, it can be verified easily that the steady-state response is also harmonic and has the same frequency $\omega$. Moreover, because Eq. (2.3) involves the function $x(t)$ and its first derivative $\dot{x}(t)$, the response must contain not only cos $\omega t$ but also sin $\omega t$. Hence, let us assume that the steady-state solution of Eq. (2.3) has the form

$$x(t) = C_1 \sin \omega t + C_2 \cos \omega t \tag{2.5}$$

where $C_1$ and $C_2$ are constants yet to be determined. Inserting solution (2.5) into Eq. (2.3), we obtain

$$\omega(C_1 \cos \omega t - C_2 \sin \omega t) + a(C_1 \sin \omega t + C_2 \cos \omega t) = Aa \cos \omega t \tag{2.6}$$

Equation (2.6) can be satisfied only if the coefficients of sin $\omega t$ on the one hand and the coefficients of cos $\omega t$ on the other hand are the same on both sides of the equation. This, in turn, requires the satisfaction of the equations

$$aC_1 - \omega C_2 = 0$$
$$\omega C_1 + aC_2 = Aa \tag{2.7}$$

which represent two algebraic equations in the unknowns $C_1$ and $C_2$. Their solution is

$$C_1 = \frac{Aa\omega}{a^2 + \omega^2} \qquad C_2 = \frac{Aa^2}{a^2 + \omega^2} \tag{2.8}$$

Introducing Eqs. (2.8) into Eq. (2.5), we obtain the steady-state solution

$$x(t) = \frac{Aa}{a^2 + \omega^2}(\omega \sin \omega t + a \cos \omega t) \tag{2.9}$$

Solution (2.9) can be expressed in a more convenient form. To this end, let us introduce the notation

$$\frac{\omega}{(a^2 + \omega^2)^{1/2}} = \sin \phi \qquad \frac{a}{(a^2 + \omega^2)^{1/2}} = \cos \phi \tag{2.10}$$

Then, Eq. (2.9) can be written as

$$x(t) = X(\omega) \cos (\omega t - \phi) \tag{2.11}$$

where

$$X(\omega) = \frac{A}{[1 + (\omega/a)^2]^{1/2}} \tag{2.12}$$

is the *amplitude* and

$$\phi(\omega) = \tan^{-1} \frac{\omega}{a} \tag{2.13}$$

is the *phase angle*. Both $X$ and $\phi$ are functions of the excitation frequency $\omega$.

The response to harmonic excitation can be obtained more conveniently by using complex vector representation of the excitation and the response. From Sec. 1.6, we recall that

$$e^{i\omega t} = \cos \omega t + i \sin \omega t \tag{2.14}$$

where $i = \sqrt{-1}$, so that Eq. (2.2) can be rewritten as

$$F(t) = kf(t) = kA \cos \omega t = \text{Re } kAe^{i\omega t} \tag{2.15a}$$

where Re denotes the real part of the function. Similarly, in the case of sinusoidal excitation we can write

$$F(t) = kf(t) = kA \sin \omega t = \text{Im } kAe^{i\omega t} \tag{2.15b}$$

where Im denotes the imaginary part of the function. Hence, we can rewrite Eq. (2.3) in the form

$$\dot{x}(t) + ax(t) = aAe^{i\omega t} \tag{2.16}$$

Then, if the excitation is given by Eq. (2.15a), we retain the real part of the response and if the excitation is given by Eq. (2.15b), we retain the imaginary part of the response.

Concentrating once again on the steady-state response, we write the solution of Eq. (2.16) in the form

$$x(t) = X(i\omega)e^{i\omega t} \tag{2.17}$$

Inserting Eq. (2.17) into Eq. (2.16), we obtain

$$Z(i\omega)X(i\omega)e^{i\omega t} = aAe^{i\omega t} \tag{2.18}$$

where

$$Z(i\omega) = a + i\omega \tag{2.19}$$

is the *impedance function* for this first-order system. Dividing Eq. (2.18) through by $e^{i\omega t}$ and solving for $X(i\omega)$, we obtain

$$X(i\omega) = \frac{aA}{Z(i\omega)} = \frac{aA}{a + i\omega} = \frac{A}{1 + i\omega\tau} \tag{2.20}$$

where $\tau = 1/a = c/k$ is the time constant. It will prove convenient to introduce the nondimensional ratio

$$G(i\omega) = \frac{X(i\omega)}{A} = \frac{1}{1 + i\omega\tau} = \frac{1 - i\omega\tau}{1 + (\omega\tau)^2} \tag{2.21}$$

where $G(i\omega)$ is known as the *frequency response*. Inserting Eq. (2.21) into Eq. (2.17), we can write the harmonic response in the general form

$$x(t) = AG(i\omega)e^{i\omega t} \tag{2.22}$$

But, the frequency response $G(i\omega)$, as any complex function, can be expressed as

$$G(i\omega) = |G(i\omega)|e^{-i\phi} \tag{2.23}$$

where $|G(i\omega)|$ is the magnitude and $\phi$ is the phase angle of $G(i\omega)$. Introducing Eq. (2.23) into Eq. (2.22), we obtain

$$x(t) = A|G(i\omega)|e^{i(\omega t - \phi)} \tag{2.24}$$

so that if the excitation is in the form of Eq. (2.15a), the response is the real part of Eq. (2.24), or

$$x(t) = A|G(i\omega)| \cos(\omega t - \phi) \tag{2.25a}$$

and if the excitation is in the form of Eq. (2.15b), the response is the imaginary part of Eq. (2.24), or

$$x(t) = A|G(i\omega)| \sin(\omega t - \phi) \tag{2.25b}$$

From Eqs. (2.25) it follows that, if the excitation is harmonic with the frequency $\omega$, the response is also harmonic and has the same frequency. Hence, in studying the nature of the response, plotting the response as a function of time will not be very rewarding. Considerably more insight into the system behavior can be gained by examining how the system responds as the driving frequency $\omega$ varies. In particular, plots of the magnitude $|G(i\omega)|$ and of the phase angle $\phi$ versus the frequency $\omega$ are very revealing. From complex algebra, if we consider Eq. (2.21), then we can write

$$|G(i\omega)| = [\text{Re}^2 G(i\omega) + \text{Im}^2 G(i\omega)]^{1/2} = \frac{1}{[1 + (\omega\tau)^2]^{1/2}} \tag{2.26}$$

and we note from Eq. (2.12) that $|G(i\omega)| = X(\omega)/A$. The plot $|G(i\omega)|$ versus $\omega\tau$ is shown in Fig. 2.1. We observe from Fig. 2.1 that for small driving frequencies the magnitude $|G(i\omega)|$ is close to 1 and for high frequencies the magnitude approaches 0. Hence, the system permits low-frequency harmonics to go through undistorted, but it attenuates greatly high-frequency harmonics. For this reason a first-order system is known as a *low-pass filter*. To obtain the phase angle, we recall first that $e^{-i\phi} = \cos\phi - i\sin\phi$. Then, using Eqs. (2.21) and (2.23), we can write

$$\phi = \tan^{-1}\left[\frac{-\text{Im } G(i\omega)}{\text{Re } G(i\omega)}\right] = \tan^{-1}\omega\tau \tag{2.27}$$

which checks with Eq. (2.13). The plot $\phi$ versus $\omega\tau$ is shown in Fig. 2.2. The plots $|G(i\omega)|$ versus $\omega\tau$ and $\phi$ versus $\omega\tau$ are known as *frequency-response plots*.

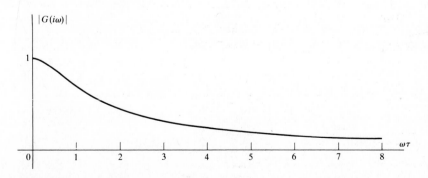

**Figure 2.1**

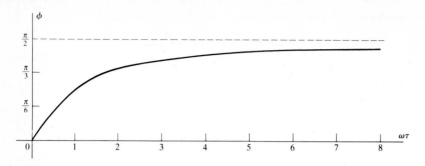

**Figure 2.2**

The magnitude $|G(i\omega)|$ of the frequency response can be interpreted geometrically by observing from Eq. (2.24) that the magnitude of the force in the spring is

$$|F_s(t)| = k|x(t)| = kA|G(i\omega)| \tag{2.28}$$

Moreover, from Eqs. (2.15), the magnitude of the harmonic excitation is

$$|F(t)| = kA \tag{2.29}$$

Hence, combining Eqs. (2.28) and (2.29), we can write

$$|G(i\omega)| = \frac{|F_s(t)|}{|F(t)|} \tag{2.30}$$

or, the magnitude of the frequency response is equal to ratio of the magnitude of the spring force $F_s(t)$ to the magnitude of the excitation force $F(t)$.

## 2.3 RESPONSE OF SECOND-ORDER SYSTEMS TO HARMONIC EXCITATION

As shown in Sec. 1.3, the differential equation of motion of a second-order system in the form of a mass-damper-spring system is

$$m\ddot{x}(t) + c\dot{x}(t) + kx(t) = F(t) \tag{2.31}$$

where all the quantities are as defined in Sec. 1.3. The free response was discussed in Sec. 1.6, so that in this section we concentrate on the forced response, and in particular on the response to harmonic excitation. This discussion follows the pattern established in Sec. 2.2 for both real and complex analysis.

Inserting Eq. (2.2) into Eq. (2.31) and dividing through by $m$, we obtain

$$\ddot{x}(t) + 2\zeta\omega_n\dot{x}(t) + \omega_n^2 x(t) = \frac{k}{m} f(t) = \omega_n^2 A \cos \omega t \tag{2.32}$$

where $\zeta$ is the viscous damping factor and $\omega_n$ the natural frequency of undamped oscillation (see Secs. 1.6 and 1.7). Letting the solution of Eq. (2.32) have the form

(2.5), we can write

$$-\omega^2(C_1 \sin \omega t + C_2 \cos \omega t) + \omega 2\zeta\omega_n(C_1 \cos \omega t - C_2 \sin \omega t)$$

$$+\omega_n^2(C_1 \sin \omega t + C_2 \cos \omega t)$$

$$= (\omega_n^2 - \omega^2)(C_1 \sin \omega t + C_2 \cos \omega t) + 2\zeta\omega\omega_n(C_1 \cos \omega t - C_2 \sin \omega t)$$

$$= \omega_n^2 A \cos \omega t \tag{2.33}$$

Equating the coefficients of $\sin \omega t$ and $\cos \omega t$, respectively, on both sides of the equation, we obtain the two algebraic equations

$$(\omega_n^2 - \omega^2)C_1 - 2\zeta\omega\omega_n C_2 = 0$$

$$2\zeta\omega\omega_n C_1 + (\omega_n^2 - \omega^2)C_2 = \omega_n^2 A \tag{2.34}$$

which have the solution

$$C_1 = \frac{\omega_n^2 A 2\zeta\omega\omega_n}{(\omega_n^2 - \omega^2)^2 + (2\zeta\omega\omega_n)^2} = \frac{2\zeta\omega/\omega_n}{[1 - (\omega/\omega_n)^2]^2 + (2\zeta\omega/\omega_n)^2} A$$

$$C_2 = \frac{\omega_n^2 A(\omega_n^2 - \omega^2)}{(\omega_n^2 - \omega^2)^2 + (2\zeta\omega\omega_n)^2} = \frac{1 - (\omega/\omega_n)^2}{[1 - (\omega/\omega_n)^2]^2 + (2\zeta\omega/\omega_n)^2} A \tag{2.35}$$

Introducing Eqs. (2.35) into Eq. (2.5), we obtain the steady-state solution

$$x(t) = \frac{A}{[1 - (\omega/\omega_n)^2]^2 + (2\zeta\omega/\omega_n)^2} \left\{ \frac{2\zeta\omega}{\omega_n} \sin \omega t + \left[ 1 - \left( \frac{\omega}{\omega_n} \right)^2 \right] \cos \omega t \right\} \tag{2.36}$$

Next let

$$\frac{2\zeta\omega/\omega_n}{\{[1 - (\omega/\omega_n)^2]^2 + (2\zeta\omega/\omega_n)^2\}^{1/2}} = \sin \phi$$

$$\frac{1 - (\omega/\omega_n)^2}{\{[1 - (\omega/\omega_n)^2]^2 + (2\zeta\omega/\omega_n)^2\}^{1/2}} = \cos \phi \tag{2.37}$$

so that the harmonic response can be written in the compact form

$$x(t) = X(\omega) \cos (\omega t - \phi) \tag{2.38}$$

where

$$X(\omega) = \frac{A}{\{[1 - (\omega/\omega_n)^2]^2 + (2\zeta\omega/\omega_n)^2\}^{1/2}} \tag{2.39}$$

is the amplitude and

$$\phi = \tan^{-1} \frac{2\zeta\omega/\omega_n}{1 - (\omega/\omega_n)^2} \tag{2.40}$$

is the phase angle.

$$G = \frac{\pm}{1 - (\omega/\omega_n)^2 + i2\zeta\frac{\omega}{\omega_n}}$$

Next, let us reproduce the above results by working with complex vectors. The motivation for this is that in future cases involving both odd-order and even-order derivatives we shall opt for the complex analysis. Hence, instead of Eq. (2.32), we consider

$$\ddot{x}(t) + 2\zeta\omega_n\dot{x}(t) + \omega_n^2 x(t) = \omega_n^2 A e^{i\omega t} \tag{2.41}$$

Then, letting the steady-state response have the form

$$x(t) = X(i\omega)e^{i\omega t} \tag{2.42}$$

Eq. (2.41) yields

$$Z(i\omega)X(i\omega)e^{i\omega t} = \omega_n^2 A e^{i\omega t} \tag{2.43}$$

where $Z(i\omega)$ is the impedance function, which in the case at hand has the expression

$$Z(i\omega) = \omega_n^2 - \omega^2 + i2\zeta\omega\omega_n \tag{2.44}$$

Inserting Eq. (2.44) into Eq. (2.43), dividing through by $e^{i\omega t}$ and solving for $X(i\omega)$, we obtain

$$X(i\omega) = \frac{\omega_n^2 A}{Z(i\omega)} = \frac{\omega_n^2 A}{\omega_n^2 - \omega^2 + i2\zeta\omega\omega_n} = \frac{A}{1 - (\omega/\omega_n)^2 + i2\zeta\omega/\omega_n} \tag{2.45}$$

so that the frequency response is

$$G(i\omega) = \frac{X(i\omega)}{A} = \frac{1}{1 - (\omega/\omega_n)^2 + i2\zeta\omega/\omega_n} \tag{2.46}$$

Following the pattern of Sec. 2.2, the harmonic response is

$$x(t) = AG(i\omega)e^{i\omega t} = A|G(i\omega)|e^{i(\omega t - \phi)} \tag{2.47}$$

where

$$|G(i\omega)| = \frac{1}{\{[1 - (\omega/\omega_n)^2]^2 + (2\zeta\omega/\omega_n)^2\}^{1/2}} \tag{2.48}$$

is the magnitude of the frequency response and is known as the *magnification factor* and $\phi$ is the phase angle and is as given by Eq. (2.40).

Considerable insight into the system behavior can be gained by examining how the magnitude and phase angle of the frequency-response function $G(i\omega)$ change with the driving frequency $\omega$. Figure 2.3 shows plots of $|G(i\omega)|$ versus $\omega/\omega_n$ for various values of $\zeta$, which permit the observation that damping tends to diminish amplitudes and to shift the peaks to the left of the vertical through $\omega/\omega_n = 1$. To find the values at which the peaks of the curves occur, we use the standard technique of calculus for finding stationary values of a function, namely, we differentiate Eq. (2.48) with respect to $\omega$ and set the result equal to zero. This leads us to the conclusion that the peaks occur at

$$\omega = \omega_n(1 - 2\zeta^2)^{1/2} \tag{2.49}$$

indicating that the maxima do not occur at the undamped natural frequency $\omega_n$ but

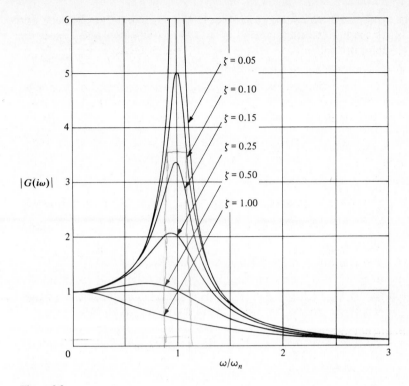

$\zeta = 0.05$

$\zeta = 0.10$

$\zeta = 0.15$

$\zeta = 0.25$

$\zeta = 0.50$

$\zeta = 1.00$

$|G(i\omega)|$

$\omega/\omega_n$

**Figure 2.3**

for $\omega/\omega_n < 1$, depending on the amount of damping. Clearly, for $\zeta > 1/\sqrt{2}$ the response has no peaks and for $\zeta = 0$ there is a discontinuity at $\omega/\omega_n = 1$. In the undamped case, $\zeta = 0$, the homogeneous differential equation reduces to that of a harmonic oscillator, leading us to the conclusion that when the driving frequency $\omega$ approaches the natural frequency $\omega_n$ the response of the harmonic oscillator tends to increase indefinitely. In such a case the harmonic oscillator is said to approach a *resonance condition* characterized by violent vibration. However, solution (2.47) is no longer valid at resonance; a new solution of Eq. (2.41) corresponding to $\omega = \omega_n$ is obtained later in this section.

We notice that for light damping, such as when $\zeta < 0.05$, the maximum of $|G(i\omega)|$ occurs in the immediate neighborhood of $\omega/\omega_n = 1$. Introducing the notation $|G(i\omega)|_{\max} = Q$, we obtain for small values of $\zeta$

$$Q \cong \frac{1}{2\zeta} \tag{2.50}$$

and the curves $|G(i\omega)|$ versus $\omega/\omega_n$ are nearly symmetric with respect to the vertical through $\omega/\omega_n = 1$ in that neighborhood. The symbol $Q$ is known as the *quality factor* because in many electrical engineering applications, such as the tuning circuit of a radio, the interest lies in an amplitude at resonance that is as large as possible. The symbol is often referred to as the "$Q$" of the circuit. The points $P_1$ and $P_2$, where the amplitude of $|G(i\omega)|$ falls to $Q/\sqrt{2}$, are called *half-power points* because the power absorbed by the resistor in an electric circuit or by the damper in a mechanical system responding harmonically at a given frequency is proportional to the square of the amplitude (see Sec. 2.10). The increment of frequency associated with the half-power points $P_1$ and $P_2$ is referred to as the *bandwidth* of the system. For light damping, it is not difficult to show that the bandwidth has the value

$$\Delta\omega = \omega_2 - \omega_1 \cong 2\zeta\omega_n \tag{2.51}$$

Moreover, comparing Eqs. (2.50) and (2.51), we conclude that

$$Q \cong \frac{1}{2\zeta} \cong \frac{\omega_n}{\omega_2 - \omega_1} \qquad \tag{2.52}$$

which can be used as a quick way of estimating $Q$ or $\zeta$.

At this point let us turn our attention to the phase angle and recall that its expression is given by Eq. (2.40). Figure 2.4 plots $\phi$ versus $\omega/\omega_n$ for selected values of $\zeta$. We notice that all curves pass through the point $\phi = \pi/2$, $\omega/\omega_n = 1$. Moreover, for $\omega/\omega_n < 1$ the phase angle tends to zero, whereas for $\omega/\omega_n > 1$ it tends to $\pi$.

For $\zeta = 0$ the plot $\phi$ versus $\omega/\omega_n$ has a discontinuity at $\omega/\omega_n = 1$, jumping from $\phi = 0$ for $\omega/\omega_n < 1$ to $\phi = \pi$ for $\omega/\omega_n > 1$. This can be easily explained by the

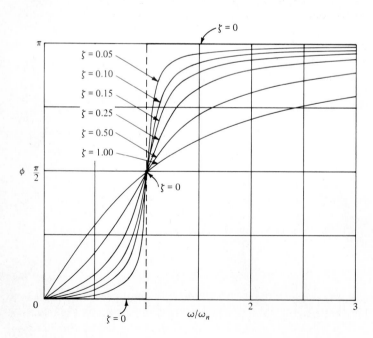

**Figure 2.4**

fact that for $\zeta = 0$ solution (2.47) reduces to

$$x(t) = \frac{1}{1 - (\omega/\omega_n)^2} A e^{i\omega t} \tag{2.53}$$

so that the response is in phase with the excitation for $\omega/\omega_n < 1$ and $180°$ out of phase for $\omega/\omega_n > 1$. Indeed, for $\omega/\omega_n < 1$ the frequency response is positive, so that the response is in the same direction as the excitation, and for $\omega/\omega_n > 1$ the frequency response becomes negative, so that the response is in a direction opposite to that of the excitation. Equation (2.53) also shows clearly that the response of a harmonic oscillator increases without bounds as the driving frequency $\omega$ approaches the natural frequency $\omega_n$.

Finally, let us consider the case of the harmonic oscillator at resonance. Because the velocity term is zero, there is no need to use the complex vector form for the excitation and response. Hence, in this case the differential equation of motion, Eq. (2.41), reduces to

$$\ddot{x}(t) + \omega_n^2 x(t) = \omega_n^2 A \cos \omega_n t \tag{2.54}$$

It is not difficult to verify by substitution that the particular solution of Eq. (2.54) is

$$x(t) = \frac{A}{2} \omega_n t \sin \omega_n t \tag{2.55}$$

which represents oscillatory response with an amplitude increasing linearly with time. This implies that the response undergoes increasingly wild fluctuations as $t$ becomes large. Physically, however, the response cannot grow indefinitely, as at a certain time the small-motions assumption implicit in linear systems is violated. Because the excitation is a cosine function and the response is a sine function, there is a $90°$ phase angle between them, as can also be concluded from Fig. 2.4. The response $x(t)$, as given by Eq. (2.55), is plotted in Fig. 2.5 as a function of time.

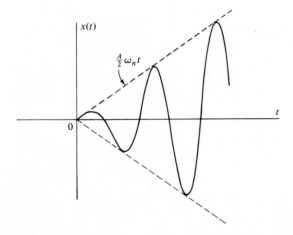

**Figure 2.5**

## 2.4 ROTATING UNBALANCED MASSES

Many mechanical systems can be represented by mathematical models of the type shown in Fig. 2.6a. The model consists of a main mass $M - m$ and two eccentric masses $m/2$ rotating in opposite directions with the constant angular velocity $\omega$. To derive the equation of motion of the system, we consider two free-body diagrams, the first for the right eccentric mass, shown in Fig. 2.6b, and the second for the main mass, shown in Fig. 2.6c. Because the effect of the two eccentric masses on the main mass can be inferred from Figs. 2.6b and c, there is no need for a free-body diagram for the left eccentric mass. Indeed, from Figs. 2.6b and c, we conclude that the reciprocating eccentric masses exert on the main mass two vertical forces $F_x$ that add up and two horizontal forces $F_y$ that cancel each other out. Because the horizontal forces cancel, the main mass undergoes no motion in the horizontal direction, so that it is only necessary to consider the vertical motion $x(t)$. As demonstrated in Sec. 1.3, by measuring the displacement $x(t)$ from the equilibrium position, the effect of the weight of the masses can be ignored in the equation of motion. We note, however, that $F_x$ contains a component equal to $mg/2$ and the force in any of the two springs contains a component equal to $Mg/2$, in addition to the values appearing in the equations of motion to be derived shortly.

From Fig. 2.6b, we observe that the vertical displacement of the eccentric mass is $x(t) + l \sin \omega t$, so that the equation of motion in the vertical direction is

$$F_x = \frac{m}{2} \frac{d^2}{dt^2} [x(t) + l \sin \omega t] = \frac{m}{2} [\ddot{x}(t) - l\omega^2 \sin \omega t] \qquad (2.56)$$

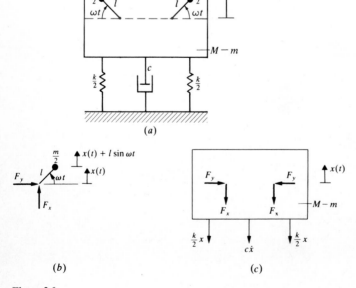

(a)

(b)                                              (c)

**Figure 2.6**

Moreover, from Fig. 2.6c, the equation of motion for the main mass is

$$-2F_x - c\dot{x}(t) - 2\frac{k}{2}x(t) = (M - m)\ddot{x}(t) \tag{2.57}$$

Substituting $F_x$ from Eq. (2.56) into Eq. (2.57) and rearranging, we obtain the equation of motion of the system in the form

$$M\ddot{x}(t) + c\dot{x}(t) + kx(t) = ml\omega^2 \sin \omega t = \text{Im}\,(ml\omega^2 e^{i\omega t}) \tag{2.58}$$

where Im denotes the imaginary part of the expression within parentheses. Hence, rotating eccentric masses exert a harmonic excitation on the system.

The solution of Eq. (2.58) can be derived directly from the results of Sec. 2.3. Indeed, from Eq. (2.47), we conclude that the response is

$$x(t) = \text{Im}\left[\frac{m}{M}l\left(\frac{\omega}{\omega_n}\right)^2 |G(i\omega)|e^{i(\omega t - \phi)}\right]$$

$$= \frac{m}{M}l\left(\frac{\omega}{\omega_n}\right)^2 |G(i\omega)| \sin(\omega t - \phi) \qquad \omega_n^2 = \frac{k}{M} \tag{2.59}$$

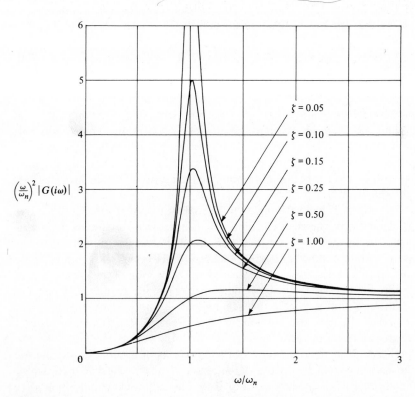

Figure 2.7

in which the phase angle $\phi$ is given by Eq. (2.40). Writing the response in the form

$$x(t) = X \sin(\omega t - \phi) \qquad (2.60)$$

we conclude that

$$X = \frac{ml}{M}\left(\frac{\omega}{\omega_n}\right)^2 |G(i\omega)| \qquad (2.61)$$

Hence, in this particular case the indicated nondimensional ratio is

$$\frac{MX}{ml} = \left(\frac{\omega}{\omega_n}\right)^2 |G(i\omega)| \qquad (2.62)$$

instead of $|G(i\omega)|$ alone, so that Fig. 2.3 is not applicable. Plots of $(\omega/\omega_n)^2|G(i\omega)|$ versus $\omega/\omega_n$ with $\zeta$ as a parameter are shown in Fig. 2.7. On the other hand, the plot $\phi$ versus $\omega/\omega_n$ remains as in Fig. 2.4.

We note that for $\omega \to 0$, $(\omega/\omega_n)^2|G(i\omega)| \to 0$, whereas for $\omega \to \infty$, $(\omega/\omega_n)^2|G(i\omega)| \to 1$. At the same time, from Eq. (2.40), we conclude that as $\omega \to \infty$, $\phi \to \pi$. Since the mass $M - m$ undergoes the displacement Im $x$, whereas the mass $m$ undergoes the displacement Im $(x + le^{i\omega t})$, it follows that for large driving frequencies $\omega$ the masses $M - m$ and $m$ move in such a way that the mass center of the system tends to remain stationary. This is true regardless of the amount of damping. Note that Im $x = -X \sin \omega t$ for large $\omega$.

## 2.5 WHIRLING OF ROTATING SHAFTS

In many mechanical applications one encounters rotating shafts carrying disks. On occasions some of these shafts experience violent vibration. To explain this phenomenon, let us consider a rotating shaft carrying a single disk. If the disk has some eccentricity, then the rotation produces a centrifugal force causing the shaft to bend. The rotation of the plane containing the bent shaft about the bearings axis is known as *whirling*.

Figure 2.8a shows a shaft rotating with the constant angular velocity $\omega$ relative to the inertial axes $x$, $y$. The shaft carries a disk of total mass $m$ at midspan and is supported elastically at both ends. Because the shaft has distributed mass, the system has an infinite number of degrees of freedom. However, if the mass of the shaft is small relative to the mass of the disk, then the motion of the system can be described approximately by the displacements $x$ and $y$ of the geometric center $S$ of the disk. Although this implies a two-degree-of-freedom system, the $x$ and $y$ motions are independent, so that the solution can be carried out as for two systems with one degree of freedom each.

As a preliminary to the derivation of the equations of motion, we denote the origin of the inertial system $x$, $y$ by $O$ and the center of mass of the disk by $C$, where $C$ is at a distance $e$ from $S$, as shown in Fig. 2.8b. The equations of motion involve the acceleration $\mathbf{a}_C$ of the mass center $C$. To compute $\mathbf{a}_C$, we first write the radius

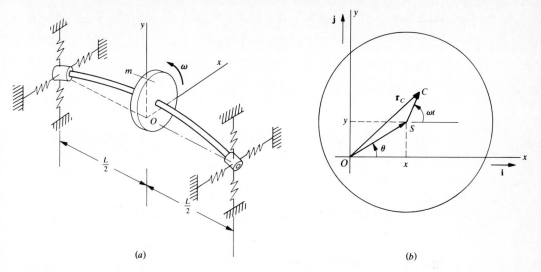

**Figure 2.8**

vector $\mathbf{r}_C$ from $O$ to $C$ in terms of cartesian components as follows:

$$\mathbf{r}_C = (x + e \cos \omega t)\mathbf{i} + (y + e \sin \omega t)\mathbf{j} \tag{2.63}$$

where $\mathbf{i}$ and $\mathbf{j}$ are constant unit vectors along axes $x$ and $y$, respectively. Then, differentiating Eq. (2.63) twice with respect to time, we obtain the acceleration of $C$ in the form

$$\mathbf{a}_C = (\ddot{x} - e\omega^2 \cos \omega t)\mathbf{i} + (\ddot{y} - e\omega^2 \sin \omega t)\mathbf{j} \tag{2.64}$$

To derive the equations of motion, we assume that the only forces acting on the disk are restoring forces due to the elastic supports and the elasticity of the shaft and resisting forces due to viscous damping, such as caused by air friction. The elastic effects are combined into equivalent spring constants $k_x$ and $k_y$ associated with the $x$ and $y$ directions, respectively. Moreover, we assume that the coefficient of viscous damping is the same in both directions and equal to $c$. The elastically restoring forces and the viscous damping forces are acting at point $S$. Considering Eq. (2.64), Newton's second law can be written in terms of $x$ and $y$ components as follows:

$$-k_x x - c\dot{x} = m(\ddot{x} - e\omega^2 \cos \omega t)$$
$$-k_y y - c\dot{y} = m(\ddot{y} - e\omega^2 \sin \omega t) \tag{2.65}$$

which can be rearranged in the form

$$\ddot{x} + 2\zeta_x \omega_{nx} \dot{x} + \omega_{nx}^2 x = e\omega^2 \cos \omega t$$
$$\ddot{y} + 2\zeta_y \omega_{ny} \dot{y} + \omega_{ny}^2 y = e\omega^2 \sin \omega t \tag{2.66}$$

where

$$\zeta_x = \frac{c}{2m\omega_{nx}} \qquad \omega_{nx} = \sqrt{\frac{k_x}{m}}$$

$$\zeta_y = \frac{c}{2m\omega_{ny}} \qquad \omega_{ny} = \sqrt{\frac{k_y}{m}} \tag{2.67}$$

are viscous damping factors and natural frequencies.

The steady-state solution of Eqs. (2.66) can be obtained by the pattern established in Sec. 2.4. Indeed, following that pattern, we can write simply

$$x(t) = X(\omega) \cos(\omega t - \phi_x) \qquad y(t) = Y(\omega) \sin(\omega t - \phi_y) \tag{2.68}$$

where the individual amplitudes are

$$X(\omega) = e\left(\frac{\omega}{\omega_{nx}}\right)^2 |G_x(i\omega)| \qquad Y(\omega) = e\left(\frac{\omega}{\omega_{ny}}\right)^2 |G_y(i\omega)| \tag{2.69}$$

in which

$$|G_x(i\omega)| = \frac{1}{\{[1 - (\omega/\omega_{nx})^2]^2 + (2\zeta_y\omega/\omega_{nx})^2\}^{1/2}}$$

$$|G_y(i\omega)| = \frac{1}{\{[1 - (\omega/\omega_{ny})^2]^2 + (2\zeta_y\omega/\omega_{ny})^2\}^{1/2}} \tag{2.70}$$

are magnification factors and

$$\phi_x = \frac{2\zeta_x\omega/\omega_{nx}}{1 - (\omega/\omega_{nx})^2} \qquad \phi_y = \frac{2\zeta_y\omega/\omega_{ny}}{1 - (\omega/\omega_{ny})^2} \tag{2.71}$$

are phase angles.

One special case of interest is that in which the stiffness is the same in both directions, $k_x = k_y = k$. In this case, the two natural frequencies coincide and so do the viscous damping factors, or

$$\omega_{nx} = \omega_{ny} = \omega = \sqrt{\frac{k}{m}} \qquad \zeta_x = \zeta_y = \zeta = \frac{c}{2m\omega_n} \tag{2.72}$$

Moreover, in view of Eqs. (2.72), we conclude from Eqs. (2.70) and (2.71) that the magnification factors on the one hand and the phase angles on the other hand are the same, or

$$|G_x(i\omega)| = |G_y(i\omega)| = |G(i\omega)| = \frac{1}{\{[1 - (\omega/\omega_n)^2]^2 + (2\zeta\omega/\omega_n)^2\}^{1/2}} \tag{2.73a}$$

$$\phi_x = \phi_y = \phi = \frac{2\zeta\omega/\omega_n}{1 - (\omega/\omega_n)^2} \tag{2.73b}$$

It follows immediately that the amplitudes of the motions $x$ and $y$ are equal to one another, or

$$X(\omega) = Y(\omega) = e\left(\frac{\omega}{\omega_n}\right)^2 |G(i\omega)| \tag{2.74}$$

But, from Fig. 2.8b and Eqs. (2.68), we can write

$$\tan\theta = \frac{y}{x} = \tan(\omega t - \phi) \tag{2.75}$$

from which we conclude that

$$\theta = \omega t - \phi \tag{2.76}$$

and that

$$\dot{\theta} = \omega \tag{2.77}$$

Hence, in this case the shaft whirls with the same angular velocity as the rotation of the disk, so that the shaft and the disk rotate together as a rigid body. This case is known as *synchronous whirl*. It is easy to verify that in synchronous whirl the radial distance from $O$ to $S$ is constant, or

$$r_{OS} = \sqrt{x^2 + y^2} = e\left(\frac{\omega}{\omega_n}\right)^2 |G(i\omega)| = \text{const} \tag{2.78}$$

so that point $S$ describes a circle about point $O$. To determine the position of $C$ relative to the whirling plane, we consider Eq. (2.76). The relation between the angles $\theta$, $\omega t$, and $\phi$ is depicted in Fig. 2.9. Indeed, from Fig. 2.9, we can interpret the phase angle $\phi$ as the angle between the radius vectors $\mathbf{r}_{OS}$ and $\mathbf{r}_{SC}$. Hence, recalling Eqs. (2.73a), we conclude that $\phi < \pi/2$ for $\omega < \omega_n$, $\phi = \pi/2$ for $\omega = \omega_n$, and $\phi > \pi/2$ for $\omega > \omega_n$. The three configurations are shown in Fig. 2.10.

As a final remark concerning synchronous whirl, we note from Eqs. (2.73) that the magnification factor and the phase angle have the same expressions as in the

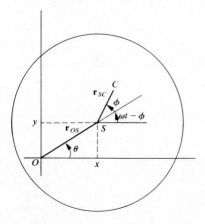

**Figure 2.9**

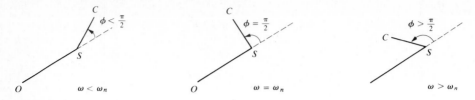

**Figure 2.10**

case of the rotating unbalanced masses discussed in Sec. 2.4. This should come as no surprise as the two phenomena are entirely analogous.

Next, let us return to the case in which the two stiffnesses are different and consider the undamped case, $c = 0$. In this case, solutions (2.68) reduce to

$$x(t) = X(\omega) \cos \omega t \qquad y(t) = Y(\omega) \sin \omega t \qquad (2.79)$$

where

$$X(\omega) = \frac{e(\omega/\omega_{nx})^2}{1 - (\omega/\omega_{nx})^2} \qquad Y(\omega) = \frac{e(\omega/\omega_{ny})^2}{1 - (\omega/\omega_{ny})^2} \qquad (2.80)$$

Dividing the first of Eqs. (2.79) by $X(\omega)$ and the second by $Y(\omega)$, squaring and adding the results, we obtain

$$\frac{x^2}{X^2} + \frac{y^2}{Y^2} = 1 \qquad (2.81)$$

which represents the equation of an ellipse. Hence, as the shaft whirls, point $S$ describes an ellipse with point $O$ as its geometric center. To gain more insight into the motion, let us consider Eqs. (2.79) and write

$$\tan \theta = \frac{y}{x} = \frac{Y}{X} \tan \omega t \qquad (2.82)$$

Differentiating both sides of Eq. (2.82) with respect to time and considering Eqs. (2.79), we obtain

$$\dot{\theta} = \frac{XY}{X^2 \cos^2 \omega t + Y^2 \sin^2 \omega t} \omega \qquad (2.83)$$

But, the denominator on the right side of Eq. (2.83) is always positive, so that the sign of $\dot{\theta}$ depends on the sign of $XY$. By convention, the sign of $\omega$ is assumed as positive, i.e., the disk rotates in the counter-clockwise sense. We can distinguish the following cases:

1. $\omega < \omega_{nx}$ and $\omega < \omega_{ny}$. In this case, we conclude from Eqs. (2.80) that $XY > 0$, so that point $S$ moves on the ellipse in the same sense as the rotation $\omega$.
2. $\omega_{nx} < \omega < \omega_{ny}$ or $\omega_{ny} < \omega < \omega_{nx}$. In either of these two cases $XY < 0$, so that $S$ moves in the opposite sense from $\omega$.

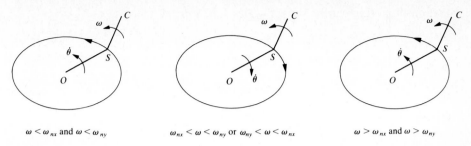

Figure captions below the three diagrams:

$\omega < \omega_{nx}$ and $\omega < \omega_{ny}$      $\omega_{nx} < \omega < \omega_{ny}$ or $\omega_{ny} < \omega < \omega_{nx}$      $\omega > \omega_{nx}$ and $\omega > \omega_{ny}$

**Figure 2.11**

3. $\omega > \omega_{nx}$ and $\omega > \omega_{ny}$. In this case $XY > 0$, so that $S$ moves in the same sense as $\omega$.

The three cases are displayed in Fig. 2.11.

Examining solutions (2.79) and (2.80) for the undamped case, we conclude that the possibility of resonance exists. In fact, there are two frequencies for which resonance is possible, namely, $\omega = \omega_{nx}$ and $\omega = \omega_{ny}$. Clearly, in the case of resonance, solutions (2.79) and (2.80) are no longer valid. It is easy to verify by substitution that the particular solutions in the two cases of resonance are

$$x(t) = \tfrac{1}{2}e\omega_{nx}t \sin \omega_{nx}t$$
$$y(t) = -\tfrac{1}{2}e\omega_{ny}t \cos \omega_{ny}t \qquad (2.84)$$

The plot $x(t)$ versus $t$ resembles that of Fig. 2.5. In fact, it is the same for $A = e$. The plot $y(t)$ versus $t$ also resembles that of Fig. 2.5 except that $\omega_{nx}$ and $\sin \omega_{nx}t$ must be replaced by $\omega_{ny}$ and $\sin (\omega_{ny}t - \pi/2)$, respectively. This is easily explained by the fact that $\sin (\omega_{ny}t - \pi/2) = -\cos \omega_{ny}t$. The two frequencies $\omega = \omega_{nx}$ and $\omega = \omega_{ny}$ are called *critical frequencies*.

## 2.6 HARMONIC MOTION OF THE SUPPORT

Another illustration of a system subjected to harmonic excitation is that in which the support undergoes harmonic motion. Considering Fig. 2.12, the differential

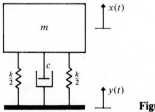

**Figure 2.12**

equation of motion can be shown to have the form

$$m\ddot{x} + c(\dot{x} - \dot{y}) + k(x - y) = 0 \tag{2.85}$$

leading to

$$\ddot{x} + 2\zeta\omega_n\dot{x} + \omega_n^2 x = 2\zeta\omega_n\dot{y} + \omega_n^2 y \tag{2.86}$$

Letting the harmonic displacement of the support be given by

$$y(t) = \text{Re}\,(Ae^{i\omega t}) \tag{2.87}$$

the response can be written as

$$x(t) = \text{Re}\left[\frac{1 + i2\zeta\omega/\omega_n}{1 - (\omega/\omega_n)^2 + i2\zeta\omega/\omega_n}\,Ae^{i\omega t}\right] \tag{2.88}$$

Following a procedure similar to that used previously, the response can be written in the form

$$x(t) = X\cos(\omega t - \phi_1) \tag{2.89}$$

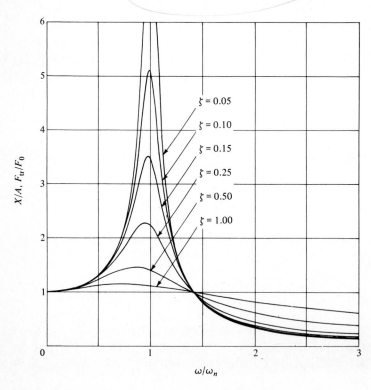

**Figure 2.13**

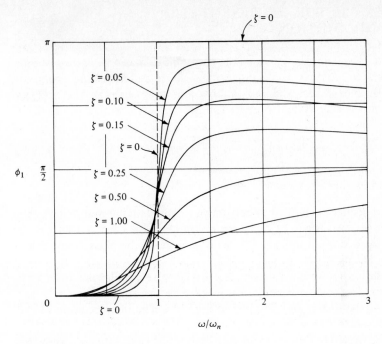

**Figure 2.14**

where

$$X = A\left(\frac{1 + (2\zeta\omega/\omega_n)^2}{[1 - \omega/\omega_n)^2]^2 + (2\zeta\omega/\omega_n)^2}\right)^{1/2} = A\left[1 + \left(\frac{2\zeta\omega}{\omega_n}\right)^2\right]^{1/2} |G(i\omega)| \quad (2.90)$$

and

$$\phi_1 = \tan^{-1}\frac{2\zeta(\omega/\omega_n)^3}{1 - (\omega/\omega_n)^2 + (2\zeta\omega/\omega_n)^2} \quad (2.91)$$

Hence, in this case the indicated nondimensional ratio is

$$\frac{X}{A} = \left[1 + \left(\frac{2\zeta\omega}{\omega_n}\right)^2\right]^{1/2} |G(i\omega)| \quad (2.92)$$

where the ratio $X/A$ is known as *transmissibility*. Curves $X/A$ versus $\omega/\omega_n$ with $\zeta$ as a parameter are plotted in Fig. 2.13. Moreover, curves $\phi_1$ versus $\omega/\omega_n$ for various values of $\zeta$ are shown in Fig. 2.14. Again, for $\zeta = 0$ the response is either in phase with the excitation for $\omega/\omega_n < 1$ or 180° out of phase with the excitation for $\omega/\omega_n > 1$.

## 2.7 COMPLEX VECTOR REPRESENTATION OF HARMONIC MOTION

The representation by complex vectors of the harmonic excitation and the response of a damped system to that excitation can be given an interesting geometric interpretation by means of a diagram in the complex plane. To this end, we consider the second-order system discussed in Sec. 2.3. Differentiating Eq. (2.47) with respect to time, we obtain

$$\dot{x}(t) = i\omega A|G(i\omega)|e^{i(\omega t - \phi)} = i\omega x(t) \tag{2.93a}$$

$$\ddot{x}(t) = (i\omega)^2 A|G(i\omega)|e^{i(\omega t - \phi)} = -\omega^2 x(t) \tag{2.93b}$$

Because $i$ can be written as $i = \cos \pi/2 + i \sin \pi/2 = e^{i\pi/2}$, we conclude that the velocity leads the displacement by the phase angle $\pi/2$ and that it is multiplied by the factor $\omega$. Moreover, because $-1$ can be expressed as $-1 = \cos \pi + i \sin \pi = e^{i\pi}$, it follows that the acceleration leads the displacement by the phase angle $\pi$ and that it is multiplied by the factor $\omega^2$.

In view of the above, we can represent Eq. (2.41) in the complex plane shown in Fig. 2.15. There is no loss of generality in regarding the amplitude $A$ as a real number, which is the assumption implied in Fig. 2.15. The interpretation of Fig. 2.15 is that the sum of the complex vectors $\ddot{x}(t)$, $2\zeta\omega_n\dot{x}(t)$, and $\omega_n^2 x(t)$ balances $\omega_n^2 Ae^{i\omega t}$, which is precisely the requirement that Eq. (2.41) be satisfied. Note that the entire diagram rotates in the complex plane with angular velocity $\omega$. It is clear that considering only the real part of the response is the equivalent of projecting the diagram on the real axis. We can just as easily retain the projections on the imaginary axis, or any other axis, without affecting the nature of the response. In view of this, it is also clear that the assumption that $A$ is real is immaterial. Choosing $A$ as a complex quantity, or considering projections on an axis other than the real axis, would merely imply the addition of a phase angle $\psi$ to all the vectors in Fig. 2.15, without changing their relative positions. This is equivalent to multiplying both sides of Eq. (2.41) by the constant factor $e^{i\psi}$.

The above geometric interpretation extends to first-order systems as well. In fact, to obtain a figure analogous to Fig. 2.15 all that is necessary is to remove the complex vector $\ddot{x}(t)$ and to adjust the magnitude of the remaining vectors, which

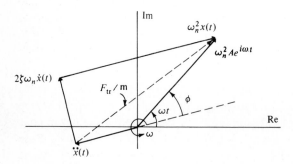

Figure 2.15

can be achieved by letting $2\zeta\omega_n = 1$ and $\omega_n^2 = a$. In the process, the rotating trapezoid of Fig. 2.15 becomes a rotating triangle.

## 2.8 VIBRATION ISOLATION

In many systems of the type shown in Fig. 1.10$a$, we are interested in transmitting as little vibration as possible to the base. This problem can become critical when the excitation is harmonic. Clearly, the force transmitted to the base is through springs and dampers. From Fig. 2.15, we conclude that the amplitude of that force is

$$F_{tr} = m[(2\zeta\omega_n \dot{x})^2 + (\omega_n^2 x)^2]^{1/2} \tag{2.94}$$

where the amplitude of the velocity is simply $\omega x$. Hence, we have

$$F_{tr} = kx\left[1 + \left(\frac{2\zeta\omega}{\omega_n}\right)^2\right]^{1/2} \tag{2.95}$$

But from Eq. (2.47), if we recall that the phase angle is of no consequence, we conclude that

$$F_{tr} = Ak\left[1 + \left(\frac{2\zeta\omega}{\omega_n}\right)^2\right]^{1/2} |G(i\omega)| \tag{2.96}$$

Because $Ak = F_0$ is the amplitude of the actual excitation force, the nondimensional ratio $F_{tr}/F_0$ is a measure of the force transmitted to the base. The ratio can be written as

$$\frac{F_{tr}}{F_0} = \left[1 + \left(\frac{2\zeta\omega}{\omega_n}\right)^2\right]^{1/2} |G(i\omega)| \tag{2.97}$$

and is recognized as the *transmissibility* given by Eq. (2.92). Hence, the plots $F_{tr}/F_0$ versus $\omega/\omega_n$ are the same as the plots $X/A$ versus $\omega/\omega_n$ shown in Fig. 2.13. It is not difficult to show that when $\omega/\omega_n = \sqrt{2}$ the full force is transmitted to the base, $F_{tr}/F_0 = 1$. For values $\omega/\omega_n > \sqrt{2}$ the force transmitted tends to decrease with increasing driving frequency $\omega$, regardless of $\zeta$. Interestingly, damping does not alleviate the situation and in fact, for $\omega/\omega_n > \sqrt{2}$, the larger the damping, the larger the transmitted force. Recalling, however, that in increasing the driving frequency we would have to go through a resonance condition for zero damping, we conclude that a small amount of damping is desirable. Moreover, the case of zero damping represents only an idealization which does not really exist, and in practice a small amount of damping is always present.

## 2.9 VIBRATION MEASURING INSTRUMENTS

There are basically three types of vibration measuring instruments, namely, those measuring accelerations, velocities, and displacements. We shall discuss the first and the third only. Many instruments consist of a case containing a mass-damper-spring system of the type shown in Fig. 2.16, and a device measuring the displacement of the mass relative to the case. The mass is constrained to move along a given axis. The displacement of the mass relative to the case is generally measured electrically. Damping may be provided by a viscous fluid inside the case.

The displacement of the case, the displacement of the mass relative to the case, and the absolute displacement of the mass are denoted by $y(t)$, $z(t)$, and $x(t)$, respectively, so that $x(t) = y(t) + z(t)$. The relative displacement $z(t)$ is the one measured, and from it we must infer the motion $y(t)$ of the case. Although we wish ultimately to determine $y(t)$, it is the response $z(t)$ which is the variable of interest. Using Newton's second law, we can write the equation of motion

$$m\ddot{x}(t) + c[\dot{x}(t) - \dot{y}(t)] + k(x(t) - y(t)) = 0 \qquad (2.98)$$

which, upon elimination of $x(t)$, can be rewritten as

$$m\ddot{z}(t) + c\dot{z}(t) + kz(t) = -m\ddot{y}(t) \qquad (2.99)$$

Assuming harmonic excitation, $y(t) = Y_0 e^{i\omega t}$, Eq. (2.99) leads to

$$m\ddot{z} + c\dot{z} + kz = Y_0 m\omega^2 e^{i\omega t} \qquad (2.100)$$

which is similar in structure to Eq. (2.58). By analogy, the response is

$$z(t) = Y_0 \left(\frac{\omega}{\omega_n}\right)^2 |G(i\omega)| e^{i(\omega t - \phi)} \qquad (2.101)$$

where the phase angle $\phi$ is given by Eq. (2.40). Introducing the notation $z(t) = Z_0 e^{i(\omega t - \phi)}$, we conclude that the plot $Z_0/Y_0$ versus $\omega/\omega_n$ is identical to that given in Fig. 2.7. The plot is shown again in Fig. 2.17 on a scale more suitable for our purposes here.

For small values of the ratio $\omega/\omega_n$ the value of the magnification factor $|G(i\omega)|$ is nearly unity and the amplitude $Z_0$ can be approximated by

$$Z_0 \cong Y_0 \left(\frac{\omega}{\omega_n}\right)^2 \qquad (2.102)$$

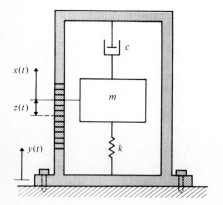

**Figure 2.16**

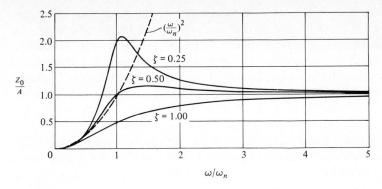

**Figure 2.17**

Because $Y_0\omega^2$ represents the acceleration of the case, the measurement $Z_0$ is proportional to the acceleration of the case, where the proportionality constant is $1/\omega_n^2$. Hence, if the frequency $\omega$ of the harmonic motion of the case is sufficiently low relative to the natural frequency of the system that the amplitude ratio $Z_0/Y_0$ can be approximated by the parabola $(\omega/\omega_n)^2$ (see Fig. 2.17), the instrument can be used as an *accelerometer*. Because the range of $\omega/\omega_n$ in which the amplitude ratio can be approximated by $(\omega/\omega_n)^2$ is the same as the range in which $|G(i\omega)|$ is approximately unity, it will prove advantageous to refer to the plot $|G(i\omega)|$ versus $\omega/\omega_n$ instead of the plot $Z_0/Y_0$ versus $\omega/\omega_n$. Figure 2.18 shows plots $|G(i\omega)|$ versus $\omega/\omega_n$ in the range $0 \leqslant \omega/\omega_n \leqslant 1$, with $\zeta$ acting as a parameter. From Fig. 2.18 we conclude that the range in which $|G(i\omega)|$ is approximately unity is very small for light damping, which implies that the natural frequency of lightly damped accelerometers must be appreciably larger than the frequency of the harmonic motion to be measured. To increase the range of utility of the instrument, larger damping is necessary. It is clear from that figure that the approximation is valid for

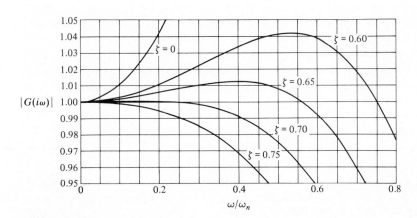

**Figure 2.18**

a larger range of $\omega/\omega_n$ if $0.65 < \zeta < 0.70$. Indeed, for $\zeta = 0.7$ the accelerometer can be used in the range $0 \leqslant \omega/\omega_n \leqslant 0.4$ with less than 1 percent error, and the range can be extended to $\omega/\omega_n \leqslant 0.7$ if proper corrections, based on the instrument calibration, are made.

The most commonly used accelerometers are the compression-type piezo-electric accelerometers. They consist of a mass resting on a piezoelectric ceramic crystal, such as quartz, barium titanate, or lead zirconium titanate, with the crystal acting both as the spring and the sensor. The accelerometers have a preload providing a compressive stress exceeding the highest dynamic stress expected. Any acceleration increases or decreases the compressive stress in the piezoelectric element, thus generating an electric charge appearing at the accelerometer terminals. Piezoelectric accelerometers have negligible damping and they typically have a frequency range from 0 to 5000 Hz (and beyond) and a natural frequency of 30,000 Hz. They tend to be very light, weighing less than 1 oz and relatively small, measuring less than 1 in in height.

Also from Fig. 2.17, we notice that for very large values of $\omega/\omega_n$, the ratio $Z_0/Y_0 = (\omega/\omega_n)^2 |G(i\omega)|$ approaches unity, regardless of the amount of damping. Hence, if the object is to measure displacements, then we should make the natural frequency of the systems very low relative to the excitation frequency, in which case the instrument is called a *seismometer*. For a seismometer, which is an instrument designed to measure earth displacements such as those caused by earthquakes or underground nuclear explosions, the requirement for a low natural frequency dictates that the spring be very soft and the mass relatively heavy, so that, in essence, the mass remains nearly stationary in inertial space while the case, being attached to the ground, moves relative to the mass. Displacement-measuring instruments are generally undamped. They typically have a frequency range from 10 to 500 Hz and a natural frequency between 2 and 5 Hz.

Because seismometers require a much larger mass than accelerometers and the relative motion of the mass in a seismometer is nearly equal in magnitude to the motion to be measured, seismometers are considerably larger in size than accelerometers. In view of this, if the interest lies in displacements, it may prove more desirable to use an accelerometer to measure the acceleration of the case, and then integrate twice with respect to time to obtain the displacement.

The above discussion has focused on the measurement of harmonic motion. In measuring more complicated motions, not only the amplitude but also the phase angle comes into play. As an example, if the motion consists of two harmonics, or

$$y(t) = Y_1 \cos \omega_1 t + Y_2 \cos \omega_2 t \qquad (2.103)$$

and the accelerometer output is

$$y_a(t) = Y_1 \cos (\omega_1 t - \phi_1) + Y_2 \cos (\omega_2 t - \phi_2) \qquad (2.104)$$

where $\phi_1$ and $\phi_2$ are two distinct phase angles, then the accelerometer fails to reproduce the motion $y(t)$, because the two harmonic components of the motion are shifted relative to one another. There are two cases in which the accelerometer output is able to reproduce the motion $y(t)$ without distortion. The first is the case of an undamped accelerometer, $\zeta = 0$, in which case the phase angle is zero. The second is the case in which the phase angle is proportional to the frequency, or

$$\phi_1 = c\omega_1 \qquad \phi_2 = c\omega_2 \qquad (2.105)$$

Indeed, introducing Eqs. (2.105) into Eq. (2.104), we obtain

$$y_a(t) = Y_1 \cos \omega_1(t - c) + Y_2 \cos \omega_2(t - c) \qquad (2.106)$$

so that both harmonics are shifted to the right on the time scale by the same time interval, thus retaining the nature of the motion $y(t)$. To explore the possibility of eliminating the phase distortion, let us consider the case of small $\omega/\omega_n$, in which case the phase angle $\phi$ is small, as can be concluded from Eq. (2.40). Then, assuming that the phase angle increases linearly with the frequency, we can write

$$\sin \phi \cong \phi = c\omega \qquad \cos \phi \cong 1 - \tfrac{1}{2}\phi^2 = 1 - \tfrac{1}{2}(c\omega)^2 \qquad (2.107)$$

Inserting Eqs. (2.107) into Eq. (2.40), we obtain

$$\tan \phi = \frac{2\zeta\omega/\omega_n}{1 - (\omega/\omega_n)^2} \cong \frac{c\omega}{1 - (c\omega)^2/2} \qquad (2.108)$$

which is satisfied provided

$$c = \sqrt{2}/\omega_n \qquad \zeta = \sqrt{2}/2 = 0.707 \qquad (2.109)$$

In general, any arbitrary motion can be regarded as a superposition of harmonic components. Hence, an accelerometer can be used for measuring arbitrary motions if the damping factor $\zeta$ is either equal to zero or equal to 0.707.

## 2.10 ENERGY DISSIPATION. STRUCTURAL DAMPING

In Sec. 2.3 we have shown that the response of a spring-damper-mass system subjected to a harmonic excitation equal to the real part of

$$F(t) = Ake^{i\omega t} \qquad (2.110)$$

is given by the real part of

$$x(t) = A|G(i\omega)|e^{i(\omega t - \phi)} = Xe^{i(\omega t - \phi)} \qquad (2.111)$$

where

$$X = A|G(i\omega)| \qquad (2.112)$$

can be interpreted as the displacement amplitude. Moreover, we have shown in Sec. 2.7 that there is no loss of generality by regarding $A$ as a real number. Clearly, because of damping, the system is not conservative, and indeed energy is dissipated. Since this energy dissipation must be equal to the work done by the external force, we can write the expression for the energy dissipated per cycle of vibration in the form

$$\Delta E_{\text{cyc}} = \int_{\text{cyc}} F \, dx = \int_0^{2\pi/\omega} F\dot{x} \, dt \tag{2.113}$$

where we recall that only the real parts of $F$ and $\dot{x}$ must be considered. Inserting Eqs. (2.93a) and (2.110) into (2.113), we obtain

$$\Delta E_{\text{cyc}} = -kA^2|G(i\omega)|\omega \int_0^{2\pi/\omega} \cos \omega t \sin(\omega t - \phi) \, dt$$

$$= m\omega_n^2 A^2|G(i\omega)|\pi \sin \phi \tag{2.114}$$

From Eqs. (2.40) and (2.48), it is not difficult to show that

$$\sin \phi = 2\zeta \frac{\omega}{\omega_n}|G(i\omega)| = \frac{c\omega}{m\omega_n^2}|G(i\omega)| \tag{2.115}$$

where it is recalled that $\zeta = c/2m\omega_n$. Inserting Eqs. (2.112) and (2.115) into (2.114), we obtain the simple expression

$$\Delta E_{\text{cyc}} = c\pi\omega X^2 \tag{2.116}$$

from which it follows that the energy dissipated per cycle is directly proportional to the damping coefficient $c$, the driving frequency $\omega$, and the square of the response amplitude.

Experience shows that energy is dissipated in all real systems, including those systems for which the mathematical model makes no specific provision for damping. For example, energy is dissipated in real springs as a result of internal friction. In contrast to viscous damping, damping due to internal friction does not depend on velocity. Experiments performed on a large variety of materials show that energy loss per cycle due to internal friction is roughly proportional to the square of the displacement amplitude,†

$$\Delta E_{\text{cyc}} = \alpha X^2 \tag{2.117}$$

where $\alpha$ is a constant independent of the frequency of the harmonic oscillation. This type of damping, called *structural damping*, is attributed to the *hysteresis phenomenon* associated with cyclic stress in elastic materials. The energy loss per cycle of stress is equal to the area inside the hysteresis loop shown in Fig. 2.19. Hence, comparing Eqs. (2.116) and (2.117), we conclude that systems possessing structural damping and subjected to harmonic excitation can be treated as if they were subjected to viscous damping with the equivalent coefficient

$$c_{\text{eq}} = \frac{\alpha}{\pi\omega} \tag{2.118}$$

† See L. Meirovitch, *Analytical Methods in Vibrations*, p. 402, The Macmillan Co., New York, 1967.

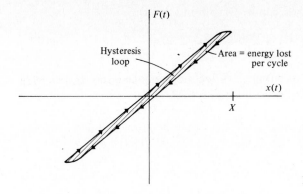

Figure 2.19

$$\ddot{x} + \frac{k}{m\omega}\gamma\dot{x} + \frac{k}{m}x = \dots$$
$$\ddot{x} + \frac{\omega_n^2}{\omega}\gamma\dot{x} + \omega_n^2 x = \dots$$

$$\frac{k}{\omega}\gamma$$

This enables us to write Eq. (1.14) in the form

$$m\ddot{x}(t) + \frac{\alpha}{\pi\omega}\dot{x}(t) + kx(t) = Ake^{i\omega t} \tag{2.119}$$

where consideration has been given to Eqs. (2.110) and (2.118). Because $\dot{x} = i\omega x$. we can rewrite Eq. (2.119) in the form

$$m\ddot{x}(t) + k(1 + i\gamma)x(t) = Ake^{i\omega t} \tag{2.120}$$

where

$$\gamma = \frac{\alpha}{\pi k} \tag{2.121}$$

is called the *structural damping factor*. The quantity $k(1 + i\gamma)$ is called *complex stiffness*, or *complex damping*.

The steady-state solution of Eq. (2.120) is the real part of

$$x(t) = \frac{Ae^{i\omega t}}{1 - (\omega/\omega_n)^2 + i\gamma} \tag{2.122}$$

and, in contrast to viscous damping, for structural damping the maximum amplitude is obtained exactly for $\omega = \omega_n$.

One word of caution is in order. *The analogy between structural and viscous damping is valid only for harmonic excitation*, because the response of a system to harmonic excitation with the driving frequency $\omega$ is implied in the foregoing development.

## 2.11 THE SUPERPOSITION PRINCIPLE

Let us consider again the second-order linear system depicted in Fig. 1.9a. In Sec. 1.3 we have shown that the differential equation for the response $x(t)$ of the system to the arbitrary excitation force $F(t)$ can be written in the form

$$m\frac{d^2x(t)}{dt^2} + c\frac{dx(t)}{dt} + kx(t) = F(t) \tag{2.123}$$

where $m$, $c$, and $k$ are the system parameters denoting the mass, the coefficient of viscous damping, and the spring constant, respectively. Quite often $x(t)$ and $F(t)$ are called the *output* and *input* of the system, respectively. The relation between the response and the excitation, or output and input, can be given an interesting and somewhat useful interpretation by introducing the *linear differential operator*

$$D = m\frac{d^2}{dt^2} + c\frac{d}{dt} + k \tag{2.124}$$

This enables us to write Eq. (2.123) in the symbolic form

$$D[x(t)] = F(t) \tag{2.125}$$

where the juxtaposition of $x(t)$ and $F(t)$ in Eq. (2.125) implies the operation $D$ on $x(t)$ in such a way as to produce Eq. (2.123). We note that an operator is linear if the differential expression $D[x(t)]$ contains the function $x(t)$ and its time derivatives to the first and zero powers only. Thus, cross products thereof and terms involving fractional powers of $x(t)$ are precluded.

The operator $D$ contains all the system characteristics because it involves all the system parameters, namely, $m$, $c$, and $k$, and it specifies the order of the derivatives multiplying each of these parameters as well. In system-analysis language, $D$ represents the "black box" of the second-order system. Relation (2.125) can be illustrated by means of the block diagram shown in Fig. 2.20, which implies that if the input $F(t)$ is fed into the black box represented by $D$, then the output is $x(t)$.

We can use the operator $D$ to define the concept of linearity of a system. To this end, we consider two excitations $F_1(t)$ and $F_2(t)$ and denote the corresponding responses by $x_1(t)$ and $x_2(t)$, so that

$$F_1(t) = D[x_1(t)] \qquad F_2(t) = D[x_2(t)] \tag{2.126}$$

Next we consider the excitation $F_3(t)$ as a linear combination of $F_1(t)$ and $F_2(t)$, namely,

$$F_3(t) = c_1F_1(t) + c_2F_2(t) \tag{2.127}$$

where $c_1$ and $c_2$ are known constants. Then, if the response $x_3(t)$ to the excitation $F_3(t)$ satisfies the relation

$$x_3(t) = c_1x_1(t) + c_2x_2(t) \tag{2.128}$$

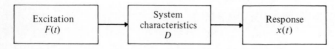

**Figure 2.20**

the system is linear; otherwise it is nonlinear. Using Eqs. (2.126) and (2.127), the above statement can be written in terms of the operator $D$ as follows:

$$D[x_3(t)] = D[c_1 x_1(t) + c_2 x_2(t)] = c_1 D[x_1(t)] + c_2 D[x_2(t)]$$

$$= c_1 F_1(t) + c_2 F_2(t) = F_3(t) \qquad (2.129)$$

Equation (2.129) represents the mathematical statement of the so-called *principle of superposition*, which clearly *applies to linear systems alone*. In words, the principle implies that for linear systems the responses to a given number of distinct excitations can be obtained separately and then combined to obtain the aggregate response. The superposition principle is a very powerful one, and has no counterpart for nonlinear systems. It is because of this principle that the theory of linear systems is so well developed compared to that of nonlinear systems. Note that we have already used the principle in Sec. 2.9 to discuss the subject of phase distortion in accelerometers. We shall use the principle again to derive the response of linear systems to periodic and nonperiodic excitation.

## 2.12 RESPONSE TO PERIODIC EXCITATION. FOURIER SERIES

In Secs. 2.2 and 2.3, we derived the steady-state response of first- and second-order systems to harmonic excitation. By virtue of the fact that the systems considered were linear, any transient response due to the initial conditions could be obtained separately and then added to the steady-state response by invoking the superposition principle.

Harmonic excitation of any arbitrary frequency $\omega$ is periodic, i.e., it repeats itself at equal intervals of time $T = 2\pi/\omega$, where $T$ is the period of excitation. In vibrations we encounter other types of periodic excitations, not necessarily harmonic. As an example, the function illustrated in Fig. 2.21 is periodic but not harmonic. Any periodic function, however, can be represented by a convergent series of harmonic functions whose frequencies are integral multiples of a certain *fundamental frequency* $\omega_0$ provided that it satisfies certain conditions to be pointed out shortly. The frequencies representing integral multiples of the fundamental frequency are called *harmonics*, with the fundamental frequency being the first harmonic. Such series of harmonic functions are known as *Fourier series*, and can

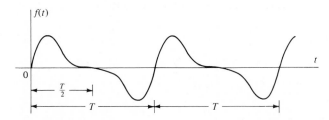

**Figure 2.21**

be written in the form (see App. A)

$$f(t) = \tfrac{1}{2}a_0 + \sum_{p=1}^{\infty} (a_p \cos p\omega_0 t + b_p \sin p\omega_0 t) \qquad \omega_0 = \frac{2\pi}{T} \qquad (2.130)$$

where $p$ are integers, $p = 1, 2, 3, \ldots$, and $T$ is the *period*. The coefficients $a_p$ and $b_p$ are given by the formulas

$$a_p = \frac{2}{T} \int_{-T/2}^{T/2} f(t) \cos p\omega_0 t \, dt \qquad p = 0, 1, 2, \ldots \qquad (2.131a)$$

$$b_p = \frac{2}{T} \int_{-T/2}^{T/2} f(t) \sin p\omega_0 t \, dt \qquad p = 1, 2, \ldots \qquad (2.131b)$$

and they represent a measure of the participation of the harmonic components $\cos p\omega_0 t$ and $\sin p\omega_0 t$, respectively, in the function $f(t)$. Note that $a_0/2$ represents the average value of $f(t)$, which in the case of Fig. 2.21 is zero. The Fourier series representation is possible provided the integrals defining $a_p$ and $b_p$ exist. We shall not pursue the subject of the integrals' existence, because for the physical problems we will be concerned with it can be safely assumed that the integrals do exist.†

There are certain cases in which the Fourier series, Eq. (2.130), can be simplified. One such case is when the function $f(t)$ is an *odd function* of time, which is defined mathematically by

$$f(t) = -f(-t) \qquad (2.132)$$

Considering Eq. (2.132), Eqs. (2.131) yield

$$a_p = \frac{2}{T} \left[ \int_{-T/2}^{0} f(t) \cos p\omega_0 t \, dt + \int_{0}^{T/2} f(t) \cos p\omega_0 t \, dt \right]$$

$$= \frac{2}{T} \left[ \int_{T/2}^{0} f(-t) \cos (-p\omega_0 t) \, d(-t) + \int_{0}^{T/2} f(t) \cos p\omega_0 t \, dt \right]$$

$$= \frac{2}{T} \left[ -\int_{0}^{T/2} f(t) \cos p\omega_0 t \, dt + \int_{0}^{T/2} f(t) \cos p\omega_0 t \, dt \right] = 0$$

$$p = 0, 1, 2, \ldots \qquad (2.133a)$$

$$b_p = \frac{2}{T} \left[ \int_{-T/2}^{0} f(t) \sin p\omega_0 t \, dt + \int_{0}^{T/2} f(t) \sin p\omega_0 t \, dt \right]$$

$$= \frac{2}{T} \left[ \int_{T/2}^{0} f(-t) \sin (-p\omega_0 t) \, d(-t) + \int_{0}^{T/2} f(t) \sin p\omega_0 t \, dt \right]$$

$$= \frac{2}{T} \left[ \int_{0}^{T/2} f(t) \sin p\omega_0 t \, dt + \int_{0}^{T/2} f(t) \sin p\omega_0 t \, dt \right]$$

$$= \frac{4}{T} \int_{0}^{T/2} f(t) \sin p\omega_0 t \, dt \qquad p = 1, 2, \ldots \qquad (2.133b)$$

† For a discussion of this subject, see A. E. Taylor, *Advanced Calculus*, p. 714, Ginn and Co., New York, 1955.

Hence, when $f(t)$ is an odd function of $t$, the Fourier series reduces to the sine series

$$f(t) = \sum_{r=1}^{\infty} b_p \sin p\omega_0 t \qquad \omega_0 = \frac{2\pi}{T} \qquad (2.134)$$

where the coefficients $b_p$ ($p = 1, 2, \ldots$) are given by Eqs. (2.133$b$). A second case is the one in which $f(t)$ is an *even function* of time, defined as

$$f(t) = f(-t) \qquad (2.135)$$

Using Eq. (2.135), Eqs. (2.131) become

$$
\begin{aligned}
a_p &= \frac{2}{T}\left[ \int_{-T/2}^{0} f(t) \cos p\omega_0 t \, dt + \int_{0}^{T/2} f(t) \cos p\omega_0 t \, dt \right] \\
&= \frac{2}{T}\left[ \int_{T/2}^{0} f(-t) \cos(-p\omega_0 t) \, d(-t) + \int_{0}^{T/2} f(t) \cos p\omega_0 t \, dt \right] \\
&= \frac{2}{T}\left[ \int_{0}^{T/2} f(t) \cos p\omega_0 t \, dt + \int_{0}^{T/2} f(t) \cos p\omega_0 t \, dt \right] \\
&= \frac{4}{T} \int_{0}^{T/2} f(t) \cos p\omega_0 t \, dt \qquad\qquad p = 0, 1, 2, \ldots \quad (2.136a)
\end{aligned}
$$

$$
\begin{aligned}
b_p &= \frac{2}{T}\left[ \int_{-T/2}^{0} f(t) \sin p\omega_0 t + \int_{0}^{T/2} f(t) \sin p\omega_0 t \, dt \right] \\
&= \frac{2}{T}\left[ \int_{T/2}^{0} f(-t) \sin(-p\omega_0 t) \, d(-t) + \int_{0}^{T/2} f(t) \sin p\omega_0 t \, dt \right] \\
&= \frac{2}{T}\left[ -\int_{0}^{T/2} f(t) \sin p\omega_0 t \, dt + \int_{0}^{T/2} f(t) \sin p\omega_0 t \, dt \right] = 0
\end{aligned}
$$

$$p = 1, 2, \ldots \quad (2.136b)$$

so that, in the case in which $f(t)$ is an even function of $t$, the Fourier series simplifies to the cosine series

$$f(t) = \tfrac{1}{2}a_0 + \sum_{p=1}^{\infty} a_p \cos p\omega_0 t \qquad \omega_0 = \frac{2\pi}{T} \qquad (2.137)$$

where the coefficients $a_p$ ($p = 0, 1, 2, \ldots$) are given by Eqs (2.136a). Expansions (2.134) and (2.137) can be easily explained by observing that $\sin p\omega_0 t$ ($p = 1, 2, \ldots$) are odd functions of time and $\cos p\omega_0 t$ ($p = 0, 1, 2, \ldots$) are even functions of time. Hence, Eq. (2.134) states that an odd periodic function cannot have even harmonic components and Eq. (2.137) states that an even periodic function cannot have odd harmonic components.

Next, we derive the response of linear systems to periodic excitation. To this end, we recognize from Secs. 2.2 and 2.3 that the response of a linear system to the excitation

$$f_{pc}(t) = a_p \cos p\omega_0 t \qquad (2.138)$$

is simply

$$x_{pc}(t) = a_p|G_p| \cos (p\omega_0 t - \phi_p) \tag{2.139}$$

where $|G_p|$ is the magnification factor and $\phi_p$ is the corresponding phase angle. Moreover, the response to the excitation

$$f_{ps}(t) = b_p \sin p\omega_0 t \tag{2.140}$$

is

$$x_{ps}(t) = b_p|G_p| \sin (p\omega_0 t - \phi_p) \tag{2.141}$$

If the excitation is in the form of a periodic function with a Fourier series expansion in the form of Eq. (2.130), by virtue of the superposition principle, the response can be written as a linear combination of responses to the individual harmonic components. Hence, considering Eqs. (2.139) and (2.141) and recognizing that the response to the constant $a_0/2$ is simply $a_0/2$, we can write the response of a mass-damper-spring system to the periodic function given by Eq. (2.130) in the form of the series

$$x(t) = \tfrac{1}{2}a_0 + \sum_{p=1}^{\infty} |G_p|[a_p \cos (p\omega_0 t - \phi_p) + b_p \sin (p\omega_0 t - \phi_p)] \tag{2.142}$$

where $a_0/2$ can be identified as the average value of the response.

For the mass-damper-spring system of Sec. 2.3, the magnification factor is

$$|G_p| = \frac{1}{\{[1 - (p\omega_0/\omega_n)^2]^2 + (2\zeta p\omega_0/\omega_n)^2\}^{1/2}} \tag{2.143}$$

and the phase angle is

$$\phi_p = \frac{2\zeta p\omega_0/\omega_n}{1 - (p\omega_0/\omega_n)^2} \tag{2.144}$$

It is clear from Eqs. (2.142) and (2.143) that if the value of one of the harmonics $p\omega_0$ in the excitation is close to the frequency $\omega_n$ of the undamped oscillation, then this particular harmonic will tend to provide a relatively larger contribution to the response, particularly for light damping. The case of zero damping has interesting implications. Specifically, we conclude from Eqs. (2.142) and (2.143) that if $p\omega_0 = \omega_n$ for a certain $p$, then a resonance condition exists. Hence, resonance can occur in undamped systems when the excitation is merely periodic, and not necessarily harmonic, provided the frequency of one of the harmonic components coincides with the system natural frequency.

In deriving the response of a linear system to harmonic excitation, we found it advantageous earlier in this chapter to represent harmonic functions in terms of complex vectors. Because periodic functions consist of series of harmonic functions, we can expect the same advantage in representing a periodic function in terms of a series of complex vectors. To this end, we recognize that the Fourier series (2.130) can also be written in what is generally known as its *complex*, or *exponential form* (see App. A)

$$f(t) = \sum_{p=-\infty}^{\infty} C_p e^{ip\omega_0 t} \tag{2.145}$$

where $C_p$ are complex coefficients having the expressions

$$C_p = \frac{1}{T} \int_{-T/2}^{T/2} f(t) e^{-ip\omega_0 t} \, dt \qquad p = 0, \pm 1, \pm 2, \dots \tag{2.146}$$

As before, the coefficient $C_0$ represents the average value of $f(t)$.

Instead of working with the negative frequencies implied in Eq. (2.145), it will prove convenient to represent the excitation in the form

$$f(t) = \tfrac{1}{2}A_0 + \mathrm{Re}\left( \sum_{p=1}^{\infty} A_p e^{ip\omega_0 t} \right) \tag{2.147}$$

where $A_0$ is a real coefficient and $A_p$ are in general complex coefficients given by

$$A_p = \frac{2}{T} \int_{-T/2}^{T/2} f(t) e^{-ip\omega_0 t} \, dt \qquad p = 0, 1, 2, \dots \tag{2.148}$$

The reason for preferring series (2.147) to (2.145) becomes obvious when we observe that every term in series (2.147) has the same form as the complex vector described by Eq. (2.15a). By analogy with the real coefficients $a_p$ and $b_p$, the complex coefficient $A_p$ represents the extent to which the harmonic component with frequency $p\omega_0$ contributes to $f(t)$.

Before attempting to obtain the response of the system to the excitation described by series (2.147), it is perhaps desirable to show that expansions (2.145) and (2.147) are indeed equivalent. Expansion (2.145) can be written as

$$f(t) = \sum_{p=-\infty}^{\infty} C_p e^{ip\omega_0 t} = C_0 + \sum_{p=1}^{\infty} C_p e^{ip\omega_0 t} + \sum_{p=-1}^{-\infty} C_p e^{ip\omega_0 t}$$

$$= C_0 + \sum_{p=1}^{\infty} (C_p e^{ip\omega_0 t} + C_p^* e^{-ip\omega_0 t}) \tag{2.149}$$

where $C_p^* = C_{-p}$ is the complex conjugate of $C_p$. Hence, using the relations $e^{ip\omega_0 t} + e^{-ip\omega_0 t} = 2\cos p\omega_0 t$, $e^{ip\omega_0 t} - e^{-ip\omega_0 t} = 2i \sin p\omega_p t$, series (2.149) reduces to

$$f(t) = C_0 + 2 \sum_{p=1}^{\infty} (\mathrm{Re}\, C_p \cos p\omega_0 t - \mathrm{Im}\, C_p \sin p\omega_0 t) \tag{2.150}$$

where $\mathrm{Re}\, C_p$ and $\mathrm{Im}\, C_p$ denote the real part and the imaginary part of $C_p$, respectively. On the other hand, series (2.147) can be written in the form

$$f(t) = \tfrac{1}{2}A_0 + \sum_{p=1}^{\infty} (\mathrm{Re}\, A_p \cos p\omega_0 t - \mathrm{Im}\, A_p \sin p\omega_0 t) \tag{2.151}$$

so that, observing from Eqs. (2.146) and (2.148) that $A_p = 2C_p$ ($p = 0, 1, 2, \dots$), we conclude that the series (2.145) and (2.147) are indeed equivalent. Note that, as for the real form of the Fourier series, the constant excitation $A_0/2$ produces a constant response also equal to $A_0/2$.

The response of a linear system to an excitation given by Re $(Ae^{i\omega t})$ was shown in Secs. 2.2 and 2.3 to have the form

$$x(t) = \text{Re}\,[AG(i\omega)e^{i\omega t}] = \text{Re}[A|G(i\omega)|e^{i(\omega t - \phi)}] \tag{2.152}$$

where $G(i\omega)$ is the frequency response. Because the system is linear, the response to the excitation $f(t)$ as given by Eq. (2.147) is simply

$$x(t) = \tfrac{1}{2}A_0 + \text{Re}\left( \sum_{p=1}^{\infty} G_p A_p e^{i p \omega_0 t} \right) \tag{2.153}$$

where, by analogy, $G_p$ is the frequency response associated with the frequency $p\omega_0$. Also by analogy with results of Secs. 2.2 and 2.3, we can rewrite the response (2.153) as

$$x(t) = \tfrac{1}{2}A_0 + \text{Re}\left[ \sum_{p=1}^{\infty} |G_p| A_p e^{i(p\omega_0 t - \phi_p)} \right] \tag{2.154}$$

where $|G_p|$ is the magnitude of $G_p$ and $\phi_p$ is the phase angle associated with the harmonic of frequency $p\omega_0$.

Note that for a mass-damper-spring system

$$G_p = \frac{1}{1 - (p\omega_0/\omega_n)^2 + i2\zeta p\omega_0/\omega_n} \tag{2.155}$$

Moreover, $|G_p|$ and $\phi_p$ are given by Eqs. (2.143) and (2.144), respectively.

**Example 2.1** Consider the excitation $f(t)$ in the form of the periodic square wave shown in Fig. 2.22, and calculate the response of an undamped single-degree-of-freedom system to that excitation. Solve the problem in two ways: first by considering a trigonometric form and then by considering a complex form of the Fourier series.

The mathematical description of the excitation is simply

$$f(t) = \begin{cases} A & \text{for} \quad 0 < t < T/2 \\ -A & \text{for} \quad -T/2 < t < 0 \end{cases} \tag{a}$$

where $T$ is the period. Observing that $f(t)$ is an odd function of time, we consider the sine series given by Eq. (2.134). Using Eqs. (2.133b), we obtain the

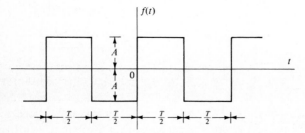

**Figure 2.22**

coefficients

$$b_p = \frac{4}{T} \int_0^{T/2} f(t) \sin p\omega_0 t \, dt = \frac{4A}{T} \int_0^{T/2} \sin p\omega_0 t \, dt$$

$$= \frac{4A}{T} \left[ -\frac{\cos p\omega_0 t}{p\omega_0} \right]_0^{T/2} = \frac{4A}{Tp\omega_0} (1 - \cos p\omega_0 T/2) \qquad (b)$$

Recalling that $\omega_0 = 2\pi/T$, we have

$$\cos \frac{p\omega_0 T}{2} = \cos p\pi = (-1)^p \qquad (c)$$

so that Eq. (b) yields

$$b_p = \frac{2A}{p\pi} [1 - (-1)^p] = \begin{cases} \dfrac{4A}{p\pi} & p = \text{odd} \\ 0 & p = \text{even} \end{cases} \qquad (d)$$

Hence, the Fourier series for the periodic square wave shown in Fig. 2.22 is

$$f(t) = \frac{4A}{\pi} \sum_{r=1,3,5,\ldots}^{\infty} \frac{1}{p} \sin p\omega_0 t \qquad \omega_0 = \frac{2\pi}{T} \qquad (e)$$

In the undamped case, $\zeta = 0$, the phase angle $\phi_p$ is $0°$ if $p\omega_0 < \omega_n$ and $180°$ if $p\omega_0 > \omega_n$, where a phase angle of $180°$ corresponds to a change in the sign of the response. As shown in Sec. 2.3, the phase angle can be taken into account automatically by replacing the magnification factor $|G_p|$ by the frequency response

$$G_p = \frac{1}{1 - (p\omega_0/\omega_n)^2} \qquad (f)$$

Hence, the response is simply

$$x(t) = \frac{4A}{\pi} \sum_{p=1,3,5,\ldots}^{\infty} \frac{\sin p\omega_0 t}{p[1 - (p\omega_0/\omega_n)^2]} \qquad \omega_0 = \frac{2\pi}{T} \qquad (g)$$

and we note that the same harmonics that participate in $f(t)$ participate also in $x(t)$, with the amplitude of the harmonics with frequencies close to $\omega_n$ gaining in magnitude relative to those with frequencies removed from $\omega_n$. It is clear from Eq. (g) that resonance occurs for $p\omega_0 = \omega_n$.

To obtain the response by means of the complex Fourier series, we insert Eqs. (a) into Eqs. (2.148) and obtain the coefficients

$$A_p = \frac{2}{T} \int_{-T/2}^{T/2} f(t) e^{-ip\omega_0 t} \, dt = \frac{2A}{T} \left( -\int_{-T/2}^0 e^{-ip\omega_0 t} \, dt + \int_0^{T/2} e^{ip\omega_0 t} \, dt \right)$$

$$= \frac{2A}{T} \left( -\int_0^{T/2} e^{ip\omega_0 t} \, dt + \int_0^{T/2} e^{-ip\omega_0 t} \, dt \right) = -\frac{4iA}{T} \int_0^{T/2} \sin p\omega_0 t \, dt$$

$$= \frac{4iA}{T} \frac{\cos p\omega_0 t}{p\omega_0} \bigg|_0^{T/2} = \frac{4iA}{Tp\omega_0} \left( \cos \frac{p\omega_0 T}{2} - 1 \right) \qquad (h)$$

Considering Eq. (*c*), we can write

$$A_p = \begin{cases} -\dfrac{4iA}{p\pi} & p = \text{odd} \\ 0 & p = \text{even} \end{cases} \tag{i}$$

Inserting Eq. (*i*) into Eq. (2.153), we obtain the response

$$x(t) = \operatorname{Re} \sum_{p=1,3,5,\dots}^{\infty} \left( -\frac{4iA}{p\pi} \right) \frac{1}{1 - (p\omega_0/\omega_n)^2} e^{ip\omega_0 t}$$

$$= \frac{4A}{\pi} \sum_{r=1,3,5,\dots}^{\infty} \frac{\sin p\omega_0 t}{p[1 - (p\omega_0/\omega_n)^2]} \qquad \omega_0 = \frac{2\pi}{T} \tag{j}$$

which is identical to the response given by Eq. (*g*), obtained by the approach based on the trigonometric form of the Fourier series.

Equation (*g*) can be used to plot $x(t)$ versus $t$, but this may not be very illuminating. Perhaps a better understanding of the system behavior can be obtained from a plot in the frequency domain instead of a plot in the time domain. Indeed, considerable information concerning the system behavior is revealed by plots showing the degree of participation of the various harmonics in the excitation $f(t)$ and in the response $x(t)$. These are plots of the amplitude of the harmonic components of the function in question versus the frequency, where such plots are known as *frequency spectra*. Figure 2.23*a* shows the

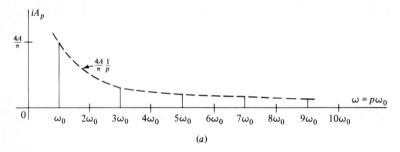

(*a*)

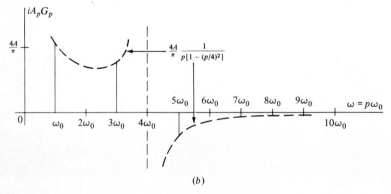

(*b*)

**Figure 2.23**

frequency spectrum for the periodic square wave $f(t)$ of Fig. 2.22, obtained by plotting the coefficients of $\sin p\omega_0 t$ in Eq. (b) versus the frequency $\omega$. Because the function $f(t)$ is periodic, its frequency spectrum consists of harmonic components with discrete frequencies, namely, $\omega = \omega_0, 3\omega_0, 5\omega_0, \ldots$. For this reason this is a *discrete frequency spectrum*. In a similar manner, Fig. 2.23b represents the frequency spectrum associated with the response $x(t)$, Eq. (g), for the case $\omega_n = 4\omega_0$. Conforming to expectation, the magnitude of the amplitudes of the harmonics $\omega_0, 3\omega_0$, and $5\omega_0$ in $x(t)$ gains relative to that of their counterparts in $f(t)$, whereas all the others are attenuated. Moreover, the amplitudes corresponding to $\omega > 4\omega_0$ are negative. Figure 2.23b also represents a discrete frequency spectrum.

## 2.13 THE UNIT IMPULSE. IMPULSE RESPONSE

In Sec. 2.12, we studied the response of a system to a periodic excitation of period $T$. The question remains as to how to obtain the response of a system to an arbitrary excitation. Clearly, in this case there is no steady-state response and the entire solution must be regarded as *transient*, although the part due to the excitation force may persist indefinitely even in the presence of damping provided, of course, that the excitation persists. Before discussing the response to arbitrary excitations, we consider the response to some special types of forcing functions.

A very important function in vibrations is the *unit impulse*, or the *Dirac delta function* (Fig. 2.24), defined mathematically as

$$\delta(t - a) = 0 \qquad \text{for } t \neq a$$
$$\int_{-\infty}^{\infty} \delta(t - a)\, dt = 1 \qquad\qquad (2.156)$$

We note that while the time interval over which the function is different from zero is by definition taken as infinitesimally small, that is, $\epsilon$ in Fig. 2.24 approaches zero in the limit and the amplitude of the function in this time interval is undefined, the area under the curve is well defined and equal to unity. We also note that the units of the Dirac delta function are $s^{-1}$, which should be immediately clear from the fact that the value of the integral in (2.156) is nondimensional. The unit impulse applied at $t = 0$ is denoted by $\delta(t)$.

The response of a system to a unit impulse applied at $t = 0$, with the initial conditions equal to zero, is called the *impulse response* of the system and is denoted

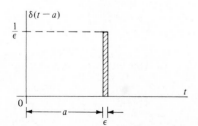

**Figure 2.24**

by $g(t)$. Clearly, the response to a unit impulse applied at a later time $t = a$ is $g(t - a)$; it can be obtained by shifting $g(t)$ to the right along the time scale by the time interval $t = a$.

**Example 2.2** Calculate the impulse response of the damper-spring system of Sec. 1.3.

Inserting $x(t) = g(t)$ and $F(t) = \delta(t)$ in Eq. (1.12), we obtain

$$c\dot{g}(t) + kg(t) = \delta(t) \tag{a}$$

where $g(0) = \dot{g}(0) = 0$ by definition. Integrating Eq. (a) with respect to time over the interval $\Delta t = \epsilon$ and taking the limit, we obtain

$$\lim_{\epsilon \to 0} \int_0^\epsilon [c\dot{g}(t) + kg(t)]\, dt = \lim_{\epsilon \to 0} \int_0^\epsilon \delta(t)\, dt = 1 \tag{b}$$

But,

$$\lim_{\epsilon \to 0} \int_0^\epsilon c\dot{g}(t)\, dt = \lim_{\epsilon \to 0} cg(t) \Big|_0^\epsilon = \lim_{\epsilon \to 0} c[g(\epsilon) - g(0)] = cg(0+)$$

$$\lim_{\epsilon \to 0} \int_0^\epsilon kg(t)\, dt = \lim_{\epsilon \to 0} kg(0)\epsilon = 0 \tag{c}$$

The notation $g(0+)$ is to be interpreted as a change in displacement at the end of the time increment $\Delta t = \epsilon$. Note that the result in the second of Eqs. (c) was obtained by invoking the mean-value theorem. Inserting Eqs. (c) into Eq. (b), we conclude that

$$g(0+) = \frac{1}{c} \tag{d}$$

The physical interpretation of Eq. (d) is that the unit impulse produces an instantaneous change in displacement, so that we can regard the effect of the unit impulse applied at $t = 0$ as being equivalent to an initial displacement $g(0) = 1/c$. In Sec. 1.5, however, we considered the response of a first-order system to an initial displacement. Hence, inserting $x(t) = g(t)$ and $x_0 = g(0) = 1/c$ in Eq. (1.31), we obtain the impulse response

$$g(t) = \begin{cases} \dfrac{1}{c} e^{-t/\tau} & t > 0 \\ 0 & t < 0 \end{cases} \tag{e}$$

where $\tau = c/k$ is the time constant.

**Example 2.3** Calculate the impulse response of the mass-damper-spring system of Sec. 1.3.

Inserting $x(t) = g(t)$ and $F(t) = \delta(t)$ into Eq. (1.14), we obtain

$$m\ddot{g}(t) + c\dot{g}(t) + kg(t) = \delta(t) \tag{a}$$

Following the same procedure as in Example 2.2, we can write

$$\lim_{\epsilon \to 0} \int_0^\epsilon (m\ddot{g} + c\dot{g} + kg) \, dt = \lim_{\epsilon \to 0} \int_0^\epsilon \delta(t) \, dt = 1 \qquad (b)$$

where

$$\lim_{\epsilon \to 0} \int_0^\epsilon m\ddot{g} \, dt = \lim_{\epsilon \to 0} m\dot{g} \Big|_0^\epsilon = \lim_{\epsilon \to 0} m[\dot{g}(\epsilon) - \dot{g}(0)] = m\dot{g}(0+)$$

$$\lim_{\epsilon \to 0} \int_0^\epsilon c\dot{g} \, dt = \lim_{\epsilon \to 0} cg \Big|_0^\epsilon = \lim_{\epsilon \to 0} c[g(\epsilon) - g(0)] = 0 \qquad (c)$$

$$\lim_{\epsilon \to 0} \int_0^\epsilon kg \, dt = 0$$

The notation $\dot{g}(0+)$ is to be interpreted as a change in velocity at the end of the time increment $\Delta t = \epsilon$. On the other hand, because the change in velocity is finite and the interval of integration $0 < t < \epsilon$ is extremely short, there is not sufficient time for displacements to develop, so that $g(\epsilon) = 0$. This fact is due to the presence of the mass $m$, which was absent in the system of Example 2.2. Combining Eqs. (b) and (c), we conclude that

$$\dot{g}(0+) = \frac{1}{m} \qquad (d)$$

The physical interpretation of Eq. (d) is that the unit impulse produces an instantaneous change in the velocity, so that we can regard the effect of a unit impulse applied at $t = 0$ as being equivalent to the effect of an initial velocity $v_0 = 1/m$. We recall, however, that in Example 1.7 we calculated the response of the system under consideration to an initial velocity $v_0$. In view of this, if we introduce $v_0 = 1/m$ into Eq. (g) of Example 1.7, we can write the impulse response in the form

$$g(t) = \begin{cases} \dfrac{1}{m\omega_d} e^{-\zeta\omega_n t} \sin \omega_d t & t > 0 \\ 0 & t < 0 \end{cases} \qquad (e)$$

where $\omega_d = (1 - \zeta^2)^{1/2}\omega_n$.

## 2.14 THE UNIT STEP FUNCTION. STEP RESPONSE

Another function of great importance in vibrations is the unit step function. The unit step function is depicted in Fig. 2.25 and is defined mathematically as follows:

$$u(t - a) = \begin{cases} 0 & \text{for } t < a \\ 1 & \text{for } t > a \end{cases} \qquad (2.157)$$

The function clearly exhibits a discontinuity at $t = a$, at which point its value jumps from 0 to 1. If the discontinuity occurs at $t = 0$ the unit step function is denoted

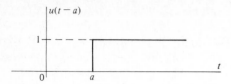

Figure 2.25

simply by $u(t)$. The unit step function is dimensionless. We notice that the multiplication of an arbitrary function $f(t)$ by the unit step function $u(t)$ sets the portion of $f(t)$ corresponding to $t < 0$ equal to zero automatically while leaving the portion for $t > 0$ unaffected.

There is a close relationship between the unit step function $u(t - a)$ and the unit impulse $\delta(t - a)$. In particular, the unit step function is the integral of the unit impulse, or

$$u(t - a) = \int_{-\infty}^{t} \delta(\xi - a) \, d\xi \tag{2.158}$$

where $\xi$ is merely a variable of integration. Conversely, the unit impulse is the time derivative of the unit step function, or

$$\delta(t - a) = \frac{du(t - a)}{dt} \tag{2.159}$$

The concept of unit step function enables us to return to some results obtained previously and express them in a more compact manner. Indeed, referring to Example 2.2, it is clear that the impulse response of a damper-spring system, Eqs. (e), can be written conveniently in the form

$$g(t) = \frac{1}{c} e^{-t/\tau} u(t) \tag{2.160}$$

Moreover, the impulse response of a mass-damper-spring system, Eqs. (e) of Example 2.3, can be expressed in the compact form

$$g(t) = \frac{1}{m\omega_d} e^{-\zeta\omega_n t} \sin \omega_d t \, u(t) \tag{2.161}$$

The response of a system to a unit step function applied at $t = 0$, with the initial conditions equal to zero, is called the *step response* of the system in question and is denoted by $\mathscr{s}(t)$. To derive the step response of a linear system, let us consider Eq. (2.125) and write the relation between the unit impulse $\delta(t)$ and the impulse response $g(t)$ in the symbolic form

$$D[g(t)] = \delta(t) \tag{2.162}$$

where $D$ is a differential operator. Integrating Eq. (2.162) with respect to time and assuming that the differentiation and integration processes are interchangeable we obtain

$$\int_{-\infty}^{t} D[g(\xi)] \, d\xi = D\left[\int_{-\infty}^{t} g(\xi) \, d\xi\right] = \int_{-\infty}^{t} \delta(\xi) \, d\xi \qquad (2.163)$$

But, according to Eq. (2.158), the right side of Eq. (2.163) is the unit step function applied at $t = a = 0$. Moreover, from Eq. (2.125), the relation between the unit step function $u(t)$ and the step response $\delta(t)$ has the symbolic form

$$D[\delta(t)] = u(t) \qquad (2.164)$$

Hence, comparing Eqs. (2.163) and (2.164), we conclude that

$$\delta(t) = \int_{-\infty}^{t} g(\xi) \, d\xi \qquad (2.165)$$

or, *the step response is the integral of the impulse response.*

The step response can be used at times to facilitate the response to relatively involved excitations. Indeed, when the excitation consists of a linear combination of step functions, the response can be expressed as a similar linear combination of step responses (see Example 2.5).

**Example 2.4** Calculate the step response of a mass-damper-spring system by integrating the impulse response, according to Eq. (2.165). Plot $\delta(t)$ versus $t$.

The impulse response of a mass-damper-spring system is given by Eq. (2.161), so that using Eq. (2.165) the step response is

$$\delta(t) = \frac{1}{m\omega_d} \int_{-\infty}^{t} e^{-\zeta\omega_n\xi} \sin \omega_d\xi \, u(\xi) \, d\xi = \frac{1}{m\omega_d} \int_{0}^{t} e^{-\zeta\omega_n\xi} \sin \omega_d\xi \, d\xi \qquad (a)$$

From Eqs. (1.37), however, it is not difficult to show that

$$\sin \omega_d\xi = \frac{e^{i\omega_d\xi} - e^{-i\omega_d\xi}}{2i} \qquad (b)$$

so that Eq. (a) can be integrated as follows:

$$\delta(t) = \frac{1}{2im\omega_d} \int_{0}^{t} e^{-\zeta\omega_n\xi}(e^{i\omega_d\xi} - e^{-i\omega_d\xi}) \, d\xi$$

$$= \frac{1}{2im\omega_d} \int_{0}^{t} [e^{-(\zeta\omega_n - i\omega_d)\xi} - e^{-(\zeta\omega_n + i\omega_d)\xi}] \, d\xi$$

$$= \frac{1}{2im\omega_d} \left[ \frac{e^{-(\zeta\omega_n - i\omega_d)\xi}}{-(\zeta\omega_n - i\omega_d)} - \frac{e^{-(\zeta\omega_n + i\omega_d)\xi}}{-(\zeta\omega_n + i\omega_d)} \right]\Bigg|_{0}^{t} \qquad (c)$$

After some algebraic operations, Eq. (c) yields the step response

$$\delta(t) = \frac{1}{k}\left[1 - e^{-\zeta\omega_n t}\left(\cos \omega_d t + \frac{\zeta\omega_n}{\omega_d} \sin \omega_d t\right)\right] u(t) \qquad (d)$$

where the unit step function $u(t)$ accounts automatically for the fact that $\delta(t) = 0$ for $t < 0$. The plot $\delta(t)$ versus $t$ is shown in Fig. 2.26.

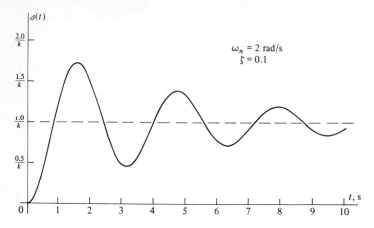

$$\omega_n = 2 \text{ rad/s}$$
$$\zeta = 0.1$$

**Figure 2.26**

**Example 2.5** Use the concept of unit step function and calculate the response $x(t)$ of an undamped single-degree-of-freedom system to the rectangular pulse shown in Fig. 2.27. Plot $x(t)$ versus $t$.

It is easy to verify that the function $F(t)$ depicted in Fig. 2.27 can be expressed conveniently in terms of unit step functions in the form

$$F(t) = F_0[\mathcal{u}(t + T) - \mathcal{u}(t - T)] \tag{a}$$

But the response of an undamped single-degree-of-freedom system to a unit step function applied at $t = 0$ can be obtained from Eq. $(e)$ of Example 2.4 by letting $\zeta = 0$ and $\omega_d = \omega_n$. The result is

$$\mathcal{s}(t) = \frac{1}{k}(1 - \cos \omega_n t)\mathcal{u}(t) \tag{b}$$

Moreover, the response to $\mathcal{u}(t + T)$ is $\mathcal{s}(t + T)$, obtained from Eq. $(b)$ of the present example by simply replacing $t$ by $t + T$. Similarly, the response to $\mathcal{u}(t - T)$ is $\mathcal{s}(t - T)$. Hence, the response to $F(t)$, as given by Eq. $(a)$, is simply

$$x(t) = F_0[\mathcal{s}(t + T) - \mathcal{s}(t - T)]$$

$$= \frac{F_0}{k}\{[1 - \cos \omega_n(t + T)]\mathcal{u}(t + T) - [1 - \cos \omega_n(t - T)]\mathcal{u}(t - T)\} \tag{c}$$

The plot $x(t)$ versus $t$ is shown in Fig. 2.28.

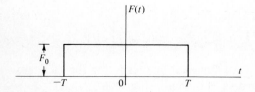

**Figure 2.27**

2,15-2,18

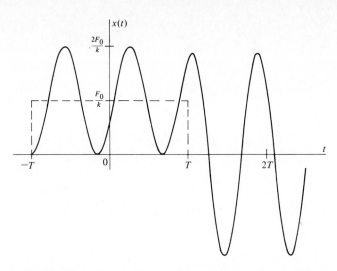

**Figure 2.28**

## 2.15 RESPONSE TO ARBITRARY EXCITATION. THE CONVOLUTION INTEGRAL

Earlier in this chapter, we studied the response of linear systems to harmonic and periodic excitations. Then, in Secs. 2.13 and 2.14, we discussed the response to a unit impulse and to a unit step function. The question remains as to how to obtain the response to arbitrary excitation.

There are various ways of deriving the response to arbitrary excitation, depending on the manner in which the excitation function is described. One way is to represent the excitation by a Fourier integral, obtained from a Fourier series through a limiting process consisting of letting the period $T$ approach infinity, so that in essence the excitation ceases to be periodic. We shall discuss this approach in Chap. 11. Another way is to regard the excitation as a superposition of impulses of varying amplitude and time of application. Similarly, the excitation can be represented by a superposition of step functions. We choose to represent the excitation as a series of impulses.

Let us consider an arbitrary excitation $F(t)$, such as that depicted in Fig. 2.29. During the small time increment $\Delta\tau$ beginning at a given time $t = \tau$, we can regard the function $F(t)$ as consisting of an impulse of magnitude $F(\tau) \Delta\tau$, as shown by the shaded area in Fig. 2.29. The impulse can be expressed mathematically as

$$\Delta F(t, \tau) = F(\tau) \Delta\tau \, \delta(t - \tau) \qquad (2.166)$$

It follows that the function $F(t)$ can be approximated by a superposition of such impulses as follows:

$$F(t) \cong \sum F(\tau) \Delta\tau \, \delta(t - \tau) \qquad (2.167)$$

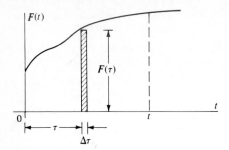

**Figure 2.29**

where the representation becomes exact as $\Delta\tau \to 0$. But, the response to the impulse described by Eq. (2.166) is

$$\Delta x(t, \tau) = F(\tau)\, \Delta\tau\, g(t - \tau) \tag{2.168}$$

so that the response to the excitation $F(t)$ is

$$x(t) \cong \sum F(\tau)g(t - \tau)\, \Delta\tau \tag{2.169}$$

Letting $\Delta\tau \to 0$, and replacing the summation by integration, we obtain

$$x(t) = \int_0^t F(\tau)g(t - \tau)\, d\tau \tag{2.170}$$

which is known as the *convolution integral*, and expresses the response as a superposition of impulse responses.

The impulse response in the integrand in (2.170) is delayed, or shifted, by the time $t = \tau$. A similar expression can be derived, however, in which the excitation function $F(t)$ is shifted instead of the impulse response. To show this, we let $t - \tau = \lambda$, $-d\tau = d\lambda$. Then, considering the limits of integration in (2.170), we observe that when $\tau = 0$, $\lambda = t$, whereas when $\tau = t$, $\lambda = 0$. Inserting these values into (2.170), it is not difficult to show that

$$x(t) = \int_0^t F(t - \lambda)g(\lambda)\, d\lambda \tag{2.171}$$

which is also referred to as a convolution integral. Recognizing that $\tau$ in (2.170) and $\lambda$ in (2.171) are dummy variables of integration, we conclude that the convolution integrals are symmetric in the excitation $F(t)$ and impulse response $g(t)$, so that we can write

$$x(t) = \int_0^t F(\tau)g(t - \tau)\, d\tau = \int_0^t F(t - \tau)g(\tau)\, d\tau \tag{2.172}$$

The question remains as to which function to shift, the excitation $F(t)$ or the impulse response $g(t)$. Logic dictates that the simpler of the two functions be shifted.

When $F(t)$ is defined for $t < 0$ the lower limit in the convolution integrals must be changed. This case is discussed in Chap. 11.

The convolution integrals are not always easy to evaluate, and in many cases they must be evaluated numerically.

The convolution integrals (2.172) can be shown to represent a special case of a broader theorem involving two arbitrary functions $f_1(t)$ and $f_2(t)$, not necessarily the excitation $F(t)$ and the impulse response $g(t)$. This can be demonstrated very conveniently by means of the Laplace transformation (see App. B, Sec. B.5).

**Example 2.6** Derive an expression for the response of a mass-damper-spring system to an arbitrary excitation $F(t)$ in terms of the convolution integral. Then, consider the undamped case and calculate the response to the one-sided harmonic excitation

$$F(t) = F_0 \sin \omega t \, \alpha(t) \tag{a}$$

by means of the convolution integral, where $\alpha(t)$ is the unit step function.

The general response of a linear system can be written in one of the two forms of the convolution integral given by Eq. (2.172), in which $F(t)$ is any arbitrary excitation and $g(t)$ is the impulse response. In the case of a mass-damper-spring system, the impulse response is given by Eq. (e) of Example 2.3. Hence, inserting Eq. (e) of Example 2.3 into Eq. (2.172), we obtain the response of a mass-damper-spring system to an arbitrary excitation in the form

$$x(t) = \frac{1}{m\omega_d} \int_0^t F(\tau) e^{-\zeta\omega_n(t-\tau)} \sin \omega_d(t - \tau) \, d\tau$$

$$= \frac{1}{m\omega_d} \int_0^t F(t - \tau) e^{-\zeta\omega_n\tau} \sin \omega_d\tau \, d\tau \tag{b}$$

Note that Eq. (b) does not include the effect of initial conditions. If in addition the system is subjected to an initial displacement and/or an initial velocity, then this response can be obtained separately and added to Eq. (b).

Letting $\zeta = 0$, $\omega_d = \omega_n$ in Eq. (b), we obtain

$$x(t) = \frac{1}{m\omega_n} \int_0^t F(\tau) \sin \omega_n(t - \tau) \, d\tau = \frac{1}{m\omega_n} \int_0^t F(t - \tau) \sin \omega_n\tau \, d\tau \tag{c}$$

Due to the nature of the excitation, it does not matter whether we shift the excitation or the impulse response. Hence, inserting Eq. (a) into Eq. (c) and recalling the trigonometric relation $\sin \alpha \sin \beta = \frac{1}{2}[\cos (\alpha - \beta) - \cos (\alpha + \beta)]$, we can write

$$x(t) = \frac{F_0}{m\omega_n} \int_0^t \sin \omega\tau \sin \omega_n(t - \tau) \, d\tau$$

$$= \frac{F_0}{2m\omega_n} \int_0^t \{\cos [(\omega + \omega_n)\tau - \omega_n t] - \cos [(\omega - \omega_n)\tau + \omega_n t]\} \, d\tau$$

$$
= \frac{F_0}{2m\omega_n} \left\{ \frac{\sin \left[ (\omega + \omega_n)\tau - \omega_n t \right]}{\omega + \omega_n} - \frac{\sin \left[ (\omega - \omega_n)\tau + \omega_n t \right]}{\omega - \omega_n} \right\} \Bigg|_0^t
$$

$$
= \frac{F_0}{k} \frac{1}{1 - (\omega/\omega_n)^2} \left( \sin \omega t - \frac{\omega}{\omega_n} \sin \omega_n t \right) \tag{d}
$$

Because there is no excitation for $t < 0$, the response should be written in the form

$$
x(t) = \frac{F_0}{k} \frac{1}{1 - (\omega/\omega_n)^2} \left( \sin \omega t - \frac{\omega}{\omega_n} \sin \omega_n t \right) \alpha(t) \tag{e}
$$

It should be noted at this point that the nature of the harmonic excitation given by Eq. (a) is distinctly different from the nature of the harmonic excitation of Sec. 2.3, as the first is defined only for $t > 0$ and the second is defined for all times. This explains the difference in the two responses.

## 2.16 SHOCK SPECTRUM

Many structures are subjected on occasions to relatively large forces applied suddenly and over periods of time that are short relative to the natural period of the structure. Such forces can produce local damage, or they can excite undesirable vibration of the structure. Indeed, at times the vibration results in large cyclic stress damaging the structure or impairing its performance. A force of this type has come to be known as a *shock*. The response of structures to shock is of vital importance in design. The severity of the shock is customarily measured in terms of the maximum value of the response. For comparison purposes, the response considered is that of an undamped single-degree-of-freedom system. The plot of the peak response of a mass-spring system to a given shock as a function of the natural frequency of the system is known as *shock spectrum*, or *response spectrum*.

A shock $F(t)$ is generally characterized by its maximum value $F_0$, its duration $T$, and its shape, or alternatively the impulse $\int_0^T F(t)\, dt$. These characteristics depend on the force-producing mechanism and on the properties of the interface material. A reasonable approximation for the force is the half-sine pulse shown in Fig. 2.30; we propose to derive the associated shock spectrum.

The mathematical definition of the half-sine pulse is

$$
F(t) = \begin{cases} F_0 \sin \omega t & 0 < t < \pi/\omega \\ 0 & t < 0 \quad \text{and} \quad t > \pi/\omega \end{cases} \tag{2.173}
$$

We must distinguish between the response during the pulse, $0 < t < \pi/\omega$, and the response subsequent to the termination of the pulse, $t > \pi/\omega$. We observe that during the pulse, the half-sine pulse has precisely the same form as the one-sided harmonic excitation of Example 2.6. Moreover, the system considered in that example is the same mass-spring system under consideration here. Hence, the

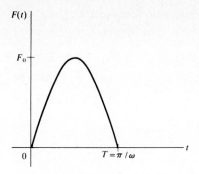

**Figure 2.30**

response *during the pulse* is as given by Eq. (*e*) of Example 2.6, or

$$x(t) = \frac{F_0}{m\omega_n} \int_0^t \sin \omega\tau \sin \omega_n(t - \tau) \, d\tau$$

$$= \frac{F_0}{k[1 - (\omega/\omega_n)^2]} \left( \sin \omega t - \frac{\omega}{\omega_n} \sin \omega_n t \right) \qquad 0 < t < \frac{\pi}{\omega} \qquad (2.174)$$

To obtain the maximum response, we must solve for the time $t_m$ at which $\dot{x} = 0$ and then substitute the value of $t_m$ in Eq. (2.174). Differentiating Eq. (2.174) with respect to time, we obtain

$$\dot{x}(t) = \frac{F_0\omega}{k[1 - (\omega/\omega_n)^2]} (\cos \omega t - \cos \omega_n t) \qquad (2.175)$$

so that, recalling the trigonometric relation

$$\cos \alpha - \cos \beta = -2 \sin \frac{\alpha + \beta}{2} \sin \frac{\alpha - \beta}{2}$$

we conclude that $t_m$ must satisfy the equation

$$\sin \frac{\omega_n + \omega}{2} t_m \sin \frac{\omega_n - \omega}{2} t_m = 0 \qquad (2.176)$$

which has two families of solutions

$$\left. \begin{array}{c} t'_m \\ t''_m \end{array} \right\} = \frac{2i\pi}{\omega_n \pm \omega} \qquad i = 1, 2, \ldots \qquad (2.177)$$

Substituting the above values in Eq. (2.174), we obtain

$$x(t'_m) = \frac{F_0}{k(1 - \omega/\omega_n)} \sin \frac{2i\pi\omega/\omega_n}{1 + \omega/\omega_n} \qquad (2.178a)$$

$$x(t''_m) = \frac{F_0}{k(1 + \omega/\omega_n)} \sin \frac{2i\pi\omega/\omega_n}{1 - \omega/\omega_n} \qquad (2.178b)$$

It is obvious from Eqs. (2.178) that the response corresponding to $t = t'_m$ achieves higher values than the response corresponding to $t = t''_m$. The question remains as to how to determine the value of the integer $i$. To answer this question, we recall that $t'_m$ must occur during the pulse, so that from Eqs. (2.177), we must have $[2i\omega/(\omega_n + \omega)] < \pi/\omega$. Hence, we conclude that for $0 < t < \pi/\omega$ we have the maximum response

$$x_{\max} = \frac{F_0 \omega_n/\omega}{k[(\omega_n/\omega) - 1]} \sin \frac{2i\pi}{1 + \omega_n/\omega} \qquad i < \frac{1}{2}\left(1 + \frac{\omega_n}{\omega}\right) \qquad (2.179)$$

To determine the response for *any time subsequent to the termination of the pulse*, $t > \pi/\omega$, we rely once again on results from Example 2.6. Replacing the upper limit in the convolution integral of Eq. (d) in Example 2.6 by $t = T$, we obtain

$$x(t) = \frac{F_0}{m\omega_n} \int_0^T \sin \omega\tau \sin \omega_n(t - \tau)\, d\tau$$

$$= \frac{F_0}{2m\omega_n} \left\{ \frac{\sin [(\omega + \omega_n)\tau - \omega_n t]}{\omega + \omega_n} - \frac{\sin [(\omega - \omega_n)\tau + \omega_n t]}{\omega - \omega_n} \right\} \Big|_0^T$$

$$= \frac{F_0 \omega_n/\omega}{k[1 - (\omega_n/\omega)^2]} [\sin \omega_n t + \sin \omega_n(t - T)] \qquad (2.180)$$

As before, to obtain the maximum response, we must first determine $t = t_m$, at which time $\dot{x}(t) = 0$. To this end, we write first

$$\dot{x}(t) = \frac{F_0 \omega_n^2/\omega}{k[1 - (\omega_n/\omega)^2]} [\cos \omega_n t + \cos \omega_n(t - T)] \qquad (2.181)$$

Then, recalling that

$$\cos \alpha + \cos \beta = 2 \cos \frac{\alpha + \beta}{2} \cos \frac{\alpha - \beta}{2}$$

we conclude that $t_m$ must satisfy the equation

$$\cos \omega_n(t_m - \tfrac{1}{2}T) \cos \tfrac{1}{2}\omega_n T = 0 \qquad (2.182)$$

which yields the solutions

$$t_m = (2i - 1)\frac{\pi}{2\omega_n} + \frac{1}{2}T \qquad i = 1, 2, \ldots \qquad (2.183)$$

Introducing $t = t_m$ in Eq. (2.180), we obtain the maximum response for $t > \pi/\omega$ in the form

$$x_{\max} = \frac{2F_0 \omega_n/\omega}{k[1 - (\omega_n/\omega)^2]} \cos \frac{\pi}{2} \frac{\omega_n}{\omega} \qquad (2.184)$$

The response spectrum is simply the plot $x_{\max}$ versus $\omega_n/\omega$, in which both Eqs. (2.179) and (2.184) must be considered. Of course, only the larger of the two values must be used. We note that for $\omega_n < \omega$ solution (2.179) is not valid, but for $\omega_n > \omega$

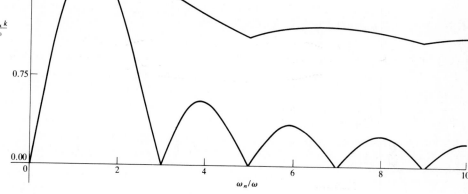

**Figure 2.31**

both solutions are valid. It turns out that the maximum response is given by Eq. (2.184) for $\omega_n < \omega$ and by Eq. (2.179) for $\omega_n > \omega$. The response spectrum is shown in Fig. 2.31 in the form of the nondimensional plot $x_{max} k/F_0$ versus $\omega_n/\omega$.

For different pulse shapes, different shock spectra can be anticipated. For a rectangular pulse, or a triangular pulse, the ratio $\omega_n/\omega$ has no meaning, because there is no $\omega$ in the definition of these pulses. However, these pulses are defined in terms of their duration $T$, so that in these cases the shock spectrum is given by $x_{max} k/F_0$ versus $T/T_n$, or $x_{max} k/F_0$ versus $2T/T_n$, where $T_n = 2\pi/\omega_n$ is the natural period of the mass-spring system.

## 2.17 SYSTEM RESPONSE BY THE LAPLACE TRANSFORMATION METHOD. TRANSFER FUNCTION

The Laplace transformation method has gained wide acceptance as a tool of analysis in the study of linear systems. In addition to providing an efficient method for solving linear differential equations with constant coefficients, the Laplace transformation permits the writing of a simple algebraic expression relating the excitation and the response of systems. In this regard, we are reminded of the operator $D$ used in Sec. 2.11 to demonstrate the principle of superposition for linear systems. However, the operator $D$ was a differential operator that merely permitted writing the relation between the excitation $F(t)$ and the response $x(t)$ in a compact form, as well as bringing the inherent characteristics of the system into sharp focus, but it did not help in any way toward the solution of the response problem. By contrast, the Laplace transformation does provide a method of solution. Significant advantages of the method are that it can treat discontinuous functions without particular difficulty and that it takes into account initial conditions automatically. Sufficient elements of the Laplace transformation to provide us with a working knowledge of the method are presented in App. B. Here we concentrate mainly on using the method to study response problems.

The *(one-sided) Laplace transformation* of $x(t)$, written symbolically as $\bar{x}(s) = \mathcal{L}x(t)$ is defined by the definite integral

$$\bar{x}(s) = \mathcal{L}x(t) = \int_0^\infty e^{-st}x(t)\,dt \qquad (2.185)$$

where $s$ is in general a complex quantity referred to as a *subsidiary variable*. The function $e^{-st}$ is known as the kernel of the transformation. Because this is a definite integral, with $t$ as the variable of integration, the transformation yields a function of $s$. To solve Eq. (2.123) by the Laplace transformation method, it is necessary to evaluate the transforms of the derivatives $dx/dt$ and $d^2x/dt^2$. A simple integration by parts leads to

$$\mathcal{L}\frac{dx(t)}{dt} = \int_0^\infty e^{-st}\frac{dx(t)}{dt}\,dt = e^{-st}x(t)\Big|_0^\infty + s\int_0^\infty e^{-st}x(t)\,dt$$

$$= s\bar{x}(s) - x(0) \qquad (2.186)$$

where $x(0)$ is the value of the function $x(t)$ at $t = 0$. Physically, it represents the initial displacement of the mass $m$. Similarly, it is not difficult to show that

$$\mathcal{L}\frac{d^2x(t)}{dt^2} = \int_0^\infty e^{-st}\frac{d^2x(t)}{dt^2}\,dt = s^2\bar{x}(s) - sx(0) - \dot{x}(0) \qquad (2.187)$$

where $\dot{x}(0)$ is the initial velocity of $m$. The Laplace transformation of the excitation function is simply

$$\bar{F}(s) = \mathcal{L}F(t) = \int_0^\infty e^{-st}F(t)\,dt \qquad (2.188)$$

Transforming both sides of Eq. (2.123), and rearranging, we obtain

$$(ms^2 + cs + k)\bar{x}(s) = \bar{F}(s) + m\dot{x}(0) + (ms + c)x(0) \qquad (2.189)$$

In the following discussion, we shall concentrate on the effect of the forcing function, although we could have just as easily regarded the right side of (2.189) as a generalized transformed excitation. Hence, ignoring the homogeneous solution, which is equivalent to letting $x(0) = \dot{x}(0) = 0$, we can write the ratio of the transformed excitation to the transformed response in the form

$$\bar{Z}(s) = \frac{\bar{F}(s)}{\bar{x}(s)} = ms^2 + cs + k \qquad (2.190)$$

where the function $\bar{Z}(s)$ is known as the *generalized impedance of the system*. We notice that $\bar{Z}(s)$ contains all the information concerning the system characteristics, in much the same way as $D$ does. By contrast, however, $\bar{Z}(s)$ is an *algebraic expression* in the $s$ domain, namely, a complex plane sometimes referred to as *the Laplace plane*, whereas $D$ is a differential operator in the time domain. The reciprocal of $\bar{Z}(s)$, denoted by

$$\bar{Y}(s) = \frac{1}{\bar{Z}(s)} \tag{2.191}$$

is called the *admittance of the system*. The concepts of impedance and admittance were used first in connection with the steady-state response of systems.

In the study of systems, we encounter a more general concept relating the transformed response to the transformed excitation. This general concept is known as the *system function*, or *transfer function*. For the special case of the second-order system described by Eq. (2.123), the transfer function has the form

$$\bar{G}(s) = \frac{\bar{x}(s)}{\bar{F}(s)} = \frac{1}{ms^2 + cs + k} = \frac{1}{m(s^2 + 2\zeta\omega_n s + \omega_n^2)} \tag{2.192}$$

where $\zeta$ and $\omega_n$ are the viscous damping factor and undamped natural frequency of the system. Note that by letting $s = i\omega$ in $\bar{G}(s)$ and multiplying by $m$, we obtain the complex frequency response $G(i\omega)$, Eq. (2.46).

Equation (2.192) can be rewritten as

$$\bar{x}(s) = \bar{G}(s)\bar{F}(s) \tag{2.193}$$

so that the transfer function can be regarded as an *algebraic operator* that operates on the transformed excitation to yield the transformed response. By analogy with the block diagram of Fig. 2.20, Eq. (2.193) can be represented by the block diagram of Fig. 2.32. In contrast to Fig. 2.20, representing a relation in the time domain in terms of the differential operator $D$, Fig. 2.32 is in the Laplace domain in terms of the algebraic operator $\bar{G}(s)$. There is another advantage of this latter approach in that it leads to the solution of the problem. Indeed, to recover the response $x(t)$ from the transformed response, we simply evaluate the *inverse Laplace transformation* of $\bar{x}(s)$, defined symbolically by

$$x(t) = \mathscr{L}^{-1}\bar{x}(s) = \mathscr{L}^{-1}\bar{G}(s)\bar{F}(s) \tag{2.194}$$

The operation $\mathscr{L}^{-1}$ involves in general a line integral in the complex domain. For our purposes, however, we need not go so deeply into the theory of the Laplace transformation method. Instead we shall look for ways of decomposing $\bar{x}(s)$ into a combination of functions whose inverse transformations are known. This is done

**Figure 2.32**

by the method of partial fractions, as presented in App. B. Also in App. B we discuss a theorem for the inversion of a function $\bar{x}(s)$ having the form of a product of two functions of $s$, such as Eq. (2.194). This is Borel's theorem, which is applicable to the product of any two functions of $s$, not necessarily $\bar{G}(s)$ and $\bar{F}(s)$.

Equation (2.194) can be used to derive the response of any linear system with constant coefficients subjected to arbitrary excitation. There are two excitations of particular interest in vibrations, namely, the unit impulse and the unit step function. These functions and the response to these functions were already discussed in Secs. 2.13 and 2.14, but in this section we wish to present the derivation of the response by means of the Laplace transformation.

The Laplace transformation of the unit impulse is

$$\bar{\delta}(s) = \int_0^\infty e^{-st}\delta(t)\,dt = e^{-st}\bigg|_{t=0} \int_0^\infty \delta(t)\,dt = 1 \qquad (2.195)$$

where use has been made of the mean-value theorem. Inserting $x(t) = g(t)$ and $\bar{F}(s) = \bar{\delta}(s) = 1$ in Eq. (2.194), we obtain the impulse response

$$g(t) = \mathcal{L}^{-1}\bar{G}(s) \qquad (2.196)$$

Hence, *the impulse response is equal to the inverse Laplace transformation of the transfer function*, so that the unit impulse and the transfer function represent a Laplace transforms pair. Clearly, they both contain all the information on the dynamic characteristics of a system, the first in an integrated form and the second in an algebraic form.

Next, we consider the Laplace transformation of the unit step function, or

$$\bar{u}(s) = \int_0^\infty e^{-st}u(t)\,dt = \int_0^\infty e^{-st}\,dt = \frac{e^{-st}}{-s}\bigg|_0^\infty = \frac{1}{s} \qquad (2.197)$$

Inserting $x(t) = \delta(t)$ and $\bar{F}(s) = \bar{u}(s) = 1/s$ in Eq. (2.194), we obtain the step response

$$\delta(t) = \mathcal{L}^{-1}\frac{\bar{G}(s)}{s} \qquad (2.198)$$

or the step response is equal to the inverse Laplace transformation of the transfer function divided by $s$.

**Example 2.7** Derive the impulse response of a damped single-degree-of-freedom system by the Laplace transformation method.

The transfer function of a damped single-degree-of-freedom system is given by Eq. (2.192), which can be rewritten in terms of partial fractions as follows:

$$\bar{G}(s) = \frac{1}{m(s^2 + 2\zeta\omega_n s + \omega_n^2)}$$

$$= \frac{1}{2i\omega_d m}\left(\frac{1}{s + \zeta\omega_n - i\omega_d} - \frac{1}{s + \zeta\omega_n + i\omega_d}\right) \qquad (a)$$

But, in general

$$\mathscr{L}^{-1} \frac{1}{s - \alpha} = e^{\alpha t} \qquad (b)$$

Hence, inserting Eq. (a) into Eq. (2.196), we obtain the impulse response

$$g(t) = \mathscr{L}^{-1} \bar{G}(s) = \mathscr{L}^{-1} \frac{1}{2i\omega_d m} \left( \frac{1}{s + \zeta\omega_n - i\omega_d} - \frac{1}{s + \zeta\omega_n + i\omega_d} \right)$$

$$= \frac{1}{2i\omega_d m} [e^{-(\zeta\omega_n - i\omega_d)t} - e^{-(\zeta\omega_n + i\omega_d)t}] = \frac{1}{m\omega_d} e^{-\zeta\omega_n t} \sin \omega_d t \qquad (c)$$

which is precisely the same as the result obtained by classical means, Eq. (e) of Example 2.3. Because there is no excitation for $t < 0$, Eq. (c) above should really be regarded as being multiplied by $u(t)$ as in Eq. (2.161).

**Example 2.8** Determine the step response of a damped single-degree-of-freedom system by the Laplace transformation method.

Introducing Eq. (2.192) into Eq. (2.198) and using the method of partial fractions, we can write

$$\mathscr{s}(t) = \mathscr{L}^{-1} \frac{\bar{G}(s)}{s} = \mathscr{L}^{-1} \frac{1}{ms(s^2 + 2\zeta\omega_n s + \omega_n^2)}$$

$$= \frac{1}{m\omega_n^2} \mathscr{L}^{-1} \left( \frac{1}{s} - \frac{\zeta\omega_n + i\omega_d}{2i\omega_d} \frac{1}{s + \zeta\omega_n - i\omega_d} + \frac{\zeta\omega_n - i\omega_d}{2i\omega_d} \frac{1}{s + \zeta\omega_n + i\omega_d} \right)$$
$$(a)$$

so that, recalling Eq. (b) of Example 2.7, the step response can be obtained as follows:

$$\mathscr{s}(t) = \frac{1}{k} \mathscr{L}^{-1} \left[ 1 - \frac{\zeta\omega_n + i\omega_d}{2i\omega_d} e^{-(\zeta\omega_n - i\omega_d)t} + \frac{\zeta\omega_n - i\omega_d}{2i\omega_d} e^{-(\zeta\omega_n + i\omega_d)t} \right]$$

$$= \frac{1}{k} \left[ 1 - \frac{1}{(1 - \zeta^2)^{1/2}} e^{-\zeta\omega_n t} \cos (\omega_d t - \psi) \right]$$
$$(b)$$

$$\psi = \tan^{-1} \frac{\zeta}{(1 - \zeta^2)^{1/2}}$$

which is essentially the same expression as that obtained in Example 2.4. In view of the fact that the excitation is zero for $t < 0$, the right side of Eq. (b) should be regarded as being multiplied by $u(t)$.

## 2.18 GENERAL SYSTEM RESPONSE

Let us consider Eq. (2.123) and obtain the response to the external excitation $F(t)$, as well as to the initial conditions $x(0) = x_0$, $\dot{x}(0) = v_0$, by the Laplace transformation method. Transforming both sides of Eq. (2.123), we obtain Eq. (2.189). Hence, using Eq. (2.189), we obtain the transformed response in the form

$$\bar{x}(s) = \frac{\bar{F}(s)}{m(s^2 + 2\zeta\omega_n s + \omega_n^2)} + \frac{s + 2\zeta\omega_n}{s^2 + 2\zeta\omega_n s + \omega_n^2} x_0 + \frac{1}{s^2 + 2\zeta\omega_n s + \omega_n^2} v_0 \tag{2.199}$$

The inverse transformation of $\bar{x}(s)$ will be carried out by considering each term on the right side of Eq. (2.199) separately. To obtain the inverse transformation of the first term on the right side of Eq. (2.199), we use Borel's theorem (see Sec. B.5). To this end, we let

$$\bar{f}_1(s) = \bar{F}(s) \qquad \bar{f}_2(s) = \frac{1}{m(s^2 + 2\zeta\omega_n s + \omega_n^2)} \tag{2.200}$$

Clearly, $f_1(t) = F(t)$. Moreover, from Sec. B.6, we conclude that

$$f_2(t) = \frac{1}{m\omega_d} e^{-\zeta\omega_n t} \sin \omega_d t \qquad \omega_d = (1 - \zeta^2)^{1/2}\omega_n \tag{2.201}$$

and we note that $f_2(t)$ is equal to the impulse response $g(t)$, as can be seen from Eq. (c) of Example 2.7. (The reader is urged to explain why.) Hence, considering Eq. (B.29), the inverse transformation of the first term on the right side of Eq. (2.199) is

$$\mathscr{L}^{-1}\bar{f}_1(s)\bar{f}_2(s) = \int_0^t f_1(\tau)f_2(\iota - \tau)\,d\tau$$

$$= \frac{1}{m\omega_d} \int_0^t F(\tau)e^{-\zeta\omega_n(t-\tau)} \sin \omega_d(t - \tau)\,d\tau \tag{2.202}$$

Also from Sec. B.6, we obtain the inverse transform of the coefficient of $x_0$ in Eq. (2.199) in the form

$$\mathscr{L}^{-1}\frac{s + 2\zeta\omega_n}{s^2 + 2\zeta\omega_n s + \omega_n^2} = \frac{1}{(1 - \zeta^2)^{1/2}} e^{-\zeta\omega_n t} \cos(\omega_d t - \psi)$$

$$\psi = \tan^{-1}\frac{\zeta}{(1 - \zeta^2)^{1/2}} \tag{2.203}$$

Moreover, the inverse transformation of the coefficient of $v_0$ can be obtained by multiplying $f_2(t)$, as given by Eq. (2.201), by $m$. Hence, considering Eqs. (2.201) through (2.203), we obtain the general response

$$x(t) = \frac{1}{m\omega_d} \int_0^t F(\tau)e^{-\zeta\omega_n(t-\tau)} \sin \omega_d(t - \tau)\,d\tau$$

$$+ \frac{x_0}{(1 - \zeta^2)^{1/2}} e^{-\zeta\omega_n t} \cos(\omega_d t - \psi) + \frac{v_0}{\omega_d} e^{-\zeta\omega_n t} \sin \omega_d t \tag{2.204}$$

and note that the Laplace transformation method permitted us to produce both the response to the initial conditions and the response to external excitation simultaneously. We shall make repeated use of Eq. (2.204) later in this text.

# PROBLEMS

**2.1** A control tab of an airplane elevator is hinged about an axis in the elevator, shown as the point $O$ in Fig. 2.33, and activated by a control linkage behaving like a torsional spring of stiffness $k_T$. The mass moment of inertia of the control tab is $I_O$, so that the natural frequency of the system is $\omega_n = \sqrt{k_T/I_O}$. Because $k_T$ cannot be calculated exactly, it is necessary to obtain the natural frequency $\omega_n$ experimentally. To this end the elevator is held fixed and the tab is excited harmonically by means of the spring $k_2$ while restrained by the spring $k_1$, as shown in Fig. 2.33, and the excitation frequency $\omega$ is varied until the resonance frequency $\omega_r$ is reached. Calculate the natural frequency $\omega_n$ of the control tab in terms of $\omega_r$ and the parameters of the experimental setup.

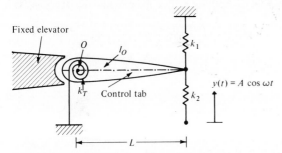

**Figure 2.33**

**2.2** A machine of mass $M$ rests on a massless elastic floor, as shown in Fig. 2.34. If a unit load is applied at midspan, the floor undergoes a deflection $x_{st}$. A shaker having total mass $m_s$ and carrying two rotating unbalanced masses (similar to the rotating masses shown in Fig. 2.6a) produces a vertical harmonic force $ml\omega^2 \sin \omega t$, where the frequency of rotation may be varied. Show how the shaker can be used to derive a formula for the natural frequency of flexural vibration of the structure.

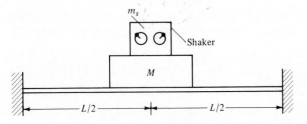

**Figure 2.34**

**2.3** Derive the differential equation of motion for the inverted pendulum of Fig. 2.35, where $A \cos \omega t$ represents a displacement excitation. Then assume small amplitudes and solve for the angle $\theta$ as a function of time.

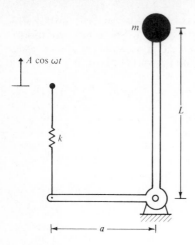

**Figure 2.35**

**2.4** One side of the tube of Prob. 1.6 is subjected to the pressure $p(t) = p_0 \cos \omega t$, where $p_0$ has units in pounds per square inch (lb/in$^2$) [newtons per square meter (N/m$^2$)]. Derive the differential equation of motion, and obtain the resonance frequency.

**2.5** The left end of the cantilever beam shown in Fig. 2.36 undergoes the harmonic motion $x(t) = A \cos \omega t$. Derive the differential equation for the motion of the mass $M$ and determine the resonance frequency. Assume that the beam is massless and that its bending stiffness $EI$ is constant.

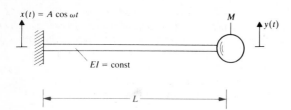

**Figure 2.36**

**2.6** The foundation of the building in Prob. 1.15 undergoes the horizontal motion $y(t) = y_0 \sin \omega t$. Derive the system response.

**2.7** Gear $A$ in Prob. 1.16 is subjected to the torque $M_A = M_0 \cos \omega t$. Derive an expression for the angular motion of gear $B$.

**2.8** Solve the differential equation

$$m\ddot{x}(t) + c\dot{x}(t) + kx(t) = kA \sin \omega t$$

describing the motion of a damped single-degree-of-freedom system subjected to a harmonic force. Assume a solution in the form $x(t) = X(\omega) \sin (\omega t - \phi)$ and derive expressions for $X$ and $\phi$ by equating coefficients of $\sin \omega t$ and $\cos \omega t$ on both sides of the equation.

**2.9** Assume a solution of Eq. (2.41) in the form $x(t) = X(\omega)e^{i(\omega t - \phi)}$ and show that this form contains the solutions to both $f(t) = A \cos \omega t$ and $f(t) = A \sin \omega t$.

**2.10** Start with Eq. (2.48) and verify Eqs. (2.49), (2.50), and (2.51).

**2.11** A mass-damper-spring system of the type shown in Fig. 1.9a has been observed to achieve a peak magnification factor $Q = 5$ at the driving frequency $\omega = 10$ rad/s. It is required to determine: (1) the damping factor, (2) the driving frequencies corresponding to the half-power points, and (3) the bandwidth of the system.

**2.12** A piece of machinery can be regarded as a rigid mass with two reciprocating rotating unbalanced masses such as in Fig. 2.6a. The total mass of the system is 12 kg and each of the unbalanced masses is equal to 0.5 kg. During normal operation, the rotation of the masses varies from zero to 600 r/min. Design a support system so that the maximum vibration amplitude will not exceed 10 percent of the rotating masses' eccentricity.

**2.13** The rotor of a turbine having the form of a disk is mounted at the midspan of a uniform steel shaft, as shown in Fig. 2.37. The mass of the disk is 15 kg and its diameter is 0.3 m. The disk has a circular hole of diameter 0.03 m at a distance of 0.12 m from the geometric center. The bending stiffness of the shaft is $EI = 1600 \text{ N} \cdot \text{m}^2$. Determine the amplitude of vibration if the turbine rotor rotates with the angular velocity of 6000 r/min. Assume that the shaft bearings are rigid.

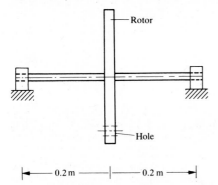

Figure 2.37

**2.14** Prove Eqs. (2.90) and (2.91).

**2.15** Consider the system of Fig. 2.16. When the support is fixed, $y = 0$, and the mass is allowed to vibrate freely, the ratio between two consecutive maximum displacement amplitudes is $x_2/x_1 = 0.8$. On the other hand, when the mass is in equilibrium, the spring is compressed by an amount $x_{st} = 0.1$ in $(2.54 \times 10^{-3} \text{ m})$. The weight of the mass is $mg = 20 \text{ lb}$ (88.96N). Let $y(t) = A \cos \omega t$, $x(t) = X \cos (\omega t - \phi)$, and plot $X/A$ versus $\omega/\omega_n$ and $\phi$ versus $\omega/\omega_n$ for $0 < \omega/\omega_n < 2$.

**2.16** The system shown in Fig. 2.38 simulates a vehicle traveling on a rough road. Let the vehicle velocity be uniform, $v = \text{const}$, and calculate the response $z(t)$, as well as the force transmitted to the vehicle.

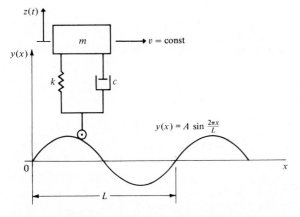

Figure 2.38

**2.17** The support of the viscously damped pendulum shown in Fig. 2.39 undergoes harmonic oscillation. Derive the differential equation of motion of the system, then assume small amplitudes and solve for $\theta(t)$.

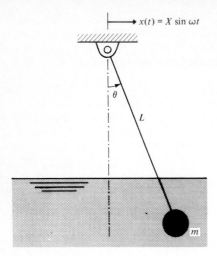

$x(t) = X \sin \omega t$

$\theta$

$L$

$m$

**Figure 2.39**

**2.18** The system of Fig. 2.6a has the following parameters: $M = 80$ kg, $m = 5$ kg, $k = 8000$ N/m, $l = 0.1$ m. Design a viscous damper so that at the rotating speed $\omega = 4\omega_n$ the force transmitted to the support does not exceed 250 N.

**2.19** It is observed that during one cycle of vibration a structurally damped single-degree-of-freedom system dissipates energy in the amount of 1.2 percent of the maximum potential energy. Calculate the structural damping factor $\gamma$.

**2.20** Refer to Eq. (2.122) and define a magnification factor $|G(i\omega)|$ and an angle $\phi$. Plot $|G(i\omega)|$ versus $\omega/\omega_n$ and $\phi$ versus $\omega/\omega_n$ for $\gamma = 0$ and $\gamma = 0.01$.

**2.21** The cam of Fig. 2.40a imparts a displacement $y(t)$ in the form of a periodic sawtooth function to the lower end of the system, where $y(t)$ is shown in Fig. 2.40b. Derive an expression for the response $x(t)$ by means of a Fourier analysis.

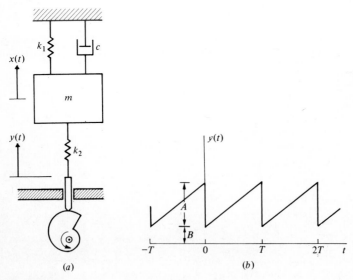

(a)

(b)

**Figure 2.40**

**2.22** Solve the differential equation

$$m\ddot{x}(t) + c\dot{x}(t) + kx(t) = kf(t)$$

by means of a Fourier analysis, where $f(t)$ is the periodic function shown in Fig. 2.41.

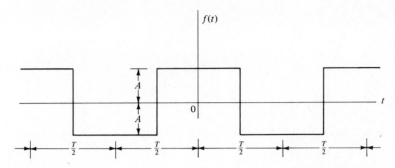

**Figure 2.41**

**2.23** Consider the system of Fig. 2.12 and use the convolution integral to solve for the response $x(t)$, where $y(t)$ has the same form as the rectangular pulse shown in Fig. 2.27. Let the system parameters be as in Prob. 2.15, and plot the response for $A = 0.4$ in (0.01 m) and $T = 10$ s for the time interval $-10$ s $< t < 20$ s. Note that for excitation functions defined for $t < 0$, the lower limit in the convolution integral must be changed (see Sec. 11.11).

**2.24** Solve the differential equation of Prob. 2.22 by means of the convolution integral for the case in which $f(t)$ is the "ramp function" given in Fig. 2.42.

$$m\ddot{x} + c\dot{x} + kx = kf$$

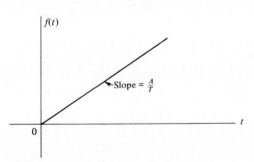

**Figure 2.42**

**2.25** Solve the differential equation of Prob. 2.22 for the case in which $f(t)$ is as given in Fig. 2.43. Regard $f(t)$ as a superposition of ramp functions.

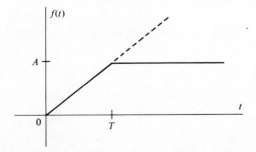

**Figure 2.43**

**2.26** Solve the differential equation of Prob. 2.22 for the case in which $f(t)$ is as given in Fig. 2.22. Regard $f(t)$ as a superposition of step functions.

**2.27** Solve the differential equation of Prob. 2.22 for the case in which $f(t)$ has the form of the triangular pulse shown in Fig. 2.44.

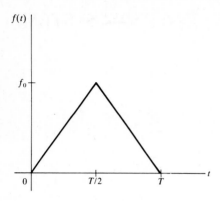

**Figure 2.44**

**2.28** Repeat Prob. 2.27 for the case in which $f(t)$ has the form of the trapezoidal pulse shown in Fig. 2.45.

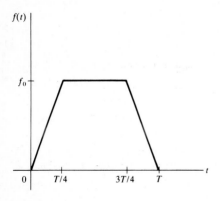

**Figure 2.45**

**2.29** Plot the shock spectrum for the system of Prob. 2.27. Compare the results with those obtained in Sec. 2.16 and draw conclusions.

**2.30** Repeat Prob. 2.29 for the system of Prob. 2.28.

**2.31** Obtain the response of the system of Prob. 2.24 by the Laplace transform method. Assume no damping and regard $f(t)$ as zero for $t < 0$.

**2.32** Obtain the response of the system of Prob. 2.26 by the Laplace transform method. Assume no damping and regard $f(t)$ as zero for $t < 0$.

3.1 – 3.5, 3.7 – 3.9

# THREE

## TWO-DEGREE-OF-FREEDOM SYSTEMS

## 3.1 INTRODUCTION

The material contained in this chapter belongs rightfully in a chapter on multi-degree-of-freedom systems, as the two-degree-of-freedom system is merely a special case of the larger class of multi-degree-of-freedom systems. Pedagogic considerations, however, tilted the balance toward a separate chapter that can serve both as an independent, more elementary treatment of two-degree-of-freedom systems and as an introduction to a more advanced study of discrete systems with an arbitrarily large number of degrees of freedom. We recall that the number of degrees of freedom of a system is defined as the number of independent coordinates necessary to describe the motion of a system completely.

If an undamped single-degree-of-freedom system is subjected to a certain initial excitation, then the ensuing motion can be described as natural vibration, in the sense that the system vibrates at the system natural frequency. What sets apart natural vibration for a multi-degree-of-freedom system from that for a single-degree-of-freedom system is that for multi-degree-of-freedom systems natural vibration implies a certain displacement configuration, or shape, assumed by the whole system during motion. Moreover, a multi-degree-of-freedom system does not possess only one natural configuration but has a finite number of natural configurations known as *natural modes of vibration*. Depending on the initial excitation, the system can vibrate in any of these modes. To each mode corresponds a unique frequency, referred to as a *natural frequency*, so that there are as many natural frequencies as there are natural modes. The natural modes possess a very important property known as *orthogonality*.

The mathematical formulation for an $n$-degree-of-freedom system consists of $n$ simultaneous ordinary differential equations. Hence, the motion of one mass depends on the motion of another. For a proper choice of coordinates, known as *principal* or *natural coordinates*, the system differential equations become independent of one another. The natural coordinates represent linear combinations of the actual displacements of the discrete masses and, conversely, the motion of the system can be regarded as a superposition of the natural coordinates. The differential equations for the natural coordinates possess the same structure as those of single-degree-of-freedom systems.

In this chapter, we begin by formulating the general equations of motion for a linear two-degree-of-freedom system and then show how to obtain the natural frequencies and modes. The response to initial excitation, as well as that to external excitation, is derived and various applications are discussed. The response to arbitrary external excitation is actually deferred to Chap. 4, when the transformation to natural coordinates is studied in a more systematic manner in conjunction with the orthogonality of natural modes for multi-degree-of-freedom systems.

## 3.2 EQUATIONS OF MOTION FOR A TWO-DEGREE-OF-FREEDOM SYSTEM

Let us consider the viscously damped system shown in Fig. 3.1$a$, and derive the associated differential equations of motion. The system is fully described by the two

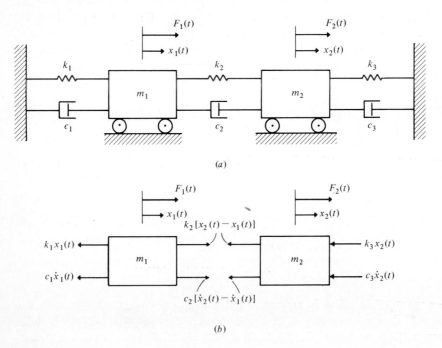

(a)

(b)

**Figure 3.1**

coordinates $x_1(t)$ and $x_2(t)$, which give the positions of the masses $m_1$ and $m_2$, respectively, for any arbitrary time $t$. The motions $x_1(t)$ and $x_2(t)$ are sufficiently small that the system operates in the linear range. To derive the differential equations of motion, we apply Newton's second law to the masses $m_1$ and $m_2$. To this end, we refer to the free-body diagrams shown in Fig. 3.1$b$. Summing forces in the horizontal direction on each mass, we can write the two equations

$$F_1(t) - c_1\dot{x}_1(t) - k_1 x_1(t) + c_2[\dot{x}_2(t) - \dot{x}_1(t)] + k_2[x_2(t) - x_1(t)] = m_1\ddot{x}_1(t)$$

$$F_2(t) - c_2[\dot{x}_2(t) - \dot{x}_1(t)] - k_2[x_2(t) - x_1(t)] - c_3\dot{x}_2(t) - k_3 x_2(t) = m_2\ddot{x}_2(t)$$

$$(3.1)$$

which can be rearranged in the form

$$m_1\ddot{x}_1(t) + (c_1 + c_2)\dot{x}_1(t) - c_2\dot{x}_2(t) + (k_1 + k_2)x_1(t) - k_2 x_2(t) = F_1(t)$$

$$m_2\ddot{x}_2(t) - c_2\dot{x}_1(t) + (c_2 + c_3)\ddot{x}_2(t) - k_2 x_1(t) + (k_2 + k_3)x_2(t) = F_2(t)$$

$$(3.2)$$

and we note that Eqs. (3.2) are not independent, because the first equation contains terms in $\dot{x}_2(t)$ and $x_2(t)$, whereas the second equation contains terms in $\dot{x}_1(t)$ and $x_1(t)$. A system described by two simultaneous second-order differential equations of the type (3.2) is known as a *two-degree-of-freedom system*. We refer to a system of simultaneous equations as *coupled*, and to the terms rendering the equations dependent on one another as *coupling terms*. In the case of Eqs. (3.2), the coupling terms are $c_2\dot{x}(t)$ and $-k_2 x_2(t)$ in the first equation and $c_2\dot{x}_1(t)$ and $-k_2 x_1(t)$ in the second equation, so that the velocity coupling terms have the coefficient $-c_2$ and the displacement coupling terms the coefficient $-k_2$. Hence, we must expect the motion of the mass $m_1$ to influence the motion of the mass $m_2$, and vice versa, except for $c_2 = k_2 = 0$ when the equations of motion (3.2) become independent of one another. The case $c_2 = k_2 = 0$ presents no interest, however, because in this case we no longer have a single two-degree-of-freedom system but two completely independent single-degree-of-freedom systems.

Equations (3.2) can be conveniently expressed in matrix form. Indeed, let us introduce the notation

$$\begin{bmatrix} m_1 & 0 \\ 0 & m_2 \end{bmatrix} = [m] \qquad \begin{bmatrix} c_1 + c_2 & -c_2 \\ -c_2 & c_2 + c_3 \end{bmatrix} = [c] \qquad \begin{bmatrix} k_1 + k_2 & -k_2 \\ -k_2 & k_2 + k_3 \end{bmatrix} = [k]$$

$$(3.3)$$

where the constant matrices $[m]$, $[c]$, and $[k]$ of the coefficients are known as the *mass matrix*, *damping matrix*, and *stiffness matrix*, respectively, and

$$\begin{Bmatrix} x_1(t) \\ x_2(t) \end{Bmatrix} = \{x(t)\} \qquad \begin{Bmatrix} F_1(t) \\ F_2(t) \end{Bmatrix} = \{F(t)\} \qquad (3.4)$$

where the $2 \times 1$ matrices $\{x(t)\}$ and $\{F(t)\}$ are the two-dimensional *displacement vector* and *force vector*, respectively. Note that rectangular matrices (of which the square matrices are a special case) are denoted by brackets and column matrices by braces. In view of Eqs. (3.3) and (3.4), Eqs. (3.2) can be written in the compact matrix form

$$[m]\{\ddot{x}(t)\} + [c]\{\dot{x}(t)\} + [k]\{x(t)\} = \{F(t)\} \qquad (3.5)$$

It is easy to see from Eqs. (3.3) that the off-diagonal elements of the matrices $[m]$, $[c]$, and $[k]$ satisfy

$$m_{12} = m_{21} = 0 \qquad c_{12} = c_{21} = -c_2 \qquad k_{12} = k_{21} = -k_2 \qquad (3.6)$$

with the implication that the matrices are symmetric, as expressed by

$$[m] = [m]^T \qquad [c] = [c]^T \qquad [k] = [k]^T \qquad (3.7)$$

where the superscript $T$ designates the transpose of the matrix. Moreover, $[m]$ is diagonal. Equation (3.5) represents a set of independent equations only when all three matrices $[m]$, $[c]$, and $[k]$ are diagonal.

The solution of Eq. (3.5) for any arbitrary force vector is difficult to obtain. The difficulty can be attributed broadly to the fact that the two equations represented by (3.5) are not independent. In the remaining part of this chapter, we discuss the free-vibration case, obtained when the forces $F_i(t)$ $(i = 1, 2)$ are zero, and the case in which the forces $F_i(t)$ are harmonic. The case in which the forces $F_i(t)$ are arbitrary will be discussed in Chap. 4, when more adequate tools for treating such problems are introduced.

## 3.3 FREE VIBRATION OF UNDAMPED SYSTEMS. NATURAL MODES

In the absence of damping and external forces, the general system of Fig. 3.1a reduces to the special case shown in Fig. 3.2, where the latter is recognized as a *conservative system* because there is no mechanism for dissipating or adding energy. The differential equations of motion for the system of Fig. 3.2 can be obtained directly from Eqs. (3.2) by letting $c_1 = c_2 = c_3 = 0$ and $F_1(t) = F_2(t) = 0$. The resulting equations are simply

$$m_1\ddot{x}_1(t) + (k_1 + k_2)x_1(t) - k_2x_2(t) = 0$$
$$m_2\ddot{x}_2(t) - k_2x_1(t) + (k_2 + k_3)x_2(t) = 0 \qquad (3.8)$$

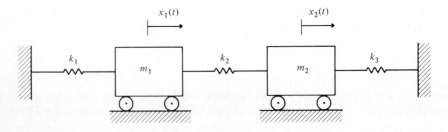

**Figure 3.2**

which represent two simultaneous homogeneous differential equations of second order. Recalling the third of Eqs. (3.3), we recognize that the coefficients

$$k_1 + k_2 = k_{11} \qquad k_2 + k_3 = k_{22} \qquad -k_2 = k_{12} = k_{21} \qquad (3.9)$$

in Eqs. (3.8) represent the elements of the stiffness matrix $[k]$, so that Eqs. (3.8) can be written in the form

$$m_1 \ddot{x}_1(t) + k_{11}x_1(t) + k_{12}x_2(t) = 0$$
$$m_2 \ddot{x}_2(t) + k_{12}x_1(t) + k_{22}x_2(t) = 0 \qquad (3.10)$$

Because Eqs. (3.10) are homogeneous, if $x_1(t)$ and $x_2(t)$ represent a solution, then $\alpha x_1(t)$ and $\alpha x_2(t)$ also represent a solution where $\alpha$ is an arbitary constant. Hence, the solution of Eqs. (3.10) can only be obtained within a constant scalar multiplier.

The interest lies in exploring the existence of a special type of solution of Eqs. (3.10), namely, one in which the coordinates $x_1(t)$ and $x_2(t)$ increase and decrease in the same proportion as time unfolds. We refer to such motion as *synchronous*. Because in this case the time dependence of $x_1(t)$ and $x_2(t)$ is the same, if synchronous motion is possible, then the ratio $x_2(t)/x_1(t)$ must be independent of time. Hence, the type of motion we are seeking is one in which *the ratio between the two displacements remains constant throughout the motion*. Referring to the displacement pattern as the system *configuration*, the implication of the preceding statement is that the shape of the system configuration does not change during motion, but the amplitude of the displacement pattern does. Denoting the time dependence of $x_1(t)$ and $x_2(t)$ by $f(t)$, the solution sought can be written in the form

$$x_1(t) = u_1 f(t) \qquad x_2(t) = u_2 f(t) \qquad (3.11)$$

where $u_1$ and $u_2$ play the role of constant amplitudes. Introducing Eqs. (3.11) into Eqs. (3.10), we obtain

$$m_1 u_1 \ddot{f}(t) + (k_{11}u_1 + k_{12}u_2)f(t) = 0$$
$$m_2 u_2 \ddot{f}(t) + (k_{12}u_1 + k_{22}u_2)f(t) = 0 \qquad (3.12)$$

For Eqs. (3.12) to possess a solution, we must have

$$-\frac{\ddot{f}(t)}{f(t)} = \frac{k_{11}u_1 + k_{12}u_2}{m_1 u_1} = \frac{k_{12}u_1 + k_{22}u_2}{m_2 u_2} = \lambda \qquad (3.13)$$

where $\lambda$ is a real constant because $m_1, m_2, k_{11}, k_{12}, u_1$, and $u_2$ are all real constants. Hence, synchronous motion is possible, provided the equations

$$\ddot{f}(t) + \lambda f(t) = 0 \qquad (3.14)$$

and

$$(k_{11} - \lambda m_1)u_1 + k_{12}u_2 = 0$$
$$k_{12}u_1 + (k_{22} - \lambda m_2)u_2 = 0 \qquad (3.15)$$

possess solutions.

It is not difficult to show that $\lambda$ must be not only real but also positive. Indeed, letting the solution of Eq. (3.14) have the exponential form

$$f(t) = Ae^{st} \tag{3.16}$$

it follows that $s$ must satisfy the equation

$$s^2 + \lambda = 0 \tag{3.17}$$

which has two roots, namely,

$$\begin{matrix} s_1 \\ s_2 \end{matrix} = \pm\sqrt{-\lambda} \tag{3.18}$$

so that solution (3.16) becomes

$$f(t) = A_1 e^{s_1 t} + A_2 e^{s_2 t} = A_1 \exp \sqrt{-\lambda}\, t + A_2 \exp -\sqrt{-\lambda}\, t \tag{3.19}$$

But if $\lambda$ is a negative number, the exponents $\sqrt{-\lambda}\, t$ and $-\sqrt{-\lambda}\, t$ are real quantities, equal in value but opposite in sign. It follows that, as $t \to \infty$, the first term of $f(t)$ tends to infinity and the second tends to zero exponentially. This, however, is inconsistent with the concept of an oscillatory system, for which the motion can neither reduce to zero nor increase without bounds. Hence, the possibility that $\lambda$ is negative must be discarded, and the one that $\lambda$ is positive must be adopted instead. Letting $\lambda = \omega^2$, where $\omega$ is real, Eq. (3.18) yields

$$\begin{matrix} s_1 \\ s_2 \end{matrix} = \pm i\omega \tag{3.20}$$

so that solution (3.19) becomes

$$f(t) = A_1 e^{i\omega t} + A_2 e^{-i\omega t} \tag{3.21}$$

where $A_1$ and $A_2$ are generally complex numbers constant in value. Recognizing that $e^{i\omega t}$ and $e^{-i\omega t}$ represent complex vectors of unit magnitude and recalling that they are related to the trigonometric functions $\cos \omega t$ and $\sin \omega t$ by

$$e^{\pm i\omega t} = \cos \omega t \pm i \sin \omega t \tag{3.22}$$

we conclude that

$$f(t) = (A_1 + A_2) \cos \omega t + i(A_1 - A_2) \sin \omega t \tag{3.23}$$

so that solution (3.23) is harmonic with the frequency $\omega$ and represents the only acceptable solution of Eq. (3.14). This implies that if synchronous motion is possible, then *the time dependence is harmonic*. But $f(t)$ is known to be a real function, so that, introducing the notation

$$A_1 + A_2 = C \cos \phi \qquad i(A_1 - A_2) = C \sin \phi \tag{3.24}$$

solution (3.23) becomes

$$f(t) = C \cos (\omega t - \phi) \tag{3.25}$$

where $C$ is an arbitrary constant, $\omega$ is the frequency of the narmonic motion, and $\phi$ is its phase angle, all three quantities being the same for both coordinates, $x_1(t)$ and $x_2(t)$.

Next we must verify whether $\lambda = \omega^2$ is arbitrary, or can take only certain values. The answer to this question lies in Eqs. (3.15). Inserting $\lambda = \omega^2$ in Eqs. (3.15), we obtain

$$(k_{11} - \omega^2 m_1)u_1 + k_{12}u_2 = 0$$
$$k_{12}u_1 + (k_{22} - \omega^2 m_2)u_2 = 0 \tag{3.26}$$

which represent two simultaneous homogeneous algebraic equations in the unknowns $u_1$ and $u_2$, with $\omega^2$ playing the role of a parameter. The problem of determining the values of the parameter $\omega^2$ for which Eqs. (3.26) admit nontrivial solutions is known as the *characteristic-value problem*, or the *eigenvalue problem*. From linear algebra, Eqs. (3.26) possess a solution only if the determinant of the coefficients of $u_1$ and $u_2$ is zero, or

$$\Delta(\omega^2) = \det \begin{bmatrix} k_{11} - \omega^2 m_1 & k_{12} \\ k_{12} & k_{22} - \omega^2 m_2 \end{bmatrix} = 0 \tag{3.27}$$

where $\Delta(\omega^2)$, known as the *characteristic determinant*, is a polynomial of second degree in $\omega^2$. Indeed, expanding Eq. (3.27), we can write

$$\Delta(\omega^2) = m_1 m_2 \omega^4 - (m_1 k_{22} + m_2 k_{11})\omega^2 + k_{11}k_{22} - k_{12}^2 = 0 \tag{3.28}$$

which represents a quadratic equation in $\omega^2$ called the *characteristic equation*, or *frequency equation*. The equation has the roots

$$\begin{matrix} \omega_1^2 \\ \omega_2^2 \end{matrix} = \frac{1}{2}\frac{m_1 k_{22} + m_2 k_{11}}{m_1 m_2} \mp \frac{1}{2}\sqrt{\left(\frac{m_1 k_{22} + m_2 k_{11}}{m_1 m_2}\right)^2 - 4\frac{k_{11}k_{22} - k_{12}^2}{m_1 m_2}} \tag{3.29}$$

so that there are only two modes in which synchronous motion is possible, one characterized by the frequency $\omega_1$ and the other by the frequency $\omega_2$, where $\omega_1$ and $\omega_2$ are known as the *natural frequencies* of the system.

It remains to determine the values of the constants $u_1$ and $u_2$. These values depend on the natural frequencies $\omega_1$ and $\omega_2$. We denote the values corresponding to $\omega_1$ by $u_{11}$ and $u_{21}$ and those corresponding to $\omega_2$ by $u_{12}$ and $u_{22}$. Hence, the first subscript identifies the position of the masses and the second subscript indicates whether the synchronous motion has the frequency $\omega_1$ or $\omega_2$. As pointed out earlier, because the problem is homogeneous, only the ratios $u_{21}/u_{11}$ and $u_{22}/u_{12}$ can be determined uniquely. Indeed, inserting $\omega_1^2$ and $\omega_2^2$ into Eqs. (3.26), we can write simply

$$\frac{u_{21}}{u_{11}} = -\frac{k_{11} - \omega_1^2 m_1}{k_{12}} = -\frac{k_{12}}{k_{22} - \omega_1^2 m_2} \tag{3.30a}$$

$$\frac{u_{22}}{u_{12}} = -\frac{k_{11} - \omega_2^2 m_1}{k_{12}} = -\frac{k_{12}}{k_{22} - \omega_2^2 m_2} \tag{3.30b}$$

The implication is that the two expressions for the ratio $u_{21}/u_{11}$ given in Eq. (3.30$a$) are equal, and a similar statement can be made concerning $u_{22}/u_{12}$ in Eq. (3.30$b$). The ratios $u_{21}/u_{11}$ and $u_{22}/u_{12}$ determine the shape assumed by the system during synchronous motion with frequencies $\omega_1$ and $\omega_2$, respectively. If one element in each ratio is assigned a certain arbitrary value, then the value of the other element follows automatically. The resulting pairs of numbers, $u_{11}$ and $u_{21}$ on the one hand and $u_{12}$ and $u_{22}$ on the other hand, are known as the *natural modes of vibration* of the system. The modes can be represented by vectors and exhibited in the form of the column matrices

$$\{u\}_1 = \begin{Bmatrix} u_{11} \\ u_{21} \end{Bmatrix} \qquad \{u\}_2 = \begin{Bmatrix} u_{12} \\ u_{22} \end{Bmatrix} \tag{3.31}$$

where $\{u\}_1$ and $\{u\}_2$ are referred to as *modal vectors*. The natural frequency $\omega_1$ and the modal vector $\{u\}_1$ constitute what is known in a broad sense as the *first mode of vibration*, and $\omega_2$ and $\{u\}_2$ constitute the *second mode of vibration*. We note that for a two-degree-of-freedom system there are two modes of vibration. We shall see in Chap. 4 that the number of modes coincides with the number of degrees of freedom. The natural modes of vibration, i.e., the natural frequencies and the modal vectors, represent a property of the system, and they are unique for a given system except for the magnitude of the modal vectors, implying that the mode shape is unique, but the amplitude is not. Indeed, because the problem is homogeneous, a modal vector multiplied by a constant scalar represents the same modal vector. It is often convenient to render a modal vector unique by assigning a given value either to one of the components of the modal vector or to the magnitude of the modal vector. This process is known as *normalization* and the resulting vector is said to represent a *normal mode*. Clearly, normalization is arbitrary and it does not affect the mode shape, as all the components of the normalized vector are changed in the same proportion.

The motion in time is obtained by recalling Eqs. (3.11) and (3.25). Hence, the two possible synchronous motions can be written in the simple vector form

$$\begin{aligned} \{x(t)\}_1 &= \{u\}_1 f_1(t) = C_1 \{u\}_1 \cos(\omega_1 t - \phi_1) \\ \{x(t)\}_2 &= \{u\}_2 f_2(t) = C_2 \{u\}_2 \cos(\omega_2 t - \phi_2) \end{aligned} \tag{3.32}$$

where we note that $f_1(t)$ and $f_2(t)$ represent the solution (3.25) corresponding to the first and second mode, respectively. We show in Sec. 3.5 that the motion of the system at any time can be obtained as a superposition of the two natural modes, namely,

$$\begin{aligned} \{x(t)\} &= \{x(t)\}_1 + \{x(t)\}_2 \\ &= C_1 \{u\}_1 \cos(\omega_1 t - \phi_1) + C_2 \{u\}_2 \cos(\omega_2 t - \phi_2) \end{aligned} \tag{3.33}$$

The amplitudes $C_1$ and $C_2$ and the phase angles $\phi_1$ and $\phi_2$ are determined by the initial displacements and initial velocities of the masses $m_1$ and $m_2$.

It is convenient to arrange the modal vectors $\{u\}_1$ and $\{u\}_2$ in a square matrix of the form

$$[u] = [\{u\}_1 \quad \{u\}_2] = \begin{bmatrix} u_{11} & u_{12} \\ u_{21} & u_{22} \end{bmatrix} \tag{3.34}$$

where $[u]$ is known as the *modal matrix*. Moreover, introducing the vector

$$\{f(t)\} = \begin{Bmatrix} f_1(t) \\ f_2(t) \end{Bmatrix} = \begin{Bmatrix} C_1 \cos(\omega_1 t - \phi_1) \\ C_2 \cos(\omega_2 t - \phi_2) \end{Bmatrix} \tag{3.35}$$

Eq. (3.33) can be written in the compact matrix form

$$\{x(t)\} = [u]\{f(t)\} \tag{3.36}$$

**Example 3.1** Consider the system of Fig. 3.2, let $m_1 = m$, $m_2 = 2m$, $k_1 = k_2 = k$, $k_3 = 2k$, and obtain the natural modes of vibration.

Using Eqs. (3.9), we obtain the elements of the stiffness matrix $[k]$ in the form

$$k_{11} = k_1 + k_2 = 2k \qquad k_{22} = k_2 + k_3 = 3k \qquad k_{12} = -k_2 = -k \quad (a)$$

so that, using Eq. (3.28), we obtain the frequency equation

$$\Delta(\omega^2) = 2m^2\omega^4 - 7mk\omega^2 + 5k^2 = 0 \tag{b}$$

which has the roots

$$\begin{aligned} \omega_1^2 \\ \omega_2^2 \end{aligned} = \left[ \frac{7}{4} \mp \sqrt{\left(\frac{7}{4}\right)^2 - \frac{5}{2}} \right] \frac{k}{m} = \begin{cases} \dfrac{k}{m} \\[2mm] \dfrac{5}{2}\dfrac{k}{m} \end{cases} \tag{c}$$

so that the natural frequencies are

$$\omega_1 = \sqrt{\frac{k}{m}} \qquad \omega_2 = 1.5811\sqrt{\frac{k}{m}} \tag{d}$$

Introducing $\omega_1^2$ and $\omega_2^2$ into Eqs. (3.30), we obtain the ratios

$$\frac{u_{21}}{u_{11}} = -\frac{k_{11} - \omega_1^2 m_1}{k_{12}} = -\frac{2k - (k/m)m}{-k} = 1$$

$$\frac{u_{22}}{u_{12}} = -\frac{k_{11} - \omega_2^2 m_1}{k_{12}} = -\frac{2k - (5k/2m)m}{-k} = -0.5 \tag{e}$$

so that the natural modes are

$$\{u\}_1 = \begin{Bmatrix} 1 \\ 1 \end{Bmatrix} \qquad \{u\}_2 = \begin{Bmatrix} 1 \\ -0.5 \end{Bmatrix} \tag{f}$$

where the constants $u_{11}$ and $u_{12}$ were taken as unity arbitrarily. This clearly

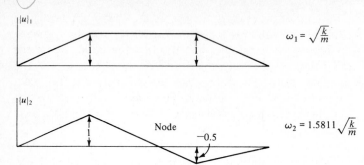

$$\omega_1 = \sqrt{\frac{k}{m}}$$

$$\omega_2 = 1.5811\sqrt{\frac{k}{m}}$$

**Figure 3.3**

does not affect the mode shapes which are plotted in Fig. 3.3. We note that the second mode possesses a point of zero displacement. Such a point is called a *node*.

## 3.4 COORDINATE TRANSFORMATIONS. COUPLING

The system of Fig. 3.4a can be regarded as an idealized mathematical model of an automobile. For simplicity, the body is represented by a rigid slab of total mass $m$ with its mass center $C$ at distances $a$ and $b$ from the springs $k_1$ and $k_2$, respectively, where the springs simulate the suspension. The body has a mass moment of inertia

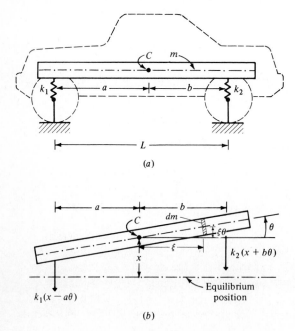

(a)

(b)

**Figure 3.4**

$I_C$ about the center $C$. Note that the mass of the tires is assumed to be negligible and their stiffness included in the stiffness of the suspension. Figure 3.4$b$ shows a free-body diagram corresponding to the body in displaced position, where the displacements consist of the vertical translation $x(t)$ of the center $C$ and the rotation $\theta(t)$ about $C$. The translation $x(t)$ is measured from the equilibrium position, so that the weight $W = mg$ of the automobile is balanced by corresponding initial compressive forces in the springs (see Sec. 1.3). The angular displacement $\theta(t)$ is assumed to be small.

There are two equations of motion, a force equation for the translation in the vertical direction and a moment equation for the rotation about the mass center $C$. To derive the equations of motion, we refer to the free-body diagram of Fig. 3.4$b$ and consider a differential element of mass $dm$ at a distance $\xi$ from the mass center $C$. Then, observing that for small angles $\theta$ the acceleration of the mass element is in the vertical direction and equal to $\ddot{x} + \xi\ddot{\theta}$, the force equation becomes

$$-k_1(x - a\theta) - k_2(x + b\theta) = \int_{\text{body}} (\ddot{x} + \xi\ddot{\theta})\, dm$$

$$= \ddot{x} \int_{\text{body}} dm + \ddot{\theta} \int_{\text{body}} \xi\, dm = m\ddot{x} \qquad (3.37a)$$

where $m = \int_{\text{body}} dm$ is the total mass of the body. Moreover, we note that the simplification on the right side of Eq. (3.37a) was possible because $\int_{\text{body}} \xi\, dm = 0$ by the definition of the mass center. Similarly, the moment equation about $C$ reduces to

$$k_1(x - a\theta)a - k_2(x + b\theta)b = \int_{\text{body}} \xi(\ddot{x} + \xi\ddot{\theta})\, dm$$

$$= \ddot{x} \int_{\text{body}} \xi\, dm + \ddot{\theta} \int_{\text{body}} \xi^2\, dm = I_C\ddot{\theta} \qquad (3.37b)$$

where $I_C = \int_{\text{body}} \xi^2\, dm$ is the mass moment of inertia of the body about the mass center. Equations (3.37) can be rearranged as

$$m\ddot{x} + (k_1 + k_2)x - (k_1a - k_2b)\theta = 0$$
$$I_C\ddot{\theta} - (k_1a - k_2b)x + (k_1a^2 + k_2b^2)\theta = 0 \qquad (3.38)$$

leading to the matrix form

$$\begin{bmatrix} m & 0 \\ 0 & I_C \end{bmatrix}\begin{Bmatrix} \ddot{x} \\ \ddot{\theta} \end{Bmatrix} + \begin{bmatrix} k_1 + k_2 & -(k_1a - k_2b) \\ -(k_1a - k_2b) & k_1a^2 + k_2b^2 \end{bmatrix}\begin{Bmatrix} x \\ \theta \end{Bmatrix} = \begin{Bmatrix} 0 \\ 0 \end{Bmatrix} \qquad (3.39)$$

Next, let us consider a point $O$ such that when a vertical force is applied at $O$ the system undergoes translation only (see Fig. 3.5$a$). Let the point $O$ be at distances $a_1$ and $b_1$ from the springs $k_1$ and $k_2$, respectively. Then, denoting by $x_1$ the vertical translation of point $O$ on the slab, we conclude from the condition of zero moment about $O$ that $k_1x_1a_1 = k_2x_1b_1$, or $k_1a_1 = k_2b_1$. Using the coordinates $x_1(t)$ and $\theta(t)$, where $\theta(t)$ denotes once again the rotation of the slab, the free-

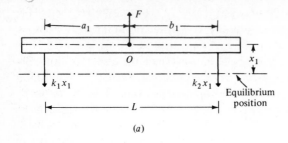

(a)

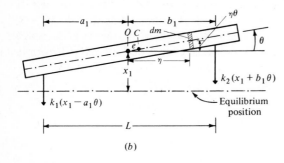

(b)                                    Figure 3.5

body diagram is now as shown in Fig. 3.5b, where $\eta$ is the distance from $O$ to $dm$. Using the same approach as before, the equations of motion are simply

$$-k_1(x_1 - a_1\theta) - k_2(x_1 + b_1\theta) = \int_{\text{body}} (\ddot{x}_1 + \eta\ddot{\theta})\, dm$$

$$= \ddot{x}_1 \int_{\text{body}} dm + \ddot{\theta} \int_{\text{body}} \eta\, dm = m\ddot{x}_1 + me\ddot{\theta}$$

$$(3.40)$$

$$k_1(x_1 - a_1\theta)a_1 - k_2(x_1 + b_1\theta)b_1 = \int_{\text{body}} \eta(\ddot{x}_1 + \eta\ddot{\theta})\, dm$$

$$= \ddot{x}_1 \int_{\text{body}} \eta\, dm + \ddot{\theta} \int_{\text{body}} \eta^2\, dm = me\ddot{x}_1 + I_0\ddot{\theta}$$

where $e = (1/m) \int_{\text{body}} \eta\, dm$ is the distance from $O$ to $C$ and $I_0 = \int_{\text{body}} \eta^2\, dm$ is the mass moment of inertia of the body about point $O$. Recalling that $k_1 a_1 = k_2 b_1$, Eqs. (3.40) reduce to

$$m\ddot{x}_1 + me\ddot{\theta} + (k_1 + k_2)x_1 = 0$$

$$me\ddot{x}_1 + I_0\ddot{\theta} + (k_1 a_1^2 + k_2 b_1^2)\theta = 0$$

$$(3.41)$$

which can be written in the matrix form

$$\begin{bmatrix} m & me \\ me & I_0 \end{bmatrix} \begin{Bmatrix} \ddot{x}_1 \\ \ddot{\theta} \end{Bmatrix} + \begin{bmatrix} k_1 + k_2 & 0 \\ 0 & k_1 a_1^2 + k_2 b_1^2 \end{bmatrix} \begin{Bmatrix} x_1 \\ \theta \end{Bmatrix} = \begin{Bmatrix} 0 \\ 0 \end{Bmatrix}$$

$$(3.42)$$

Examining Eq. (3.39), we conclude that by using as coordinates the translation $x(t)$ of the mass center $C$ and the rotation $\theta(t)$ about $C$, the equations of motion are coupled through the stiffness terms. Such coupling is referred to as *elastic coupling*. On the other hand, the equations of motion described by (3.42) are coupled through the mass terms. Such coupling is called *inertial coupling*. The preceding statements demonstrate that the nature of coupling depends only on the choice of coordinates, i.e., on how we describe the system mathematically, rather than on the system itself. Indeed, we have the freedom to describe the motion of the system in terms of any pair of independent coordinates, and the interest lies in a pair of coordinates offering the greatest simplification. In particular, the most desirable system of coordinates is that for which the equations of motion are uncoupled both elastically and inertially, i.e., for which the mass and stiffness matrices are both diagonal. We show in Sec. 3.5 that such coordinates do indeed exist, and they are known as *natural coordinates*, or *principal coordinates*. The general motion of an undamped linear system can be represented as a linear combination of the natural modes multiplied by these natural coordinates.

## 3.5 ORTHOGONALITY OF MODES. NATURAL COORDINATES

The modal vectors $\{u\}_1$ and $\{u\}_2$ possess a very useful property known as orthogonality. We propose to demonstrate first the orthogonality property and then to show how it can be used to uncouple the equations of motion, thus enabling us to solve the equations with the same ease as solving the equations of single-degree-of-freedom systems.

Considering Eqs. (3.30), we can write the modal vectors as follows:

$$\{u\}_1 = u_{11} \left\{ \begin{array}{c} 1 \\ -\dfrac{k_{11} - \omega_1^2 m_1}{k_{12}} \end{array} \right\} \qquad \{u\}_2 = u_{12} \left\{ \begin{array}{c} 1 \\ -\dfrac{k_{11} - \omega_2^2 m_1}{k_{12}} \end{array} \right\} \qquad (3.43)$$

where $\omega_1^2$ and $\omega_2^2$ are given by Eq. (3.29). Next, form the matrix product

$$\{u\}_2^T [m]\{u\}_1 = u_{11} u_{12} \left\{ \begin{array}{c} 1 \\ -\dfrac{k_{11} - \omega_2^2 m_1}{k_{12}} \end{array} \right\}^T \begin{bmatrix} m_1 & 0 \\ 0 & m_2 \end{bmatrix} \left\{ \begin{array}{c} 1 \\ -\dfrac{k_{11} - \omega_1^2 m_1}{k_{12}} \end{array} \right\}$$

$$= u_{11} u_{12} \left[ m_1 + \frac{m_2}{k_{12}^2} (k_{11} - \omega_1^2 m_1)(k_{11} - \omega_2^2 m_1) \right] \qquad (3.44)$$

which is a scalar. Inserting $\omega_1^2$ and $\omega_2^2$ from Eq. (3.29) into Eq. (3.44), we conclude that

$$\{u\}_2^T [m]\{u\}_1 = 0 \qquad (3.45)$$

so that *the modal vectors* $\{u\}_1$ *and* $\{u\}_2$ *are orthogonal.* Because the matrix product in Eq. (3.45) contains the mass matrix $[m]$ as a weighting matrix, this is not an ordinary orthogonality but *orthogonality with respect to* $[m]$.

Next, rewrite Eqs. (3.26) in the matrix form

$$[k]\{u\} = \omega^2[m]\{u\} \tag{3.46}$$

But, both modes $\omega_1^2, \{u\}_1$ and $\omega_2^2, \{u\}_2$ must satisfy Eq. (3.46), so that we can write

$$[k]\{u\}_1 = \omega_1^2[m]\{u\}_1 \tag{3.47a}$$

$$[k]\{u\}_2 = \omega_2^2[m]\{u\}_2 \tag{3.47b}$$

Multiplying Eq. (3.47a) on the left by $\{u\}_2^T$ and considering Eq. (3.45), we obtain

$$\{u\}_2^T[k]\{u\}_1 = 0 \tag{3.48}$$

so that *the modal vectors* $\{u\}_1$ *and* $\{u\}_2$ *are orthogonal with respect to the stiffness matrix* $[k]$ *as well.* Of course, because the matrices $[m]$ and $[k]$ are symmetric, Eqs. (3.45) and (3.48) are valid also when the positions of $\{u\}_1$ and $\{u\}_2$ are interchanged. This statement can be verified by taking the transpose of Eqs. (3.45) and (3.48), and recalling that the transpose of a product of matrices is equal to the product of the transposed matrices in reversed order. It is worth noting that, multiplying Eq. (3.47a) by $\{u\}_1^T$ and Eq. (3.47b) by $\{u\}_2^T$, we can write

$$\{u\}_i^T[k]\{u\}_i = \omega_i^2\{u\}_i^T[m]\{u\}_i \qquad i = 1, 2 \tag{3.49}$$

where $\omega_i$ ($i = 1, 2$) are the natural frequencies of the system.

The orthogonality property can be used to uncouple the equations of motion both elastically and inertially. To justify this statement, we first rewrite the equations of motion, Eqs. (3.10), in the matrix form

$$[m]\{\ddot{x}(t)\} + [k]\{x(t)\} = \{0\} \tag{3.50}$$

Then, we seek a solution of Eq. (3.50) as the linear combination

$$\{x(t)\} = \{u\}_1 q_1(t) + \{u\}_2 q_2(t) \tag{3.51}$$

where $q_1(t)$ and $q_2(t)$ are two functions of time that remain to be determined. Introducing Eq. (3.51) into Eq. (3.50), we obtain

$$[m](\{u\}_1\ddot{q}_1(t) + \{u\}_2\ddot{q}_2(t)) + [k](\{u\}_1 q_1(t) + \{u\}_2 q_2(t)) = \{0\} \tag{3.52}$$

Multiplying Eq. (3.52) on the left by $\{u\}_1^T$ and considering the orthogonality of $\{u\}_1$ and $\{u\}_2$ with respect to $[m]$ and $[k]$, as well as Eqs. (3.49), we conclude that $q_1(t)$ must satisfy the equation

$$\ddot{q}_1(t) + \omega_1^2 q_1(t) = 0 \tag{3.53a}$$

Similarly, multiplying Eq. (3.52) on the left by $\{u\}_2^T$ and considering Eqs. (3.45), (3.48), and (3.49), we conclude that $q_2(t)$ satisfies the equation

$$\ddot{q}_2(t) + \omega_2^2 q_2(t) = 0 \tag{3.53b}$$

We observe that Eqs. (3.53) represent two independent equations, one for $q_1(t)$ and one for $q_2(t)$. This is in contrast with Eqs. (3.10) for $x_1(t)$ and $x_2(t)$, which are simultaneous equations. Hence, the linear transformation (3.51), when used in conjunction with the orthogonality of the modal vectors, does indeed permit simultaneous elastic and inertial uncoupling of the equations of motion. Coordinates such as $q_1(t)$ and $q_2(t)$, for which the system of equations of motion are independent, are called *natural coordinates*, or *principal coordinates*.

Equations (3.53) describe two independent harmonic oscillators (Sec. 1.6). Their solutions are simply

$$q_1(t) = C_1 \cos(\omega_1 t - \phi_1) \qquad (3.54a)$$

$$q_2(t) = C_2 \cos(\omega_2 t - \phi_2) \qquad (3.54b)$$

where $C_i$ and $\phi_i$ $(i = 1, 2)$ are amplitudes and phase angles, respectively, so that Eq. (3.51) becomes

$$\{x(t)\} = C_1\{u\}_1 \cos(\omega_1 t - \phi_1) + C_2\{u\}_2 \cos(\omega_2 t - \phi_2) \qquad (3.55)$$

The amplitudes $C_1$ and $C_2$ and the phase angles $\phi_1$ and $\phi_2$ depend on the initial displacements and velocities of $m_1$ and $m_2$. Equation (3.55) is identical to Eq. (3.33), thus justifying the statement made in Sec. 3.3 that *the motion of the system at any time can be expressed as a superposition of the natural modes of vibration multiplied by the natural coordinates.*

The orthogonality of the modal vectors and the process of uncoupling the equations of motion will be presented in a more formal manner in Chap. 4.

**Example 3.2** Consider the automobile of Sec. 3.4, let the system parameters have the values $m = 100$ lb·s²/ft, $I_C = 1600$ lb·s²·ft, $k_1 = 2400$ lb/ft, $k_2 = 2700$ lb/ft, $a = 4.40$ ft, $b = 5.60$ ft, calculate the natural modes of the system and write an expression for the response.

To determine the natural coordinates, it is necessary to find the natural modes first. Inserting the values of the parameters given into the equations of motion, Eq. (3.39), we obtain

$$\begin{bmatrix} 100 & 0 \\ 0 & 1600 \end{bmatrix} \begin{Bmatrix} \ddot{x} \\ \ddot{\theta} \end{Bmatrix} + \begin{bmatrix} 5100 & 4560 \\ 4560 & 131{,}136 \end{bmatrix} \begin{Bmatrix} x \\ \theta \end{Bmatrix} = \begin{Bmatrix} 0 \\ 0 \end{Bmatrix} \qquad (a)$$

leading to the eigenvalue problem

$$-\omega^2 \begin{bmatrix} 100 & 0 \\ 0 & 1600 \end{bmatrix} \begin{Bmatrix} X \\ \Theta \end{Bmatrix} + \begin{bmatrix} 5100 & 4560 \\ 4560 & 131{,}136 \end{bmatrix} \begin{Bmatrix} X \\ \Theta \end{Bmatrix} = \begin{Bmatrix} 0 \\ 0 \end{Bmatrix} \qquad (b)$$

where $X$ and $\Theta$ are the amplitudes of $x(t)$ and $\theta(t)$, respectively. Hence, the system characteristic equation is

$$\det \begin{bmatrix} 5100 - 100\omega^2 & 4560 \\ 4560 & 131{,}136 - 1600\omega^2 \end{bmatrix}$$

$$= 160{,}000(\omega^4 - 132.96\omega^2 + 4050.00) = 0 \qquad (c)$$

having the solutions

$$\begin{matrix} \omega_1^2 \\ \omega_2^2 \end{matrix} = 66.48 \mp \sqrt{66.48^2 - 4050.00} = 66.48 \mp 19.22$$

$$= \begin{cases} 47.26 \ (\text{rad/s})^2 \\ 85.70 \ (\text{rad/s})^2 \end{cases} \tag{d}$$

so that the natural frequencies are $\omega_1 = 6.88$ rad/s and $\omega_2 = 9.26$ rad/s. Inserting $\omega_1^2$ from Eqs. (d) into the first row of Eq. (b), we obtain

$$-47.26 \times 100X_1 + 5100X_1 + 4560\Theta_1 = 0$$

yielding

$$\frac{\Theta_1}{X_1} = -\frac{374}{4560} = -0.0820 \ \text{rad/ft} \tag{e}$$

Moreover, introducing $\omega_2^2$ from Eqs. (d) into the first row of Eq. (b), we have

$$-85.70 \times 100X_2 + 5100X_2 + 4560\Theta_2 = 0$$

from which we obtain

$$\frac{\Theta_2}{X_2} = \frac{3470}{4560} = 0.7610 \ \text{rad/ft} \tag{f}$$

Hence, letting arbitrarily $X_1 = 1$ and $X_2 = 1$, the natural modes become

$$\{u\}_1 = \begin{Bmatrix} X_1 \\ \Theta_1 \end{Bmatrix} = \begin{bmatrix} 1 \\ -0.0820 \end{bmatrix} \qquad \{u\}_2 = \begin{Bmatrix} X_2 \\ \Theta_2 \end{Bmatrix} = \begin{bmatrix} 1 \\ 0.7610 \end{bmatrix} \tag{g}$$

Note that the same results would have been obtained had we used the second row of Eq. (b) instead of the first. The modes are plotted in Fig. 3.6.

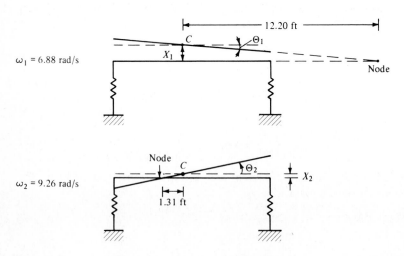

**Figure 3.6**

The natural modes can be verified to be orthogonal with respect to both the mass matrix and the stiffness matrix. The orthogonality property can be used to uncouple the equations of motion and to determine the natural coordinates, as outlined in Eqs. (3.50)–(3.54). It is not really necessary to carry out these steps. Instead, we can insert the computed modal vectors, Eqs. ($g$), directly into Eq. (3.55) and obtain the system response

$$\begin{Bmatrix} x(t) \\ \theta(t) \end{Bmatrix} = C_1 \begin{Bmatrix} 1 \\ -0.0820 \end{Bmatrix} \cos(6.88t - \theta_1)$$

$$+ C_2 \begin{Bmatrix} 1 \\ 0.7610 \end{Bmatrix} \cos(9.26t - \phi_2) \qquad (h)$$

The coefficients $C_1$ and $C_2$ and the phase angles $\phi_1$ and $\phi_2$ depend on the initial conditions $x(0)$, $\dot{x}(0)$, $\theta(0)$, and $\dot{\theta}(0)$. Their calculation is shown in Sec. 3.6.

# 3.6 RESPONSE OF A TWO-DEGREE-OF-FREEDOM SYSTEM TO INITIAL EXCITATION

It was indicated in Sec. 3.5 that the motion of a system at any time can be regarded as the superposition of the natural modes multiplied by the natural coordinates. More specifically, the motion of a two-degree-of-freedom system can be written in the vector form (3.55). The natural frequencies $\omega_1$ and $\omega_2$ are unique for a given system and the modal vectors $\{u\}_1$ and $\{u\}_2$ can be rendered unique through normalization. On the other hand, the amplitudes $C_1$ and $C_2$ and the phase angles $\phi_1$ and $\phi_2$ play the role of constants of integration and their values depend on the initial conditions. Letting the initial conditions have the values $x_1(0) = x_{10}$, $x_2(0) = x_{20}$, $\dot{x}_1(0) = v_{10}$, $\dot{x}_2(0) = v_{20}$, and inserting these values into Eq. (3.55) with $t = 0$, we obtain

$$\begin{aligned} x_{10} &= C_1 u_{11} \cos\phi_1 + C_2 u_{12} \cos\phi_2 \\ x_{20} &= C_1 u_{21} \cos\phi_1 + C_2 u_{22} \cos\phi_2 \\ v_{10} &= C_1 \omega_1 u_{11} \sin\phi_1 + C_2 \omega_2 u_{12} \sin\phi_2 \\ v_{20} &= C_1 \omega_1 u_{21} \sin\phi_1 + C_2 \omega_2 u_{22} \sin\phi_2 \end{aligned} \qquad (3.56)$$

which can be regarded as two pairs of algebraic equations, the first pair in the unknowns $C_1 \cos\phi_1$, $C_2 \cos\phi_2$ and the second pair in the unknowns $C_1 \sin\phi_1$, $C_2 \sin\phi_2$. Equations (3.56) have the solution

$$C_1 \cos\phi_1 = \frac{1}{\det[u]}(u_{22}x_{10} - u_{12}x_{20})$$

$$C_2 \cos\phi_2 = \frac{1}{\det[u]}(u_{11}x_{20} - u_{21}x_{10}) \qquad (3.57)$$

$$C_1 \sin \phi_1 = \frac{1}{\omega_1 \det [u]} (u_{22}v_{10} - u_{12}v_{20})$$

$$C_2 \sin \phi_2 = \frac{1}{\omega_2 \det [u]} (u_{11}v_{20} - u_{21}v_{10})$$

From Eqs. (3.57), we obtain

$$C_1 = \frac{1}{\det [u]} \sqrt{(u_{22}x_{10} - u_{12}x_{20})^2 + \frac{(u_{22}v_{10} - u_{12}v_{20})^2}{\omega_1^2}}$$

$$C_2 = \frac{1}{\det [u]} \sqrt{(u_{11}x_{20} - u_{21}x_{10})^2 + \frac{(u_{11}v_{20} - u_{21}v_{10})^2}{\omega_2^2}} \qquad (3.58)$$

$$\phi_1 = \tan^{-1} \frac{u_{22}v_{10} - u_{12}v_{20}}{\omega_1(u_{22}x_{10} - u_{12}x_{20})}$$

$$\phi_2 = \tan^{-1} \frac{u_{11}v_{20} - u_{21}v_{10}}{\omega_2(u_{11}x_{20} - u_{21}x_{10})}$$

Equations (3.55) and (3.58) define the response of a two-degree-of-freedom system to initial excitation completely. We shall see in Chap. 4 that the response can be obtained in matrix form in a more systematic way.

**Example 3.3** Consider the system of Example 3.1 and obtain the response to the initial excitation $x_1(0) = x_{10} = 1.2$, $x_2(0) = x_{20} = 0$, $\dot{x}_1(0) = v_{10} = 0$, $\dot{x}_2(0) = v_{20} = 0$.

From Eqs. (d) of Example 3.1, we have $\omega_1 = \sqrt{k/m}$, $\omega_2 = 1.5811 \sqrt{k/m}$. Moreover, choosing arbitrarily $u_{11} = 1, u_{12} = 1$ in Example 3.1, we obtained the modal matrix

$$[u] = \begin{bmatrix} u_{11} & u_{12} \\ u_{21} & u_{22} \end{bmatrix} = \begin{bmatrix} 1 & 1 \\ 1 & -0.5 \end{bmatrix} \qquad (a)$$

which has the determinant

$$\det [u] = \begin{vmatrix} 1 & 1 \\ 1 & -0.5 \end{vmatrix} = -0.5 - 1 = -1.5 \qquad (b)$$

Inserting the initial conditions listed above and the values given by Eqs. (a) and (b) into Eqs. (3.58), we obtain

$$C_1 = \frac{u_{22}x_{10}}{\det [u]} = \frac{-0.5 \times 1.2}{-1.5} = 0.4$$

$$C_2 = \frac{-u_{21}x_{10}}{\det [u]} = \frac{-1.2}{-1.5} = 0.8 \qquad (c)$$

$$\phi_1 = \phi_2 = 0$$

Hence, introducing Eqs. (*c*) into Eq. (3.55), we obtain the response

$$\begin{Bmatrix} x_1(t) \\ x_2(t) \end{Bmatrix} = 0.4 \begin{Bmatrix} 1 \\ 1 \end{Bmatrix} \cos \sqrt{\frac{k}{m}} t + 0.8 \begin{Bmatrix} 1 \\ -0.5 \end{Bmatrix} \cos 1.5811 \sqrt{\frac{k}{m}} t \qquad (d)$$

It must be pointed out that the arbitrary choice $u_{11} = 1$, $u_{12} = 1$ did not affect the final outcome. Indeed, any other choice would have resulted in such values for $C_1$ and $C_2$ as to keep Eq. (*d*) unchanged.

## 3.7 BEAT PHENOMENON

A very interesting phenomenon is encountered when the natural frequencies of a two-degree-of-freedom system are very close in value. To illustrate the phenomenon, let us consider two identical pendulums connected by a spring, as shown in Fig. 3.7*a*. The corresponding free-body diagrams are shown in Fig. 3.7*b*, in which the assumption of small angles $\theta_1$ and $\theta_2$ is implied. The moment equations about the points $O$ and $O'$, respectively, yield the differential equations of motion

$$\begin{aligned} mL^2\ddot{\theta}_1 + mgL\theta_1 + ka^2(\theta_1 - \theta_2) &= 0 \\ mL^2\ddot{\theta}_2 + mgL\theta_2 - ka^2(\theta_1 - \theta_2) &= 0 \end{aligned} \qquad (3.59)$$

which can be arranged in the matrix form

$$\begin{bmatrix} mL_2 & 0 \\ 0 & mL^2 \end{bmatrix} \begin{Bmatrix} \ddot{\theta}_1 \\ \ddot{\theta}_2 \end{Bmatrix} + \begin{bmatrix} mgL + ka^2 & -ka^2 \\ -ka^2 & mgL + ka^2 \end{bmatrix} \begin{Bmatrix} \theta_1 \\ \theta_2 \end{Bmatrix} = \begin{Bmatrix} 0 \\ 0 \end{Bmatrix} \qquad (3.60)$$

indicating that the system is coupled elastically. As expected, when the spring stiffness $k$ reduces to zero the coupling disappears and the two pendulums reduce to independent simple pendulums with identical natural frequencies equal to $\sqrt{g/L}$. For $k \neq 0$, Eq. (3.60) yields the eigenvalue problem

$$-\omega^2 \begin{bmatrix} mL^2 & 0 \\ 0 & mL^2 \end{bmatrix} \begin{Bmatrix} \Theta_1 \\ \Theta_2 \end{Bmatrix} + \begin{bmatrix} mgL + ka^2 & -ka^2 \\ -ka^2 & mgL + ka^2 \end{bmatrix} \begin{Bmatrix} \Theta_1 \\ \Theta_2 \end{Bmatrix} = \begin{Bmatrix} 0 \\ 0 \end{Bmatrix}$$

$$(3.61)$$

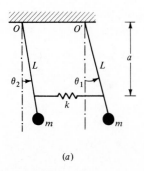

(*a*)

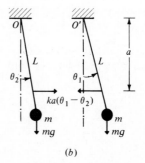

(*b*)

**Figure 3.7**

leading to the characteristic equation

$$\det \begin{bmatrix} mgL + ka^2 - \omega^2 mL^2 & -ka^2 \\ -ka^2 & mgL + ka^2 - \omega^2 mL^2 \end{bmatrix}$$

$$= (mgL + ka^2 - \omega^2 mL^2)^2 - (ka^2)^2 = 0 \quad (3.62,$$

which is equivalent to

$$mgL + ka^2 - \omega^2 mL^2 = \pm ka^2 \qquad (3.63)$$

Hence, the two natural frequencies are

$$\omega_1 = \sqrt{\frac{g}{L}} \qquad \omega_2 = \sqrt{\frac{g}{L} + 2\frac{k}{m}\frac{a^2}{L^2}} \qquad (3.64)$$

The natural modes are obtained from the equations

$$-\omega_i^2 \begin{bmatrix} mL^2 & 0 \\ 0 & mL^2 \end{bmatrix} \begin{Bmatrix} \Theta_1 \\ \Theta_2 \end{Bmatrix}_i + \begin{bmatrix} mgL + ka^2 & -ka^2 \\ -ka & mgL + ka^2 \end{bmatrix} \begin{Bmatrix} \Theta_1 \\ \Theta_2 \end{Bmatrix}_i = \begin{Bmatrix} 0 \\ 0 \end{Bmatrix} \quad i = 1, 2$$

$$(3.65)$$

Inserting $\omega_1^2 = g/L$ and $\omega_2^2 = g/L + 2(k/m)(a^2/L^2)$ into Eqs. (3.65), and solving for the ratios $\Theta_{21}/\Theta_{11}$ and $\Theta_{22}/\Theta_{12}$, we obtain

$$\frac{\Theta_{21}}{\Theta_{11}} = 1 \qquad \frac{\Theta_{22}}{\Theta_{12}} = -1 \qquad (3.66)$$

so that in the first natural mode the two pendulums move like a single pendulum with the spring $k$ unstretched, which can also be concluded from the fact that the first natural frequency of the system is that of the simple pendulum, $\omega_1 = \sqrt{g/L}$ On the other hand, in the second natural mode the two pendulums are 180° out of phase. The two modes are shown in Fig. 3.8.

As was pointed out in Sec. 3.5, the general motion of the system can be expressed as a superposition of the two natural modes multiplied by the associated

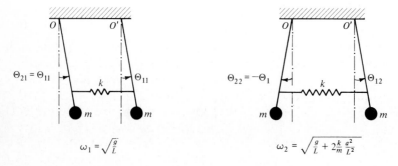

Figure 3.8

natural coordinates, or

$$\begin{Bmatrix} \theta_1(t) \\ \theta_2(t) \end{Bmatrix} = C_1 \begin{Bmatrix} \Theta_1 \\ \Theta_2 \end{Bmatrix}_1 \cos(\omega_1 t - \phi_1) + C_2 \begin{Bmatrix} \Theta_1 \\ \Theta_2 \end{Bmatrix}_2 \cos(\omega_2 t - \phi_2) \quad (3.67)$$

Choosing $\Theta_{11} = \Theta_{12} = 1$ and using Eqs. (3.66), Eqs. (3.67) can be rewritten in the scalar form

$$\theta_1(t) = C_1 \cos(\omega_1 t - \phi_1) + C_2 \cos(\omega_2 t - \phi_2)$$
$$\theta_2(t) = C_1 \cos(\omega_1 t - \phi_1) - C_2 \cos(\omega_2 t - \phi_2) \quad (3.68)$$

Letting the initial conditions be $\theta_1(0) = \theta_0$, $\theta_2(0) = \dot{\theta}_1(0) = \dot{\theta}_2(0) = 0$, Eqs. (3.68) become

$$\theta_1(t) = \tfrac{1}{2}\theta_0 \cos \omega_1 t + \tfrac{1}{2}\theta_0 \cos \omega_2 t$$

$$= \theta_0 \cos \frac{\omega_2 - \omega_1}{2} t \cos \frac{\omega_2 + \omega_1}{2} t$$

$$\theta_2(t) = \tfrac{1}{2}\theta_0 \cos \omega_1 t - \tfrac{1}{2}\theta_0 \cos \omega_2 t \quad (3.69)$$

$$= \theta_0 \sin \frac{\omega_2 - \omega_1}{2} t \sin \frac{\omega_2 + \omega_1}{2} t$$

Note that, in deriving Eqs. (3.69), we used the trigonometric relations $\cos(\alpha \pm \beta) = \cos\alpha\cos\beta \mp \sin\alpha\sin\beta$, in which $\alpha = (\omega_2 - \omega_1)t/2$, $\beta = (\omega_2 + \omega_1)t/2$.

Next let us consider the case in which $ka^2$ is very small in value compared with $mgL$. Examining Eq. (3.60), we conclude that this statement is equivalent to saying that the coupling provided by the spring $k$ is very weak. In this case, Eqs. (3.69) can be written in the form

$$\theta_1(t) \cong \theta_0 \cos \tfrac{1}{2}\omega_B t \cos \omega_{\text{ave}} t$$
$$\theta_2(t) \cong \theta_0 \sin \tfrac{1}{2}\omega_B t \sin \omega_{\text{ave}} t \quad (3.70)$$

where $\omega_B/2$ and $\omega_{\text{ave}}$ are approximated by

$$\frac{\omega_B}{2} = \frac{\omega_2 - \omega_1}{2} \cong \frac{1}{2}\frac{k}{m}\frac{a^2}{\sqrt{gL^3}} \qquad \omega_{\text{ave}} = \frac{\omega_2 + \omega_1}{2} \cong \sqrt{\frac{g}{L}} + \frac{1}{2}\frac{k}{m}\frac{a^2}{\sqrt{gL^3}} \quad (3.71)$$

Hence, $\theta_1(t)$ and $\theta_2(t)$ can be regarded as being harmonic functions with frequency $\omega_{\text{ave}}$ and with amplitudes varying slowly according to $\theta_0 \cos \tfrac{1}{2}\omega_B t$ and $\theta_0 \sin \tfrac{1}{2}\omega_B t$, respectively. The plots $\theta_1(t)$ versus $t$ and $\theta_2(t)$ versus $t$ are shown in Fig. 3.9, with the slowly varying amplitudes indicated by the dashed-line envelopes. Geometrically, Fig. 3.9a (or Fig. 3.9b) implies that if two harmonic functions possessing equal amplitudes and nearly equal frequencies are added, then the resulting function is an *amplitude-modulated* harmonic function with a frequency equal to the average frequency. At first, when the two harmonic waves reinforce each other, the amplitude is doubled, and later, as the two waves cancel each other,

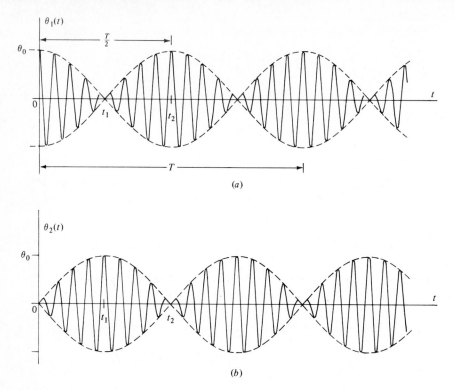

**Figure 3.9**

the amplitude reduces to zero. The phenomenon is known as the *beat phenomenon*, and the frequency of modulation $\omega_B$, which in this particular case is equal to $ka^2/m\sqrt{gL^3}$, is called the *beat frequency*. From Fig. 3.9a, we conclude that the time between two maxima is $T/2 = 2\pi/\omega_B$, whereas the period of the amplitude-modulated envelope is $T = 4\pi/\omega_B$.

Although in our particular case the beat phenomenon resulted from the weak coupling of two pendulums, the phenomenon is not exclusively associated with two-degree-of-freedom systems. Indeed, the beat phenomenon is purely the result of adding two harmonic functions of equal amplitudes and nearly equal frequencies. For example, the phenomenon occurs in twin-engine propeller airplanes, in which the propeller noise grows and diminishes in intensity as the sound waves generated by the two propellers reinforce and cancel each other in turn.

We observe from Fig. 3.9 that there is a 90° phase angle between $\theta_1(t)$ and $\theta_2(t)$. At $t = 0$ the first pendulum (right pendulum in Fig. 3.7a) begins to swing with the amplitude $\theta_0$ while the second pendulum is at rest. Soon thereafter the second pendulum is entrained, gaining amplitude while the amplitude of the first decreases. At $t_1 = \pi/\omega_B$ the amplitude of the first pendulum becomes zero, whereas the amplitude of the second pendulum reaches $\theta_0$. At $t_2 = 2\pi/\omega_B$ the amplitude of the

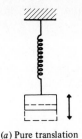

(a) Pure translation      (b) Pure torsion      **Figure 3.10**

first pendulum reaches $\theta_0$ once again and that of the second pendulum reduces to zero. The motion keeps repeating itself, so that every interval of time $T/4 = \pi/\omega_B$ there is a complete transfer of energy from one pendulum to the other.

Another example of a system exhibiting the beat phenomenon is the "Wilberforce spring", consisting of a mass of finite dimensions suspended by a helical spring such that the frequency of vertical translation and the frequency of torsional motion are very close in value. In this case, the kinetic energy changes from pure translational in the vertical direction to pure rotational about the vertical axis, as shown in Fig. 3.10.

## 3.8 RESPONSE OF A TWO-DEGREE-OF-FREEDOM SYSTEM TO HARMONIC EXCITATION

Let us return to the damped system of Sec. 3.2 and write Eq. (3.5) in the expanded form

$$m_{11}\ddot{x}_1 + m_{12}\ddot{x}_2 + c_{11}\dot{x}_1 + c_{12}\dot{x}_2 + k_{11}x_1 + k_{12}x_2 = F_1(t)$$
$$m_{12}\ddot{x}_1 + m_{22}\ddot{x}_2 + c_{12}\dot{x}_1 + c_{22}\dot{x}_2 + k_{12}x_1 + k_{22}x_2 = F_2(t) \tag{3.72}$$

where the diagonal mass matrix has been replaced by a more general nondiagonal but symmetric matrix. Next, let us consider the following harmonic excitation:

$$F_1(t) = F_1 e^{i\omega t} \qquad F_2(t) = F_2 e^{i\omega t} \tag{3.73}$$

and write the steady-state response as

$$x_1(t) = X_1 e^{i\omega t} \qquad x_2(t) = X_2 e^{i\omega t} \tag{3.74}$$

where $X_1$ and $X_2$ are in general complex quantities depending on the driving frequency $\omega$ and the system parameters. Inserting Eqs. (3.73) and (3.74) into (3.72), we obtain the two algebraic equations

$$(-\omega^2 m_{11} + i\omega c_{11} + k_{11})X_1 + (-\omega^2 m_{12} + i\omega c_{12} + k_{12})X_2 = F_1$$
$$(-\omega^2 m_{12} + i\omega c_{12} + k_{12})X_1 + (-\omega^2 m_{22} + i\omega c_{22} + k_{22})X_2 = F_2 \tag{3.75}$$

Introducing the notation

$$Z_{ij}(\omega) = -\omega^2 m_{ij} + i\omega c_{ij} + k_{ij} \qquad i,j = 1, 2 \tag{3.76}$$

where the functions $Z_{ij}(\omega)$ are known as *impedances*, Eqs. (3.75) can be written in the compact matrix form

$$[Z(\omega)]\{X\} = \{F\} \tag{3.77}$$

where $[Z(\omega)]$ is called the *impedance matrix*, $\{X\}$ is the column matrix of the displacement amplitudes, and $\{F\}$ is the column matrix of the excitation amplitudes.

The solution of Eq. (3.77) can be obtained by premultiplying both sides of the equation by the inverse $[Z(\omega)]^{-1}$ of the impedance matrix $[Z(\omega)]$, with the result

$$\{X\} = [Z(\omega)]^{-1}\{F\} \tag{3.78}$$

where the inverse $[Z(\omega)]^{-1}$ can be shown to have the form (see App. C)

$$[Z(\omega)]^{-1} = \frac{1}{\det[Z(\omega)]}\begin{bmatrix} Z_{22}(\omega) & -Z_{12}(\omega) \\ -Z_{12}(\omega) & Z_{11}(\omega) \end{bmatrix}$$

$$= \frac{1}{Z_{11}(\omega)Z_{22}(\omega) - Z_{12}^2(\omega)}\begin{bmatrix} Z_{22}(\omega) & -Z_{12}(\omega) \\ -Z_{12}(\omega) & Z_{11}(\omega) \end{bmatrix} \tag{3.79}$$

Introducing Eq. (3.79) into (3.78), and performing the multiplication, we can write

$$X_1(\omega) = \frac{Z_{22}(\omega)F_1 - Z_{12}(\omega)F_2}{Z_{11}(\omega)Z_{22}(\omega) - Z_{12}^2(\omega)} \qquad X_2(\omega) = \frac{-Z_{12}(\omega)F_1 + Z_{11}(\omega)F_2}{Z_{11}(\omega)Z_{22}(\omega) - Z_{12}^2(\omega)}$$
$$\tag{3.80}$$

and we note that the functions $X_1(\omega)$ and $X_2(\omega)$ are analogous to the frequency response of Sec. 2.3.

Next, let us confine ourselves to the undamped system of Fig. 3.2. Moreover, let $F_2 = 0$, so that Eqs. (3.76) yield

$$Z_{11}(\omega) = k_{11} - \omega^2 m_1 \qquad Z_{22}(\omega) = k_{22} - \omega^2 m_2 \qquad Z_{12}(\omega) = k_{12} \quad (3.81)$$

Introducing Eqs. (3.81) into (3.80), we obtain

$$X_1(\omega) = \frac{(k_{22} - \omega^2 m_2)F_1}{(k_{11} - \omega^2 m_1)(k_{22} - \omega^2 m_2) - k_{12}^2}$$

$$X_2(\omega) = \frac{-k_{12}F_1}{(k_{11} - \omega^2 m_1)(k_{22} - \omega^2 m_2) - k_{12}^2} \tag{3.82}$$

For a given set of system parameters, Eqs. (3.82) can be used to plot $X_1(\omega)$ versus $\omega$ and $X_2(\omega)$ versus $\omega$, thus obtaining the magnitude of the response for any excitation frequency $\omega$.

**Example 3.4** Let us consider the system of Example 3.1 and plot the frequency-response curves.

Using the parameter values of Example 3.1, Eqs. (3.82) become

$$X_1(\omega) = \frac{(3k - 2m\omega^2)F_1}{2m^2\omega^4 - 7mk\omega^2 + 5k^2} \qquad X_2(\omega) = \frac{kF_1}{2m^2\omega^4 - 7mk\omega^2 + 5k^2} \quad (a)$$

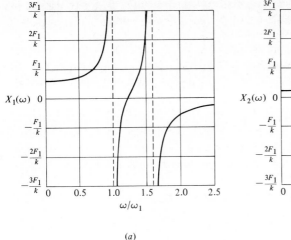
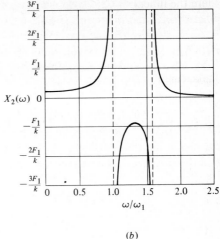

(a)

(b)

**Figure 3.11**

But the denominator of $X_1$ and $X_2$ is recognized as the characteristic determinant, which can be written as

$$\Delta(\omega^2) = 2m^2\omega^4 - 7mk\omega^2 + 5k^2 = 2m^2(\omega^2 - \omega_1^2)(\omega^2 - \omega_2^2) \qquad (b)$$

where

$$\omega_1^2 = \frac{k}{m} \qquad \omega_2^2 = \frac{5}{2}\frac{k}{m} \qquad (c)$$

are the squares of the system's natural frequencies. Hence, Eqs. (a) can be written in the form

$$X_1(\omega) = \frac{2F_1}{5k} \frac{(3/2) - (\omega/\omega_1)^2}{[1 - (\omega/\omega_1)^2][1 - (\omega/\omega_2)^2]}$$

$$X_2(\omega) = \frac{F_1}{5k} \frac{1}{[1 - (\omega/\omega_1)^2][1 - (\omega/\omega_2)^2]} \qquad (d)$$

The frequency response curves $X_1(\omega)$ versus $\omega/\omega_1$ and $X_2(\omega)$ versus $\omega/\omega_1$ are plotted in Fig. 3.11.

## 3.9 UNDAMPED VIBRATION ABSORBERS

When rotating machinery operates at a constant frequency close to resonance, violent vibrations are induced. Assuming that the system can be represented by a single-degree-of-freedom system subjected to harmonic excitation, the situation may be alleviated by changing either the mass or the spring. At times, however, this is not possible. In such a case, a second mass and spring can be added to the system, where the added mass and spring are so designed as to produce a two-degree-of-freedom system whose frequency response is zero at the excitation frequency. We note from Fig. 3.11a that a point for which the frequency response is zero does exist. The new two-degree-of-freedom system has two resonant frequencies, but these frequencies generally present no problem because they differ from the operating frequency.

Let us consider the system of Fig. 3.12, where the original single-degree-of-freedom system, referred to as the main system, consists of the mass $m_1$ and the spring $k_1$, and the added system, referred to as the absorber, consists of the mass $m_2$ and the spring $k_2$. The equations of motion of the combined system can be shown to be

$$m_1\ddot{x}_1 + (k_1 + k_2)x_1 - k_2 x_2 = F_1 \sin \omega t$$
$$m_2\ddot{x}_2 - k_2 x_1 + k_2 x_2 = 0$$

(3.83)

Letting the solution of Eqs. (3.83) be

$$x_1(t) = X_1 \sin \omega t \qquad x_2(t) = X_2 \sin \omega t$$

(3.84)

we obtain two algebraic equations for $X_1$ and $X_2$ having the matrix form

$$\begin{bmatrix} k_1 + k_2 - \omega^2 m_1 & -k_2 \\ -k_2 & k_2 - \omega^2 m_2 \end{bmatrix} \begin{Bmatrix} X_1 \\ X_2 \end{Bmatrix} = \begin{Bmatrix} F_1 \\ 0 \end{Bmatrix}$$

(3.85)

Following the pattern of Sec. 3.8, the solution of Eq. (3.85) can be shown to be

$$X_1 = \frac{(k_2 - \omega^2 m_2)F_1}{(k_1 + k_2 - \omega^2 m_1)(k_2 - \omega^2 m_2) - k_2^2}$$

$$X_2 = \frac{k_2 F_1}{(k_1 + k_2 - \omega^2 m_1)(k_2 - \omega^2 m_2) - k_2^2}$$

(3.86)

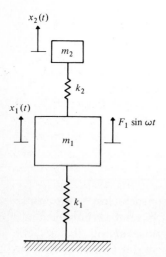

**Figure 3.12**

It is customary to introduce the notation:

$\omega_n = \sqrt{k_1/m_1}$ = the natural frequency of the main system alone

$\omega_a = \sqrt{k_2/m_2}$ = the natural frequency of the absorber alone

$x_{st} = F_1/k_1$ = the static deflection of the main system

$\mu = m_2/m_1$ = the ratio of the absorber mass to the main mass

With this notation, Eqs. (3.86) can be rewritten as

$$X_1 = \frac{[1 - (\omega/\omega_a)^2]x_{st}}{[1 + \mu(\omega_a/\omega_n)^2 - (\omega/\omega_n)^2][1 - (\omega/\omega_a)^2] - \mu(\omega_a/\omega_n)^2} \qquad (3.87a)$$

$$X_2 = \frac{x_{st}}{[1 + \mu(\omega_a/\omega_n)^2 - (\omega/\omega_n)^2][1 - (\omega/\omega_a)^2] - \mu(\omega_a/\omega_n)^2} \qquad (3.87b)$$

From Eq. (3.87a), we conclude that for $\omega = \omega_a$, the amplitude $X_1$ of the main mass reduces to zero. Hence, the absorber can indeed perform the task for which it is designed, namely, to eliminate the vibration of the main mass, provided the natural frequency of the absorber is the same as the operating frequency of the machinery. Moreover, for $\omega = \omega_a$, Eq. (3.87b) reduces to

$$X_2 = -\left(\frac{\omega_n}{\omega_a}\right)^2 \frac{x_{st}}{\mu} = -\frac{F_1}{k_2} \qquad (3.88)$$

so that, inserting Eq. (3.88) into the second of Eqs. (3.84), we obtain

$$x_2(t) = -\frac{F_1}{k_2} \sin \omega t \qquad (3.89)$$

from which we conclude that the force in the absorber spring at any time is

$$k_2 x_2(t) = -F_1 \sin \omega t \qquad (3.90)$$

Hence, the absorber exerts on the main mass a force $-F_1 \sin \omega t$ which balances exactly the applied force $F_1 \sin \omega t$. Because the same effect is obtained by any absorber provided its natural frequency is equal to the operating frequency, there is a wide choice of absorber parameters. The actual choice is generally dictated by limitations placed on the amplitude $X_2$ of the absorber motion.

Although a vibration absorber is designed for a given operating frequency $\omega$, the absorber can perform satisfactorily for operating frequencies that vary slightly from $\omega$. In this case, the motion of $m_1$ is not zero, but its amplitude is very small. This statement can be verified by using Eq. (3.87a) and plotting $X_1(\omega)/x_{st}$ versus $\omega/\omega_a$. Figure 3.13 shows such a plot for $\mu = 0.2$ and $\omega_n = \omega_a$. The shaded area indicates the domain in which the performance of the absorber can be regarded as satisfactory. As pointed out earlier, one disadvantage of the vibration absorber is that two new resonant frequencies are created, as can be seen from Fig. 3.13. To reduce the amplitude at the resonant frequencies, damping can be added, but this results in an increase in amplitude in the neighborhood of the operating frequency

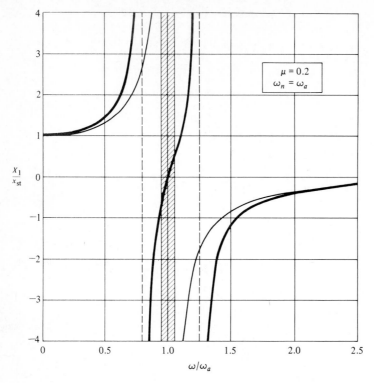

**Figure 3.13**

$\omega = \omega_a$. It should be recalled that any rotating machinery builds up its frequency from rest, so that the system is likely to go through the first resonant frequency. As a matter of interest, the plot $X_1/x_{st}$ versus $\omega/\omega_n$ corresponding to the main system alone is also shown in Fig. 3.13.

## PROBLEMS

3.1 The system of Fig. 3.14 consists of two point masses $m_1$ and $m_2$ carried by a weightless string subjected to the constant tension $T$. Assume small transverse displacements $y_1(t)$ and $y_2(t)$ and derive the differential equations of motion.

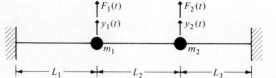

**Figure 3.14**

**3.2** Two disks of mass polar moments of inertia $I_1$ and $I_2$ are mounted on a circular massless shaft consisting of two segments of torsional stiffness $GJ_1$ and $GJ_2$, respectively (see Fig. 3.15). Derive the differential equations for the rotational motion of the disks.

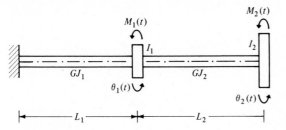

Figure 3.15

**3.3** A rigid bar of mass per unit length $m$ carries a point mass $M$ at its right end. The bar is supported by two springs, as shown in Fig. 3.16. Derive the differential equations for the translation of point $A$ and rotation about $A$. Assume small motions.

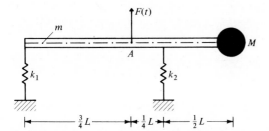

Figure 3.16

**3.4** Derive the differential equations of motion for the double pendulum shown in Fig. 3.17. The angles $\theta_1$ and $\theta_2$ can be arbitrarily large.

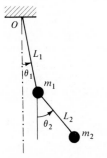

Figure 3.17

**3.5** Derive the differential equations of motion for the system shown in Fig. 3.18. Let the angle $\theta$ be small.

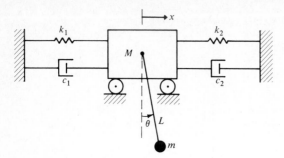

**Figure 3.18**

**3.6** Consider the system of Fig. 3.19 and show that it can be reduced to a single-degree-of-freedom system of equivalent mass $m_{eq} = m_1 m_2/(m_1 + m_2)$. Note that an unrestrained system such as this is known as a semidefinite system (see Sec. 4.10).

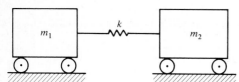

**Figure 3.19**

**3.7** The system of Fig. 3.20 represents an airfoil section being tested in a wind tunnel. Let the airfoil have total mass $m$ and mass moment of inertia $I_C$ about the mass center $C$, and derive the differential equations of motion.

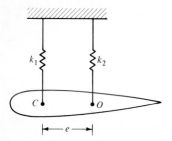

**Figure 3.20**

**3.8** A uniform thin rod is suspended by a string, as shown in Fig. 3.21. Derive the differential equations of motion of the system for arbitrarily large angles.

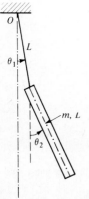

**Figure 3.21**

**3.9** A rigid bar of mass per unit length $\rho(\eta) = \rho_0(1 + \eta/L)$ is supported by two springs, as shown in Fig. 3.22. Assume small motions and derive the differential equations of motion.

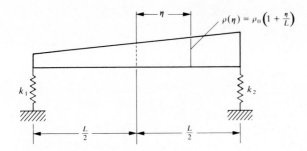

Figure 3.22

**3.10** Figure 3.23 depicts a two-story building. Assume the horizontal members to be rigid and the columns to be massless and derive the differential equations for the horizontal translation of the masses.

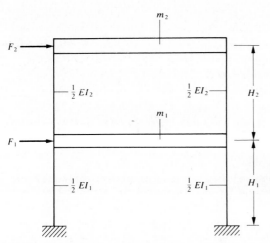

Figure 3.23

$HW\ 7$
$(1/18$

**3.11** A rigid uniform bar is supported by two translational springs and one torsional spring (Fig. 3.24). Derive the differential equations of motion.

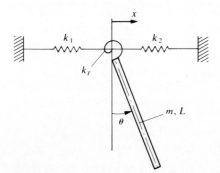

Figure 3.24

**3.12** Figure 3.25 shows a system of gears mounted on shafts. The radii of gears $A$ and $B$ are related by $R_A/R_B = n$. Derive the differential equations for the torsional motion of the system.

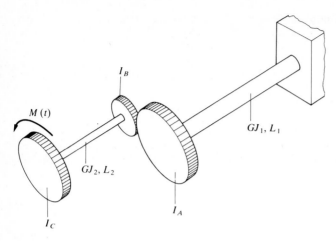

**Figure 3.25**

**3.13** Consider the system of Prob. 3.1, let $m_1 = m_2 = m$ and $L_1 = L_2 = L_3 = L$, and calculate the natural frequencies and natural modes. Plot the modes.

**3.14** Use Eq. (3.29) and prove that the two ratios in Eq. (3.30a) are identical. Repeat the problem for the two ratios in Eq. (3.30b).

**3.15** Consider the system of Prob. 3.2, let $I_1 = I_2 = I$, $GJ_1 = GJ_2 = GJ$, $L_1 = L_2 = L$, and calculate the natural frequencies and natural modes. Plot the modes.

**3.16** Consider the system of Prob. 3.3, let $k_1 = k$, $k_2 = 2k$, $M = mL$, and calculate the natural frequencies and natural modes. Plot the modes.

**3.17** Consider the double pendulum of Prob. 3.4 and linearize the equations of motion by assuming that $\theta_1(t)$ and $\theta_2(t)$ are small. Then let $m_1 = m_2 = m$, $L_1 = L_2 = L$, and calculate the natural frequencies and natural modes. Plot the modes.

**3.18** Linearize the equations of motion for the system of Prob. 3.8 and calculate the natural frequencies and natural modes.

**3.19** Obtain the natural frequencies and modes of vibration for the building of Prob. 3.10. Plot the modes. Let $m_1 = m_2 = m$, $H_1 = H_2 = H$ and $I_1 = I_2 = I$.

**3.20** Repeat Prob. 3.19, but for the system of Prob. 3.12. Let $n = 2$, $I_A = 5I$, $I_B = 2I$, $I_C = I$, and $k_1 = k_2 = k$.

**3.21** Consider the system of Prob. 3.3 and find a set of coordinates for which the system is elastically uncoupled. Then let $k_1 = k$, $k_2 = 2k$, $M = mL$, and calculate the natural frequencies and natural modes. Plot the modes. Compare the results with those obtained in Prob. 3.16 and draw conclusions.

**3.22** Consider Example 3.2 and use Eqs. (d) in conjunction with the second row of Eq. (b) to derive the natural modes.

**3.23** Verify that the natural modes in Example 3.2 are orthogonal both with respect to the mass matrix and the stiffness matrix.

**3.24** Obtain the response of the system of Prob. 3.15 to the initial excitation $\theta_1(0) = 0$, $\theta_2(0) = 1.5$, $\dot{\theta}_1(0) = 1.8\sqrt{GJ/IL}$, $\dot{\theta}_2(0) = 0$.

**3.25** Obtain the response of the system of Prob. 3.13 to the initial excitation $y_1(0) = 1.0$, $y_2(0) = -1.0$, $\dot{y}_1(0) = \dot{y}_2(0) = 0$. Explain your results.

**3.26** Consider the system of Fig. 3.2, let the excitation have the form

$$F_1(t) = F_1 \cos \omega t \qquad F_2(t) = 0$$

and derive Eqs. (3.82) by assuming the solution in the form of trigonometric functions.

**3.27** Let the system of Prob. 3.15 be acted upon by the torques

$$M_1(t) = 0 \qquad M_2(t) = M_2 e^{i\omega t}$$

and obtain expressions for the frequency responses $\Theta_1(\omega)$ and $\Theta_2(\omega)$. Plot $\Theta_1(\omega)$ versus $\omega$ and $\Theta_2(\omega)$ versus $\omega$.

**3.28** The foundation of the building of Prob. 3.19 undergoes the horizontal motion $y(t) = Y_0 \sin \omega t$. Derive expressions for the displacements of $m_1$ and $m_2$.

**3.29** A piece of machinery weighing 4800 lb ($2.1352 \times 10^4$ N) is observed to deflect 1.2 in ($3.05 \times 10^{-2}$ m) when at rest. A harmonic force of 100 lb (444.8 N) amplitude induces resonance. Design a vibration absorber undergoing a maximum deflection of 0.1 in ($2.54 \times 10^{-3}$ m). What is the value of the mass ratio $\mu$?

CHAPTER
# FOUR

## MULTI-DEGREE-OF-FREEDOM SYSTEMS

### 4.1 INTRODUCTION

The systems with one and two degrees of freedom discussed in the first three chapters represented simple mathematical models of complex physical systems. These simple models were able to explain the dynamic behavior of the complex systems. Quite often, however, such idealizations are not possible, and mathematical models with a larger number of degrees of freedom must be considered.

Most vibrational systems encountered in physical situations have distributed properties, such as mass and stiffness. Systems of this type are said to possess an infinite number of degrees of freedom, because the system is fully described only when the motion is known at every point of the system. In many cases, the mass and stiffness distributions are highly nonuniform, and for such systems it may be more feasible to construct discrete mathematical models, which need only a finite number of parameters to describe the mass and stiffness properties. Moreover, a description of the motion of such discrete models requires only a finite number of coordinates. In this manner, systems with an infinite number of degrees of freedom are reduced to systems with only a finite number of degrees of freedom. For example, Fig. 4.1a represents a nonuniform shaft with torsional stiffness $GJ(x)$ and mass moment of inertia $I(x)$ per unit length at any arbitrary point $x$, as shown. The continuous shaft can be approximated by the discrete model depicted in Fig. 4.1b, obtained by dividing the actual shaft into six segments and "lumping" the mass associated with each of these segments into six rigid disks of mass moments of inertia $I_i$ ($i = 1, 2, \ldots, 6$) connected by seven massless shafts of torsional rigidity $GJ_i$ ($i = 1, 2, \ldots, 7$). The parameters $I_i$ and $GJ_i$ are assigned values so as to simulate the

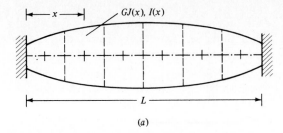

(a)

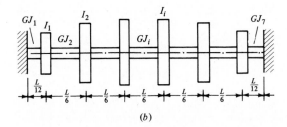

(b)                                                   **Figure 4.1**

continuous model as closely as possible. Note that in this case the continuous system is represented by a six-degree-of-freedom system, but this representation was merely to illustrate the lumping process and quite often the number of degrees of freedom is dictated by the nature of the problem.

As shown in Chap. 3, two-degree-of-freedom systems represent a significant departure from single-degree-of-freedom systems in the sense that the natural modes of vibration of the former have no counterpart in the latter. On the other hand, there is no basic difference between two- and many-degree-of-freedom systems, except that the latter require more efficient treatment. Such treatment is made possible by the use of concepts of linear algebra in conjunction with matrix methods.

The motion of multi-degree-of-freedom systems is generally described by a finite set of simultaneous second-order ordinary differential equations. The solution of such sets of equations is not an easy task, even when the equations are linear and they possess constant coefficients, because the coupling terms require that the equations be solved simultaneously. Such a solution is not feasible, however, so that the most indicated approach is to remove the coupling by means of a coordinate transformation producing a set of independent second-order ordinary differential equations of motion. Then, the solution of the independent equations can be carried out individually by the methods of Chap. 2. As demonstrated in Chap. 3, the coordinate transformation decoupling the equations of motion is based on the modal vectors of the system, and the coordinates describing the independent equations are the natural coordinates. Finally, the solution of the simultaneous equations of motion is obtained by simply inserting the expressions for the natural coordinates into the equations describing the coordinate transformation in question. The process whereby the solution of a set of simultaneous equations of motion is carried out by transforming the simultaneous equations into a set of independent equations for the natural coordinates, solving the independent equations, and expressing the solution of the simultaneous equations as a linear combination of the modal vectors multiplied by the natural coordinates is known as modal analysis. In addition to permitting efficient solutions of otherwise difficult problems, modal analysis affords a great deal of insight into the behavior of complex vibrating systems.

The emphasis in this chapter is placed on systematic ways of treating vibration problems associated with $n$-degree-of-freedom systems and on modern methods for obtaining numerical results by using high-speed electronic computers. The chapter generalizes and extends the material of Chap. 3. It begins by deriving the differential equations of motion for an $n$-degree-of-freedom system. Concentrating on linear systems, the equations are conveniently expressed in matrix form. To reduce the system of simultaneous equations of motion to uncoupled form by means of a linear transformation, we must first obtain the modal matrix. This leads naturally to the eigenvalue problem and its solution, where the latter consists of the system natural frequencies and modal vectors. The understanding of the eigenvalue problem and the properties of its solution are greatly enhanced by the use of concepts from linear algebra. A special appendix is devoted to reviewing these concepts. Two methods for the solution of the eigenvalue problem are presented, the first based on the characteristic determinant and the second based on matrix iteration. The responses of an $n$-degree-of-freedom system to initial excitation and externally applied forces are derived by modal analysis.

## 4.2 NEWTON'S EQUATIONS OF MOTION. GENERALIZED COORDINATES

Let us consider the system of particles of Fig. 4.2, where the particles have constant masses $m_i$ $(i = 1, 2, \ldots, N)$. The particles may be connected by springs, not necessarily linear, and are acted on by forces given by the vectors $\mathbf{F}_i$ $(i = 1, 2, \ldots, N)$, which could be external to the system or forces in the springs connecting $m_i$ with all or some of the remaining masses. We write the forces $\mathbf{F}_i$ in the form

$$\mathbf{F}_i = F_{xi}\mathbf{i} + F_{yi}\mathbf{j} + F_{zi}\mathbf{k} \qquad i = 1, 2, \ldots, N \tag{4.1}$$

where $F_{xi}, F_{yi}, F_{zi}$ are the cartesian components of the vector $\mathbf{F}_i$ in the directions $x$, $y$, and $z$, respectively, and $\mathbf{i}, \mathbf{j}, \mathbf{k}$ are corresponding unit vectors. We note that the vector notation is merely a way of writing all three components of a vector quantity by means of only one mathematical symbol. In addition to the applied forces $\mathbf{F}_i$, we assume that there are constraint forces $\mathbf{f}_i$ acting on the masses $m_i$. Such forces can occur if the motion of mass $m_i$ is restricted in some fashion. The constraint forces can be written as

$$\mathbf{f}_i = f_{xi}\mathbf{i} + f_{yi}\mathbf{j} + f_{zi}\mathbf{k} \qquad i = 1, 2, \ldots, N \tag{4.2}$$

where $f_{xi}, f_{yi}, f_{zi}$ are their cartesian components. Because the motion of $m_i$ is in

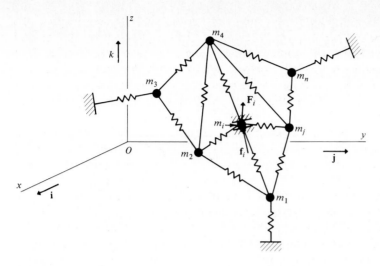

**Figure 4.2**

general three-dimensional, the displacement of $m_i$ can be written in the form of the vector

$$\mathbf{r}_i = x_i \mathbf{i} + y_i \mathbf{j} + z_i \mathbf{k} \qquad i = 1, 2, \ldots, N \tag{4.3}$$

where $x_i$, $y_i$, $z_i$ are the cartesian components of the displacement vector. Using Newton's second law for each particle, we can write the equations of motion in terms of cartesian coordinates as follows:

$$F_{xi} + f_{xi} = m_i \ddot{x}_i$$
$$F_{yi} + f_{yi} = m_i \ddot{y}_i \qquad i = 1, 2, \ldots, N \tag{4.4}$$
$$F_{zi} + f_{zi} = m_i \ddot{z}_i$$

which can be rewritten in the compact vector notation

$$\mathbf{F}_i + \mathbf{f}_i = m_i \ddot{\mathbf{r}}_i \qquad i = 1, 2, \ldots, N \tag{4.5}$$

Equations (4.4), or (4.5), represent a system of $3N$ second-order differential equations of motion. They can be linear or nonlinear, according to whether the forces $\mathbf{F}_i$ and $\mathbf{f}_i$ are linear or nonlinear functions of the displacements $\mathbf{r}_i$ and their time rates of change $\dot{\mathbf{r}}_i$.

In most cases, the constraint forces $\mathbf{f}_i$ are not given explicitly but implicitly through constraint equations placing restrictions on the motion of any of the masses $m_i$. As a result, not all coordinates $x_i$, $y_i$, $z_i$ ($i = 1, 2, \ldots, N$) are independent. Indeed, one constraint equation can be used, at least in principle, to eliminate one coordinate from the problem formulation. If there are $c$ constraint equations then the number of independent coordinates describing the system is only

$$n = 3N - c \tag{4.6}$$

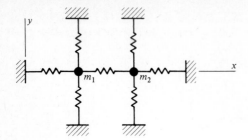

Figure 4.3

In this case the system is said to possess *n degrees of freedom*. Quite often, however, the constraints are taken into account automatically. As a simple illustration, we consider the system of Fig. 4.3, in which the motion of the masses $m_1$ and $m_2$ is restricted to the plane $xy$, so that the constraint equations are $z_1 = z_2 = 0$. Because in this case $N = 2$, we conclude from Eq. (4.6) that the number of degrees of freedom of the system is $n = 4$. Indeed, the motion of the system can be described by the cartesian coordinates $x_1, y_1, x_2, y_2$. This set of four coordinates is not the only possible one, but any four-degree-of-freedom system requires a minimum of four coordinates to describe its motion fully. We refer to a set of coordinates that describes the motion of a system completely as *generalized coordinates*, and denote them by $q_k$ $(k = 1, 2, \ldots, n)$. In the case of Fig. 4.3, the generalized coordinates can be taken as $q_1 = x_1$, $q_2 = y_1$, $q_3 = x_2$, $q_4 = y_2$. We shall see later that the generalized coordinates are not unique for a system, and that any $n$ coordinates capable of describing completely the motion of the system can serve as a set of generalized coordinates.

Another simple illustration is the double pendulum shown in Fig. 4.4. The positions of the masses $m_1$ and $m_2$ can be given by the cartesian coordinates $x_1, y_1$, and $x_2, y_2$, respectively. This is not a four-degree-of-freedom system, however,

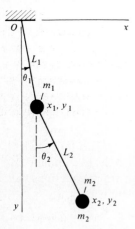

Figure 4.4

because we have two constraint equations, namely,

$$x_1^2 + y_1^2 = L_1^2 \qquad (x_2 - x_1)^2 + (y_2 - y_1)^2 = L_2^2 \qquad (4.7)$$

so that the system has only two degrees of freedom. Indeed, a convenient set of generalized coordinates for the system is $q_1 = \theta_1, q_2 = \theta_2$, where $\theta_1$ and $\theta_2$ are the angles shown in Fig. 4.4. Note that the first of the constraint equations (4.7) can be interpreted as confining the mass $m_1$ to a circle of radius $L_1$ and with the center at $O$.

## 4.3 EQUATIONS OF MOTION FOR LINEAR SYSTEMS. MATRIX FORMULATION

We are interested in the motion of a multi-degree-of-freedom system in the neighborhood of an equilibrium position, where the equilibrium position is as defined in Sec. 1.4. Without loss of generality, we assume that the equilibrium position is given by the trivial solution $q_1 = q_2 = \cdots = 0$. Moreover, we assume that the generalized displacements from the equilibrium position are sufficiently small that the force-displacement and force-velocity relations are linear, so that the generalized coordinates and their time derivatives appear in the differential equations of motion at most to the first power. This represents, in essence, the so-called *small-motions assumption*, leading to a linear system of equations. In this section, we derive the differential equations of motion by applying Newton's second law. Another procedure for deriving the equations of motion is the Lagrangian approach. Because the Lagrangian approach requires additional mathematical tools, its presentation is deferred to Chap. 6.

Let us consider the linear system consisting of $n$ masses $m_i$ $(i = 1, 2, \ldots, n)$ connected by springs and dampers, as shown in Fig. 4.5$a$, and draw the free-body diagram associated with the typical mass $m_i$ (see Fig. 4.5$b$). Because the motion

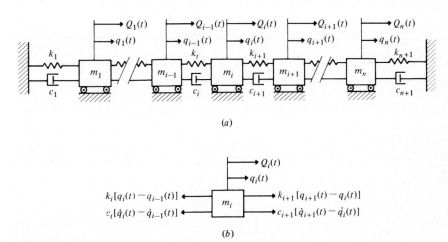

(a)

(b)

**Figure 4.5**

takes place in one dimension, the total number of degrees of freedom of the system coincides with the number of masses $n$. In view of this, we can dispense with the vector notation, and denote the generalized coordinates representing the displacements of the masses $m_i$ by $q_i(t)$ $(i = 1, 2, ..., n)$. Applying Newton's second law to a typical mass $m_i$ we can write the differential equation of motion

$$Q_i(t) + c_{i+1}[\dot{q}_{i+1}(t) - \dot{q}_i(t)] + k_{i+1}[q_{i+1}(t) - q_i(t)]$$

$$- c_i[\dot{q}_i(t) - \dot{q}_{i-1}(t)] - k_i[q_i(t) - q_{i-1}(t)] = m_i\ddot{q}_i(t) \quad (4.8)$$

where $Q_i(t)$ represents the externally impressed force. Equation (4.8) can be rearranged in the form

$$m_i\ddot{q}_i(t) - c_{i+1}\dot{q}_{i+1}(t) + (c_i + c_{i+1})\dot{q}_i(t) - c_i\dot{q}_{i-1}(t)$$

$$- k_{i+1}q_{i+1}(t) + (k_i + k_{i+1})q_i(t) - k_iq_{i-1}(t) = Q_i(t) \quad (4.9)$$

Next, let us introduce the notation

$$m_{ij} = \delta_{ij}m_i$$

$$
\begin{array}{llll}
c_{ij} = 0 & k_{ij} = 0 & j = 1, 2, ..., i - 2, i + 2, ..., n & \\
c_{ij} = -c_i & k_{ij} = -k_i & j = i - 1 & (4.10) \\
c_{ij} = c_i + c_{i+1} & k_{ij} = k_i + k_{i+1} & j = i & \\
c_{ij} = -c_{i+1} & k_{ij} = -k_{i+1} & j = i + 1 &
\end{array}
$$

where $m_{ij}$, $c_{ij}$, and $k_{ij}$ are referred to as *mass, damping,* and *stiffness coefficients,* respectively, and $\delta_{ij}$ is the *Kronecker delta,* defined as being equal to unity for $i = j$ and equal to zero for $i \neq j$. In view of notation (4.10), Eq. (4.9) can be used to express the complete set of equations of motion of the system as follows:

$$\sum_{j=1}^{n} [m_{ij}\ddot{q}_j(t) + c_{ij}\dot{q}_j(t) + k_{ij}q_j(t)] = Q_i(t) \qquad i = 1, 2, ..., n \quad (4.11)$$

which constitutes a set of $n$ simultaneous second-order ordinary differential equations for the generalized coordinates $q_i(t)$ $(i = 1, 2, ..., n)$. We note that Eqs. (4.11) are quite general, and indeed they can accommodate other end conditions as well. For example, if the right end is free instead of fixed, then we can simply set $c_{n+1} = k_{n+1} = 0$ in Eqs. (4.10). Although at this particular point the notation (4.10) appears as an undesirable complication, its advantage lies in the fact that the use of double index for the coefficients permits writing Eqs. (4.11) in matrix notation. We shall have ample opportunity to work with the coefficients $m_{ij}$, $c_{ij}$, and $k_{ij}$ and to study their interesting and useful properties. In particular, it will be shown that the mass, damping, and stiffness coefficients are symmetric

$$m_{ij} = m_{ji} \qquad c_{ij} = c_{ji} \qquad k_{ij} = k_{ji} \qquad i, j = 1, 2, ..., n \quad (4.12)$$

and that these coefficients control the system behavior, especially in the case of free vibration. Note that we encountered these coefficients for the first time in Sec. 3.2.

In spite of the fact that Eqs. (4.11) possess constant coefficients, the general closed-form solution of the equations is extremely difficult to obtain, particularly because of the coupling introduced by the damping coefficients $c_{ij}$. Under special circumstances, however, the solution of Eqs. (4.11) is possible. In attempting a solution, it will prove convenient to write Eqs. (4.11) in matrix form. To this end, we arrange the coefficients $m_{ij}$, $c_{ij}$, and $k_{ij}$ in the following square matrices:

$$[m_{ij}] = [m] \qquad [c_{ij}] = [c] \qquad [k_{ij}] = [k] \tag{4.13}$$

and we note that the symmetry of the coefficients is expressed by the relations

$$[m] = [m]^T \qquad [c] = [c]^T \qquad [k] = [k]^T \tag{4.14}$$

where the superscript $T$ denotes the transpose of the matrix in question. Moreover, we can arrange the generalized coordinates $q_i(t)$ and generalized impressed forces $Q_i(t)$ in the column matrices

$$\{q_i(t)\} = \{q(t)\} \qquad \{Q_i(t)\} = \{Q(t)\} \tag{4.15}$$

so that, using simple rules of matrix multiplication, Eqs. (4.11) can be written in the compact form

$$[m]\{\ddot{q}(t)\} + [c]\{\dot{q}(t)\} + [k]\{q(t)\} = \{Q(t)\} \tag{4.16}$$

As in Sec. 3.2, the matrices $[m]$, $[c]$, and $[k]$ are called the *mass*, or *inertia, damping,* and *stiffness matrices,* respectively. The matrix $[m]$ is diagonal because of our particular choice of coordinates. For a different set of generalized coordinates $[m]$ is not necessarily diagonal

The remainder of this chapter is devoted primarily to ways of obtaining the response of multi-degree-of-freedom systems.

**Example 4.1** Consider the three-degree-of-freedom system of Fig. 4.6a and derive the system differential equations of motion by using Newton's second law. The springs exhibit linear behavior and the dampers are viscous.

As shown in Fig. 4.6a, the generalized coordinates $q_1(t)$, $q_2(t)$, and $q_3(t)$ represent the horizontal translations of masses $m_1$, $m_2$, and $m_3$, respectively, and $Q_1(t)$, $Q_2(t)$, and $Q_3(t)$ are the associated generalized externally applied forces. To derive the equations of motion by Newton's second law, we draw three free-body diagrams, associated with masses $m_1$, $m_2$, and $m_3$, respectively. They are all shown in Fig. 4.6b, where the forces in the springs and dampers between masses $m_1$ and $m_2$ on the one hand and $m_2$ and $m_3$ on the other hand are the same in magnitude but opposite in direction. Application of Newton's second law for masses $m_i$ ($i = 1, 2, 3$) leads to the equations of motion

$$Q_1 + c_2(\dot{q}_2 - \dot{q}_1) + k_2(q_2 - q_1) - c_1\dot{q}_1 - k_1 q_1 = m_1\ddot{q}_1$$

$$Q_2 + c_3(\dot{q}_3 - \dot{q}_2) + k_3(q_3 - q_2) - c_2(\dot{q}_2 - \dot{q}_1) - k_2(q_2 - q_1) = m_2\ddot{q}_2 \tag{a}$$

$$Q_3 - c_3(\dot{q}_3 - \dot{q}_2) - k_3(q_3 - q_2) = m_3\ddot{q}_3$$

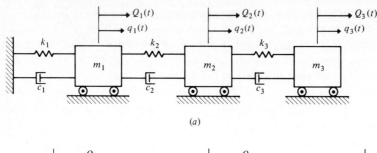

(a)

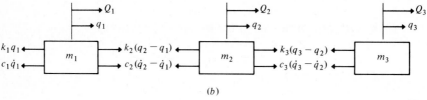

(b)

**Figure 4.6**

which can be rearranged in the form

$$m_1\ddot{q}_1 + (c_1 + c_2)\dot{q}_1 - c_2\dot{q}_2 + (k_1 + k_2)q_1 - k_2q_2 = Q_1$$

$$m_2\ddot{q}_2 - c_2\dot{q}_1 + (c_2 + c_3)\dot{q}_2 - c_3\dot{q}_3 - k_2q_1 + (k_2 + k_3)q_2 - k_3q_3 = Q_2 \quad (b)$$

$$m_3\ddot{q}_3 - c_3\dot{q}_2 + c_3\dot{q}_3 - k_3q_2 + k_3q_3 = Q_3$$

It is not difficult to see that Eqs. (b) can be expressed in the matrix form (4.16), where matrices $[m]$, $[c]$, and $[k]$ are given by

$$[m] = \begin{bmatrix} m_1 & 0 & 0 \\ 0 & m_2 & 0 \\ 0 & 0 & m_3 \end{bmatrix} \quad (c)$$

$$[c] = \begin{bmatrix} c_1 + c_2 & -c_2 & 0 \\ -c_2 & c_2 + c_3 & -c_3 \\ 0 & -c_3 & c_3 \end{bmatrix} \quad (d)$$

$$[k] = \begin{bmatrix} k_1 + k_2 & -k_2 & 0 \\ -k_2 & k_2 + k_3 & -k_3 \\ 0 & -k_3 & k_3 \end{bmatrix} \quad (e)$$

which are clearly symmetric. Moreover, $[m]$ is diagonal.

## 4.4 INFLUENCE COEFFICIENTS

In the study of discrete linear systems of the type treated in Example 4.1, it is of vital importance to know not only the inertia properties but also the stiffness properties of the system. These properties are implicit in the differential equations of motion in the form of the mass coefficients $m_{ij}$ and stiffness coefficients $k_{ij}$ introduced in Sec. 4.3. The latter coefficients can be obtained by other means, not necessarily involving the equations of motion. In fact, the stiffness coefficients are more properly known as *stiffness influence coefficients*, and can be derived by using a definition to be introduced shortly. There is one more type of influence coefficients, namely, *flexibility influence coefficients*. They are intimately related to the stiffness influence coefficients, which is to be expected, because both types of coefficients can be used to describe the manner in which the system deforms under forces.

In Sec. 1.2 we examined springs exhibiting linear behavior. In particular, we introduced the spring constant concept for a single spring and the equivalent spring constant for a given combination of springs. In this section we introduce the concept of influence coefficients by expanding on the approach of Sec. 1.2.

Let us consider a simple discrete system, as shown in Fig. 4.7, which is similar to that of Fig. 4.5a except that it has no damping. The system consists of $n$ point masses $m_i$ occupying the positions $x = x_i$ $(i = 1, 2, \ldots, n)$ when in equilibrium. In general, there are forces $F_i$ $(i = 1, 2, \ldots, n)$ acting upon each point mass $m_i$, respectively, so that the masses undergo displacements $u_i$. In the following, we propose to establish relations between the forces acting upon the system and the resulting displacements in terms of both flexibility and stiffness influence coefficients.

Let us first assume that the system is acted upon by a single force $F_j$ at $x = x_j$, and consider the displacement at any arbitrary point $x = x_i$ $(i = 1, 2, \ldots, n)$ due to the force $F_j$. With this in mind, we *define the flexibility influence coefficient $a_{ij}$ as the displacement of point $x = x_i$ due to a unit force, $F_j = 1$, applied at $x = x_j$.* Because the system is linear, displacements increase proportionally with forces, so that the displacement corresponding to a force of arbitrary magnitude $F_j$ is $a_{ij}F_j$. Moreover, for a linear system, we can invoke the principle of superposition and obtain the displacement $u_i$ at $x = x_i$ resulting from all forces $F_j$ $(j = 1, 2, \ldots, n)$ by simply summing up the individual contributions, with the result

$$u_i = \sum_{j=1}^{n} a_{ij}F_j \tag{4.17}$$

Note that in this particular case the coefficients $a_{ij}$ have units $LF^{-1}$, where $L$ and $F$ represent length and force, respectively. In other cases, involving torques and angular displacements, they can have different units.

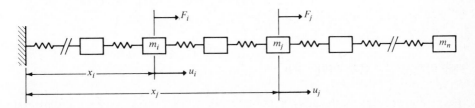

**Figure 4.7**

By analogy, we can *define the stiffness influence coefficient $k_{ij}$ as the force required at $x = x_i$ to produce a unit displacement, $u_j = 1$, at point $x = x_j$, and such that the displacements at all points for which $x \neq x_j$ are zero.* To obtain zero displacements at all points defined by $x \neq x_j$, the forces must simply hold these points fixed. Hence, the force at $x = x_i$ producing a displacement of arbitrary magnitude $u_j$ at $x = x_j$ is $k_{ij}u_j$. In reality the points for which $x \neq x_j$ are not fixed, so that, invoking once again the superposition principle, the force at $x = x_i$ producing displacements $u_j$ at $x = x_j$ ($j = 1, 2, \ldots, n$) is simply

$$F_i = \sum_{j=1}^{n} k_{ij}u_j \qquad (4.18)$$

It should be pointed out here that the stiffness coefficients as defined above, represent a special type of coefficient given in a more general form in Chap. 6. The coefficients $k_{ij}$ defined here have units $L^{-1}F$.

We note that for a single-degree-of-freedom system with only one spring the stiffness influence coefficient is merely the spring constant, whereas the flexibility influence coefficient is its reciprocal. A similar conclusion can be reached in a more general context for multi-degree-of-freedom systems. In this regard, matrix notation turns out to be most useful. Introducing square matrices whose elements are the flexibility and stiffness influence coefficients, respectively,

$$[a_{ij}] = [a] \qquad [k_{ij}] = [k] \qquad (4.19)$$

where $[a]$ is known as the *flexibility matrix* and $[k]$ as the *stiffness matrix*, and using simple rules of matrix multiplication, Eqs. (4.17) and (4.18) can be written in the compact matrix form

$$\{u\} = [a]\{F\} \qquad (4.20)$$

and

$$\{F\} = [k]\{u\} \qquad (4.21)$$

in which $\{u\}$ and $\{F\}$ are column matrices representing the $n$-dimensional displacement and force vectors with components $u_i$ ($i = 1, 2, \ldots, n$) and $F_j$ ($j = 1, 2, \ldots, n$), respectively. Equation (4.20) represents a linear transformation, with matrix $[a]$ playing the role of an operator that operates on $\{F\}$ to produce the column matrix $\{u\}$. In view of this, Eq. (4.21) can be regarded as the inverse transformation leading from $\{u\}$ to $\{F\}$. Because (4.21) and (4.20) relate the same vectors $\{u\}$ and $\{F\}$, matrices $[a]$ and $[k]$ must clearly be related. Indeed, introducing (4.21) into (4.20), we obtain

$$\{u\} = [a]\{F\} = [a][k]\{u\} \qquad (4.22)$$

with the obvious conclusion that

$$[a][k] = [1] \qquad (4.23)$$

where $[1] = [\delta_{ij}]$ is the identity or unit matrix of order $n$, with all its elements equal to the Kronecker delta $\delta_{ij}$ $(i, j = 1, 2, \ldots, n)$. Equation (4.23) implies that

$$[a] = [k]^{-1} \qquad [k] = [a]^{-1} \qquad (4.24)$$

or the *flexibility and stiffness matrices are the inverse of each other*.

It should be pointed out that, although the definition of the stiffness coefficients $k_{ij}$ sounds forbidding, these coefficients are often easier to evaluate than the flexibility coefficients $a_{ij}$, as can be concluded from Example 4.2. Moreover, quite frequently many of the stiffness coefficients have zero values. Nevertheless, the calculation of the stiffness coefficients by the definition given above is not the most efficient. More often than not it is possible to calculate the stiffness coefficients in a much simpler manner, namely, by means of the potential energy, as demonstrated in Sec. 4.5.

**Example 4.2** Consider the three-degree-of-freedom system shown in Fig. 4.8a and use the definitions to calculate the flexibility and stiffness matrices.

To calculate the flexibility influence coefficients $a_{ij}$, we apply unit forces $F_j = 1$ $(j = 1, 2, 3)$, in sequence, as shown in Figs. 4.8b, c, and d, respectively. In each case, the same unit force is acting everywhere to the left of the point of application $x = x_j$ of the unit force. On the other hand, the force is zero to the right of $x = x_j$. It follows that the elongation of every spring is equal to the reciprocal of the spring constant to the left of $x_j$ and to zero to the right of $x_j$. Hence, displacements are equal to the sum of the elongations of the springs to the left of $x_j$, and including $x_j$, and to $u_j$ to the right of $x_j$, so that, from Figs. 4.8b, c, and d we conclude that

$$a_{11} = u_1 = \frac{1}{k_1} \qquad a_{21} = u_2 = u_1 = \frac{1}{k_1} \qquad a_{31} = u_3 = u_2 = u_1 = \frac{1}{k_1}$$

$$a_{12} = u_1 = \frac{1}{k_1} \qquad a_{22} = u_2 = \frac{1}{k_1} + \frac{1}{k_2} \qquad a_{32} = u_3 = u_2 = \frac{1}{k_1} + \frac{1}{k_2} \qquad (a)$$

$$a_{13} = u_1 = \frac{1}{k_1} \qquad a_{23} = u_2 = \frac{1}{k_1} + \frac{1}{k_2} \qquad a_{33} = u_3 = \frac{1}{k_1} + \frac{1}{k_2} + \frac{1}{k_3}$$

The coefficients given by $(a)$ can be exhibited in the matrix form

$$[a] = \begin{bmatrix} \dfrac{1}{k_1} & \dfrac{1}{k_1} & \dfrac{1}{k_1} \\[2mm] \dfrac{1}{k_1} & \dfrac{1}{k_1} + \dfrac{1}{k_2} & \dfrac{1}{k_1} + \dfrac{1}{k_2} \\[2mm] \dfrac{1}{k_1} & \dfrac{1}{k_1} + \dfrac{1}{k_2} & \dfrac{1}{k_1} + \dfrac{1}{k_2} + \dfrac{1}{k_3} \end{bmatrix} \qquad (b)$$

and we note that the flexibility matrix $[a]$ is symmetric. This is no coincidence, as we shall have the oportunity to learn in Sec. 4.5.

The stiffness influence coefficients $k_{ij}$ are obtained from Fig. 4.8e, f, and g, in which the coefficients are simply the shown forces, where forces opposite in

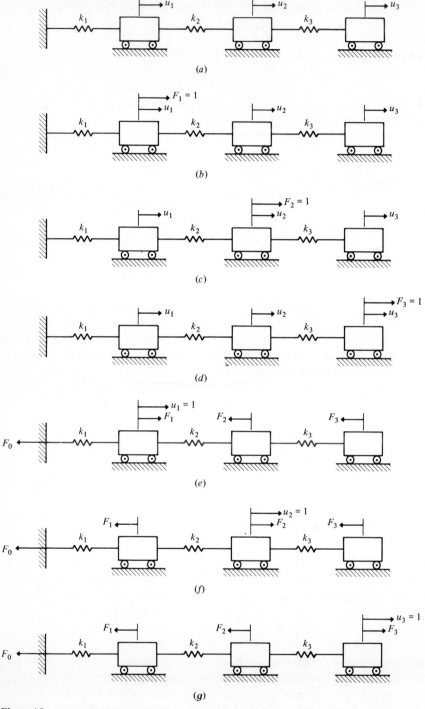

**Figure 4.8**

direction to the unit displacements must be assigned negative signs. From Fig. 4.8e we conclude that, corresponding to $u_1 = 1$, $u_2 = u_3 = 0$, there are the reaction forces $F_0 = -k_1$ and $F_2 = -k_2$ and, because for equilibrium we must have $F_0 + F_1 + F_2 = 0$, it follows that

$$k_{11} = F_1 = k_1 + k_2 \qquad k_{21} = F_2 = -k_2 \qquad k_{31} = F_3 = 0 \qquad (c)$$

where $F_3 = 0$ because no force is needed to keep the third mass in place. Similarly, from Fig. 4.8f and g we obtain

$$k_{12} = F_1 = -k_2 \qquad k_{22} = F_2 = k_2 + k_3 \qquad k_{32} = F_3 = -k_3 \qquad (d)$$

and

$$k_{13} = F_1 = 0 \qquad k_{23} = F_2 = -k_3 \qquad k_{33} = F_3 = k_3 \qquad (e)$$

The coefficients $k_{ij}$, Eqs. (e), (f), and (g), lead to the stiffness matrix

$$[k] = \begin{bmatrix} k_1 + k_2 & -k_2 & 0 \\ -k_2 & k_2 + k_3 & -k_3 \\ 0 & -k_3 & k_3 \end{bmatrix} \qquad (f)$$

where $[k]$ is also symmetric, as expected. Examining Eq. (f), it is easy to see that it is identical to Eq. (e) of Example 4.1.

We note from Eq. (f) that the elements $k_{13}$ and $k_{31}$ are zero. For systems such as that of Fig. 4.7, many more stiffness coefficients are equal to zero. In fact, it is easy to verify by inspection that the only coefficients which are not zero are those on the main diagonal and those immediately above and below the main diagonal. A matrix whose nonzero elements are clustered around the main diagonal is referred to as *banded*.

Using matrix algebra, it is not difficult to verify that $[a]$ and $[k]$, as given in Eqs. (b) and (f), are the inverse of one another. The verification is left to the reader as an exercise.

## 4.5 PROPERTIES OF THE STIFFNESS AND INERTIA COEFFICIENTS

Consider a single linear spring acted upon by a given force. The force in the spring corresponding to a displacement $\zeta$ is proportional to $\zeta$ and has the form $F_\zeta = -k\zeta$, where $k$ is the spring constant. If the spring is initially unstretched, then the potential energy corresponding to a final displacement $u$ is defined as (see Sec. 6.2)

$$V = \int_u^0 F_\zeta \, d\zeta = \int_u^0 (-k\zeta) \, d\zeta = \tfrac{1}{2}ku^2 = \tfrac{1}{2}Fu \qquad (4.25)$$

where $F$ is the final applied force. Equation (4.25) is quadratic in $u$, with the spring constant $k$ playing the role of a coefficient. It is reasonable to expect that for multi-degree-of-freedom linear systems the potential energy due to elastic effects alone can also be written in a quadratic form similar to Eq. (4.25). This is indeed the case, and the coefficients turn out to be the stiffness coefficients introduced in Sec. 4.4.

With reference to Fig. 4.7, let us focus our attention on the point mass $m_i$. When subjected to a force $F_i$, the mass undergoes a displacement $u_i$. Because for linear systems the displacement increases proportionally with the force, by analogy with Eq. (4.25), the elastic potential energy associated with the displacement of the point mass $m_i$ is

$$V_i = \tfrac{1}{2}F_i u_i \tag{4.26}$$

Note that the elastic potential energy is often referred to as *strain energy*. Assuming that there are $n$ forces $F_i$ $(i = 1, 2, \ldots, n)$ present, the strain energy for the entire system is simply

$$V = \sum_{i=1}^{n} V_i = \frac{1}{2}\sum_{i=1}^{n} F_i u_i \tag{4.27}$$

But the force $F_i$ is related to the displacements $u_j$ $(j = 1, 2, \ldots, n)$ according to Eq. (4.18). Inserting Eq. (4.18) into (4.27), we obtain

$$V = \frac{1}{2}\sum_{i=1}^{n} u_i \left( \sum_{j=1}^{n} k_{ij}u_j \right) = \frac{1}{2}\sum_{i=1}^{n}\sum_{j=1}^{n} k_{ij}u_i u_j \tag{4.28}$$

where $k_{ij}$ $(i, j = 1, 2, \ldots, n)$ are the stiffness influence coefficients. On the other hand, Eq. (4.17) relates the displacement $u_i$ to the forces $F_j$ $(j = 1, 2, \ldots, n)$, so that inserting Eq. (4.17) into (4.27), we arrive at

$$V = \frac{1}{2}\sum_{i=1}^{n} F_i \left( \sum_{j=1}^{n} a_{ij}F_j \right) = \frac{1}{2}\sum_{i=1}^{n}\sum_{j=1}^{n} a_{ij}F_i F_j \tag{4.29}$$

where $a_{ij}$ $(i, j = 1, 2, \ldots, n)$ are the flexibility influence coefficients.

The flexibility coefficients $a_{ij}$ and the stiffness coefficients $k_{ij}$ have a very important property, namely, they are symmetric. This statement is true for the flexibility and stiffness coefficients for any linear multi-degree-of-freedom mechanical system. The proof is based on the principle of superposition. Considering Fig. 4.7, let us assume that only the force $F_i$ is acting on the system, and denote by $u_i' = a_{ii}F_i$ the displacement produced at $x = x_i$ and by $u_j' = a_{ji}F_i$ that produced at $x = x_j$, where primes indicate that the displacements are produced by $F_i$ alone. It follows that the potential energy due to the force $F_i$ is

$$\tfrac{1}{2}F_i u_i' = \tfrac{1}{2}a_{ii}F_i^2 \tag{4.30}$$

Next let us apply a force $F_j$ at $x = x_j$, resulting in additional displacements $u_i'' = a_{ij}F_j$ and $u_j'' = a_{jj}F_j$ at $x = x_i$ and $x = x_j$, respectively, where double primes denote displacements due to $F_j$ alone. Because the force $F_i$ does not change during the application of $F_j$, the total potential energy has the expression

$$\tfrac{1}{2}F_i u_i' + F_i u_i'' + \tfrac{1}{2}F_j u_j'' = \tfrac{1}{2}a_{ii}F_i^2 + a_{ij}F_i F_j + \tfrac{1}{2}a_{jj}F_j^2 \tag{4.31}$$

Now let us apply the same forces $F_i$ and $F_j$ but in reverse order. Applying first a force $F_j$ at $x = x_j$, and denoting by $u''_j = a_{jj}F_j$ the displacement produced at $x = x_j$ and by $u''_i = a_{ij}F_j$ that produced at $x = x_i$, the potential energy due to $F_j$ alone is

$$\tfrac{1}{2}F_j u''_j = \tfrac{1}{2}a_{jj}F_j^2 \qquad (4.32)$$

Next we apply a force $F_i$ at $x = x_i$ and denote the resulting displacement at $x = x_i$ by $u'_i = a_{ii}F_i$ and that at $x = x_j$ by $u'_j = a_{ji}F_i$. This time we recognize that it is $F_j$ that does not change during the application of $F_i$, so that the potential energy is

$$\tfrac{1}{2}F_j u''_j + F_j u'_j + \tfrac{1}{2}F_i u'_i = \tfrac{1}{2}a_{jj}F_j^2 + a_{ji}F_jF_i + \tfrac{1}{2}a_{ii}F_i^2 \qquad (4.33)$$

But the potential energy must be the same regardless of the order in which the forces $F_i$ and $F_j$ are applied. Hence, Eqs. (4.32) and (4.33) must have the same value, which yields

$$a_{ij}F_iF_j = a_{ji}F_jF_i \qquad (4.34)$$

with the obvious conclusion that the flexibility influence coefficients are symmetric,

$$a_{ij} = a_{ji} \qquad (4.35)$$

Equation (4.35) is the statement of *Maxwell's reciprocity theorem* and can be proved for more general linear systems than that of Fig. 4.7.

In matrix notation, Eq. (4.35) takes the form

$$[a] = [a]^T \qquad (4.36)$$

where the superscript $T$ denotes the transpose of the matrix in question. Considering Eq. (4.36), and using Eq. (4.23), it is not difficult to show that the stiffness influence coefficients are also symmetric, as expressed by the matrix equation

$$[k] = [k]^T \qquad (4.37)$$

The potential energy can be written in the form of a triple matrix product. Indeed, in matrix notation, Eq. (4.28) has the form

$$V = \tfrac{1}{2}\{u\}^T[k]\{u\} \qquad (4.38)$$

whereas Eq. (4.29) can be written as

$$V = \tfrac{1}{2}\{F\}^T[a]\{F\} \qquad (4.39)$$

where $\{u\}$ and $\{F\}$ are column matrices representing the $n$-dimensional displacement and force vectors.

Another matrix of special interest in vibrations is the mass matrix. It turns out that the mass matrix is associated with the kinetic energy. For a single mass $m$ moving with the velocity $\dot{u}$, the kinetic energy is defined as (see Sec. 6.2)

$$T = \tfrac{1}{2}m\dot{u}^2 \qquad (4.40)$$

Considering a multi-degree-of-freedom system and denoting by $\dot{u}_i$ the velocity of the mass $m_i$ ($i = 1, 2, \ldots, n$), the kinetic energy is simply

$$T = \frac{1}{2} \sum_{i=1}^{n} m_i \dot{u}_i^2 \tag{4.41}$$

which can be written in the form of the triple matrix product

$$T = \tfrac{1}{2}\{\dot{u}\}^T[m]\{\dot{u}\} \tag{4.42}$$

in which $[m]$ is the mass (or inertia) matrix. In this particular case, the matrix $[m]$ is diagonal. In general $[m]$ need not be diagonal (see Sec. 6.6), although it is symmetric. We assume that this is the case with $[m]$ in Eq. (4.42). It is worth pointing out here that the matrices $[k]$ and $[m]$ in Eqs. (4.38) and (4.42), respectively, are precisely the stiffness and mass matrices appearing in the differential equations of motion for a discrete linear system, as derived in Sec. 4.3.

Equations (4.38) and (4.42) are merely quadratic forms in matrix notation, the first in terms of generalized coordinates and the second in terms of generalized velocities. It will prove of interest to study some of the properties of quadratic forms, as from this it is possible to infer certain motion characteristics of multi-degree-of-freedom systems. Quadratic forms represent a special type of functions, so that we first present certain definitions concerning functions in general and then apply these definitions to quadratic functions of particular interest in vibrations.

A function of several variables is said to be *positive (negative) definite* if it is never negative (positive) and is equal to zero if and only if all the variables are zero. A function of several variables is said to be *positive (negative) semidefinite* if it is never negative (positive) and can be zero even when some or all the variables are not zero. A function of several variables is said to be *sign-variable* if it can take either positive or negative values. A criterion for testing the positive definiteness of a function, known as Sylvester's criterion, is discussed in Sec. 9.7.

For quadratic forms, the sign definiteness is governed by the corresponding constant coefficients. In the particular case of the kinetic energy $T$ and the potential energy $V$, these coefficients are $m_{ij}$ and $k_{ij}$, respectively. In view of the definitions for the sign definiteness of functions, we can define a matrix whose elements are the coefficients of a positive (negative) definite quadratic form as a *positive (negative) definite matrix*. Likewise, a matrix whose elements are the coefficients of a positive (negative) semidefinite quadratic form is said to be a *positive (negative) semidefinite matrix*. Sometimes a positive (negative) semidefinite function is known as merely *positive (negative)*.

The kinetic energy is always positive definite, so that $[m]$ is always positive definite. The question remains as to the sign properties of the potential energy and the associated matrix $[k]$. Two cases of particular interest in the area of vibrations are that in which $[k]$ is positive definite and that in which $[k]$ is only positive semidefinite. When both $[m]$ and $[k]$ are positive definite, the system is said to be a *positive definite system* and the motion is that of *undamped free vibration*. This case is discussed in Sec. 4.7. When $[m]$ is positive definite and $[k]$ is only positive semidefinite, the system is referred to as a *positive semidefinite system*, and the motion is again undamped free vibration but rigid-body motion is possible because semidefinite systems are unrestrained, that is to say, such systems are supported in a manner in which rigid-body motion can take place. This case is discussed in Sec. 4.12.

**Example 4.3** Derive the stiffness matrix for the system of Example 4.2 by means of the potential energy.

Considering Fig. 4.8a and recognizing that the elongations of the springs $k_1$, $k_2$, and $k_3$ are $u_1$, $u_2 - u_1$, and $u_3 - u_2$, respectively, the potential energy is simply

$$V = \tfrac{1}{2}[k_1 u_1^2 + k_2(u_2 - u_1)^2 + k_3(u_3 - u_2)^2]$$
$$= \tfrac{1}{2}[(k_1 + k_2)u_1^2 + (k_2 + k_3)u_2^2 + k_3 u_3^2 - 2k_2 u_1 u_2 - 2k_3 u_2 u_3] \qquad (a)$$

which can be rewritten in the matrix form

$$V = \tfrac{1}{2}\{u\}^T[k]\{u\} \qquad (b)$$

where

$$\{u\} = \begin{Bmatrix} u_1 \\ u_2 \\ u_3 \end{Bmatrix} \qquad [k] = \begin{bmatrix} k_1 + k_2 & -k_2 & 0 \\ -k_2 & k_2 + k_3 & -k_3 \\ 0 & -k_3 & k_3 \end{bmatrix} \qquad (c)$$

are the displacement vector and stiffness matrix, respectively. Clearly, the stiffness matrix is the same as that obtained in Example 4.2. It is also clear that the derivation of the stiffness matrix via the potential energy is appreciably more expeditious than through the use of the definition. This is often the case, and not merely in this particular example.

## 4.6 LINEAR TRANSFORMATIONS. COUPLING

As demonstrated in Sec. 3.4, coupling depends on the coordinates used to describe the motion and is not a basic characteristic of the system. In this section, we discuss the ideas of coordinate transformations and coupling in broader terms.

Focusing our attention on the undamped case, we set $[c] = [0]$ in Eq. (4.16), where $[0]$ is the null square matrix of order $n$, and obtain the corresponding system of differential equations of motion

$$[m]\{\ddot{q}(t)\} + [k]\{q(t)\} = \{Q(t)\} \qquad (4.43)$$

where $\{Q(t)\}$ is a column matrix whose elements are the $n$ generalized externally impressed forces. For the purpose of this discussion, we consider the matrices $[m]$ and $[k]$ as arbitrary, except that they are symmetric and their elements constant. The column matrices $\{q\}$ and $\{Q\}$ represent $n$-dimensional vectors of generalized coordinates and forces, respectively.

It is clear from Eq. (4.43) that if $[m]$ is not diagonal, then the equations of motion are coupled through the inertial forces. On the other hand, if $[k]$ is not diagonal, the equations are coupled through the elastically restoring forces. In general (4.43) represents a set of $n$ simultaneous linear second-order ordinary differential equations with constant coefficients. The solution of such a set of equations is not a simple task, and we wish to explore means of facilitating it. To this end, we express the equations of motion in a different set of generalized coordinates $\eta_j(t)$ $(j = 1, 2, \ldots, n)$ such that any coordinate $q_i(t)$ $(i = 1, 2, \ldots, n)$ is a linear combination of the coordinates $\eta_j(t)$. Hence, let us consider the linear transformation

$$\{q(t)\} = [u]\{\eta(t)\} \tag{4.44}$$

in which $[u]$ is a constant nonsingular square matrix, referred to as a *transformation matrix*. The matrix $[u]$ can be regarded as an operator transforming the vector $\{\eta\}$ into the vector $\{q\}$. Because $[u]$ is constant, we also have

$$\{\dot{q}(t)\} = [u]\{\dot{\eta}(t)\} \qquad \{\ddot{q}(t)\} = [u]\{\ddot{\eta}(t)\} \tag{4.45}$$

so that the same transformation matrix $[u]$ connects the velocity vectors $\{\dot{\eta}\}$ and $\{\dot{q}\}$ and the acceleration vectors $\{\ddot{\eta}\}$ and $\{\ddot{q}\}$. Inserting Eqs. (4.44) and (4.45) into (4.43), we arrive at

$$[m][u]\{\ddot{\eta}(t)\} + [k][u]\{\eta(t)\} = \{Q(t)\} \tag{4.46}$$

Next, we premultiply both sides of Eq. (4.46) by $[u]^T$ and obtain

$$[M]\{\ddot{\eta}(t)\} + [K]\{\eta(t)\} = \{N(t)\} \tag{4.47}$$

where the matrices

$$[M] = [u]^T[m][u] = [M]^T \qquad [K] = [u]^T[k][u] = [K]^T \tag{4.48}$$

are symmetric because $[m]$ and $[k]$ are symmetric. Moreover,

$$\{N(t)\} = [u]^T\{Q(t)\} \tag{4.49}$$

is an $n$-dimensional vector whose elements are the generalized forces $N_i$ associated with the generalized coordinates $\eta_i$. Note that $N_i$ are linear combinations of $Q_j$ $(j = 1, 2, \ldots, n)$.

The derivation of the matrices $[M]$ and $[K]$ can be effected in a more natural manner by considering the kinetic and potential energy. Indeed, recalling Eqs. (4.44) and (4.45) and recognizing that $\{q(t)\}^T = \{\eta(t)\}^T[u]^T$, $\{\dot{q}(t)\}^T = \{\dot{\eta}(t)\}^T[u]^T$, the kinetic and potential energy, Eqs. (4.42) and (4.38), can be expressed in the form

$$T = \tfrac{1}{2}\{\dot{\eta}(t)\}^T[M]\{\dot{\eta}(t)\} \tag{4.50}$$

$$V = \tfrac{1}{2}\{\eta(t)\}^T[K]\{\eta(t)\} \tag{4.51}$$

where $[M]$ and $[K]$ are the mass and stiffness matrices corresponding to the coordinates $\eta_j(t)$ $(j = 1, 2, \ldots, n)$ and are as given by Eqs. (4.48). The derivation of the column matrix $\{N(t)\}$ can be carried out by means of the virtual work expression (see Sec. 6.5).

At this point we wish to return to the concept of coupling. If matrix $[M]$ is diagonal, then system (4.47) is said to be *inertially uncoupled*. On the other hand, if $[K]$ is diagonal, then the system is said to be *elastically uncoupled*. *The object of the transformation* (4.44) *is to produce diagonal matrices* $[M]$ *and* $[K]$ *simultaneously, because only then does the system consist of independent equations of motion.* Hence, if such a transformation matrix $[u]$ can be found, then Eq. (4.47) represents a set of $n$ independent equations of the type

$$M_j \ddot{n}_j(t) + K_j n_j(t) = N_j(t) \qquad j = 1, 2, \ldots, n \qquad (4.52)$$

where one of the two subscripts in $M_{jj}$ and $K_{jj}$ has been dropped because they are identical. Equations (4.52) have precisely the same structure as that of an undamped single-degree-of-freedom system [see Eq. (1.14) with $c = 0$], and can be readily solved by the methods of Chap. 2.

We state here (and prove later) that a linear transformation matrix $[u]$ diagonalizing $[m]$ and $[k]$ simultaneously does indeed exist. This particular matrix $[u]$ is known as the *modal matrix*, because it consists of the *modal vectors* or *characteristic vectors*, representing the *natural modes* of the system, and the coordinates $n_j(t)$ $(j = 1, 2, \ldots, n)$ are called *natural*, or *principal, coordinates*. The procedure of solving the system of simultaneous differential equations of motion by transforming them into a set of independent equations by means of the modal matrix is generally referred to as *modal analysis*.

It is perhaps appropriate to pause at this point and reflect on the coordinate transformation (4.44), leading from equations of motion in terms of the coordinates $q_i(t)$ $(i = 1, 2, \ldots, n)$ to equations of motion in terms of the coordinates $n_j(t)$ $(j = 1, 2, \ldots, n)$. The new mass and stiffness matrices $[M]$ and $[K]$ are related to the original mass and stiffness matrices $[m]$ and $[k]$ by Eqs. (4.48). In the special case in which $[u]$ is the modal matrix, the matrices $[M]$ and $[K]$ become diagonal simultaneously and the matrix $[u]$ is said to be *orthogonal* (with respect to both $[m]$ and $[k]$). Moreover, in this case Eqs. (4.48) represent an *orthogonal transformation*, which is a special case of a similarity transformation and, as shown in App. C, *the nature of the system does not change in similarity transformations*. But, because the new mass and stiffness matrices $[M]$ and $[K]$ are both diagonal, the equations of motion in terms of the coordinates $n_j(t)$ $(j = 1, 2, \ldots, n)$ become independent and very easy to solve. Hence, the linear transformation (4.44), in which $[u]$ is the modal matrix, permits an expeditious solution of the equations of motion.

It remains to find a way of determining the modal matrix $[u]$ for a given system. This can be accomplished by solving the eigenvalue problem associated with the matrices $[m]$ and $[k]$, a subject discussed in Sec. 4.7. It should be pointed out that we already used a linear transformation of the type (4.44) to uncouple the equations of motion. Indeed, the vectors $\{u\}_1$ and $\{u\}_2$ multiplying the principal coordinates $q_1(t)$ and $q_2(t)$ in Sec. 3.5 were the modal vectors, and hence the columns of the modal matrix $[u]$. But, as pointed out in Sec. 3.3, the modal vectors satisfy homogeneous algebraic equations, so that their magnitudes cannot be determined uniquely; only the ratios of the components of the modal vectors can. It is often convenient to choose the magnitude of the modal vectors so as to reduce the matrix $[M]$ to the identity matrix, which automatically reduces the matrix $[K]$

to the diagonal matrix of natural frequencies squared. This process is known as *normalization* and, under these circumstances, the modal matrix $[u]$ is said to be *orthonormal* (with respect to $[m]$ and $[k]$). In addition, the natural, or principal coordinates $\eta_j(t)$ $(j = 1, 2, ..., n)$ become *normal coordinates*.

**Example 4.4** The modal matrix associated with the mass and stiffness matrices

$$[m] = m \begin{bmatrix} 1 & 0 & 0 \\ 0 & 1 & 0 \\ 0 & 0 & 2 \end{bmatrix} \qquad [k] = k \begin{bmatrix} 2 & -1 & 0 \\ -1 & 3 & -2 \\ 0 & -2 & 2 \end{bmatrix} \qquad (a)$$

can be shown to be (see Example 4.7)

$$[u] = m^{-1/2} \begin{bmatrix} 0.2691 & -0.8782 & 0.3954 \\ 0.5008 & -0.2231 & -0.8363 \\ 0.5817 & 0.2992 & 0.2685 \end{bmatrix} \qquad (b)$$

Show that, when used as a transformation matrix, the matrix $[u]$ diagonalizes $[m]$ and $[k]$ simultaneously.

Inserting Eqs. $(a)$ and $(b)$ into Eqs. (4.48), we obtain the matrices

$$[M] = [u]^T[m][u]$$

$$= \begin{bmatrix} 0.2691 & 0.5008 & 0.5817 \\ -0.8782 & -0.2231 & 0.2992 \\ 0.3954 & -0.8363 & 0.2685 \end{bmatrix} \begin{bmatrix} 1 & 0 & 0 \\ 0 & 1 & 0 \\ 0 & 0 & 2 \end{bmatrix}$$

$$\times \begin{bmatrix} 0.2691 & -0.8782 & 0.3954 \\ 0.5008 & -0.2231 & -0.8363 \\ 0.5817 & 0.2992 & 0.2685 \end{bmatrix}$$

$$= \begin{bmatrix} 1 & 0 & 0 \\ 0 & 1 & 0 \\ 0 & 0 & 1 \end{bmatrix} \qquad (c)$$

and

$$[K] = [u]^T[k][u]$$

$$= \frac{k}{m} \begin{bmatrix} 0.2691 & 0.5008 & 0.5817 \\ -0.8782 & -0.2231 & 0.2992 \\ 0.3954 & -0.8363 & 0.2685 \end{bmatrix} \begin{bmatrix} 2 & -1 & 0 \\ -1 & 3 & -2 \\ 0 & -2 & 2 \end{bmatrix}$$

$$\times \begin{bmatrix} 0.2691 & -0.8782 & 0.3954 \\ 0.5008 & -0.2231 & -0.8363 \\ 0.5817 & 0.2992 & 0.2685 \end{bmatrix}$$

$$= \frac{k}{m} \begin{bmatrix} 0.1392 & 0 & 0 \\ 0 & 1.7458 & 0 \\ 0 & 0 & 4.1152 \end{bmatrix} \qquad (d)$$

which are clearly diagonal. Moreover, $[M]$ is the identity matrix, so that the modal matrix $[u]$ is orthonormal. Consistent with this, the diagonal elements of $[K]$ are equal to the natural frequencies squared, as we shall verify later.

## 4.7 UNDAMPED FREE VIBRATION. EIGENVALUE PROBLEM

In Sec. 4.6, we pointed out that, in the absence of damping, the equations of motion can be decoupled by using a transformation of coordinates, with the modal matrix acting as the transformation matrix. To determine the modal matrix, we must solve the so-called eigenvalue problem, a problem associated with free vibration, i.e., vibration in which the external forces are zero. In this section, we show how the free vibration problem leads directly to the eigenvalue problem, the solution of the latter yielding the natural modes of vibration. Then, we show that the natural motions, defined as motions in which the system vibrates in any one of the natural modes, can be identified as special cases of free vibration. Finally, we show that in the general case of free vibration, the motion can be regarded as a linear combination of the natural motions.

In the absence of external forces, $\{Q(t)\} = \{0\}$, Eq. (4.43) reduces to

$$[m]\{\ddot{q}(t)\} + [k]\{q(t)\} = \{0\} \tag{4.53}$$

which represents a set of $n$ simultaneous homogeneous differential equations of the type

$$\sum_{j=1}^{n} m_{ij}\ddot{q}_j(t) + \sum_{j=1}^{n} k_{ij}q_j(t) = 0 \qquad i = 1, 2, \dots, n \tag{4.54}$$

We are interested in a special type of solution of the set (4.54), namely, that in which all the coordinates $q_j(t)$ $(j = 1, 2, \dots, n)$ execute synchronous motion. Physically, this implies a motion in which all the coordinates have the same time dependence, and the general configuration of the motion does not change, except for the amplitude, so that the ratio between any two coordinates $q_i(t)$ and $q_j(t)$, $i \neq j$, remains constant during the motion. Mathematically, this type of motion is expressed by

$$q_j(t) = u_j f(t) \qquad j = 1, 2, \dots, n \tag{4.55}$$

where $u_j$ $(j = 1, 2, \dots, n)$ are constant amplitudes and $f(t)$ is a function of time that is the same for all the coordinates $q_j(t)$. We are interested in the case in which the coordinates $q_j(t)$ represent stable oscillation, which implies that $f(t)$ must be bounded.

Inserting Eqs. (4.55) into (4.54), and recognizing that the function $f(t)$ does not depend on the index $j$, we obtain

$$\ddot{f}(t) \sum_{j=1}^{n} m_{ij}u_j + f(t) \sum_{j=1}^{n} k_{ij}u_j = 0 \qquad i = 1, 2, \dots, n \tag{4.56}$$

Equations (4.56) can be written in the form

$$-\frac{\ddot{f}(t)}{f(t)} = \frac{\sum_{j=1}^{n} k_{ij} u_j}{\sum_{j=1}^{n} m_{ij} u_j} \qquad i = 1, 2, \ldots, n \qquad (4.57)$$

with the implication that the time dependence and the positional dependence are separable, which is akin to the separation of variables for partial differential equations. Using the standard argument, we observe that the left side of (4.57) does not depend on the index $i$, whereas the right side does not depend on time, so that the two ratios must be equal to a constant. Assuming that $f(t)$ is a real function, the constant must be a real number. Denoting the constant by $\lambda$, the set (4.57) yields

$$\ddot{f}(t) + \lambda f(t) = 0 \qquad (4.58)$$

$$\sum_{j=1}^{n} (k_{ij} - \lambda m_{ij}) u_j = 0 \qquad i = 1, 2, \ldots, n \qquad (4.59)$$

Let us consider a solution of Eq. (4.58) in the exponential form

$$f(t) = A e^{st} \qquad (4.60)$$

Introducing solution (4.60) into (4.58), we conclude that $s$ must satisfy the equation

$$s^2 + \lambda = 0 \qquad (4.61)$$

which has two roots

$$\begin{matrix} s_1 \\ s_2 \end{matrix} = \pm \sqrt{-\lambda} \qquad (4.62)$$

If $\lambda$ is a negative number (we have already concluded that it must be real), then $s_1$ and $s_2$ are real numbers, equal in magnitude but opposite in sign. In this case, Eq. (4.58) has two solutions, one decreasing and the other increasing exponentially with time. These solutions, however, are inconsistent with stable motion, so that the possibility that $\lambda$ is negative must be discarded and the one that $\lambda$ is positive considered. Letting $\lambda = \omega^2$, where $\omega$ is real, Eq. (4.62) yields

$$\begin{matrix} s_1 \\ s_2 \end{matrix} = \pm i\omega \qquad (4.63)$$

so that the solution of Eq. (4.58) becomes

$$f(t) = A_1 e^{i\omega t} + A_2 e^{-i\omega t} \qquad (4.64)$$

where $A_1$ and $A_2$ are generally complex numbers constant in value. Recognizing that $e^{i\omega t}$ and $e^{-i\omega t}$ represent complex vectors of unit magnitude, we conclude that solution (4.64) is harmonic with the frequency $\omega$, and that it is the only acceptable solution of Eq. (4.58). This implies that if synchronous motion is possible, then the time dependence is harmonic. Because $f(t)$ is a real function, $A_2$ is the complex conjugate of $A_1$. It is easy to verify that solution (4.64) can be expressed in the form

$$f(t) = C \cos (\omega t - \phi) \qquad (4.65)$$

where $C$ is an arbitrary constant, $\omega$ is the frequency of the harmonic motion, and $\phi$ its phase angle, all three quantities being the same for every coordinate $q_j(t)$ $(j = 1, 2, \ldots, n)$.

To complete the solution of Eqs. (4.54), we must determine the amplitudes $u_j$ $(j = 1, 2, \ldots, n)$. To this end, we turn to Eqs. (4.59), which constitute a set of $n$ homogeneous algebraic equations in the unknowns $u_j$, with $\lambda = \omega^2$ playing the role of a parameter. Not any arbitrary value of $\omega^2$ permits a solution of Eqs. (4.59), but only a select set of $n$ values. The problem of determining the values of $\omega^2$ for which a nontrivial solution $u_j$ $(j = 1, 2, \ldots, n)$ of Eqs. (4.59) exists is known as the *characteristic-value*, or *eigenvalue problem*.

It will prove convenient to write Eqs. (4.59) in the matrix form

$$[k]\{u\} = \omega^2 [m]\{u\} \qquad (4.66)$$

Equation (4.66) represents the eigenvalue problem associated with matrices $[m]$ and $[k]$ and it possesses a nontrivial solution if and only if the determinant of the coefficients of $u_j$ vanishes. This can be expressed in the form

$$\Delta(\omega^2) = |k_{ij} - \omega^2 m_{ij}| = 0 \qquad (4.67)$$

where $\Delta(\omega^2)$ is called the *characteristic determinant*, with Eq. (4.67) itself being known as the *characteristic equation*, or *frequency equation*. It is an equation of degree $n$ in $\omega^2$, and it possesses in general $n$ distinct roots, referred to as *characteristic values*, or *eigenvalues*. The $n$ roots are denoted $\omega_1^2, \omega_2^2, \ldots, \omega_n^2$ and the square roots of these quantities are the *system natural frequencies* $\omega_r$ $(r = 1, 2, \ldots, n)$. The natural frequencies can be arranged in order of increasing magnitude, namely, $\omega_1 \leqslant \omega_2 \leqslant \cdots \leqslant \omega_n$. The lowest frequency $\omega_1$ is referred to as the *fundamental frequency*, and for many practical problems it is the most important one. In general all frequencies $\omega_r$ are distinct and the equality sign never holds, except in *degenerate* cases (see discussion of such cases in Sec. 4.8). It follows that there are $n$ frequencies $\omega_r$ $(r = 1, 2, \ldots, n)$ in which harmonic motion of the type (4.65) is possible.

Associated with every one of the frequencies $\omega_r$ there is a certain nontrivial vector $\{u\}_r$, whose elements $u_{ir}$ are real numbers, where $\{u\}_r$ is a solution of the eigenvalue problem, such that

$$[k]\{u\}_r = \omega_r^2 [m]\{u\}_r \qquad r = 1, 2, \ldots, n \qquad (4.68)$$

The vectors $\{u\}_r$ $(r = 1, 2, \ldots, n)$ are known as *characteristic vectors*, or *eigenvectors*. The eigenvectors are also referred to as *modal vectors* and represent physically the so-called *natural modes*. These vectors are unique only in the sense that the ratio between any two elements $u_{ir}$ and $u_{jr}$ is constant. The value of the elements themselves is arbitrary, however, because Eq. (4.66) is homogeneous, so that if $\{u\}_r$ is a solution of the equation, then $\alpha_r\{u\}_r$ is also a solution, where $\alpha_r$ is an arbitrary constant. Hence, we can say that *the shape of the natural modes is unique, but the amplitude is not.*

If one of the elements of the eigenvector $\{u\}_r$ is assigned a certain value, then the eigenvector is rendered unique in an absolute sense, because this automatically causes an adjustment in the values of the remaining $n-1$ elements by virtue of the fact that the ratio between any two elements is constant. The process of adjusting the elements of the natural modes to render their amplitude unique is called *normalization,* and the resulting vectors are referred to as *normal modes.* A very convenient normalization scheme consists of setting

$$\{u\}_r^T[m]\{u\}_r = 1 \qquad r = 1, 2, \ldots, n \tag{4.69}$$

which has the advantage that it yields

$$\{u\}_r^T[k]\{u\}_r = \omega_r^2 \qquad r = 1, 2, \ldots, n \tag{4.70}$$

This can be easily shown by premultiplying both sides of (4.68) by $\{u\}_r^T$. Note that if this normalization scheme is used, then the elements of $\{u\}_r$ have units of $M^{-1/2}$, where $M$ represents symbolically the units of the elements $m_{ij}$ of the inertia matrix $[m]$. This, in turn, establishes the units of the constant $C$ in Eq. (4.65), as can be concluded from Eqs. (4.55).

Another normalization scheme consists of setting the value of the largest element of the modal vector $\{u\}_r$ equal to 1, which may be convenient for plotting the modes. Clearly, *the normalization process is devoid of physical significance and should be regarded as a mere convenience.*

In view of Eqs. (4.55) and (4.65), we conclude that Eq. (4.53) has the solutions

$$\{q(t)\}_r = \{u\}_r f_r(t) \qquad r = 1, 2, \ldots, n \tag{4.71}$$

where

$$f_r(t) = C_r \cos(\omega_r t - \phi_r) \qquad r = 1, 2, \ldots, n \tag{4.72}$$

in which $C_r$ and $\phi_r$ are constants of integration representing amplitudes and phase angles, respectively. Hence, the free vibration problem admits special independent solutions in which the system vibrates in any one of the natural modes. These solutions are referred to as *natural motions.* Because for a linear system the general solution is the sum of the individual solutions, we can write the general solution of Eq. (4.53) as a linear combination of the natural motions, or

$$\{q(t)\} = \sum_{r=1}^{n} \{q(t)\}_r = \sum_{r=1}^{n} \{u\}_r f_r(t) = [u]\{f(t)\} \tag{4.73}$$

where

$$[u] = [\{u\}_1 \quad \{u\}_2 \quad \ldots \quad \{u\}_n] \tag{4.74}$$

is the *modal matrix* and $\{f(t)\}$ is a vector whose components $f_r(t)$ are given by Eqs. (4.72). The constants $C_r$ and $\phi_r$ $(r = 1, 2, \ldots, n)$ entering into $\{f(t)\}$ depend on the initial conditions $\{q(0)\}$ and $\{\dot{q}(0)\}$. In Sec. 4.9, we obtain solution (4.73), together with the evaluation of the constants of integration, by a more formal approach, namely, by modal analysis.

It should be pointed out that motion characteristics as described above are typical of positive definite systems, i.e., system for which the mass and stiffness matrices are real, symmetric, and positive definite.

**Example 4.5** Derive the equations of motion for the two-degree-of-freedom system shown in Fig. 4.9, obtain the natural frequencies and natural modes and write the general solution to the free-vibration problem.

From Fig. 4.9, we can write the equations of motion

$$m_1\ddot{x}_1(t) + (k_1 + k_2)x_1(t) - k_2x_2(t) = 0$$
$$m_2\ddot{x}_2(t) - k_2x_1(t) + (k_2 + k_3)x_2(t) = 0$$

(a)

so that the mass and stiffness matrices have the form

$$[m] = \begin{bmatrix} m_1 & 0 \\ 0 & m_2 \end{bmatrix} = \begin{bmatrix} m & 0 \\ 0 & 2m \end{bmatrix}$$

$$[k] = \begin{bmatrix} k_1 + k_2 & -k_2 \\ -k_2 & k_2 + k_3 \end{bmatrix} = \begin{bmatrix} 2k & -k \\ -k & 2k \end{bmatrix}$$

(b)

Introducing matrices (b) into Eq. (4.67), we arrive at the characteristic equation

$$\Delta(\omega^2) = \begin{vmatrix} 2k - \omega^2 m & -k \\ -k & 2k - 2\omega^2 m \end{vmatrix} = 2m^2\omega^4 - 6km\omega^2 + 3k^2 = 0 \quad (c)$$

Letting $k/m = \Omega^2$, Eq. (c) reduces to

$$\left(\frac{\omega}{\Omega}\right)^4 - 3\left(\frac{\omega}{\Omega}\right)^2 + \frac{3}{2} = 0$$

(d)

which has the roots

$$\begin{matrix} \left(\dfrac{\omega_1}{\Omega}\right)^2 \\ \left(\dfrac{\omega_2}{\Omega}\right)^2 \end{matrix} = \frac{3}{2} \mp \left[\left(\frac{3}{2}\right)^2 - \frac{3}{2}\right]^{1/2} = \frac{3}{2}\left(1 \mp \frac{1}{\sqrt{3}}\right)$$

(e)

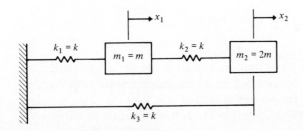

**Figure 4.9**

so that the natural frequencies are

$$\omega_1 = \left[\frac{3}{2}\left(1 - \frac{1}{\sqrt{3}}\right)\right]^{1/2} \Omega = 0.7962\sqrt{\frac{k}{m}}$$

$$\omega_2 = \left[\frac{3}{2}\left(1 + \frac{1}{\sqrt{3}}\right)\right]^{1/2} \Omega = 1.5382\sqrt{\frac{k}{m}}$$

$(f)$

To obtain the natural modes, we write Eq. (4.68) in the explicit form

$$(k_{11} - \omega_r^2 m_{11})u_{1r} + (k_{12} - \omega_r^2 m_{12})u_{2r} = 0$$
$$(k_{21} - \omega_r^2 m_{21})u_{1r} + (k_{22} - \omega_r^2 m_{22})u_{2r} = 0 \qquad r = 1, 2, \ldots, n \qquad (g)$$

which in our case reduce to

$$\left[2 - \left(\frac{\omega_r}{\Omega}\right)^2\right]u_{1r} - u_{2r} = 0$$

$$-u_{1r} + 2\left[1 - \left(\frac{\omega_r}{\Omega}\right)^2\right]u_{2r} = 0 \qquad r = 1, 2, \ldots, n \qquad (h)$$

Because the problem is homogeneous, we can only solve for one element of a given modal vector in terms of the other. To this end, it is sufficient to solve only one of Eqs. (h) for each value of r. Which equation is solved is immaterial, because both yield the same result. We choose to solve the first equation. Letting $r = 1$, and using the value of $(\omega_1/\Omega)^2$ from Eq. (e), we obtain

$$u_{21} = \left[2 - \left(\frac{\omega_1}{\Omega}\right)^2\right]u_{11} = \left[2 - \frac{3}{2}\left(1 - \frac{1}{\sqrt{3}}\right)\right]u_{11} = 1.3660u_{11} \qquad (i)$$

so that the first mode can be written in the form

$$\{u\}_1 = \begin{Bmatrix} 1.0000 \\ 1.3660 \end{Bmatrix} \qquad (j)$$

where we normalized the mode by setting $u_{11} = 1.0000$. In a similar fashion, we have

$$u_{22} = \left[2 - \left(\frac{\omega_2}{\Omega}\right)^2\right]u_{12} = \left[2 - \frac{3}{2}\left(1 + \frac{1}{\sqrt{3}}\right)\right]u_{12} = -0.3360u_{12} \qquad (k)$$

leading to the second mode

$$\{u\}_2 = \begin{Bmatrix} 1.0000 \\ -0.3660 \end{Bmatrix} \qquad (l)$$

where we set $u_{12} = 1.0000$. Note that the second mode has a sign change, so that at some point between masses $m_1$ and $m_2$ the displacement is zero. Such a point is called a *node*. The modes are plotted in Fig. 4.10.

According to Eq. (4.73), the solution of the free-vibration problem asso-

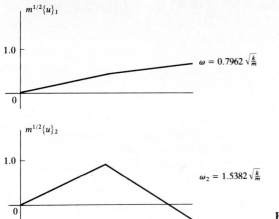

**Figure 4.10**

ciated with Eqs. (*a*) can be written in the form

$$\begin{Bmatrix} x_1(t) \\ x_2(t) \end{Bmatrix} = C_1 \begin{Bmatrix} 1.0000 \\ 1.3660 \end{Bmatrix} \cos\left(0.7962 \sqrt{\frac{k}{m}}\, t - \phi_1\right)$$

$$+ C_2 \begin{Bmatrix} 1.0000 \\ -0.3660 \end{Bmatrix} \cos\left(1.5382 \sqrt{\frac{k}{m}}\, t - \phi_2\right) \qquad (m)$$

where $C_1$, $C_2$, $\phi_1$, and $\phi_2$ are determined from the initial conditions $x_1(0)$, $x_2(0)$, $\dot{x}_1(0)$, and $\dot{x}_2(0)$, as shown in Sec. 4.9.

## 4.8 ORTHOGONALITY OF MODAL VECTORS. EXPANSION THEOREM

The natural modes possess a very important and useful property known as *orthogonality*. This is not an ordinary orthogonality, but an orthogonality with respect to the inertia matrix $[m]$ (and also with respect to the stiffness matrix $[k]$). Following is a proof of the orthogonality of the modal vectors $\{u\}_r$ ($r = 1, 2, \ldots, n$).

Let us consider two distinct solutions $\omega_r^2$, $\{u\}_r$, and $\omega_s^2$, $\{u\}_s$ of the eigenvalue problem (4.66). These solutions can be written in the form

$$[k]\{u\}_r = \omega_r^2[m]\{u\}_r \qquad (4.75)$$

$$[k]\{u\}_s = \omega_s^2[m]\{u\}_s \qquad (4.76)$$

Premultiplying both sides of (4.75) by $\{u\}_s^T$ and both sides of (4.76) by $\{u\}_r^T$, we obtain

$$\{u\}_s^T[k]\{u\}_r = \omega_r^2\{u\}_s^T[m]\{u\}_r \qquad (4.77)$$

$$\{u\}_r^T[k]\{u\}_s = \omega_s^2\{u\}_r^T[m]\{u\}_s \qquad (4.78)$$

Next, let us transpose Eq. (4.78), recall from Sec. 4.5 that matrices $[m]$ and $[k]$ are symmetric, and subtract the result from (4.77) to obtain

$$(\omega_r^2 - \omega_s^2)\{u\}_s^T[m]\{u\}_r = 0 \tag{4.79}$$

Because in general the natural frequencies are distinct, $\omega_r \neq \omega_s$, Eq. (4.79) is satisfied provided

$$\{u\}_s^T[m]\{u\}_r = 0 \qquad r \neq s \tag{4.80}$$

which is the statement of the *orthogonality condition* of the modal vectors. We note that the orthogonality is with respect to the inertia matrix $[m]$, which plays the role of a weighting matrix. Inserting Eq. (4.80) into (4.77), it is easy to see that the modal vectors are also orthogonal with respect to the stiffness matrix $[k]$,

$$\{u\}_s^T[k]\{u\}_r = 0 \qquad r \neq s \tag{4.81}$$

We stress again that the orthogonality relations (4.80) and (4.81) are valid only if $[m]$ and $[k]$ are symmetric. In many problems of practical interest the inertia matrix $[m]$ is diagonal, so that in these cases orthogonality condition (4.80) is simpler to use. Regardless of whether $[m]$ is diagonal or not, however, condition (4.80) is ordinarily used rather than condition (4.81).

If the modes are normalized, then they are called *orthonormal*, and if the normalization scheme is according to Eq. (4.69), the modes satisfy the relation

$$\{u\}_r^T[m]\{u\}_s = \delta_{rs} \qquad r, s = 1, 2, \ldots, n \tag{4.82}$$

where $\delta_{rs}$ is the Kronecker delta (see definition in Sec. 4.3).

The question remains as to the case in which $p$ natural frequencies are equal, where $p$ is an integer such that $2 \leqslant p \leqslant n$. In this case, the modal vectors associated with the repeated eigenvalue are orthogonal to the remaining $n - p$ vectors, but in general they may not be orthogonal to one another. Fortunately, when the eigenvalue problem is defined in terms of two real symmetric matrices, such as the matrices $[m]$ and $[k]$ in the case at hand, the modal vectors corresponding to the repeated eigenvalue are orthogonal to one another. Indeed, according to a theorem of linear algebra,† if an eigenvalue of a real symmetric matrix is repeated $p$ times, then the matrix has $p$ but not more than $p$ mutually orthogonal eigenvectors corresponding to the repeated eigenvalue. The eigenvectors are not uniquely determined because, for repeated eigenvalues, any linear combination of the associated eigenvectors is also an eigenvector. In general, however, it is possible to choose $p$ linear combinations of the eigenvectors corresponding to the repeated eigenvalue such that these combinations constitute mutually orthogonal eigenvectors, thus determining uniquely the eigenvectors in question. The above theorem is equally valid for the case in which the eigenvalue problem is defined in terms of two real symmetric matrices instead of one, if such a problem can be transformed into one in terms of a single real symmetric matrix by means of a linear transformation. The fact that the mass matrix $[m]$ is positive definite guarantees that a

† See D. C. Murdoch, *Linear Algebra*, sec. 6.5, John Wiley & Sons, Inc., New York, 1970.

transformation to a single real symmetric matrix is always possible.† Hence, *all the system eigenvectors are orthogonal*, regardless of whether the system possesses repeated eigenvalues or not. A system with repeated eigenvalues is referred to as *degenerate*.

The modal vectors can be conveniently arranged in a square matrix of order $n$, known as the *modal matrix* and having the form

$$[u] = [\{u\}_1 \quad \{u\}_2 \quad \cdots \quad \{u\}_n] \tag{4.83}$$

where $[u]$ is in fact the transformation matrix introduced in Sec. 4.6. In view of definition (4.83), all $n$ solutions of the eigenvalue problem, Eq. (4.68), can be written in the compact matrix equation

$$[k][u] = [m][u][\omega^2] \tag{4.84}$$

where $[\omega^2]$ is a diagonal matrix of the natural frequencies squared. The fact that the modal matrix $[u]$ can be used as the transformation matrix uncoupling the system differential equations of motion is due to the orthogonality property of the natural modes. If the modes are normalized so as to satisfy Eqs. (4.82), then we can write

$$[u]^T[m][u] = [1] \qquad [u]^T[k][u] = [\omega^2] \tag{4.85}$$

where $[1]$ is the unit matrix. Note that the second of Eqs. (4.85) follows directly from (4.84).

The eigenvectors $\{u\}_r$ $(r = 1, 2, \ldots, n)$ form a *linearly independent set*, implying that any $n$-dimensional vector can be constructed as a linear combination of these eigenvectors. Physically this implies that any motion of the system can be regarded at any given time as a superposition of the natural modes multiplied by appropriate constants, where the constants are a measure of the degree of participation of each mode in the motion. The normal mode representation of the motion permits the transformation of a simultaneous set of differential equations of motion into an independent set, where the transformation matrix is the modal matrix $[u]$.

To prove that the set of vectors $\{u\}_r$ is linearly independent, we assume that the vectors are linearly dependent and arrive at a contradiction. For the vectors $\{u\}_r$ to be linearly dependent they must satisfy an equation of the type

$$c_1\{u\}_1 + c_2\{u\}_2 + \cdots + c_n\{u\} = \sum_{r=1}^{n} c_r\{u\}_r = \{0\} \tag{4.86}$$

where $c_r$ $(r = 1, 2, \ldots, n)$ are nonzero constants. Premultiplying Eq. (4.86) by $\{u\}_s^T[m]$, we obtain

$$\sum_{r=1}^{n} c_r\{u\}_s^T[m]\{u\}_r = 0 \tag{4.87}$$

But the triple matrix product $\{u\}_s^T[m]\{u\}_r$ is equal to zero for $r \neq s$ and is different from zero for $r = s$. It follows that Eq. (4.87) can be satisfied only if $c_s = 0$.

† See L. Meirovitch, *Computational Methods in Structural Dynamics*, sec. 3.3, Sijthoff & Noordhoff International Publishers, The Netherlands, 1980.

Repeating the operation $n$ times, for $s = 1, 2, \ldots, n$, we conclude that Eq. (4.87) can be satisfied only in the *trivial case* defined by $c_1 = c_2 = \cdots = c_n = 0$. Hence, the eigenvectors $\{u\}_r$ cannot satisfy any equation of the type (4.86), with the obvious conclusion that the system modal vectors are linearly independent.

Because the modal vectors $\{u\}_r$ cannot satisfy any equation of the type (4.86), we must have

$$\{u\} = c_1\{u\}_1 + c_2\{u\}_2 + \cdots + c_n\{u\}_n \neq \{0\} \qquad (4.88)$$

where $\{u\}$ is called a *linear combination* of $\{u\}_1, \{u\}_2, \ldots, \{u\}_n$, with coefficients $c_1, c_2, \ldots, c_n$ (see App. C, Sec. C.3). The totality of linear combinations obtained by letting the coefficients $c_1, c_2, \ldots, c_n$ vary forms the *vector space* $\{u\}$, which is said to be *spanned* by $\{u\}_1, \{u\}_2, \ldots, \{u\}_n$. The set of vectors $\{u\}_r, (r = 1, 2, \ldots, n)$ is called a *generating system* of $\{u\}$ and, because the vectors are independent, the generating system is said to be a *basis* of $\{u\}$. Hence, any vector belonging to the space $\{u\}$ can be generated in the form of the linear combination (4.88). Physically this implies that *any possible motion of the system can be described as a linear combination of the modal vectors*. Considering Eq. (4.88) and the orthogonality condition in the form (4.82), the coefficients $c_r$ can be obtained by writing simply

$$c_r = \{u\}_r^T[m]\{u\} \qquad r = 1, 2, \ldots, n \qquad (4.89)$$

where the coefficients $c_r$ are a measure of the contribution of the associated modes $\{u\}_r$ to the motion $\{u\}$. Equations (4.88) and (4.89) are known in vibrations under the name of the *expansion theorem*. The derivation of the response of a system by modal analysis is based on the expansion theorem.

The natural frequencies $\omega_r$ and associated natural modes $\{u\}_r, (r = 1, 2, \ldots, n)$ are paired together and represent a unique characteristic of the system. Their values depend solely on the matrices $[m]$ and $[k]$. Every one of the pairs $\omega_r, \{u\}_r$ can be excited independently of any other pair $\omega_s, \{u\}_s, r \neq s$. For example, if the system is excited by a harmonic forcing function with frequency $\omega_r$, then the system configuration will resemble the natural mode $\{u\}_r$. Of course, this represents a resonance condition, and the motion will tend to increase without bounds until the small-motions assumption is violated. On the other hand, if the system is imparted an initial excitation resembling the natural mode $\{u\}_r$, then the ensuing motion will be synchronous harmonic oscillation with the natural frequency $\omega_r$. We shall devote ample time to the relation between the system response and the normal modes.

## 4.9 RESPONSE OF SYSTEMS TO INITIAL EXCITATION. MODAL ANALYSIS

Let us consider once again the free vibration of an undamped system. From Sec. 4.7, we can write the equations of motion in the matrix form

$$[m]\{\ddot{q}(t)\} + [k]\{q(t)\} = \{0\} \qquad (4.90)$$

where $\{q(t)\}$ is the vector of the generalized coordinates $q_i(t)$ $(i = 1, 2, ..., n)$. We seek now a formal solution of Eq. (4.90).

At some arbitrary time $t = t_1$ the solution of Eq. (4.90) is $\{q(t_1)\}$. But by the expansion theorem, Eq. (4.88), the solution $\{q(t_1)\}$ can be regarded as a superposition of the normal modes. Denoting the coefficients $c_r$ for this particular configuration by $\eta_r(t_1)$ $(r = 1, 2, ..., n)$, we can write

$$\{q(t_1)\} = \eta_1(t_1)\{u\}_1 + \eta_2(t_1)\{u\}_2 + \cdots + \eta_n(t_1)\{u\}_n \tag{4.91}$$

where, according to Eq. (4.89), the coefficients have the values

$$\eta_r(t_1) = \{u\}_r^T[m]\{q(t_1)\} \qquad r = 1, 2, ..., n \tag{4.92}$$

But $t_1$ is arbitrary, so that its value can be changed at will. Because Eq. (4.92) must hold for all values of time, we can replace $t_1$ by $t$, and write in general

$$\eta_r(t) = \{u\}_r^T[m]\{q(t)\} \qquad r = 1, 2, ..., n \tag{4.93}$$

where the coefficients $\eta_r(t)$ can be regarded as linear combinations of the generalized coordinates $q_i(t)$, and hence as functions of time. In view of this, a formal solution of Eq. (4.90) can be written in the form

$$\{q(t)\} = \eta_1(t)\{u\}_1 + \eta_2(t)\{u\}_2 + \cdots + \eta_n(t)\{u\}_n$$

$$= \sum_{r=1}^{n} \eta_r(t)\{u\}_r = [u]\{\eta(t)\} \tag{4.94}$$

where $[u]$ is recognized as the modal matrix and $\{\eta(t)\}$ is the vector of the functions $\eta_r(t)$ $(r = 1, 2, ..., n)$. Equation (4.94) can be regarded as a linear transformation relating the vectors $\{q(t)\}$ and $\{\eta(t)\}$, where the transformation matrix $[u]$ is constant. It follows immediately from Eq. (4.95) that

$$\{\ddot{q}(t)\} = [u]\{\ddot{\eta}(t)\} \tag{4.95}$$

so that, inserting Eqs. (4.94) and (4.95) into Eq. (4.90), premultiplying the result by $[u]^T$, and considering Eqs. (4.85), we arrive at the independent set of equations

$$\ddot{\eta}_r(t) + \omega_r^2\eta_r(t) = 0 \qquad r = 1, 2, ..., n \tag{4.96}$$

where the variables $\eta_r(t)$ are identified as the *normal coordinates* of the system. By analogy with the free-vibration solution of an undamped single-degree-of-freedom system, Eq. (1.39), the solution of (4.96) is simply

$$\eta_r(t) = C_r \cos(\omega_r t - \phi_r) \qquad r = 1, 2, ..., n \tag{4.97}$$

where $C_r$ and $\phi_r$ $(r = 1, 2, ..., n)$ are constants of integration representing the amplitudes and phase angles of the normal coordinates. Inserting Eqs. (4.97) back into transformation (4.94), we obtain

$$\{q(t)\} = [u]\{\eta(t)\} = \sum_{r=1}^{n} \eta_r(t)\{u\}_r = \sum_{r=1}^{n} C_r\{u\}_r \cos(\omega_r t - \phi_r) \tag{4.98}$$

so that the free vibration of a multi-degree-of-freedom system consists of a superposition of $n$ modal vectors multiplied by harmonic functions with frequencies equal to the system natural frequencies and with amplitudes and phase angles depending on the initial conditions.

Letting $\{q(0)\}$ and $\{\dot{q}(0)\}$ be the initial displacement and velocity vectors, respectively, Eq. (4.98) leads to

$$\{q(0)\} = \sum_{r=1}^{n} C_r\{u\}_r \cos \phi_r$$

$$\{\dot{q}(0)\} = \sum_{r=1}^{n} C_r \omega_r \{u\}_r \sin \phi_r$$

(4.99)

Premultiplying Eqs. (4.99) by $\{u\}_s^T[m]$, and considering the orthonormality relations, Eqs. (4.82), we can write

$$C_r \cos \phi_r = \{u\}_r^T[m]\{q(0)\}$$

$$\qquad r = 1, 2, \ldots, n \qquad (4.100)$$

$$C_r \sin \phi_r = \frac{1}{\omega_r}\{u\}_r^T[m]\{\dot{q}(0)\}$$

so that, introducing Eqs. (4.100) into (4.98), we obtain the general expression

$$\{q(t)\} = \sum_{r=1}^{n} (\{u\}_r^T[m]\{q(0)\} \cos \omega_r t + \{u\}_r^T[m]\{\dot{q}(0)\} \frac{1}{\omega_r} \sin \omega_r t)\{u\}_r$$

(4.101)

which represents the response of the system to the initial displacement vector $\{q(0)\}$ and the initial velocity vector $\{\dot{q}(0)\}$.

Next, let us assume that the initial displacement vector resembles a given normal mode, say $\{u\}_s$, whereas the initial velocity vector is zero. Introducing $\{q(0)\} = q_0\{u\}_s$ and $\{\dot{q}(0)\} = \{0\}$ into Eq. (4.101), and considering Eqs. (4.82), the response is simply

$$\{q(t)\} = \sum_{r=1}^{n} (q_0\{u\}_r^T[m]\{u\}_s \cos \omega_r t)\{u\}_r$$

$$= \sum_{r=1}^{n} q_0 \delta_{rs}\{u\}_r \cos \omega_r t = q_0\{u\}_s \cos \omega_s t \qquad (4.102)$$

which represents synchronous harmonic oscillation at the natural frequency $\omega_s$ with the system configuration resembling the $s$th mode at all times, thus justifying the statement made at the end of Sec. 4.8 that the natural modes can be excited independently of one another.

**Example 4.6** Consider the system of Example 4.5 and verify that the natural modes are orthogonal. Then obtain the response to the initial conditions $\dot{x}_1(0) = v_0$, $x_1(0) = x_2(0) = \dot{x}_2(0) = 0$.

Inserting the modal vectors $\{u\}_1$ and $\{u\}_2$, Eqs. ($j$) and ($l$) of Example 4.5, into Eq. (4.82), we obtain

$$\{u\}_1^T[m]\{u\}_2 = \begin{Bmatrix} 1.0000 \\ 1.3660 \end{Bmatrix}^T \begin{bmatrix} m & 0 \\ 0 & 2m \end{bmatrix} \begin{Bmatrix} 1.0000 \\ -0.3660 \end{Bmatrix}$$

$$= m(1.0000 - 2 \times 1.3660 \times 0.3660) = 0 \qquad (a)$$

so that the modes are verified as being orthogonal with respect to the mass matrix.

The general response of a multi-degree-of-freedom system to initial excitation is given by Eq. (4.101). Of course, we must change the notation from $\{q(t)\}$, $\{q(0)\}$, and $\{\dot{q}(0)\}$ to $\{x(t)\}$, $\{x(0)\}$, and $\{\dot{x}(0)\}$, respectively. Because in our case $\{x(0)\} = \{0\}$, the response becomes

$$\{x(t)\} = \sum_{r=1}^{2} \left( \{u\}_r^T[m]\{\dot{x}(0)\} \frac{1}{\omega_r} \sin \omega_r t \right) \{u\}_r \qquad (b)$$

where

$$\{\dot{x}(0)\} = \begin{Bmatrix} v_0 \\ 0 \end{Bmatrix} \qquad (c)$$

Before using Eq. (b), however, we recall that the modal vectors must be normalized according to Eq. (4.69). Hence, let us assume that the normalized modal vectors have the form

$$\{u\}_1 = \alpha_1 \begin{Bmatrix} 1.0000 \\ 1.3660 \end{Bmatrix} \qquad \{u\}_2 = \alpha_2 \begin{Bmatrix} 1.0000 \\ -0.3660 \end{Bmatrix} \qquad (d)$$

where the constants $\alpha_1$ and $\alpha_2$ are evaluated by using Eq. (4.69). Indeed, we can write

$$\{u\}_1^T[m]\{u\}_1 = \alpha_1^2 \begin{Bmatrix} 1.0000 \\ 1.3660 \end{Bmatrix}^T \begin{bmatrix} m & 0 \\ 0 & 2m \end{bmatrix} \begin{Bmatrix} 1.0000 \\ 1.3660 \end{Bmatrix} = 4.7320m\alpha_1^2 = 1$$

$$\{u\}_2^T[m]\{u\}_2 = \alpha_2^2 \begin{Bmatrix} 1.0000 \\ -0.3660 \end{Bmatrix}^T \begin{bmatrix} m & 0 \\ 0 & 2m \end{bmatrix} \begin{Bmatrix} 1.0000 \\ -0.3660 \end{Bmatrix} = 1.2679m\alpha_2^2 = 1$$

$$(e)$$

yielding the constants

$$\alpha_1 = \frac{0.4597}{\sqrt{m}} \qquad \alpha_2 = \frac{0.8881}{\sqrt{m}} \qquad (f)$$

Hence, inserting the above values into (d), we obtain the normal modes

$$\{u\}_1 = \frac{1}{\sqrt{m}} \begin{Bmatrix} 0.4597 \\ 0.6280 \end{Bmatrix} \qquad \{u\}_2 = \frac{1}{\sqrt{m}} \begin{Bmatrix} 0.8881 \\ -0.3251 \end{Bmatrix} \qquad (g)$$

Next, let us recall from Example 4.5 that the system natural frequencies are

$$\omega_1 = 0.7962 \sqrt{\frac{k}{m}} \qquad \omega_2 = 1.5382 \sqrt{\frac{k}{m}} \qquad (h)$$

and form

$$\frac{1}{\omega_1} \{u\}_1^T [m] \{\dot{x}(0)\} = \frac{1}{0.7962\sqrt{k/m}} \frac{1}{\sqrt{m}} \begin{Bmatrix} 0.4597 \\ 0.6280 \end{Bmatrix}^T \begin{bmatrix} m & 0 \\ 0 & 2m \end{bmatrix} \begin{Bmatrix} v_0 \\ 0 \end{Bmatrix}$$

$$= 0.5774 \frac{mv_0}{\sqrt{k}}$$

$$\frac{1}{\omega_2} \{u\}_2^T [m] \{\dot{x}(0)\} = \frac{1}{1.5382\sqrt{k/m}} \frac{1}{\sqrt{m}} \begin{Bmatrix} 0.8880 \\ -0.3251 \end{Bmatrix}^T \begin{bmatrix} m & 0 \\ 0 & 2m \end{bmatrix} \begin{Bmatrix} v_0 \\ 0 \end{Bmatrix} \qquad (i)$$

$$= 0.5774 \frac{mv_0}{\sqrt{k}}$$

so that, introducing Eqs. (g) through (i) into (b), we obtain the response

$$\{x(t)\} = \left(0.5774 \frac{mv_0}{\sqrt{k}} \sin 0.7962 \sqrt{\frac{k}{m}} t\right) \frac{1}{\sqrt{m}} \begin{Bmatrix} 0.4597 \\ 0.6277 \end{Bmatrix}$$

$$+ \left(0.5774 \frac{mv_0}{\sqrt{k}} \sin 1.5382 \sqrt{\frac{k}{m}} t\right) \frac{1}{\sqrt{m}} \begin{Bmatrix} 0.8881 \\ -0.3251 \end{Bmatrix}$$

$$= v_0 \sqrt{\frac{m}{k}} \begin{Bmatrix} 0.2654 \\ 0.3626 \end{Bmatrix} \sin 0.7962 \sqrt{\frac{k}{m}} t$$

$$+ v_0 \sqrt{\frac{m}{k}} \begin{Bmatrix} 0.5127 \\ -0.1877 \end{Bmatrix} \sin 1.5382 \sqrt{\frac{k}{m}} t \qquad (j)$$

and note that the elements of $\{x(t)\}$ have units of length, as should be expected.

As a matter of interest, let us calculate the velocity vector $\{\dot{x}(t)\}$. Differentiating Eq. (j) with respect to time, we have simply

$$\{\dot{x}(t)\} = v_0 \begin{Bmatrix} 0.2113 \\ 0.2887 \end{Bmatrix} \cos 0.7962 \sqrt{\frac{k}{m}} t + v_0 \begin{Bmatrix} 0.7887 \\ -0.2887 \end{Bmatrix} \cos 1.5382 \sqrt{\frac{k}{m}} t \quad (k)$$

Letting $t = 0$ in Eq. (k), we obtain $\{\dot{x}(0)\} = \{v_0 \ \ 0\}^T$, thus verifying the validity of the solution.

## 4.10 SOLUTION OF THE EIGENVALUE PROBLEM BY THE CHARACTERISTIC DETERMINANT

In Sec. 4.7, we showed that the eigenvalue problem has a solution provided the parameter $\omega^2$ satisfies the $n$th-degree algebraic equation (4.67), known as the characteristic equation. In this section, we expand on the subject.

Let us write the eigenvalue problem (4.66) in the form

$$[m]\{u\} - \lambda[k]\{u\} = \{0\} \qquad \lambda = \frac{1}{\omega^2} \qquad (4.103)$$

Premultiplying Eq. (4.103) through by $[k]^{-1} = [a]$, where $[a]$ is the flexibility matrix, the eigenvalue problem becomes

$$([D] - \lambda[1])\{u\} = \{0\} \qquad (4.104)$$

where

$$[D] = [k]^{-1}[m] = [a][m] \qquad (4.105)$$

is known as the *dynamical matrix*. Note that, in general, $[D]$ is *not* symmetric. In view of this, the characteristic equation can be written as

$$\Delta(\lambda) = \det([D] - \lambda[1]) = |([D] - \lambda[1])| = 0 \qquad (4.106)$$

where $\Delta(\lambda)$ is a polynomial of degree $n$ in $\lambda$. In general, Eq. (4.106) possesses $n$ distinct real and positive roots $\lambda_r$, related to the system natural frequencies by $\lambda_r = 1/\omega_r^2$ $(r = 1, 2, ..., n)$. Note that the value of $\lambda$ in this section corresponds to the reciprocal of $\lambda$ defined in Sec. 4.7.

If $\{u\}_r$ represents the eigenvector corresponding to the eigenvalue $\lambda_r$, then the $n$ solutions of the eigenvalue problem (4.104) can be written as follows:

$$([D] - \lambda_r[1])\{u\}_r = \{0\} \qquad r = 1, 2, ..., n \qquad (4.107)$$

For a given eigenvalue $\lambda_r$, Eq. (4.107) represents $n$ homogeneous algebraic equations in the unknowns $u_{ir}$ $(i = 1, 2, ..., n)$, so that the values of $u_{ir}$ can be obtained only within a constant scalar multiplier. If one of the unknowns is assigned an arbitrary value, such as unity, then any $n - 1$ of the equations can be regarded as constituting a nonhomogeneous set and solved for the remaining $n - 1$ unknowns by any method for the solution of algebraic equations, such as Gaussian elimination in conjunction with back substitution.[†]

The question remains as to how to obtain the eigenvalues $\lambda_r$ $(r = 1, 2, ..., n)$. When the number of degrees of freedom is three or higher, it is advisable to obtain the eigenvalues by a computational algorithm, such as the $QR$ method or the one based on Sturm's theorem. Both algorithms can be found in the reference cited above.[‡]

**Example 4.7** Consider the three-degree-of-freedom system of Example 4.2 and obtain the solution of the eigenvalue problem by the method employing the characteristic determinant. Let $m_1 = m_2 = m$, $m_3 = 2m$, $k_1 = k_2 = k$, and $k_3 = 2k$.

The inertia matrix of the system is simply

$$[m] = \begin{bmatrix} m_1 & 0 & 0 \\ 0 & m_2 & 0 \\ 0 & 0 & m_3 \end{bmatrix} = m \begin{bmatrix} 1 & 0 & 0 \\ 0 & 1 & 0 \\ 0 & 0 & 2 \end{bmatrix} \qquad (a)$$

† L. Meirovitch, *Computational Methods in Structural Dynamics*, sec. 5.2, Sijthoff & Noordhoff International Publishers, The Netherlands, 1980.

‡ L. Meirovitch, op. cit., secs. 5.12 and 5.13.

whereas from Example 4.2 we obtain the flexibility matrix

$$
[a] = \begin{bmatrix} \dfrac{1}{k_1} & \dfrac{1}{k_1} & \dfrac{1}{k_1} \\ \dfrac{1}{k_1} & \dfrac{1}{k_1}+\dfrac{1}{k_2} & \dfrac{1}{k_1}+\dfrac{1}{k_2} \\ \dfrac{1}{k} & \dfrac{1}{k_1}+\dfrac{1}{k_2} & \dfrac{1}{k_1}+\dfrac{1}{k_2}+\dfrac{1}{k_3} \end{bmatrix} = \dfrac{1}{k}\begin{bmatrix} 1 & 1 & 1 \\ 1 & 2 & 2 \\ 1 & 2 & 2.5 \end{bmatrix}
\tag{b}
$$

In view of definition (4.105), the dynamical matrix is

$$
[D] = [a][m] = \dfrac{m}{k}\begin{bmatrix} 1 & 1 & 1 \\ 1 & 2 & 2 \\ 1 & 2 & 2.5 \end{bmatrix}\begin{bmatrix} 1 & 0 & 0 \\ 0 & 1 & 0 \\ 0 & 0 & 2 \end{bmatrix} = \dfrac{m}{k}\begin{bmatrix} 1 & 1 & 2 \\ 1 & 2 & 4 \\ 1 & 2 & 5 \end{bmatrix}
\tag{c}
$$

The eigenvalue problem can be written in the form

$$
\dfrac{m}{k}\begin{bmatrix} 1 & 1 & 2 \\ 1 & 2 & 4 \\ 1 & 2 & 5 \end{bmatrix}\begin{Bmatrix} u_1 \\ u_2 \\ u_3 \end{Bmatrix} = \dfrac{1}{\omega^2}\begin{Bmatrix} u_1 \\ u_2 \\ u_3 \end{Bmatrix}
\tag{d}
$$

and, introducing the notation

$$
\lambda = \dfrac{k}{m}\dfrac{1}{\omega^2}
\tag{e}
$$

we obtain the characteristic equation

$$
\Delta(\lambda) = \begin{vmatrix} 1-\lambda & 1 & 2 \\ 1 & 2-\lambda & 4 \\ 1 & 2 & 5-\lambda \end{vmatrix} = -(\lambda^3 - 8\lambda^2 + 6\lambda - 1) = 0
\tag{f}
$$

which has the solutions

$$
\lambda_1 = 7.1842 \qquad \lambda_2 = 0.5728 \qquad \lambda_3 = 0.2430
\tag{g}
$$

Inserting the above values of $\lambda_1$, $\lambda_2$, and $\lambda_3$ into $[D] - \lambda_r[1]$ $(r = 1, 2, 3)$, we can write the matrices

$$
[D] - \lambda_1[1] = \begin{bmatrix} 1-\lambda_1 & 1 & 2 \\ 1 & 2-\lambda_1 & 4 \\ 1 & 2 & 5-\lambda_1 \end{bmatrix}
$$

$$
= \begin{bmatrix} -6.1842 & 1 & 2 \\ 1 & -5.1842 & 4 \\ 1 & 2 & -2.1842 \end{bmatrix}
$$

$$
[D] - \lambda_2[1] = \begin{bmatrix} 1-\lambda_2 & 1 & 2 \\ 1 & 2-\lambda_2 & 4 \\ 1 & 2 & 5-\lambda_2 \end{bmatrix}
$$

$$
= \begin{bmatrix} 0.4272 & 1 & 2 \\ 1 & 1.4272 & 4 \\ 1 & 2 & 4.4272 \end{bmatrix}
$$

$$
\tag{h}
$$

$$[D] - \lambda_3[1] = \begin{bmatrix} 1 - \lambda_3 & 1 & 2 \\ 1 & 2 - \lambda_3 & 4 \\ 1 & 2 & 5 - \lambda_3 \end{bmatrix}$$

$$= \begin{bmatrix} 0.7570 & 1 & 2 \\ 1 & 1.7570 & 4 \\ 1 & 2 & 4.7570 \end{bmatrix}$$

Retaining the first two equations from the set described by Eq. (4.107), we can write for the first mode

$$-6.1842u_{11} + u_{21} + 2u_{31} = 0$$
$$u_{11} - 5.1842u_{21} + 4u_{31} = 0$$

$$(i)$$

Letting $u_{11} = 1.0000$ arbitrarily, the solution of Eqs. (i) is

$$u_{11} = 1.0000 \qquad u_{21} = 1.8608 \qquad u_{31} = 2.1617 \qquad (j)$$

Similarly, the equations for the second mode are

$$0.4272u_{12} + u_{22} = 2u_{32} = 0$$
$$u_{12} + 1.4272u_{22} + 4u_{32} = 0$$

$$(k)$$

having the solution

$$u_{12} = 1.0000 \qquad u_{22} = 0.2542 \qquad u_{32} = -0.3407 \qquad (l)$$

Finally, the equations for the third mode are

$$0.7570u_{13} + u_{23} + 2u_{33} = 0$$
$$u_{12} + 1.7570u_{23} + 4u_{33} = 0$$

$$(m)$$

so that

$$u_{13} = 1.0000 \qquad u_{23} = -2.1152 \qquad u_{33} = 0.6791 \qquad (n)$$

Using the normalization scheme (4.69), we obtain the normal modes

$$\{u\}_1 = m^{-1/2} \begin{Bmatrix} 0.2691 \\ 0.5008 \\ 0.5817 \end{Bmatrix}$$

$$\{u\}_2 = m^{-1/2} \begin{Bmatrix} 0.8781 \\ 0.2232 \\ -0.2992 \end{Bmatrix} \qquad (o)$$

$$\{u\}_3 = m^{-1/2} \begin{Bmatrix} 0.3954 \\ -0.8363 \\ 0.2685 \end{Bmatrix}$$

It is typical of the normal modes that $\{u\}_1$ should exhibit no sign change, $\{u\}_2$ should exhibit one sign change, and $\{u\}_3$ should exhibit two sign changes.

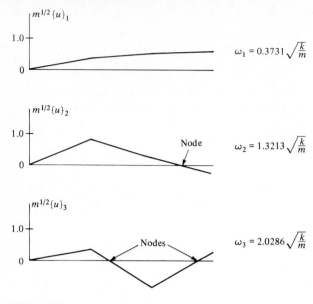

**Figure 4.11**

Correspondingly, the modes have no nodes, one node, and two nodes, where a node is defined as a point of zero displacement. Using $(e)$, the associated natural frequencies are

$$\omega_1 = \sqrt{\frac{k}{m\lambda_1}} = 0.3731 \sqrt{\frac{k}{m}}$$

$$\omega_2 = \sqrt{\frac{k}{m\lambda_2}} = 1.3213 \sqrt{\frac{k}{m}} \qquad (p)$$

$$\omega_3 = \sqrt{\frac{k}{m\lambda_3}} = 2.0286 \sqrt{\frac{k}{m}}$$

The modes are plotted in Fig. 4.11.

## 4.11 SOLUTION OF THE EIGENVALUE PROBLEM BY MATRIX ITERATION. POWER METHOD USING MATRIX DEFLATION

There are various matrix iteration schemes for the solution of the eigenvalue problem, such as the Jacobi method, the $QR$ method, and the method based on Sturm's theorem. The Jacobi method yields all the eigenvalues and eigenvectors simultaneously. On the other hand, the $QR$ method and the method based on Sturm's theorem yield only the eigenvalues, so that the eigenvectors must be computed separately. To this end, the method based on the characteristic determinant discussed in Sec. 4.10 comes immediately to mind. A more efficient technique for computing the eigenvectors corresponding to known eigenvalues is inverse iteration. All these techniques lie beyond the scope of this text and can be found in another text by Meirovitch.† In this text, we present perhaps the simplest of the matrix iteration schemes, namely, the *power method using matrix deflation*.

The power method is based on the expansion theorem (see Sec. 4.8). The implication of the theorem is that the solution of the eigenvalue problem (4.104) consists of $n$ linearly independent eigenvectors $\{u\}_r$ $(r = 1, 2, ..., n)$ of $[D]$, where $[D]$ is the *dynamical matrix* given by Eq. (4.105). The expansion theorem implies further that these eigenvectors span the $n$-dimensional vector space $\{u\}$, where $\{u\}$ represents a possible motion of the system. Hence, any such vector $\{u\}$ can be expressed as a linear combination of the eigenvectors $\{u\}_r$, where $\{u\}_r$ satisfy the equations

$$[D]\{u\}_r = \lambda_r\{u\}_r \qquad \lambda_r = \frac{1}{\omega_r^2} \qquad r = 1, 2, ..., n \qquad (4.108)$$

Solutions (4.108) can be given an interesting interpretation in terms of linear transformations (see Sec. C.3). Specifically, matrix $[D]$ can be regarded as representing a linear transformation that transforms any eigenvector $\{u\}_r$ into itself, within the constant scalar multiplier $\lambda_r = 1/\omega_r^2$ $(r = 1, 2, ..., n)$. On the other hand, if an arbitrary vector $\{v\}_1$, other than an eigenvector, is premultiplied by $[D]$, then the vector will not duplicate itself but will be transformed into another vector $\{v\}_2$, generally different from $\{v\}_1$. However, by the expansion theorem, Eq. (4.88), we can write

$$\{v\}_1 = c_1\{u\}_1 + c_2\{u\}_2 + \cdots + c_n\{u\}_n = \sum_{r=1}^{n} c_r\{u\}_r \qquad (4.109)$$

where $c_r$ are constant coefficients depending on the basis $\{u\}_1, \{u\}_2, ..., \{u\}_n$ and on the vector $\{v\}_1$. Next let us premultiply $\{v\}_1$ by $[D]$, consider Eqs. (4.108), and obtain the vector $\{v\}_2$ in the form

$$\{v\}_2 = [D]\{v\}_1 = \sum_{r=1}^{n} c_r[D]\{u\}_r = \lambda_1 \sum_{r=1}^{n} c_r\frac{\lambda_r}{\lambda_1}\{u\}_r \qquad (4.110)$$

In contrast to $\{v\}_1$, in which the eigenvectors $\{u\}_r$ are multiplied by the constants $c_r$, the eigenvectors $\{u\}_r$ in the vector $\{v\}_2$ are multiplied by $c_r\lambda_r/\lambda_1$, where the constant multiplier $\lambda_1$ in front of the series is immaterial and can be ignored because the problem is homogeneous. But the eigenvalues $\lambda_r$ are such that $\lambda_1 \geqslant \lambda_2 \geqslant \cdots \geqslant \lambda_n$. Moreover, we confine ourselves to the case in which all the eigenvalues are distinct, $\lambda_1 > \lambda_2 > \cdots > \lambda_n$. Because $\lambda_r/\lambda_1 < 1$ $(r = 2, 3, ..., n)$, and the ratios decrease with increasing $r$, the participation of the higher modes in $\{v\}_2$ tends to decrease, as opposed to their participation in $\{v\}_1$. Hence if $\{v\}_1$ is regarded as a trial vector toward obtaining the modal vector $\{u\}_1$, then $\{v\}_2$ must be regarded as an

---

† Meirovitch, op. cit., chap. 5.

improved trial vector. Of course, the procedure can be repeated with $\{v\}_2$ as a new trial vector, so that if $\{v\}_2$ divided by $\lambda_1$ is premultiplied by $[D]$, we obtain

$$\{v\}_3 = \frac{1}{\lambda_1} [D]\{v\}_2 = \frac{1}{\lambda_1} [D]^2\{v\}_1$$

$$= \sum_{r=1}^{n} c_r \frac{\lambda_r}{\lambda_1} [D]\{u\}_1 = \lambda_1 \sum_{r=1}^{n} c_r \left(\frac{\lambda_r}{\lambda_1}\right)^2 \{u\}_r \qquad (4.111)$$

and it is clear that $\{v\}_3$ is an even better trial vector for $\{u\}_1$ than $\{v\}_2$, so that by premultiplying the newly obtained vectors repeatedly by $[D]$ we are establishing an iteration procedure converging to the first eigenvalue and eigenvector. Hence, in general we have

$$\{v\}_p = \frac{1}{\lambda_1} [D]\{v\}_{p-1} = \cdots = \frac{1}{\lambda_1^{p-2}} [D]^{p-1}\{v\}_1$$

$$= \lambda_1 \sum_{r=1}^{n} c_r \left(\frac{\lambda_r}{\lambda_1}\right)^{p-1} \{u\}_1 \qquad (4.112)$$

so that for a sufficiently large integer $p$ the first term in the series (4.112) becomes the dominant one, for which reason $\lambda_1$ is sometimes referred to as the *dominant eigenvalue*. It follows that

$$\lim_{p \to \infty} \frac{1}{\lambda_1} \{v\}_p = \lim_{p \to \infty} \frac{1}{\lambda_1^{p-1}} [D]^{p-1}\{v\}_1 = c_1\{u\}_1 \qquad (4.113)$$

Moreover, when convergence is achieved the vectors $\{v\}_{p-1}$ and $\{v\}_p$ satisfy Eq. (4.108) because they can both be regarded as $\{u\}_1$. Hence, at this point $\{v\}_{p-1}$ and $\{v\}_p$ are proportional to one another, the constant of proportionality being $\lambda_1 = 1/\omega_1^2$, so that the lowest natural frequency can be obtained from

$$\lim_{p \to \infty} \frac{v_{i,p-1}}{v_{i,p}} = \omega_1^2 \qquad (4.114)$$

where $v_{i,p-1}$ and $v_{i,p}$ are the elements in the $i$th row of the vectors $\{v\}_{p-1}$ and $\{v\}_p$, respectively. Although we let $p$ approach infinity in Eqs. (4.113) and (4.114), in practice only a finite number of iterations will suffice to reach a desired level of accuracy.

It appears from the above that the rate of convergence of the iteration process depends on how fast the ratios $(\lambda_r/\lambda_1)^{p-1}$ $(r = 2, 3, \ldots, n)$ go to zero. There are two factors affecting the number of iteration steps necessary to achieve satisfactory accuracy. The first factor depends on the system itself, and in particular on how much larger $\lambda_1$ is than $\lambda_2$, because the effect of one iteration step is to multiply the second term in the series for the trial vectors by $\lambda_2/\lambda_1$. Clearly, the larger $\lambda_1$ is compared to $\lambda_2$, the faster the separation of the eigenvectors $\{u\}_1$ and $\{u\}_2$ is, with the implication that the number of steps necessary for convergence is relatively small. The second factor depends on the skill and experience of the analyst, because the closer the first trial vector resembles the first modal vector $\{u\}_1$, the faster the convergence tends to be. In general, for a given system, there are certain clues as to the selection of the first trial vector based on physical considerations. Specifically, from the nature of the system, it is possible to make a rough guess of the displacement pattern in the first mode. Quite often, however, the convergence acceleration is not sufficiently significant to justify the effort in trying to guess the first mode.

This iteration scheme has a major advantage in that it is "errorproof" in the sense that if an error is made in one of the iteration steps, this only sets back the iteration process but does not affect the final result. The error amounts to beginning a new iteration sequence with a new trial vector, which is likely to delay convergence but definitely not destroy it. In general, convergence is achieved regardless of how poor the first trial vector is. Clearly, the iteration process is errorproof only if the matrix $[D]$ is correct. The iteration leads to the first mode, with the only exception being the case in which the trial vector coincides exactly with one of the higher-ordered modes $\{u\}_s$ ($s = 2, 3, \ldots, n$), that is, the case in which the coefficients in the series (4.109) are such that $c_r = c_r \delta_{rs}$ ($r = 1, 2, \ldots, n$), where $\delta_{rs}$ is the Kronecker delta. In this case, a premultiplication of $\{u\}_s$ by $[D]$ merely reproduces the vector $\{u\}_s$. If the iteration process is programmed for electronic computation, then even this choice will not prevent convergence to the dominant mode, because roundoff tends to introduce a $\{u\}_1$ component in $\{v\}_1$, however small. This small component is sufficient to cause the iteration process to converge to the dominant mode. The convergence begins slowly, but ultimately the rate of convergence depends on the ratio $\lambda_2/\lambda_1$.

The question remains as to how to obtain the higher modes. The lower eigenvalues $\lambda_r$ ($r = 2, 3, \ldots, n$) corresponding to the higher frequencies $\omega_r$, are sometimes referred to as *subdominant eigenvalues*. One possibility is to construct a trial vector that is entirely free of the eigenvector $\{u\}_1$, otherwise the iteration process using the dynamical matrix $[D]$ leads invariably to $\{u\}_1$. If we can make sure that the trial vector is free of the first eigenvector, then the iteration leads automatically to the second mode. Such a trial vector can be obtained by using two vectors corresponding to two consecutive iterations. Indeed, from Eq. (4.112), we can write

$$\{v\}_p - \{v\}_{p-1} = \frac{1}{\lambda_1^{p-1}} \sum_{r=1}^{n} c_r(\lambda_r^{p-1} - \lambda_1 \lambda_r^{p-2})\{u\}_r$$

$$= \frac{1}{\lambda_1^{p-1}} \sum_{r=2}^{n} c_r(\lambda_r - \lambda_1)\lambda_r^{p-2}\{u\}_r \tag{4.115}$$

where the vectors $\{v\}_{p-1}$ and $\{v\}_p$ have already been computed. But the vectors $\{v\}_p$ and $\{v\}_{p-1}$ are nearly equal, so that in general it is very difficult to retain significance in $\{v\}_p - \{v\}_{p-1}$. Hence, this method, known as *vector deflation*, does not appear suitable. We shall consider instead another technique not suffering from this drawback, where the method is called *matrix deflation*. The method will now be described.

If $\lambda_1$ and $\{u\}_1$ are the first eigenvalue and eigenvector associated with the dynamical matrix $[D]$, and $\{u\}_1$ is normalized so as to satisfy $\{u\}_1^T[m]\{u\}_1 = 1$, then the matrix

$$[D]_2 = [D] - \lambda_1\{u\}_1\{u\}_1^T[m] \qquad (4.116)$$

has the same eigenvalues as $[D]$ except that $\lambda_1$ is replaced by zero. Indeed, postmultiplying Eq. (4.116) by any arbitrary vector, such as the one given by Eq. (4.109), we obtain

$$[D]_2\{v\}_1 = \sum_{r=1}^{n} c_r[D]_2\{u\}_r$$

$$= \sum_{r=1}^{n} c_r[D]\{u\}_r - \lambda_1\{u\}_1 \sum_{r=1}^{n} c_r\{u\}_1^T[m]\{u\}_r \qquad (4.117)$$

Recalling Eqs. (4.82) and (4.108), however, Eq. (4.117) reduces to

$$[D]_2\{v\}_1 = \sum_{r=2}^{n} c_r\lambda_r\{u\}_r \qquad (4.118)$$

where the right side of Eq. (4.118) is completely free of the first eigenvector. Hence, we conclude that an iteration using any arbitrary trial vector in conjunction with the matrix $[D]_2$ given by Eq. (4.116) iterates automatically to the second eigenvalue $\lambda_2$ and eigenvector $\{u\}_2$ in the same way as $[D]$ iterates to the first eigenvalue and eigenvector. The matrix $[D]_2$ is called the *deflated matrix* corresponding to the second eigenvalue, or the first subdominant eigenvalue.

Because the dominant eigenvalue of $[D]_2$ is $\lambda_2$, the deflation process can be repeated by using

$$[D]_3 = [D]_2 - \lambda_2\{u\}_2\{u\}_2^T[m] \qquad (4.119)$$

to obtain the third eigenvalue $\lambda_3$ and eigenvector $\{u\}_3$. The procedure can be generalized by writing

$$[D]_s = [D]_{s-1} - \lambda_{s-1}\{u\}_{s-1}\{u\}_{s-1}^T[m] \qquad s = 2, 3, \ldots, n \qquad (4.120)$$

The iteration processes to the higher modes are also errorproof. However, one word of caution is in order. The iterations are errorproof only if the matrices $[D]_s$ $(s = 1, 2, \ldots, n)$, where $[D]_1 = [D]$, are correct. If an error is made in calculating any of the matrices $[D]_s$, no convergence to the corresponding modes is to be expected. Moreover, if the eigenvectors $\{u\}_1$, $\{u\}_2, \ldots$ are not computed with sufficient accuracy, $[D]_2, [D]_3, \ldots$ become progressively inaccurate, thus propagating the error.

Actually the power method using matrix deflation works also for the case of repeated eigenvalues, provided the eigenvectors corresponding to a repeated eigenvalue are orthogonal to one another. As pointed out in Sec. 4.8, this is always the case for eigenvalue problems that can be expressed in terms of a single real symmetric matrix, which is guaranteed by a positive definite mass matrix.

**Example 4.8**   Solve the eigenvalue problem of Example 4.7 by the power method using matrix deflation.

Equation (d) of Example 4.7 can be written in the form

$$
\begin{bmatrix} 1 & 1 & 2 \\ 1 & 2 & 4 \\ 1 & 2 & 5 \end{bmatrix} \begin{Bmatrix} u_1 \\ u_2 \\ u_3 \end{Bmatrix} = \lambda \begin{Bmatrix} u_1 \\ u_2 \\ u_3 \end{Bmatrix} \qquad \lambda = \frac{k}{m\omega^2} \qquad (a)
$$

Letting the first trial vector have the elements $v_1 = 1$, $v_2 = 2$, $v_3 = 3$, the first iteration is simply

$$
\begin{bmatrix} 1 & 1 & 2 \\ 1 & 2 & 4 \\ 1 & 2 & 5 \end{bmatrix} \begin{Bmatrix} 1.0000 \\ 2.0000 \\ 3.0000 \end{Bmatrix} = \begin{Bmatrix} 1.0000 + 2.0000 + 6.0000 \\ 1.0000 + 4.0000 + 12.0000 \\ 1.0000 + 4.0000 + 15.0000 \end{Bmatrix}
$$

$$
= 20.0000 \begin{Bmatrix} 0.4500 \\ 0.8500 \\ 1.0000 \end{Bmatrix}
$$

where the resulting vector has been normalized by letting $v_3 = 1$. Using that vector as an improved trial vector, we obtain

$$
\begin{bmatrix} 1 & 1 & 2 \\ 1 & 2 & 4 \\ 1 & 2 & 5 \end{bmatrix} \begin{Bmatrix} 0.4500 \\ 0.8500 \\ 1.0000 \end{Bmatrix} = \begin{Bmatrix} 0.4500 + 0.8500 + 2.0000 \\ 0.4500 + 1.7000 + 4.0000 \\ 0.4500 + 1.7000 + 5.0000 \end{Bmatrix}
$$

$$
= 7.1500 \begin{Bmatrix} 0.4515 \\ 0.8601 \\ 1.0000 \end{Bmatrix}
$$

The third iteration yields

$$
\begin{bmatrix} 1 & 1 & 2 \\ 1 & 2 & 4 \\ 1 & 2 & 5 \end{bmatrix} \begin{Bmatrix} 0.4615 \\ 0.8601 \\ 1.0000 \end{Bmatrix} = \begin{Bmatrix} 0.4615 + 0.8601 + 2.0000 \\ 0.4615 + 1.7203 + 4.0000 \\ 0.4615 + 1.7203 + 5.0000 \end{Bmatrix}
$$

$$
= 7.1818 \begin{Bmatrix} 0.4625 \\ 0.8608 \\ 1.0000 \end{Bmatrix}
$$

Convergence is achieved at the sixth iteration in the form

$$
\begin{bmatrix} 1 & 1 & 2 \\ 1 & 2 & 4 \\ 1 & 2 & 5 \end{bmatrix} \begin{Bmatrix} 0.4626 \\ 0.8608 \\ 1.0000 \end{Bmatrix} = \begin{Bmatrix} 0.4626 + 0.8608 + 2.0000 \\ 0.4626 + 1.7216 + 4.0000 \\ 0.4626 + 1.7216 + 5.0000 \end{Bmatrix}
$$

$$
= 7.1842 \begin{Bmatrix} 0.4626 \\ 0.8608 \\ 1.0000 \end{Bmatrix}
$$

with the conclusion that $\lambda_1 = 7.1842$ and $\{u\}_1$ is the vector on the right side.

Normalizing the eigenvector so that $\{u\}_1^T[m]\{u\}_1 = 1$, where $[m]$ is given by Eq. (a) of Example 4.7, we obtain the first normal mode and natural frequency

$$\{u\}_1 = m^{-1/2} \begin{Bmatrix} 0.2691 \\ 0.5008 \\ 0.5817 \end{Bmatrix} \qquad \omega_1 = \frac{1}{\sqrt{7.1842}}\sqrt{\frac{k}{m}} = 0.3731\sqrt{\frac{k}{m}} \qquad (b)$$

To obtain the second mode, we use Eq. (4.116) and form the matrix

$$[D]_2 = [D] - \lambda_1\{u\}_1\{u\}_1^T[m]$$

$$= \begin{bmatrix} 1 & 1 & 2 \\ 1 & 2 & 4 \\ 1 & 2 & 5 \end{bmatrix} - 7.1842 \begin{Bmatrix} 0.2691 \\ 0.5008 \\ 0.5817 \end{Bmatrix} \begin{Bmatrix} 0.2691 \\ 0.5008 \\ 0.5817 \end{Bmatrix}^T \begin{bmatrix} 1 & 0 & 0 \\ 0 & 1 & 0 \\ 0 & 0 & 2 \end{bmatrix}$$

$$= \begin{bmatrix} 0.4797 & 0.0319 & -0.2494 \\ 0.0319 & 0.1985 & -0.1856 \\ -0.1247 & -0.0928 & 0.1376 \end{bmatrix} \qquad (c)$$

Expecting a node, we use as the elements of the first trial vector for the second mode $v_1 = 1$, $v_2 = 1$, $v_3 = -1$, so that the first iteration to the second mode is

$$\begin{bmatrix} 0.4797 & 0.0319 & -0.2494 \\ 0.0319 & 0.1985 & -0.1856 \\ -0.1247 & -0.0928 & 0.1376 \end{bmatrix} \begin{Bmatrix} 1.0000 \\ 1.0000 \\ -1.0000 \end{Bmatrix} = 0.7610 \begin{Bmatrix} 1.0000 \\ 0.5467 \\ -0.4666 \end{Bmatrix}$$

whereas the second iteration is

$$\begin{bmatrix} 0.4797 & 0.0319 & -0.2494 \\ 0.0319 & 0.1985 & -0.1856 \\ -0.1247 & -0.0928 & 0.1376 \end{bmatrix} \begin{Bmatrix} 1.0000 \\ 0.5467 \\ -0.4666 \end{Bmatrix} = 0.6135 \begin{Bmatrix} 1.0000 \\ 0.3700 \\ -0.3905 \end{Bmatrix}$$

The fourteenth iteration yields

$$\begin{bmatrix} 0.4797 & 0.0319 & -0.2494 \\ 0.0319 & 0.1985 & -0.1856 \\ -0.1247 & -0.0928 & 0.1376 \end{bmatrix} \begin{Bmatrix} 1.0000 \\ 0.2541 \\ -0.3407 \end{Bmatrix} = 0.5728 \begin{Bmatrix} 1.0000 \\ 0.2541 \\ -0.3407 \end{Bmatrix}$$

at which point we conclude that convergence has been achieved. The second normal mode and natural frequency are

$$\{u\}_2 = m^{-1/2} \begin{Bmatrix} 0.8782 \\ 0.2231 \\ -0.2992 \end{Bmatrix} \qquad \omega_2 = \frac{1}{\sqrt{0.5728}}\sqrt{\frac{k}{m}} = 1.3213\sqrt{\frac{k}{m}} \qquad (d)$$

For the third mode, we use Eq. (4.119) and write

$$[D]_3 = [D]_2 - \lambda_2\{u\}_2\{u\}_2^T[m] = \begin{bmatrix} 0.0380 & -0.0804 & 0.0516 \\ -0.0804 & 0.1700 & -0.1091 \\ 0.0258 & -0.0546 & 0.0350 \end{bmatrix} \qquad (e)$$

Using the same procedure as above, we obtain the third normal mode and natural frequency

$$\{u\}_3 = m^{-1/2} \begin{Bmatrix} 0.3954 \\ -0.8363 \\ 0.2685 \end{Bmatrix} \qquad \omega_3 = \frac{1}{\sqrt{0.2430}} \sqrt{\frac{k}{m}} = 2.0285 \sqrt{\frac{k}{m}} \qquad (f)$$

The results compare favorably with those obtained in Example 4.7 by using the characteristic determinant method. It should be pointed out that in actuality the above computations were carried out using six decimal places, but to save space only four decimal places were given.

## 4.12 SYSTEMS ADMITTING RIGID-BODY MOTIONS

The undamped free vibration of a multi-degree-of-freedom linear system, in which the system is capable of harmonic oscillation in any one or all of the modes of vibration, is typical of positive definite systems, i.e., systems defined by real symmetric positive definite mass and stiffness matrices. The behavior is somewhat different when the stiffness matrix $[k]$ is only positive semidefinite.

As indicated in Sec. 4.5, when $[m]$ is positive definite and $[k]$ is only positive semidefinite, the system is *positive semidefinite*. Physically this implies that the system is supported in such a manner that rigid-body motion is possible. When the potential energy is due to elastic effects alone, if the body undergoes rigid-body motion, i.e., if there are no elastic deformations, then the potential energy is zero without all the coordinates being identically equal to zero. Such a semidefinite system is shown in Fig. 4.12, where the system consists of three disks of mass polar moments of inertia $I_1$, $I_2$, and $I_3$ connected by two massless shafts of lengths $L_1$ and $L_2$ and torsional stiffnesses $GJ_1$ and $GJ_2$, respectively. The system is supported at both ends by means of frictionless sleeves in such a way that the entire system can rotate freely as a whole. Of course, torsional deformations can also be present, so that in general the motion of the system is a combination of rigid and elastic motions. Denoting by $\theta_i(t)$ ($i = 1, 2, 3$) the angular displacements and velocities on the three disks, the kinetic energy becomes

$$T = \tfrac{1}{2}(I_1\dot{\theta}_1^2 + I_2\dot{\theta}_2^2 + I_3\dot{\theta}_3^2) = \tfrac{1}{2}\{\dot{\theta}\}^T[I]\{\dot{\theta}\} \tag{4.121}$$

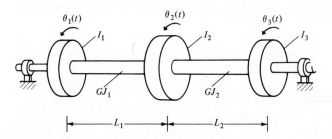

**Figure 4.12**

where the inertia matrix $[I]$ is diagonal

$$[I] = \begin{bmatrix} I_1 & 0 & 0 \\ 0 & I_2 & 0 \\ 0 & 0 & I_3 \end{bmatrix} \tag{4.122}$$

On the other hand, the potential energy has the expression

$$V = \tfrac{1}{2}[k_1(\theta_2 - \theta_1)^2 + k_2(\theta_3 - \theta_2)^2] = \tfrac{1}{2}\{\theta\}^T[k]\{\theta\} \tag{4.123}$$

where the stiffness matrix has the form

$$[k] = \begin{bmatrix} k_1 & -k_1 & 0 \\ -k_1 & k_1 + k_2 & -k_2 \\ 0 & -k_2 & k_2 \end{bmatrix} \tag{4.124}$$

in which we have used the notation $k_i = GJ_i/L_i$ $(i = 1, 2)$. Assuming synchronous motion

$$\theta_i(t) = \Theta_i f(t) \qquad i = 1, 2, 3 \tag{4.125}$$

where $\Theta_i$ $(i = 1, 2, 3)$ are constants and $f(t)$ is harmonic, we arrive at the eigenvalue problem

$$\omega^2[I]\{\Theta\} = [k]\{\Theta\} \tag{4.126}$$

Premultiplying both sides of (4.126) by $\{\Theta\}^T$, we obtain

$$\omega^2\{\Theta\}^T[I]\{\Theta\} = \{\Theta\}^T[k]\{\Theta\} \tag{4.127}$$

Considering the matrix (4.124), we conclude by inspection that the rigid-body motion

$$\{\Theta\} = \{\Theta\}_0 = \Theta_0\{1\} \tag{4.128}$$

where $\{1\}$ is a column matrix with all its elements equal to unity and $\Theta_0 = \text{const} \neq 0$, renders the right side of Eq. (4.127) equal to zero. But the triple matrix product on the left side of (4.127) is always positive,

$$\{\Theta\}_0^T[I]\{\Theta\}_0 > 0 \tag{4.129}$$

except when the vector $\{\Theta\}_0$ is identically zero, a case ruled out as trivial. It follows that the only possible way of satisfying Eq. (4.127) is for the frequency corresponding to $\{\Theta\}_0$ to be zero, $\omega_0 = 0$. Hence, for a semidefinite system there is at least one zero eigenvalue. We refer to the mode defined by $\omega_0$, $\{\Theta\}_0$ as the *rigid-body mode*, or *zero mode*. The fact that a semidefinite system possesses a zero eigenvalue is consistent with the fact that the stiffness matrix is singular, i.e., its determinant is equal to zero, as can be verified from Eq. (4.124).

Because the rigid-body mode, defined by a constant eigenvector $\{\Theta\}_0$ and a zero natural frequency $\omega_0$, is a solution of the eigenvalue problem (4.126), it follows that any other eigenvector must be orthogonal to it, namely, it must satisfy the condition

$$\{\Theta\}_0^T[I]\{\Theta\} = \Theta_0(I_1\Theta_1 + I_2\Theta_2 + I_3\Theta_3) = 0 \tag{4.130}$$

where $\Theta_i$ $(i = 1, 2, 3)$ are the components of $\{\Theta\}$. Because $\Theta_0$ is nonzero by definition, Eq. (4.130) implies that

$$I_1\Theta_1 + I_2\Theta_2 + I_3\Theta_3 = 0 \qquad (4.131)$$

In view of Eqs. (4.125), Eq. (4.131) can also be written in the form

$$I_1\dot{\theta}_1(t) + I_2\dot{\theta}_2(t) + I_3\dot{\theta}_3(t) = 0 \qquad (4.132)$$

which implies physically that the system angular momentum associated with the elastic motion is equal to zero, where the momentum is about an axis coinciding with the axis of the shaft. Hence, *the orthogonality of the rigid-body mode to the elastic modes is equivalent to the preservation of zero angular momentum* in pure elastic motion.

The general motion of an unrestrained system consists of a combination of elastic modes and the rigid-body motion. Clearly, this type of motion is possible only for unrestrained systems, such as that shown in Fig. 4.12, because if one of the ends were to be clamped, then the reactive torque at that end would prevent rigid-body rotation from taking place. From Eq. (4.131), we conclude that the elastic motion must be such that the weighted average rotation of the system is zero, where the weighting factors are the moments of inertia $I_i$ $(i = 1, 2, 3)$. The equivalent statement for an unrestrained discrete system in translational motion is that the system mass center is at rest at all times.

As pointed out earlier, det $[k]$ is equal to zero, so that $[k]$ is a singular matrix, with the implication that the inverse matrix $[k]^{-1}$ does not exist. Recalling that $[k]^{-1} = [a]$ is the flexibility matrix, this fact can be easily explained physically by recognizing that for an unrestrained system it is not possible to define flexibility influence coefficients. If the interest lies in solving a positive definite eigenvalue problem, then one can remove the singularity of $[k]$ by transforming the eigenvalue problem associated with the unrestrained system into one for the elastic modes alone, as shown in the following.

Although there are three disks involved in the system of Fig. 4.12, as far as the elastic motion alone is concerned, this is not truly a three-degree-of-freedom system, because Eq. (4.132) can be regarded as a constraint equation that can be used to eliminate one coordinate from the problem formulation. Indeed, if we write

$$\theta_3 = -\frac{I_1}{I_3}\theta_1 - \frac{I_2}{I_3}\theta_2 \qquad (4.133)$$

then, we can express the relation between the constrained vector $\{\theta\}_c$ and the arbitrary vector $\{\theta\}$ in the form

$$\left\{ \begin{matrix} \theta_1 \\ \theta_2 \\ \theta_3 \end{matrix} \right\}_c = \begin{bmatrix} 1 & 0 & 0 \\ 0 & 1 & 0 \\ -\dfrac{I_1}{I_3} & -\dfrac{I_2}{I_3} & 0 \end{bmatrix} \left\{ \begin{matrix} \theta_1 \\ \theta_2 \\ \theta_3 \end{matrix} \right\} = \begin{bmatrix} 1 & 0 \\ 0 & 1 \\ -\dfrac{I_1}{I_3} & -\dfrac{I_2}{I_3} \end{bmatrix} \left\{ \begin{matrix} \theta_1 \\ \theta_2 \end{matrix} \right\} \qquad (4.134)$$

where we note that the coordinate $\theta_3$ is not really needed in the solution for the system elastic motion, because it is automatically determined as soon as $\theta_1$ and $\theta_2$ are known. There is nothing unique about $\theta_3$, as we could have eliminated either $\theta_1$ or $\theta_2$ from the problem formulation without affecting the final results. An expression similar to (4.134) exists for the angular velocities $\dot{\theta}_i$ ($i = 1, 2, 3$), so that we can write

$$\{\theta\}_c = [c]\{\theta\} \qquad \{\dot{\theta}\}_c = [c]\{\dot{\theta}\} \tag{4.135}$$

where

$$[c] = \begin{bmatrix} 1 & 0 \\ 0 & 1 \\ -\dfrac{I_1}{I_3} & -\dfrac{I_2}{I_3} \end{bmatrix} \tag{4.136}$$

plays the role of a constraint matrix. We note again that whereas the constrained vectors $\{\theta\}_c$ and $(\dot{\theta})_c$ possess three components, the arbitrary vectors $\{\theta\}$ and $\{\dot{\theta}\}$ in Eqs. (4.135) possess only two components. The linear transformations (4.135) can be used to reduce the kinetic and potential energy to expressions in $\theta_1$ and $\theta_2$ alone. Indeed, inserting Eq. (4.135) into (4.121) and (4.123), and recognizing that the vectors in (4.121) and (4.123) are constrained, we obtain

$$T = \tfrac{1}{2}\{\dot{\theta}\}_c^T[I]\{\dot{\theta}\}_c = \tfrac{1}{2}\{\dot{\theta}\}^T[c]^T[I][c]\{\dot{\theta}\} = \tfrac{1}{2}\{\dot{\theta}\}^T[I']\{\dot{\theta}\} \tag{4.137}$$

and

$$V = \tfrac{1}{2}\{\theta\}_c^T[k]\{\theta\}_c = \tfrac{1}{2}\{\theta\}^T[c]^T[k][c]\{\theta\} = \tfrac{1}{2}\{\theta\}^T[k']\{\theta\} \tag{4.138}$$

where

$$[I'] = [c]^T[I][c] = \frac{1}{I_3}\begin{bmatrix} I_1(I_1 + I_3) & I_1 I_2 \\ I_1 I_2 & I_2(I_2 + I_3) \end{bmatrix} \tag{4.139}$$

and

$$[k'] = [c]^T[k][c]$$

$$= \frac{1}{I_3^2}\begin{bmatrix} k_1 I_3^2 + k_2 I_1^2 & -k_1 I_3^2 + k_2 I_1(I_2 + I_3) \\ -k_1 I_3^2 + k_2 I_1(I_2 + I_3) & (k_1 + k_2)I_3^2 + k_2 I_2(2I_2 + I_3) \end{bmatrix} \tag{4.140}$$

are $2 \times 2$ symmetric positive definite matrices.

The eigenvalue problem associated with the transformed system is

$$\omega^2[I']\{\Theta\} = [k']\{\Theta\} \tag{4.141}$$

which possesses all the characteristics associated with a positive definite system. Its solution consists of the natural modes $\{\Theta\}_1$, $\{\Theta\}_2$ and the associated natural frequencies $\omega_1, \omega_2$, respectively. The modes $\{\Theta\}_1$ and $\{\Theta\}_2$ give only the rotations of disks 1 and 2. The rotations of disk 3 in these elastic modes are obtained by considering the first of Eqs. (4.135), and writing

$$\{\Theta\}_{1c} = [c]\{\Theta\}_1 \qquad \{\Theta\}_{2c} = [c]\{\Theta\}_2 \tag{4.142}$$

where the elements of the constrained modes $\{\Theta\}_{1c}$ and $\{\Theta\}_{2c}$ are such that Eq. (4.131) is satisfied automatically.

We stress again that Eqs. (4.142) represent only the elastic modes. In addition, for this semidefinite system, we have the rigid-body mode $\{\Theta\}_0 = \Theta_0\{1\}$ with the natural frequency $\omega_0 = 0$.

**Example 4.9** Consider the unrestrained system of Fig. 4.12, let $k_1 = k_2 = k$ and $I_1 = I_2 = I_3 = I$ and obtain the natural modes of the system by solving a positive definite eigenvalue problem.

The natural modes are obtained by solving the eigenvalue problem (4.141), where the matrices $[I']$ and $[k']$ are given by Eqs. (4.139) and (4.140). Using the data given above, the two matrices have the explicit form

$$[I'] = \frac{1}{I}\begin{bmatrix} 2I^2 & I^2 \\ I^2 & 2I^2 \end{bmatrix} = I\begin{bmatrix} 2 & 1 \\ 1 & 2 \end{bmatrix} \tag{a}$$

and

$$[k'] = \frac{1}{I^2}\begin{bmatrix} 2kI^2 & kI^2 \\ kI^2 & 5kI^2 \end{bmatrix} = k\begin{bmatrix} 2 & 1 \\ 1 & 5 \end{bmatrix} \tag{b}$$

It is clear that the matrix $[k']$ is not singular because its determinant is different from zero. In fact, $[k']$ is positive definite.

The eigenvalue problem for the system is obtained by inserting Eqs. (a) and (b) into Eq. (4.141). The solution of the eigenvalue problem is

$$\omega_1 = \sqrt{\frac{k}{I}} \qquad \{\Theta\}_1 = \begin{Bmatrix} 1 \\ 0 \end{Bmatrix}$$

$$\tag{c}$$

$$\omega_2 = \sqrt{\frac{3k}{I}} \qquad \{\Theta\}_2 = \begin{Bmatrix} 0.5 \\ -1 \end{Bmatrix}$$

Using Eq. (4.136), we can write the constraint matrix

$$[c] = \begin{bmatrix} 1 & 0 \\ 0 & 1 \\ -1 & -1 \end{bmatrix} \tag{d}$$

so that, from Eqs. (4.142), the constrained eigenvectors corresponding to the elastic modes are

$$\{\Theta\}_{1c} = \begin{bmatrix} 1 & 0 \\ 0 & 1 \\ -1 & -1 \end{bmatrix}\begin{Bmatrix} 1 \\ 0 \end{Bmatrix} = \begin{Bmatrix} 1 \\ 0 \\ -1 \end{Bmatrix}$$

$$\tag{e}$$

$$\{\Theta\}_{2c} = \begin{bmatrix} 1 & 0 \\ 0 & 1 \\ -1 & -1 \end{bmatrix}\begin{Bmatrix} 0.5 \\ -1 \end{Bmatrix} = \begin{Bmatrix} 0.5 \\ -1 \\ 0.5 \end{Bmatrix}$$

$$\omega_0 = 0$$

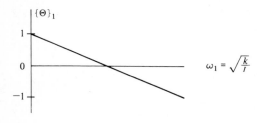

$$\omega_1 = \sqrt{\frac{k}{I}}$$

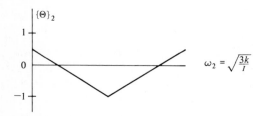

$$\omega_2 = \sqrt{\frac{3k}{I}}$$

**Figure 4.13**

In addition, we have the rigid-body mode

$$\omega_0 = 0 \qquad \{\Theta\}_0 = \begin{Bmatrix} 1 \\ 1 \\ 1 \end{Bmatrix} \qquad (f)$$

It can be verified that the three modes are orthogonal with respect to the inertia matrix $[I]$ and the stiffness matrix $[k]$. The modes are plotted in Fig. 4.13.

From Fig. 4.13, we observe that in the first elastic mode the first and third disks have displacements equal in magnitude but opposite in sense, while the center disk is at rest at all times, as it coincides with a node. This mode is what is generally called an *antisymmetric mode*. On the other hand, in the second elastic mode, the first and third disks have displacements equal in magnitude and in the same sense, while the center disk moves in opposite sense. This is a *symmetric mode*. In fact, the rigid-body mode is also a symmetric mode. Symmetric and antisymmetric modes are common occurrences in systems with symmetrical parameter distributions, such as the system of Fig. 4.12, and we shall have ample opportunity later in this text to verify this statement.

## 4.13 RAYLEIGH'S QUOTIENT

Rayleigh's quotient occupies a unique position in vibrations. It is not only fundamental to vibration theory, but it also has practical value, as it can be used as a means of estimating the fundamental frequency of a system or as a tool in speeding up convergence to the solution of the eigenvalue problem in matrix iteration. Moreover, the concept enhances the understanding of the nature of vibrating systems. To introduce the concept, let us return to the eigenvalue problem

$$\lambda[m]\{u\} = [k]\{u\} \qquad \lambda = \omega^2 \qquad (4.143)$$

where $[m]$ and $[k]$ are symmetric matrices. (Note that in earlier discussions the parameter $\lambda$ was defined as the reciprocal of $\omega^2$.) The inertia matrix $[m]$ is always positive definite, whereas for the systems considered in this chapter the stiffness matrix $[k]$ can be positive definite or positive semidefinite. Equation (4.143) can be written in terms of its solutions $\lambda_r$, $\{u\}_r$ $(r = 1, 2, \ldots, n)$ as follows:

$$\lambda_r[m]\{u\}_r = [k]\{u\}_r \qquad r = 1, 2, \ldots, n \qquad (4.144)$$

Premultiplying both sides of (4.144) by $\{u\}_r^T$ and dividing by the scalar $\{u\}_r^T[m]\{u\}_r$, we obtain

$$\lambda_r = \omega_r^2 = \frac{\{u\}_r^T[k]\{u\}_r}{\{u\}_r^T[m]\{u\}_r} \qquad r = 1, 2, \ldots, n \qquad (4.145)$$

so that the eigenvalue $\lambda_r = \omega_r^2$ can be written in the form of a quotient of two triple matrix products representing quadratic forms, where the numerator is related to the potential energy and the denominator to the kinetic energy in a given mode.

Next, consider any arbitrary vector $\{u\}$, premultiply both sides of Eq. (4.143) by $\{u\}^T$, divide the resulting equation through by $\{u\}^T[m]\{u\}$, and obtain

$$\lambda = \omega^2 = R(\{u\}) = \frac{\{u\}^T[k]\{u\}}{\{u\}^T[m]\{u\}} \qquad (4.146)$$

where $R(\{u\})$ is a scalar whose value depends not only on the matrices $[m]$ and $[k]$ but also on the vector $\{u\}$. Whereas matrices $[m]$ and $[k]$ reflect the system characteristics, the vector $\{u\}$ is arbitrary, so that for a given system $R(\{u\})$ depends on the vector $\{u\}$ alone. The scalar $R(\{u\})$ is called *Rayleigh's quotient* and it possesses very interesting properties. Clearly, if the arbitrary vector $\{u\}$ coincides with one of the system eigenvectors, then the quotient reduces to the associated eigenvalue. Moreover, the quotient has stationary values in the neighbourhood of the system eigenvectors. To show this, let us consider the expansion theorem of Sec. 4.8 and represent the arbitrary vector $\{u\}$ as a linear combination of the system eigenvectors in the form

$$\{u\} = \sum_{r=1}^{n} c_r\{u\}_r = [u]\{c\} \qquad (4.147)$$

where $[u]$ is the modal matrix and $\{c\}$ a vector with its elements consisting of the coefficients $c_r$. Let the eigenvectors be normalized so that the modal matrix satisfies

$$[u]^T[m][u] = [1] \qquad [u]^T[k][u] = [\lambda] \qquad (4.148)$$

where $[1]$ is the unit matrix and $[\lambda]$ is the diagonal matrix of the eigenvalues $\lambda_r$. Introducing transformation (4.147) into Eq. (4.146) and considering Eqs. (4.148), we obtain

$$R(\{u\}) = \frac{\{c\}^T[u]^T[k][u]\{c\}}{\{c\}^T[u]^T[m][u]\{c\}} = \frac{\{c\}^T[\lambda]\{c\}}{\{c\}^T[1]\{c\}} = \frac{\displaystyle\sum_{i=1}^{n} \lambda_i c_i^2}{\displaystyle\sum_{i=1}^{n} c_i^2} \qquad (4.149)$$

Next, assume that the trial vector $\{u\}$ differs only slightly from the eigenvector $\{u\}_r$. Mathematically, this implies that the coefficients $c_i$ ($i \neq r$) are very small compared to $c_r$, or

$$c_i = \epsilon_i c_r \qquad i = 1, 2, \ldots, n; \, i \neq r \qquad (4.150)$$

where $\epsilon_i$ are small numbers, $\epsilon_i \ll 1$. Dividing the numerator and denominator of (4.149) by $c_r^2$, we obtain

$$R(\{u\}) = \frac{\lambda_r + \displaystyle\sum_{i=1}^{n} (1 - \delta_{ir})\epsilon_i^2}{1 + \displaystyle\sum_{i=1}^{n} (1 - \delta_{ir})\epsilon_i^2} \cong \lambda_r + \sum_{i=1}^{n} (\lambda_i - \lambda_r)\epsilon_i^2 \qquad (4.151)$$

where $\delta_{ir}$ is the Kronecker delta. The use of $(1 - \delta_{ir})$ excludes automatically the terms corresponding to $i = r$ from the series in the numerator and denominator. We note that the series on the right side of (4.151) is a quantity of second order. Hence, if the trial vector $\{u\}$ differs from the eigenvector $\{u\}_r$ by a small quantity of first order, then $R(\{u\})$ differs from the eigenvalue $\lambda_r$ by a small quantity of second order. The implication is that *Rayleigh's quotient has a stationary value in the neighborhood of an eigenvector*, where *the stationary value is the corresponding eigenvalue.*

In the neighborhood of the fundamental mode, Rayleigh's quotient has not merely a stationary value but a minimum. Indeed, if we let $r = 1$ in Eq. (4.151), we obtain

$$R(\{u\}) \cong \lambda_1 + \sum_{i=2}^{n} (\lambda_i - \lambda_1)\epsilon_i^2 \qquad (4.152)$$

Because in general $\lambda_i > \lambda_1$ ($i = 2, 3, \ldots, n$), it follows that

$$R(u) \geqslant \lambda_1 \qquad (4.153)$$

where the equality sign holds only if all $\epsilon_i$ ($i = 2, 3, \ldots, n$) are identically zero. Hence, *Rayleigh's quotient is never lower than the first eigenvalue, and the minimum value it can take is that of the first eigenvalue itself.* In view of the above, we conclude that a practical application of Rayleigh's quotient is to obtain estimates for the fundamental frequency of the system. To this end, a very good estimate can be obtained by using as a trial vector $\{u\}$ the vector of static displacements obtained by subjecting the masses to forces proportional to their weights.

In the preceding discussion we assumed that matrices $[m]$ and $[k]$ were given, and we used Eq. (4.146) to examine the behavior of $R(\{u\})$ with changing $\{u\}$. Equation (4.146), however, can be used also to examine how Rayleigh's quotient changes with $[m]$ and $[k]$ for a given $\{u\}$. It is clear that when the elements of $[k]$ increase in value the quotient increases, whereas when the elements of $[m]$ increase in value the quotient decreases. Physically this implies that the natural frequencies increase if the system is made stiffer, and they decrease if the system is made more massive.

**Example 4.10** Consider the system of Fig. 4.8a, use the data of Example 4.7 and obtain an estimate of the fundamental frequency by means of Rayleigh's quotient.

From Examples 4.2 and 4.7, we have the mass and stiffness matrices

$$[m] = m \begin{bmatrix} 1 & 0 & 0 \\ 0 & 1 & 0 \\ 0 & 0 & 2 \end{bmatrix} \qquad [k] = k \begin{bmatrix} 2 & -1 & 0 \\ -1 & 3 & -2 \\ 0 & -2 & 2 \end{bmatrix} \qquad (a)$$

As a trial vector, we will use a vector of static displacements as described above. To this end, we subject the system of Fig. 4.8a to forces proportional to the weight of the masses, or

$$\{F\} = mg \begin{Bmatrix} 1 \\ 1 \\ 2 \end{Bmatrix} \qquad (b)$$

The vector of static displacements can be obtained by using Eq. (4.20), where $[a]$ is the flexibility matrix. In the case at hand

$$[a] = [k]^{-1} = \frac{1}{k} \begin{bmatrix} 1 & 1 & 1 \\ 1 & 2 & 2 \\ 1 & 2 & 2.5 \end{bmatrix} \qquad (c)$$

Hence, inserting Eqs. (b) and (c) into Eq. (4.20), we obtain the trial vector

$$\{u\} = [a]\{F\} = \frac{mg}{k} \begin{bmatrix} 1 & 1 & 1 \\ 1 & 2 & 2 \\ 1 & 2 & 2.5 \end{bmatrix} \begin{Bmatrix} 1 \\ 1 \\ 2 \end{Bmatrix} = \frac{mg}{k} \begin{Bmatrix} 4 \\ 7 \\ 8 \end{Bmatrix} \qquad (d)$$

Ignoring the scalar multiplier $mg/k$, the triple matrix products involved in Eq. (4.146) can be computed as follows:

$$\{u\}^T[m]\{u\} = m \begin{Bmatrix} 4 \\ 7 \\ 8 \end{Bmatrix}^T \begin{bmatrix} 1 & 0 & 0 \\ 0 & 1 & 0 \\ 0 & 0 & 2 \end{bmatrix} \begin{Bmatrix} 4 \\ 7 \\ 8 \end{Bmatrix} = 193m$$

$$\qquad (e)$$

$$\{u\}^T[k]\{u\} = k \begin{Bmatrix} 4 \\ 7 \\ 8 \end{Bmatrix}^T \begin{bmatrix} 2 & -1 & 0 \\ -1 & 3 & -2 \\ 0 & -2 & 2 \end{bmatrix} \begin{Bmatrix} 4 \\ 7 \\ 8 \end{Bmatrix} = 27k$$

Inserting Eqs. (e) into Eq. (4.146), we obtain

$$\omega^2 = R(\{u\}) = \frac{27k}{193m} = 0.1399 \frac{k}{m} \qquad (f)$$

so that the estimated fundamental frequency is

$$\omega = 0.3740 \sqrt{\frac{k}{m}} \qquad (g)$$

It was shown in Example 4.7 that the first natural frequency has the value

$$\omega_1 = 0.3731 \sqrt{\frac{k}{m}} \qquad (h)$$

so that the percentage error is

$$\frac{\omega - \omega_1}{\omega} = \frac{0.3740 - 0.3731}{0.3740} = 0.2406\% \qquad (i)$$

Hence, the estimated first frequency differs from the calculated one by less than one quarter of one percent, which is a remarkable result. Of course, the estimate is so good because the trial vector $\{u\}$ resembles the first eigenvector $\{u\}_1$ very closely, as can be verified by means of results from Example 4.7.

## 4.14 GENERAL RESPONSE OF DISCRETE LINEAR SYSTEMS. MODAL ANALYSIS

Until now the discussion has been confined to the free vibration of discrete linear systems, placing the emphasis on the role of the natural modes in the construction of the system response. Indeed, in Sec. 4.9 we have shown how to determine the response of an undamped $n$-degree-of-freedom system to initial excitation by means of modal analysis. However, modal analysis can be used to derive the response of undamped systems to any arbitrary excitation, whether in the form of initial excitation or externally impressed forces, and under certain circumstances also the response of viscously damped systems.

Considering first the response of an undamped system, we recall Eq. (4.43), representing the system differential equations of motion in the matrix form

$$[m]\{\ddot{q}(t)\} + [k]\{q(t)\} = \{Q(t)\} \qquad (4.154)$$

where $[m]$ and $[k]$ are $n \times n$ symmetric matrices, called correspondingly the inertia and stiffness matrix, and $\{q(t)\}$ and $\{Q(t)\}$ are the $n$-dimensional generalized coordinate and force vectors, respectively. Equation (4.154) constitutes a system of $n$ simultaneous ordinary differential equations with constant coefficients. The equations are linear, and a solution can be obtained by the Laplace transformation method, at least in principle. In practice, however, the solution can be quite laborious, even for a two-degree-of-freedom system, so that a different method is advised. Indeed, a solution by modal analysis is appreciably less laborious. The basic idea behind modal analysis is to transform the simultaneous set of equations represented by (4.154) into an independent set of equations, where the transformation matrix is the modal matrix.

To obtain the solution of Eq. (4.154) by modal analysis, we must first solve the eigenvalue problem associated with matrices $[m]$ and $[k]$. The solution can be written in the general matrix form

$$[m][u][\omega^2] = [k][u] \tag{4.155}$$

where $[u]$ is the modal matrix and $[\omega^2]$ the diagonal matrix of the natural frequencies squared. The modal matrix can be normalized so as to satisfy

$$[u]^T[m][u] = [1] \qquad [u]^T[k][u] = [\omega^2] \tag{4.156}$$

Next, we consider the linear transformation

$$\{q(t)\} = [u]\{\eta(t)\} \tag{4.157}$$

relating the vectors $\{q(t)\}$ and $\{\eta(t)\}$, where the vectors represent two different sets of generalized coordinates. Because $[u]$ is a constant matrix, a transformation similar to (4.157) exists between $\{\ddot{q}(t)\}$ and $\{\ddot{\eta}(t)\}$. Introducing transformation (4.157) into (4.154), premultiplying the result by $[u]^T$, and considering Eqs. (4.156), we obtain

$$\{\ddot{\eta}(t)\} + [\omega^2]\{\eta(t)\} = \{N(t)\} \tag{4.158}$$

where

$$\{N(t)\} = [u]^T\{Q(t)\} \tag{4.159}$$

is an $n$-dimensional vector of generalized forces associated with the vector of generalized coordinates $\{\eta(t)\}$.

Equation (4.158) represents a set of $n$ independent equations of the form

$$\ddot{\eta}_r(t) + \omega_r^2\eta_r(t) = N_r(t) \qquad r = 1, 2, \ldots, n \tag{4.160}$$

where $\eta_r(t)$ are recognized as the system *normal coordinates*, introduced in Sec. 4.6, and $N_r(t)$ are associated generalized forces. Equations (4.160) have the same structure as the differential equation of motion of a single-degree-of-freedom system of unit mass, natural frequency $\omega_r$, and impressed force $N_r(t)$. Hence, the solution of Eqs. (4.160) can be obtained by the methods of Chap. 2. Indeed, letting $m = 1$, $\zeta = 0$, and $\omega_d = \omega_r$ in Eq. (2.204), we can write the complete solution

$$\eta_r(t) = \frac{1}{\omega_r} \int_0^t N_r(\tau) \sin \omega_r(t - \tau) \, d\tau + \eta_r(0) \cos \omega_r t$$

$$+ \frac{\dot{\eta}_r(0)}{\omega_r} \sin \omega_r t \qquad r = 1, 2, \ldots, n \tag{4.161}$$

where $\eta_r(0)$ and $\dot{\eta}_r(0)$ are the initial generalized displacements and velocities, respectively. But Eq. (4.157) can also be expressed in the form

$$\{q(t)\} = [u]\{\eta(t)\} = \sum_{r=1}^{n} \{u\}_r \eta_r(t) \tag{4.162}$$

where $\{u\}_r$ $(r = 1, 2, \ldots, n)$ are the normalized modal vectors. Hence, the complete response of an undamped $n$-degree-of-freedom system can be obtained by inserting the normal coordinates (4.161) into Eq. (4.162). Note that the normal coordinates are sometimes called *modal coordinates*.

The expressions for the normal coordinates, Eqs. (4.161), contain the initial generalized displacements $\eta_r(0)$ and velocities $\dot{\eta}_r(0)$ $(r = 1, 2, \ldots, n)$, which are related to the actual initial displacements $q_i(0)$ and velocities $\dot{q}_i(0)$ $(i = 1, 2, \ldots, n)$. To establish this relation, we let $t = 0$ in Eq. (4.162) and write

$$\{q(0)\} = [u]\{\eta(0)\} = \sum_{r=1}^{n} \{u\}_r \eta_r(0) \tag{4.163}$$

where $\{q(0)\}$ is the vector of initial displacements. Premultiplying Eq. (4.163) by $\{u\}_s^T[m]$ and considering the orthonormality of the modal vectors, Eqs. (4.82), we obtain the initial modal displacements

$$\eta_r(0) = \{u\}_r^T[m]\{q(0)\} \qquad r = 1, 2, \ldots, n \tag{4.164}$$

Similarly, the initial modal velocities have the form

$$\dot{\eta}_r(0) = \{u\}_r^T[m]\{\dot{q}(0)\} \qquad r = 1, 2, \ldots, n \tag{4.165}$$

where $\{\dot{q}(0)\}$ is the initial modal velocity vector. Note that the response to initial conditions derived here is the same as that obtained in Sec. 4.9.

The response of a general viscously damped $n$-degree-of-freedom system represents a much more difficult problem. The difficulty can be traced to the coupling introduced by damping. To show this, we recall from Sec. 4.3 that the differential equations of motion of a viscously damped $n$-degree-of-freedom system can be written in the matrix form

$$[m]\{\ddot{q}(t)\} + [c]\{\dot{q}(t)\} + [k]\{q(t)\} = \{Q(t)\} \tag{4.166}$$

where $[c]$ is the $n \times n$ symmetric damping matrix. The remaining quantities are as defined in Eq. (4.154). Using the transformation (4.157), Eq. (4.166) can be reduced to

$$\{\ddot{\eta}(t)\} + [C]\{\dot{\eta}(t)\} + [\omega^2]\{\eta(t)\} = \{N(t)\} \tag{4.167}$$

where

$$[C] = [u]^T[c][u] \tag{4.168}$$

is an $n \times n$ symmetric matrix, *generally nondiagonal*. Hence, in general the classical modal analysis does not lead to an independent system of differential equations of motion. Here, we shall consider some special cases in which $[C]$ is diagonal, or at least it can be treated approximately as diagonal.

In the special case in which $[c]$ is a linear combination of the matrices $[m]$ and

$[k]$, namely, when

$$[c] = \alpha[m] + \beta[k] \qquad (4.169)$$

where $\alpha$ and $\beta$ are constants, matrix $[C]$ does indeed become diagonal,

$$[C] = \alpha[1] + \beta[\omega^2] \qquad (4.170)$$

so that the set (4.167) reduces to an independent set of equations. The case described by Eq. (4.169) is known as *proportional damping*. Introducing the notation

$$[C] = [2\zeta\omega] \qquad (4.171)$$

the $n$ independent sets of equations can be written in the form

$$\ddot{\eta}_r(t) + 2\zeta_r\omega_r\dot{\eta}_r(t) + \omega_r^2\eta_r(t) = N_r(t) \qquad r = 1, 2, \therefore, n \qquad (4.172)$$

where the notation has been chosen so as to render the structure of the equations identical to that of a viscously damped single-degree-of-freedom system of the type studied in Chap. 2.

There are other special cases in which matrix $[C]$ becomes diagonal. They do not occur very frequently, however, and a discussion of these cases lies beyond the scope of this text.†

A case occurring frequently is that in which damping is very small. In such a case, the coupling introduced by the off-diagonal terms of $[C]$ can be regarded as being a second-order effect, and a reasonable approximation can be obtained by discarding these off-diagonal terms. This amounts to regarding $[C]$ as diagonal, although in fact it is not.

When damping is not small, matrix $[C]$ is generally not diagonal, nor can it be regarded as diagonal. This case is treated in Sec. 12.5.

Returning to Eqs. (4.172), we wish to obtain a solution by using the results of Sec. 2.18. Letting $m = 1$ and $\omega_d = \omega_{dr}$ in Eq. (2.204), and converting the notation to that used here, we can write simply

$$\eta_r(t) = \frac{1}{\omega_{dr}} \int_0^t N_r(\tau)e^{-\zeta_r\omega_r(t-\tau)} \sin \omega_{dr}(t - \tau)\, d\tau$$

$$+ e^{-\zeta_r\omega_r t}\left[\frac{\eta_r(0)}{(1 - \zeta_r^2)^{1/2}} \cos (\omega_{dr}t - \psi_r) + \frac{\dot{\eta}_r(0)}{\omega_{dr}} \sin \omega_{dr}t\right] \qquad r = 1, 2, \ldots, n \qquad (4.173)$$

where

$$\omega_{dr} = (1 - \zeta_r^2)^{1/2}\omega_r \qquad (4.174)$$

---

† For a discussion of these cases, see T. K. Caughey, "Classical Normal Modes in Damped Linear Dynamic Systems," *Journal of Applied Mechanics*, vol. 27, pp. 269–271, 1960.

is the damped frequency in the $r$th mode, and

$$\psi_r = \tan^{-1} \frac{\zeta_r}{(1 - \zeta_r^2)^{1/2}} \tag{4.175}$$

is a phase angle associated with the $r$th mode. Hence, the solution of Eq. (4.166) is obtained by introducing Eqs. (4.173) into Eq. (4.162).

**Example 4.11** Let the system shown in Fig. 4.9 be acted upon by the forces

$$F_1(t) = 0 \qquad F_2(t) = F_0 \alpha(t) \tag{a}$$

where $\alpha(t)$ is the unit step function, and derive the system response.

From Example 4.5, we can write the differential equations of motion

$$m\ddot{x}_1(t) + 2kx_1(t) - kx_2(t) = 0$$
$$2m\ddot{x}_2(t) - kx_1(t) + 2kx_2(t) = F_0 \alpha(t) \tag{b}$$

which can be expressed in the matrix form

$$[m]\{\ddot{x}(t)\} + [k]\{x(t)\} = \{F(t)\} \tag{c}$$

where

$$[m] = m\begin{bmatrix} 1 & 0 \\ 0 & 2 \end{bmatrix} \qquad [k] = k\begin{bmatrix} 2 & -1 \\ -1 & 2 \end{bmatrix} \tag{d}$$

are the inertia and stiffness matrices for the system and

$$\{x(t)\} = \begin{Bmatrix} x_1(t) \\ x_2(t) \end{Bmatrix} \qquad \{F(t)\} = \begin{Bmatrix} 0 \\ F_0 \alpha(t) \end{Bmatrix} \tag{e}$$

are the two-dimensional displacement and force vectors, respectively.

To solve the problem by modal analysis, we must first solve the eigenvalue problem associated with $[m]$ and $[k]$. This was actually done in Example 4.5, from which we obtain the natural frequencies and natural modes

$$\omega_1 = 0.7962\sqrt{\frac{k}{m}} \qquad \{u\}_1 = \frac{1}{\sqrt{m}}\begin{Bmatrix} 0.4597 \\ 0.6280 \end{Bmatrix}$$

$$\omega_2 = 1.5382\sqrt{\frac{k}{m}} \qquad \{u\}_2 = \frac{1}{\sqrt{m}}\begin{Bmatrix} 0.8881 \\ -0.3251 \end{Bmatrix} \tag{f}$$

where the modes were normalized in Example 4.6 according to Eq. (4.69). The modal vectors can be arranged in the modal matrix

$$[u] = \frac{1}{\sqrt{m}}\begin{bmatrix} 0.4597 & 0.8881 \\ 0.6280 & -0.3251 \end{bmatrix} \tag{g}$$

Following the procedure outlined earlier, we make use of the linear transformation

$$\{x(t)\} = [u]\{\eta(t)\} \tag{h}$$

where $\{\eta(t)\}$ is a two-dimensional vector of generalized coordinates, and obtain Eq. (4.158) in which $\{N(t)\}$ is the two-dimensional vector of generalized forces having the form

$$\{N(t)\} = [u]^T \{F(t)\} = \frac{1}{\sqrt{m}} \begin{bmatrix} 0.4597 & 0.6280 \\ 0.8881 & -0.3251 \end{bmatrix} \begin{Bmatrix} 0 \\ F_0 u(t) \end{Bmatrix}$$

$$= \frac{F_0}{\sqrt{m}} \begin{Bmatrix} 0.6280 \\ -0.3251 \end{Bmatrix} u(t) \tag{i}$$

Inserting the elements of (i) into Eq. (4.161), we obtain

$$\eta_1(t) = 0.6280 \frac{F_0}{\sqrt{m}} \frac{1}{\omega_1} \int_0^t u(\tau) \sin \omega_1(t - \tau) \, d\tau$$

$$= 0.6280 \frac{F_0}{\omega_1^2 \sqrt{m}} (1 - \cos \omega_1 t)$$

$$\eta_2(t) = -0.3251 \frac{F_0}{\sqrt{m}} \frac{1}{\omega_2} \int_0^t u(t) \sin \omega_2(t - \tau) \, d\tau \tag{j}$$

$$= -0.3251 \frac{F_0}{\omega_2^2 \sqrt{m}} (1 - \cos \omega_2 t)$$

Finally, introducing Eqs. (j) into (h), and considering Eqs. (f) and (g), we can write explicitly

$$x_1(t) = \frac{F_0}{m} \left[ 0.4597 \times 0.6280 \frac{1}{\omega_1^2} (1 - \cos \omega_1 t) \right.$$

$$\left. -0.8881 \times 0.3251 \frac{1}{\omega_2^2} (1 - \cos \omega_2 t) \right]$$

$$= \frac{F_0}{k} \left[ 0.4553 \left( 1 - \cos 0.7962 \sqrt{\frac{k}{m}} t \right) \right.$$

$$\left. - 0.1220 \left( 1 - \cos 1.5383 \sqrt{\frac{k}{m}} t \right) \right]$$

$$x_2(t) = \frac{F_0}{m} \left[ 0.6280^2 \frac{1}{\omega_1^2} (1 - \cos \omega_1 t) \right. \tag{k}$$

$$\left. + 0.3251^2 \frac{1}{\omega_2^2} (1 - \cos \omega_2 t) \right]$$

$$= \frac{F_0}{k} \left[ 0.6219 \left( 1 - \cos 0.7962 \sqrt{\frac{k}{m}} t \right) \right.$$

$$\left. + 0.0447 \left( 1 - \cos 1.5382 \sqrt{\frac{k}{m}} t \right) \right]$$

## PROBLEMS

**4.1** Four discrete masses $m_i$ ($i = 1, 2, 3, 4$) are connected to an inextensible string, as shown in Fig. 4.14. Assume that the tension in the string is constant and that the displacements are small (so that the sine and tangent of an angle can be approximated by the angle itself) and derive Newton's equations of motion by summing up forces acting in the vertical direction on each of the masses.

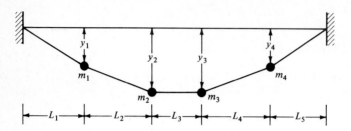

**Figure 4.14**

**4.2** Derive Newton's equations of motion for the system shown in Fig. 4.15 and write the equations in matrix form.

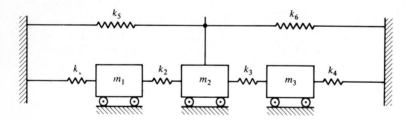

**Figure 4.15**

**4.3** Repeat Prob. 4.2 for the system of Fig. 4.16.

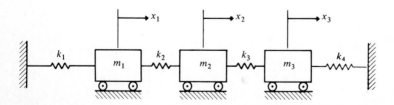

**Figure 4.16**

**4.4** Derive the equations of motion for the $n$-story building shown in Fig. 4.17. Make use of the concept of equivalent springs (Sec. 1.2).

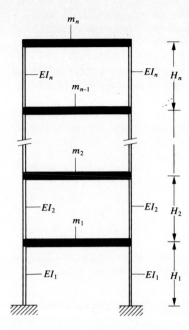

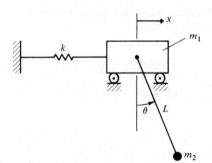

**Figure 4.17**

**4.5** Derive Newton's equations of motion for the system of Fig. 4.18. The angle $\theta$ is arbitrarily large.

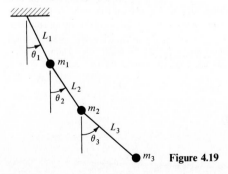

$m_2$     **Figure 4.18**

**4.6** Derive Newton's equations of motion for the triple pendulum shown in Fig. 4.19. The angles $\theta_i$ $(i = 1, 2, 3)$ are arbitrarily large.

**4.7** Consider the system of Fig. 4.15 and calculate the flexibility and stiffness influence coefficients by using the definitions. Make use of the concept of equivalent springs (Sec. 1.2). Arrange the coefficients in matrix form, let $k_1 = k_2 = k_3 = k$, $k_4 = k_5 = k_6 = 2k$, and check your results by inverting the stiffness matrix to obtain the flexibility matrix.

**4.8** Consider the system of Prob. 4.1, let $L_i = L$ ($i = 1, 2, \ldots, 5$) and determine the flexibility influence coefficients by using the definition. Check your results by inverting the stiffness matrix of Prob. 4.1 for the same special case.

**4.9** Consider a cantilever bar supporting three point masses $m_i$ ($i = 1, 2, 3$), as depicted in Fig. 4.20. The segments between the support and the point masses are massless and possess corresponding flexural rigidities $EI_i$ ($i = 1, 2, 3$). Derive the flexibility matrix for the system.

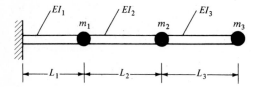

Figure 4.20

**4.10** Consider the system of Prob. 4.3 and determine the mass and stiffness matrices by writing the kinetic and potential energy expressions. Then, consider the linear transformation

$$x_1 = y_1 \qquad x_2 - x_1 = y_2 \qquad x_3 - x_2 = y_3$$

write the kinetic and potential energy expressions in terms of the new coordinates, and determine the associated mass and stiffness matrices. Compare the mass and stiffness matrices corresponding to the two sets of coordinates and draw conclusions concerning the nature of coupling for both sets of coordinates. Explain the reasons for the different types of coupling.

**4.11** The system shown in Fig. 4.21 consists of four masses connected by three springs. Show how the system can be reduced to a three-degree-of-freedom system for the elastic motion.

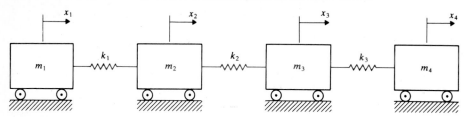

**Figure 4.21**

**4.12** Consider a bar hinged at the left end and free at the right end, as shown in Fig. 4.22. In this case the system is positive semidefinite and there exists a rigid-body mode in the form of rigid-body rotation of the bar about point $O$. Derive the eigenvalue problem for the elastic motion of the system. *Hint:* Assume that the displacements of the masses consist of a rigid part and an elastic part, where the first is due to the rigid rotation about $O$ and the second is due to flexure, as measured relative to the line of rotation. For the kinetic energy use absolute velocities (consisting of the sum of the rigid and elastic parts), whereas for the potential energy use only the elastic part of the displacements. Then, use the conservation of the angular momentum about $O$ to eliminate the rigid-body rotation from the kinetic energy.

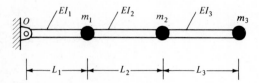

Figure 4.22

**4.13** Consider the system of Prob. 4.3, let $m_1 = m_2 = m_3 = m$, $k_1 = k_2 = k_3 = k_4 = k$ and use Rayleigh's quotient to produce estimates of all three natural frequencies.

**4.14** Use Rayleigh's quotient to produce estimates of the two lowest natural frequencies of the system of Prob. 4.8. Let $m_i = m$ $(i = 1, 2, 3, 4)$.

**4.15** Use Rayleigh's quotient to estimate the lowest natural frequency of the system of Prob. 4.9. Let $L_i = L$, $m_i = m$, $EI_i = EI$ $(i = 1, 2, 3)$.

**4.16** Solve the eigenvalue problem corresponding to the system of Fig. 4.15 by the characteristic determinant method and obtain the natural frequencies and natural modes for the case $m_1 = m_2 = m$, $m_3 = 2m$, $k_1 = k_2 = k_3 = k$, $k_4 = k_5 = k_6 = 2k$. Plot the natural modes.

**4.17** Consider the triple pendulum of Prob. 4.6, linearize the equations of motion by assuming small angles $\theta_i$ $(i = 1, 2, 3)$, solve the associated eigenvalue problem by the characteristic determinant method, and obtain the natural frequencies and natural modes for the case $L_1 = L_2 = L_3 = L$, $m_1 = m_2 = m_3 = m$. Plot the natural modes.

**4.18** Check the orthogonality with respect to the inertia matrix and stiffness matrix of the eigenvectors computed in Probs. 4.16 and 4.17.

**4.19** Consider the system of Prob. 4.3, let $k_1 = k_2 = k$, $k_3 = k_4 = 2k$, $m_1 = m$, $m_2 = m_3 = 2m$ and solve the eigenvalue problem by matrix iteration. Plot the natural modes.

**4.20** Solve the eigenvalue problem of Prob. 4.16 by matrix iteration. Plot the natural modes.

**4.21** Repeat Prob. 4.20 for the system of Prob. 4.17.

**4.22** Repeat Prob. 4.20 for the system of Prob. 4.4 with $n = 3$. Let $EI_i = EI$, $m_i = m$, $H_i = H$ $(i = 1, 2, 3)$.

**4.23** Repeat Prob. 4.20 for the system of Prob. 4.9. Let $EI_i = EI$, $m_i = m$, $L_i = L$ $(i = 1, 2, 3)$.

**4.24** Repeat Prob. 4.23 for the system of Prob. 4.12. Note that the modes must be in terms of absolute displacements and not elastic displacements alone. *Hint:* To determine the contribution of the rotation to the modes, use the same equation as that used to eliminate the rotation from the kinetic energy.

**4.25** Determine the response of the three-story building of Prob. 4.22 to the horizontal ground motion $y(t) = A \sin \omega t$, where $A$ is a displacement amplitude.

**4.26** Determine the response of the system of Prob. 4.15 to the excitation $F_1 = F_3 = 0$, $F_2 = F_0 \alpha(t)$, where $\alpha(t)$ is the unit step function.

**4.27** Determine the response of the triple pendulum of Prob. 4.17 to a horizontal force in the form of an impulse of amplitude $\hat{F}_0$ applied to the mass $m_3$ at $t = 0$.

**4.28** Determine the response of the system of Prob. 4.19 to a horizontal periodic force applied to the mass $m_3$. The periodic force is as shown in Fig. 2.22.

**4.29** The system shown in Fig. 4.23 is the same as that in Prob. 4.19 but with the viscous damping $c_1 = c_2 = c$, $c_3 = c_4 = 2c$ added. Determine the response to the initial excitation $x_1(0) = x_2(0) = 0$, $x_3(0) = x_0$, $\dot{x}_i(0) = 0$ $(i = 1, 2, 3)$.

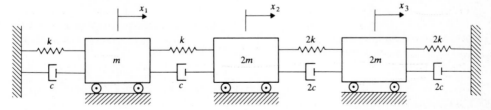

**Figure 4.23**

**4.30** Solve Prob. 4.27 under the assumption that the bobs of the triple pedulum are subjected to resisting forces proportional to their velocities, where the proportionality constants are $c_1 = c_2 = c_3 = c$.

# FIVE

# CONTINUOUS SYSTEMS. EXACT SOLUTIONS

## 5.1 GENERAL DISCUSSION

Chapter 4 was devoted exclusively to the vibration of discrete systems, whereas this chapter is devoted to continuous systems. This fact should not be interpreted, however, as an indication that discrete and continuous systems represent different types of systems exhibiting dissimilar dynamical characteristics. In reality the opposite is true, as discrete and continuous systems represent merely two mathematical models of identical physical systems. The basic difference between discrete and continuous systems is that discrete systems have a finite number of degrees of freedom and continuous systems have an infinite number of degrees of freedom. This results from the fact that the index $i$ identifying a typical lumped mass has as counterpart an independent spatial variable $x$ identifying the nominal position of an infinitesimal mass element. Consistent with this, discrete systems are governed by ordinary differential equations and continuous systems by partial differential equations. Nevertheless, because discrete and continuous systems represent in general models of identical physical systems, they display similar dynamical behavior.

This chapter begins by stressing the intimate relation between discrete and continuous systems. In fact, the mathematical formulation for a given continuous system is derived as a limiting case of that of a discrete system. The discussion continues by showing that various concepts introduced in our study of discrete systems have their counterparts in continuous systems. Indeed, to a finite set of eigenvalues and finite-dimensional eigenvectors corresponds an infinite set of eigenvalues and space-dependent eigenfunctions. Concepts such as the orthogonality of natural modes of vibration and the ensuing expansion theorem can be defined for continuous systems in a manner analogous to that for discrete systems, and the same can be said about Rayleigh's quotient.

In this chapter, a number of continous systems are discussed, such as strings in transverse vibration, rods in axial vibration, shafts in torsion, and bars in bending. Strings, rods, and shafts are governed by second-order differential equations in space and are analogous in nature. On the other hand, bars are governed by fourth-order differential equations. Exact solutions for the vibration of continuous systems can be obtained only in special cases, mainly when the system parameters are uniformly distributed. In this case, second-order differential equations in space reduce to the so-called "wave equation." A discussion of the wave equation enables us to demonstrate the connection between traveling and standing waves. Finally, expressions for the kinetic and potential energy of continuous systems are derived, thus completing the analogy with discrete systems.

## 5.2 RELATION BETWEEN DISCRETE AND CONTINUOUS SYSTEMS. BOUNDARY-VALUE PROBLEM

As pointed out in Sec. 5.1, there is a very intimate relation between discrete and continuous systems, as they generally represent two distinct mathematical models of the same physical system. To demonstrate this, we derive the differential equation for the transverse vibration of a string first by regarding it as a discrete system and letting it approach a continuous model in the limit. Then, we formulate the problem by regarding the system as continuous from the beginning.

Let us consider a system of discrete masses $m_i$ ($i = 1, 2, \ldots, n$) connected by massless strings, where the masses $m_i$ are subjected to the external forces $F_i$, as shown in Fig. 5.1a. To derive the differential equation of motion for a typical mass $m_i$, we concentrate our attention on the three adjacent masses $m_{i-1}$, $m_i$, and $m_{i+1}$ of Fig. 5.1b. The tensions in the string segments connecting $m_i$ to $m_{i-1}$ and $m_{i+1}$ are denoted by $T_{i-1}$ and $T_i$, and the horizontal projections of these segments by $\Delta x_{i-1}$ and $\Delta x_i$, respectively. The displacements $y_i(t)$ ($i = 1, 2, \ldots, n$) of the masses $m_i$ are assumed to be small, so that the projections $\Delta x_i$ remain essentially unchanged during motion. Moreover, the angles between the string segments and the horizontal are sufficiently small that the sine and tangent of the angles are approximately equal to one another. Hence, using Newton's second law, the equation of motion of the mass $m_i$ in the vertical direction has the form

$$T_i \frac{y_{i+1} - y_i}{\Delta x_i} - T_{i-1} \frac{y_i - y_{i-1}}{\Delta x_{i-1}} + F_i = m_i \frac{d^2 y_i}{dt^2} \tag{5.1}$$

Equation (5.1) is applicable to any mass $m_i$ ($i = 2, 3, \ldots, n - 1$). The equation can also be used for $i = 1$ and $i = n$, but certain provisions must be made to reflect the

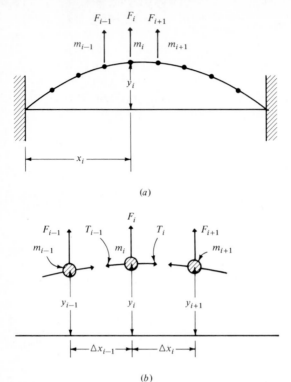

(a)

(b)                                                    Figure 5.1

way the system is supported, as we shall see shortly. Rearranging Eq. (5.1), we obtain the set of simultaneous ordinary differential equations

$$\frac{T_i}{\Delta x_i} y_{i+1} - \left(\frac{T_i}{\Delta x_i} + \frac{T_{i-1}}{\Delta x_{i-1}}\right) y_i + \frac{T_{i-1}}{\Delta x_{i-1}} y_{i-1} + F_i = m_i \frac{d^2 y_i}{dt^2}$$

$$i = 1, 2, \dots, n \quad (5.2)$$

in the variables $y_i$ ($i = 1, 2, \dots, n$), and we notice that the equations for $i = 1$ and $i = n$ contain the displacements $y_0$ and $y_{n+1}$, respectively. If the string is fixed at both ends, as is the case with the system shown in Fig. 5.1a, then we must set

$$y_0(t) = y_{n+1}(t) = 0 \quad (5.3)$$

in Eqs. (5.2). In other cases different conditions are possible. Indeed, if the ends $x = 0$ and $x = L$ are attached to vertical springs, or if they are free to move along a vertical line, the end conditions must reflect the fact that there is a force proportional to the stretching of the spring, or that the vertical component of the force at that particular end is zero. We shall not pursue this subject any further at this time but return to Eq. (5.1), because our object is to draw the analogy between discrete and continuous systems.

If we introduce the notation $y_{i+1} - y_i = \Delta y_i$, $y_i - y_{i-1} = \Delta y_{i-1}$, Eq. (5.1) becomes

$$T_i \frac{\Delta y_i}{\Delta x_i} - T_{i-1} \frac{\Delta y_{i-1}}{\Delta x_{i-1}} + F_i = m_i \frac{d^2 y_i}{dt^2} \qquad i = 1, 2, \ldots, n \qquad (5.4)$$

But the first two terms on the left side of Eq. (5.4) constitute the incremental change in the vertical force component between the left and right sides of $m_i$. In view of this, we can write Eq. (5.4) as

$$\Delta \left( T_i \frac{\Delta y_i}{\Delta x_i} \right) + F_i = m_i \frac{d^2 y_i}{dt^2} \qquad i = 1, 2, \ldots, n \qquad (5.5)$$

Moreover, dividing both sides of (5.5) by $\Delta x_i$, we arrive at

$$\frac{\Delta}{\Delta x_i} \left( T_i \frac{\Delta y_i}{\Delta x_i} \right) + \frac{F_i}{\Delta x_i} = \frac{m_i}{\Delta x_i} \frac{d^2 y_i}{dt^2} \qquad i = 1, 2, \ldots, n \qquad (5.6)$$

At this time we let the number $n$ of masses $m_i$ increase indefinitely, while the masses themselves and the distance between them decrease correspondingly, and replace the indexed position $x_i$ by the independent spatial variable $x$, so that in the limit, as $\Delta x_i \to 0$, Eq. (5.6) reduces to

$$\frac{\partial}{\partial x} \left[ T(x) \frac{\partial y(x, t)}{\partial x} \right] + f(x, t) = \rho(x) \frac{\partial^2 y(x, t)}{\partial t^2} \qquad (5.7)$$

which must be satisfied over the domain $0 < x < L$, where

$$f(x, t) = \lim_{\Delta x_i \to 0} \frac{F_i(t)}{\Delta x_i} \qquad \rho(x) = \lim_{\Delta x_i \to 0} \frac{m_i}{\Delta x_i} \qquad (5.8)$$

are the distributed transverse force on the string and the mass density at point $x$, respectively. We note that, by virtue of the fact that the indexed position $x_i$ is replaced by the independent spatial variable $x$, total derivatives with respect to the time $t$ become partial derivatives with respect to $t$, whereas ratios of increments are replaced directly by partial derivatives with respect to $x$. Equation (5.7) represents the *partial differential equation of the string*. Similarly, conditions (5.3) must be replaced by

$$y(0, t) = y(L, t) = 0 \qquad (5.9)$$

which are generally known as the *boundary conditions* of the problem. Equations (5.7) and (5.9) constitute what is referred to as a *boundary-value problem*. In fact, the transverse displacement $y(x, t)$ is also subject to the initial conditions

$$y(x, 0) = y_0(x) \qquad \left. \frac{\partial y(x, t)}{\partial t} \right|_{t=0} = v_0(x) \qquad (5.10)$$

where $y_0(x)$ is the initial displacement and $v_0(x)$ the initial velocity at every point $x$ of the string, so that Eqs. (5.7), (5.9), and (5.10) represent a *boundary-value and initial-value problem* simultaneously.

As mentioned above, the problem can be formulated more directly by considering the string as a continuous system, as shown in Fig. 5.2a, where $f(x, t)$, $\rho(x)$, and $T(x)$ are respectively the distributed force, mass density, and tension at point $x$. Figure 5.2b represents the free-body diagram corresponding to an element of string of length $dx$. Again writing Newton's second law for the force component in the vertical direction, we obtain

$$
\left[ T(x) + \frac{\partial T(x)}{\partial x} dx \right] \left[ \frac{\partial y(x, t)}{\partial x} + \frac{\partial^2 y(x, t)}{\partial x^2} dx \right]
$$

$$
- T(x) \frac{\partial y(x, t)}{\partial x} + f(x, t) dx = \rho(x) dx \frac{\partial^2 y(x, t)}{\partial t^2} \quad (5.11)
$$

Canceling appropriate terms and ignoring second-order terms in $dx$, Eq. (5.11) reduces to

$$
\frac{\partial T(x)}{\partial x} \frac{\partial y(x, t)}{\partial x} dx + T(x) \frac{\partial^2 y(x, t)}{\partial x^2} dx + f(x, t) dx = \rho(x) dx \frac{\partial^2 y(x, t)}{\partial t^2} \quad (5.12)
$$

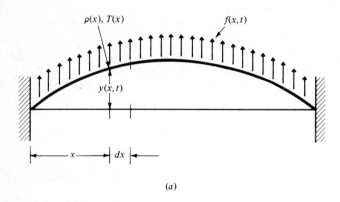

(a)

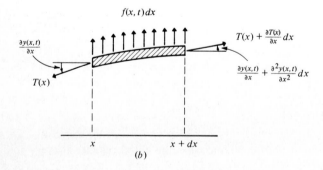

(b)

**Figure 5.2**

and, after dividing both sides by $dx$, we can write Eq. (5.12) in the more compact form

$$\frac{\partial}{\partial x}\left[T(x)\frac{\partial y(x, t)}{\partial x}\right] + f(x, t) = \rho(x)\frac{\partial^2 y(x, t)}{\partial t^2} \qquad 0 < x < L \qquad (5.13)$$

which is identical to Eq. (5.7) in every respect. Moreover, from Fig. 5.2b, we recognize that the displacement of the string at the two ends must be zero, $y(0, t) = y(L, t) = 0$, thus duplicating boundary conditions (5.9). This completes the mathematical analogy between the discrete and continuous models.

The unmistakable conclusion is that Figs. 5.1a and 5.2a, although different in appearance, represent two intimately related mathematical models. In this particular section we made the transition from the discrete system, Fig. 5.1a, to the continuous one, Fig. 5.2a, through a limiting process equivalent to spreading the masses over the entire string. In many practical applications, particularly if the string is nonuniform, it is more common to follow the opposite path and lump a continuous system into discrete masses. This can be done by using the second of Eqs. (5.8) and writing $m_i = \rho(x_i)\,\Delta x_i$. Regardless of which mathematical model is ultimately chosen, it is clear that we should expect similar vibrational characteristics.

It turns out that the longitudinal vibration of a thin rod and the torsional vibration of a shaft of circular cross section satisfy similar boundary-value problems. In fact, the corresponding problems can be derived from the associated discrete models (see Probs. 5.1 and 5.2). For longitudinal vibration, the parameters $\rho(x)$ and $T(x)$ must be replaced by the mass per unit length $m(x)$ and the axial stiffness $EA(x)$, respectively, where $E$ is the modulus of elasticity and $A(x)$ the cross-sectional area. For torsional vibration, they must be replaced by the mass polar moment of inertia per unit length $I(x)$ and the torsional stiffness $GJ(x)$, respectively, where $G$ is the shear modulus and $J(x)$ the polar moment of inertia of the cross-sectional area.

## 5.3 FREE VIBRATION. THE EIGENVALUE PROBLEM

Let us consider the vibrating string of Sec. 5.2. In the case of free vibration, namely, when the distributed force is zero, $f(x, t) = 0$, the boundary-value problem reduces to the differential equation

$$\frac{\partial}{\partial x}\left[T(x)\frac{\partial y(x, t)}{\partial x}\right] = \rho(x)\frac{\partial^2 y(x, t)}{\partial t^2} \qquad 0 < x < L \qquad (5.14)$$

and the boundary conditions

$$y(0, t) = y(L, t) = 0 \qquad (5.15)$$

Although the free-vibration problem for a continuous system, Eqs. (5.14) and (5.15), differs in appearance from that of a discrete system, Eqs. (4.54), the general approach to the solution is the same. Hence, we wish to explore the possibility of synchronous motion, that is to say, a motion in which the general shape of the string displacement does not change with time, while the amplitude of this general shape does change with time. Stating it differently, every point of the string executes the same motion in time, passing through the equilibrium position at the same time and reaching its maximum excursion at the same time. In mathematical terminology, this implies that the displacement $y(x, t)$ is separable in space and time, so that we wish to examine the possibility that the solution of the boundary-value problem can be written in the form

$$y(x, t) = Y(x)F(t) \qquad (5.16)$$

where $Y(x)$ represents the general string configuration and depends on the spatial variable $x$ alone, and where $F(t)$ indicates the type of motion the string configuration executes with time and depends on $t$ alone. Consistent with the approach used in Sec. 4.7 for discrete systems, we confine ourselves to the case in which $y(x, t)$ undergoes stable harmonic oscillation, which implies that $F(t)$ must be bounded for all times.

Introducing Eq. (5.16) into (5.14), and dividing through by $\rho(x)Y(x)F(t)$, we obtain

$$\frac{1}{\rho(x)Y(x)} \frac{d}{dx}\left[ T(x) \frac{dY(x)}{dx} \right] = \frac{1}{F(t)} \frac{d^2F(t)}{dt^2} \qquad (5.17)$$

where, because $Y$ depends only on $x$ and $F$ only on $t$, partial derivatives have been replaced by total derivatives. Moreover, the variables have been separated so that the left side of Eq. (5.17) depends on $x$ alone, whereas the right side depends on $t$ alone. Using the standard argument employed in conjunction with the separation of variables method (see also Sec. 4.7), we conclude that the only way Eq. (5.17) can be satisfied for every $x$ and $t$ is that both sides be constant. In view of the results derived in Sec. 4.7, we denote the constant by $-\omega^2$, so that Eq. (5.17) leads to

$$\frac{d^2F(t)}{dt^2} + \omega^2 F(t) = 0 \qquad (5.18)$$

$$-\frac{d}{dx}\left[ T(x) \frac{dY(x)}{dx} \right] = \omega^2 \rho(x) Y(x) \qquad 0 < x < L \qquad (5.19)$$

We recall from Sec. 4.7 that the reason for selecting the constant as negative is for Eq. (5.18) to represent the equation of a harmonic oscillator, whose solution consists of trigonometric functions. Had we chosen a positive constant, the solution of the resulting equation would have been in terms of exponential functions, one with a positive exponent and the other with a negative one. Because the solution with the positive exponent diverges with time and that with the negative exponent decays with time, these solutions are inconsistent with the stable oscillation considered here, for which the motion amplitude must remain finite. It follows that, if synchronous motion is possible, then the function $F(t)$ expressing the time dependence must be harmonic. Hence, as in Sec. 4.7, we can write the solution of Eq. (5.18) in the form

$$F(t) = C \cos(\omega t - \phi) \tag{5.20}$$

where $C$ is an arbitrary constant, $\omega$ the frequency of the harmonic motion, and $\phi$ its phase angle, all three quantities being the same for any function $Y(x)$ that is a solution of Eq. (5.19).

The question remains as to the displacement configuration, namely, the function $Y(x)$. Clearly, $Y(x)$ must satisfy Eq. (5.19) over the domain $0 < x < L$. Moreover, from Eqs. (5.15), it must also satisfy the boundary conditions

$$Y(0) = Y(L) = 0 \tag{5.21}$$

Following is a general discussion of the solution of Eqs. (5.19) and (5.21). It parallels the discussion of the eigenvalue problem for discrete systems (see Sec. 4.7), the various concepts being entirely analogous.

We note that Eq. (5.19) contains the parameter $\omega^2$, as yet undetermined. The problem of determining the values of the parameter $\omega^2$ for which nontrivial solutions $Y(x)$ of Eq. (5.19) exist, where the solutions are subject to boundary conditions (5.21), is called the *characteristic-value*, or *eigenvalue*, *problem*. The corresponding values of the parameter are known as *characteristic values*, or *eigenvalues*, and the associated functions $Y(x)$ as *characteristic functions*, or *eigenfunctions*. Equation (5.19) is a second-order ordinary differential equation and contains the parameter $\omega^2$. Hence, we must determine two constants of integration, in addition to $\omega^2$, but we have at our disposal only two boundary conditions. Because Eq. (5.19) is homogeneous, however, we conclude that only the shape of the function $Y(x)$ can be determined uniquely and that the amplitude of the function is arbitrary. Indeed, if $Y(x)$ is a solution of Eq. (5.19), then $\alpha Y(x)$ is also a solution, where $\alpha$ is a constant multiplier. It follows that one of the two boundary conditions (5.21) can be used to solve for one constant of integration in terms of the other, thus determining the general shape of $Y(x)$ but not its amplitude. The other boundary condition can be used to produce the so-called *characteristic equation*, or *frequency equation*; the values of the parameter $\omega^2$ are obtained by solving this equation. The solution of the characteristic equation consists of a denumerably infinite set of discrete characteristic values, the square roots of which are the system *natural frequencies* $\omega_r$ $(r = 1, 2, \ldots)$. To each characteristic value, or natural frequency, corresponds an eigenfunction, or *natural mode*, $Y_r(x)$. As mentioned above, because the problem is homogeneous, $A_r Y_r(x)$ represents the same natural mode, where $A_r$ is an arbitrary constant, so that the amplitudes of the natural modes are undetermined. The constants $A_r$, and hence the amplitudes, can be determined uniquely if a certain normalization process is used, in which case the natural modes become *normal modes*. The natural frequencies $\omega_r$ and associated natural modes $Y_r(x)$ $(r = 1, 2, \ldots)$ depend on the system parameters $\rho(x)$ and $T(x)$, as well as on the boundary conditions; thus they are a characteristic of the system. Note that the modes $Y_r(x)$ can be regarded as infinite-dimensional eigenvectors, obtained as limiting cases of finite-dimensional eigenvectors in a process that replaces the discrete indexed position $x_i$ by the continuous spatial variable $x$.

We recall that for discrete systems we also identified a set of natural frequencies and natural modes representing a characteristic of the system. Another characteristic common to discrete and continuous systems is the *orthogonality of modes*, a property to be discussed later. Hence, the analogy between discrete and continuous systems is complete, with the exception that for discrete systems the set of natural frequencies and modes is finite, whereas for continuous systems the set is infinite. The orthogonality condition can be written as

$$\int_0^L \rho(x) Y_r(x) Y_s(x) \, dx = 0 \qquad r \neq s \tag{5.22}$$

where $Y_r(x)$ and $Y_s(x)$ are two distinct eigenfunctions. For convenience, the modes can be normalized by writing

$$\int_0^L \rho(x) Y_r(x) Y_s(x) \, dx = \delta_{rs} \qquad r, s = 1, 2, \ldots \tag{5.23}$$

where $\delta_{rs}$ is the Kronecker delta. Moreover, we shall verify later that the eigenfunctions $Y_r(x)$ satisfy also the relation

$$\int_0^L T(x) \frac{dY_r(x)}{dx} \frac{dY_s(x)}{dx} \, dx = \omega_r^2 \delta_{rs} \qquad r, s = 1, 2, \ldots \tag{5.24}$$

In view of the above, the free-vibration solution of Eq. (5.14) can be represented by an infinite series of the system eigenfunctions in the form

$$y(x, t) = \sum_{r=1}^{\infty} Y_r(x)\eta_r(t) \tag{5.25}$$

Introducing Eq. (5.25) into (5.14), multiplying the result by $Y_s(x)$, integrating over the domain $0 < x < L$, recalling that the system eigenfunctions satisfy Eq. (5.19) and assuming that they are normalized so as to satisfy conditions (5.23) and (5.24), we arrive at the infinite set of harmonic equations

$$\ddot{\eta}_r(t) + \omega_r^2 \eta_r(t) = 0 \qquad r = 1, 2, \ldots \tag{5.26}$$

where the time-dependent functions $\eta_r(t)$ are the system *natural coordinates*, which in this case are also *normal coordinates*. As in Sec. 4.7, the solution of Eqs. (5.26) can be written as

$$\eta_r(t) = C_r \cos(\omega_r t - \phi_r) \qquad r = 1, 2, \ldots \tag{5.27}$$

where the constants $C_r$ and $\phi_r$ are the amplitude and phase angle, respectively, quantities which depend on the initial conditions. The response of the system to initial conditions can be obtained by inserting (5.27) into (5.25). We shall not pursue the subject any further at this point, but return to it in Sec. 5.9, where the response to both initial excitation and forcing functions is presented.

A simple illustration of the solution of the eigenvalue problem is furnished in Example 5.1. Further elaboration, including a proof of the orthogonality property, is provided in subsequent sections.

**Example 5.1** Solve the eigenvalue problem associated with a uniform string fixed at $x = 0$ and $x = L$ (see Fig. 5.3), and plot the first three eigenfunctions. The tension $T$ in the string is constant.

Inserting $\rho(x) = \rho = $ const, $T(x) = T = $ const in Eq. (5.19), we conclude that the transverse displacement $Y(x)$ must satisfy the differential equation

$$\frac{d^2 Y(x)}{dx^2} + \beta^2 Y(x) = 0 \qquad \beta^2 = \frac{\omega^2 \rho}{T} \qquad (a)$$

over the domain $0 < x < L$. Moreover, because the ends are fixed, the displacement must be zero at $x = 0$ and $x = L$. Hence, the solution $Y$ of Eq. $(a)$ is subject to the boundary conditions

$$Y(0) = 0 \qquad Y(L) = 0 \qquad (b)$$

Equation $(a)$ is harmonic in $x$, and its solution can be written in the form

$$Y(x) = A \sin \beta x + B \cos \beta x \qquad (c)$$

where $A$ and $B$ are constants of integration. Inserting the first of boundary conditions $(b)$ into $(c)$, we conclude that $B = 0$, so that the solution reduces to

$$Y(x) = A \sin \beta x \qquad (d)$$

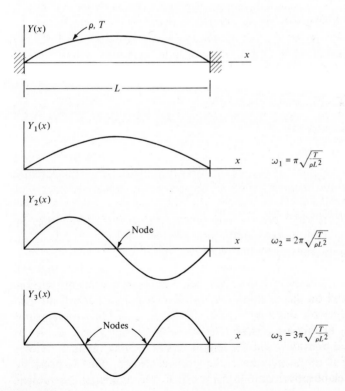

$$\omega_1 = \pi \sqrt{\frac{T}{\rho L^2}}$$

$$\omega_2 = 2\pi \sqrt{\frac{T}{\rho L^2}}$$

$$\omega_3 = 3\pi \sqrt{\frac{T}{\rho L^2}}$$

**Figure 5.3**

On the other hand, introducing the second of boundary conditions (*b*) into Eq. (*d*), we obtain

$$Y(L) = A \sin \beta L = 0 \qquad (e)$$

There are two ways in which Eq. (*e*) can be satisfied, namely, $A = 0$ and $\sin \beta L = 0$. But $A = 0$ must be ruled out, because this would yield the trivial solution $Y(x) = 0$. It follows that we must have

$$\sin \beta L = 0 \qquad (f)$$

which is recognized as the *characteristic equation*. Its solution consists of the infinite set of *characteristic values*.

$$\beta_r L = r\pi \qquad r = 1, 2, \ldots \qquad (g)$$

to which corresponds the infinite set of *eigenfunctions*

$$Y_r(x) = A_r \sin \frac{r\pi x}{L} \qquad (h)$$

where $A_r$ are undetermined amplitudes, with the implication that only the mode shapes can be determined uniquely. The first three natural modes are plotted in Fig. 5.3, where the modes have been normalized by letting $A_r = 1$. We note that the first mode has no nodes, the second has one node and the third has two nodes. In general the *r*th mode has $r - 1$ nodes ($r = 1, 2, \ldots$). From the second of Eqs. (*a*) we conclude that the system *natural frequencies* are

$$\omega_r = \beta_r \sqrt{\frac{T}{\rho}} = r\pi \sqrt{\frac{T}{\rho L^2}} \qquad r = 1, 2, \ldots \qquad (i)$$

The frequency $\omega_1$ is called the *fundamental frequency* and the higher frequencies $\omega_r$ ($r = 2, 3, \ldots$) are referred to as *overtones*. The overtones are integral multiples of the fundamental frequency, for which reason the fundamental frequency is called the *fundamental harmonic* and the overtones are known as *higher harmonics*.

Vibrating systems which possess harmonic overtones are distinguished by the fact that under certain excitations they produce pleasant sounds. Such systems are not commonly encountered in nature but can be manufactured, particularly for use in musical instruments. It is a well-known fact that the string is the major ingredient in a large number of musical instruments, such as the violin, the piano, the guitar and many other instruments related to them. For example, the violin has four strings which possess four fundamental frequencies. From Eq. (*i*), we observe that these frequencies depend on the tension $T$, the mass density $\rho$ and the length $L$. The violinist tuning a violin merely ensures that the strings have the proper tension. This is done by comparing the pitch of a given note to that produced by a different instrument known to be tuned correctly. One must not infer from this, however, that the violin yields only four fundamental frequencies and their higher harmonics. Indeed, whereas $\rho$ and $T$ are constant for each string, the violinist can change the pitch by adjusting the length of the strings. Hence, when fingers are run on the fingerboard, the artist merely adjusts the length $L$ of the strings. Thus, there is a large variety of frequencies at the violinist's disposal. Generally the sounds consist of a combination of harmonics, with the lower harmonics being the predominant ones. However, a talented performer excites the proper array of higher harmonics to produce a pleasing sound.

**Example 5.2** Consider the eigenvalue problem of Example 5.1 and verify that the eigenfunctions satisfy the orthogonality relations, Eqs. (5.23) and (5.24).

Before verifying the satisfaction of Eqs. (5.23) and (5.24), we must normalize the modes according to

$$\int_0^L \rho(x) Y_r^2(x)\, dx = 1 \qquad r = 1, 2, \ldots \tag{a}$$

Hence, inserting Eqs. (h) of Example 5.1 into Eq. (a), and recalling that $\rho(x) = \rho = \text{const}$, we can write

$$\rho A_r^2 \int_0^L \sin^2 \frac{r\pi x}{L}\, dx = 1 \qquad r = 1, 2, \ldots \tag{b}$$

But $\sin^2 \alpha = \frac{1}{2}(1 - \cos 2\alpha)$, so that

$$\int_0^L \sin^2 \frac{r\pi x}{L}\, dx = \frac{1}{2} \int_0^L \left(1 - \cos \frac{2r\pi x}{L}\right) dx$$

$$= \frac{1}{2} \left(x - \frac{\sin 2r\pi x/L}{2r\pi/L}\right)\Bigg|_0^L = \frac{L}{2} \tag{c}$$

Inserting Eq. (c) into (b), we conclude that

$$A_r = \sqrt{\frac{2}{\rho L}} \qquad r = 1, 2, \ldots \tag{d}$$

so that the normal modes become

$$Y_r(x) = \sqrt{\frac{2}{\rho L}} \sin \frac{r\pi x}{L} \qquad r = 1, 2, \ldots \tag{e}$$

Using Eq. (e), we can form

$$\int_0^L \rho\, Y_r(x) Y_s(x)\, dx = \frac{2}{L} \int_0^L \sin \frac{r\pi x}{L} \sin \frac{s\pi x}{L}\, dx \tag{f}$$

Recalling that $\sin \alpha \sin \beta = \frac{1}{2}[\cos(\alpha - \beta) - \cos(\alpha + \beta)]$, we can write

$$\int_0^L \sin\frac{r\pi s}{L}\sin\frac{s\pi x}{L}\,dx = \frac{1}{2}\int_0^L\left[\cos\frac{(r-s)\pi x}{L} - \cos\frac{(r+s)\pi x}{L}\right]dx$$

$$= \frac{1}{2}\left[\frac{\sin[r-s]\pi x/L}{(r-s)\pi/L} - \frac{\sin(r+s)\pi x/L}{(r+s)\pi/L}\right]\Bigg|_0^L$$

$$= \begin{cases} 0 & r \neq s \\ \dfrac{L}{2} & r = s \end{cases} \qquad (g)$$

Hence, inserting Eq. ($g$) into ($f$), we can write

$$\int_0^L \rho Y_r(x)Y_s(x)\,dx = \delta_{rs} \qquad r,s = 1,2,\ldots \qquad (h)$$

where $\delta_{rs}$ is the Kronecker delta, thus verifying Eq. (5.23).

To verify Eq. (5.24), we follow a procedure similar to that above, recall Eq. ($i$) of Example 5.1, and write

$$\int_0^L T\frac{dY_r(x)}{dx}\frac{dY_s(x)}{dx}\,dx = T\frac{2}{\rho L}\frac{r\pi}{L}\frac{s\pi}{L}\int_0^L \cos\frac{r\pi x}{L}\cos\frac{s\pi x}{L}\,dx$$

$$= T\frac{2}{\rho L}\frac{r\pi}{L}\frac{s\pi}{L}\frac{L}{2}\delta_{rs}$$

$$= \omega_r^2\,\delta_{rs} \qquad r,s = 1,2,\ldots \qquad (i)$$

which is identical to Eq. (5.24).

Note that the fact that the eigenfunctions $Y_r(x)$ in this particular case satisfy relations (5.23) and (5.24) is a mere reiteration of the ordinary orthogonality of trigonometric functions. However, we shall have the opportunity to establish that the orthogonality of the eigenfunctions is much more general in nature, as the eigenfunctions of a system are trigonometric functions only in very special cases.

## 5.4 CONTINUOUS VERSUS DISCRETE MODELS FOR THE AXIAL VIBRATION OF RODS

To bring the parallel between continuous and discrete models into sharper focus, we consider a specific system and compare the solutions of the eigenvalue problem obtained by regarding the same system first as continuous and then as discrete. A system that lends itself readily to such an analysis is the rod in axial vibration.

As indicated in Sec. 5.2, the boundary-value problem for the axial vibration of a thin rod has the same structure as that for the transverse vibration of a string (see Prob. 5.1). To obtain the first from the second, we must replace the system parameters $\rho(x)$ and $T(x)$ by $m(x)$ and $EA(x)$, respectively, where $m(x)$ is the mass per unit length of rod and $EA(x)$ the axial stiffness, in which $E$ is the modulus of elasticity and $A(x)$ the cross-sectional area. It also follows that the structure of the eigenvalue problems is similar, subject to the same parameter substitution.

Let us consider the axial vibration of a thin rod fixed at both ends (see Fig. 5.4). In view of the above discussion, if we use Eqs. (5.19) and (5.21) and assume that the axial displacement $u(x, t)$ is separable in space and time, or

$$u(x, t) = U(x)F(t) \tag{5.28}$$

in which $F(t)$ is harmonic, we can write the eigenvalue problem directly in the form

$$-\frac{d}{dx}\left[EA(x)\frac{dU(x)}{dx}\right] = \omega^2 m(x)U(x) \qquad 0 < x < L \tag{5.29}$$

where $U(x)$ is subject to the boundary conditions

$$U(0) = U(L) = 0 \tag{5.30}$$

The differential equation (5.29) possesses space-dependent coefficients, so that in general no closed-form solution can be expected. A closed-form solution can be obtained in the special case of a *uniform rod*, $m(x) = m = \text{const}$, $EA(x) = EA = \text{const}$. Considering that case, Eq. (5.29) reduces to

$$\frac{d^2U(x)}{dx^2} + \beta^2 U(x) = 0 \qquad \beta^2 = \omega^2 \frac{m}{EA} \tag{5.31}$$

which must be satisfied over the domain $0 < x < L$. Of course, boundary conditions (5.30) remain the same.

The eigenvalue problem defined by the differential equation (5.31) and the boundary conditions (5.30) has precisely the same structure as that for the string fixed at both ends discussed in Example 5.1. It follows that the solution has the same structure, subject to the parameter substitution pointed out above. Hence, using the results of Example 5.1, we can write directly the system natural frequencies

$$\omega_r = \beta_r \sqrt{\frac{EA}{m}} = r\pi \sqrt{\frac{EA}{mL^2}} \qquad r = 1, 2, \ldots \tag{5.32}$$

Moreover, if the modes are normalized by letting $A_r = 1$ $(r = 1, 2, \ldots)$, we obtain

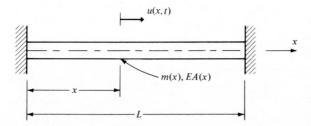

Figure 5.4

the normal modes

$$U_r(x) = \sin\frac{r\pi x}{L} \qquad r = 1, 2, \ldots \tag{5.33}$$

The first five normal modes are plotted in Fig. 5.5 in solid lines.

Next, let us solve the same problem by regarding the system as discrete. An equivalent discrete system can be obtained by dividing the rod into five equal segments, lumping the mass of the segments in the center as shown in Fig. 5.6 and regarding the lumped masses $M$ as being connected by springs of equivalent stiffnesses $k$ and $2k$, where $k$ is such that the springs undergo the same elongations as the corresponding rod segments would under identical loading. Hence, the lumped masses have the value $M = mL/5$ and the spring constant is $k = 5EA/L$. Accordingly, the eigenvalue problem can be written as

$$[k]\{u\} = \omega^2[m]\{u\} \tag{5.34}$$

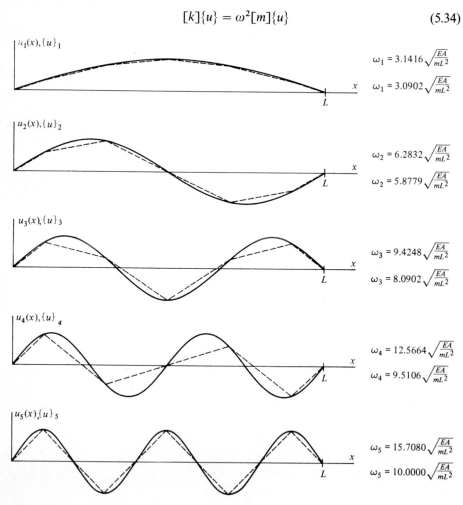

$\omega_1 = 3.1416\sqrt{\dfrac{EA}{mL^2}}$

$\omega_1 = 3.0902\sqrt{\dfrac{EA}{mL^2}}$

$\omega_2 = 6.2832\sqrt{\dfrac{EA}{mL^2}}$

$\omega_2 = 5.8779\sqrt{\dfrac{EA}{mL^2}}$

$\omega_3 = 9.4248\sqrt{\dfrac{EA}{mL^2}}$

$\omega_3 = 8.0902\sqrt{\dfrac{EA}{mL^2}}$

$\omega_4 = 12.5664\sqrt{\dfrac{EA}{mL^2}}$

$\omega_4 = 9.5106\sqrt{\dfrac{EA}{mL^2}}$

$\omega_5 = 15.7080\sqrt{\dfrac{EA}{mL^2}}$

$\omega_5 = 10.0000\sqrt{\dfrac{EA}{mL^2}}$

**Figure 5.5**

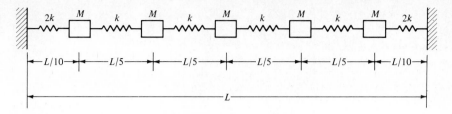

**Figure 5.6**

where the stiffness matrix has the form

$$[k] = \frac{5EA}{L} \begin{bmatrix} 3 & -1 & 0 & 0 & 0 \\ -1 & 2 & -1 & 0 & 0 \\ 0 & -1 & 2 & -1 & 0 \\ 0 & 0 & -1 & 2 & -1 \\ 0 & 0 & 0 & -1 & 3 \end{bmatrix} \qquad (5.35)$$

whereas the mass matrix is simply

$$[m] = \frac{mL}{5} \begin{bmatrix} 1 & 0 & 0 & 0 & 0 \\ 0 & 1 & 0 & 0 & 0 \\ 0 & 0 & 1 & 0 & 0 \\ 0 & 0 & 0 & 1 & 0 \\ 0 & 0 & 0 & 0 & 1 \end{bmatrix} \qquad (5.36)$$

The solution of the eigenvalue problem (5.34), in conjunction with matrices (5.35) and (5.36), was obtained by means of a computer program, with the results

$$\{u\}_1 = \begin{Bmatrix} 0.3090 \\ 0.8090 \\ 1.0000 \\ 0.8090 \\ 0.3090 \end{Bmatrix} \qquad \omega_1 = 3.0902 \sqrt{\frac{EA}{mL^2}} \qquad (5.37a)$$

$$\{u\}_2 = \begin{Bmatrix} 0.5878 \\ 0.9511 \\ 0 \\ -0.9511 \\ -0.5878 \end{Bmatrix} \qquad \omega_2 = 5.8779 \sqrt{\frac{EA}{mL^2}} \qquad (5.37b)$$

$$\{u\}_3 = \begin{Bmatrix} 0.8090 \\ 0.3090 \\ -1.0000 \\ 0.3090 \\ 0.8090 \end{Bmatrix} \qquad \omega_3 = 8.0902 \sqrt{\frac{EA}{mL^2}} \qquad (5.37c)$$

$$\{u\}_4 = \left\{ \begin{array}{c} 0.9511 \\ -0.5878 \\ 0 \\ 0.5878 \\ -0.9511 \end{array} \right\} \qquad \omega_4 = 9.5106 \sqrt{\frac{EA}{mL^2}} \qquad (5.37d)$$

$$\{u\}_5 = \left\{ \begin{array}{c} 1.0000 \\ -1.0000 \\ 1.0000 \\ -1.0000 \\ 1.0000 \end{array} \right\} \qquad \omega_5 = 10.0000 \sqrt{\frac{EA}{mL^2}} \qquad (5.37e)$$

where the modes have been normalized so as to match the amplitudes of the normal modes (5.33) of the continuous system. The modes are plotted in Fig. 5.5 in dashed lines. The natural frequencies of the continuous model are given on the corresponding top lines and those of the discrete model on the bottom lines. It is easy to see that, whereas the first mode and natural frequency are relatively close to those of the continuous model, accuracy is lost rapidly for higher modes in the discrete model, in the sense that the displacements are not very representative and the frequencies are not good approximations of those of the continuous system.

We note that the natural frequencies of the discrete system are lower than those of the corresponding continuous model. The reason is that, although the total mass is the same in both systems, in the case of the discrete model the mass is shifted toward the center of the system instead of being uniformly distributed. This tends to increase the effect of the system inertia relative to its stiffness, resulting in lower natural frequencies. Of course, accuracy can be improved by increasing the number of degrees of freedom of the discrete system.

## 5.5 BENDING VIBRATION OF BARS. BOUNDARY CONDITIONS

The transverse vibration of a string, axial vibration of a thin rod and torsional vibration of a circular shaft all lead to the same form of boundary-value problem, namely, one consisting of a partial differential equation of second order in both space and time and two boundary conditions, one at each end. Of course, the system parameters are different in each case (see Sec. 5.2). By contrast, the boundary-value problem for a bar in flexure is defined by a fourth-order differential equation in space requiring two boundary conditions at each end. In this section, we derive this boundary-value problem and use the opportunity to discuss the nature of various types of boundary conditions.

Let us consider the bar in flexure shown in Fig. 5.7a. The transverse displacement at any point $x$ and time $t$ is denoted by $y(x, t)$ and the transverse force per unit length by $f(x, t)$. The system parameters are the mass per unit length $m(x)$

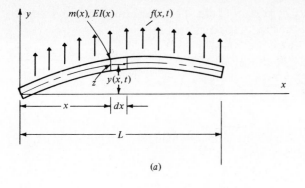

(a)

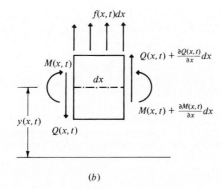

(b)                                    **Figure 5.7**

and the flexural rigidity $EI(x)$, where $E$ is Young's modulus of elasticity and $I(x)$ the cross-sectional area moment of inertia about an axis normal to $x$ and $y$ and passing through the center of the cross-sectional area. Figure 5.7b shows the free-body diagram corresponding to a bar element of length $dx$, where $Q(x, t)$ denotes the shearing force and $M(x, t)$ the bending moment. We use the so-called "simple-beam theory," according to which the rotation of the element is insignificant compared to the vertical translation, and the shear deformation is small in relation to the bending deformation. This theory is valid if the ratio between the length of the bar and its height is relatively large (say more than 10), and if the bar does not become too "wrinkled" because of flexure. In the area of vibrations the above statements imply ignoring the rotatory inertia and shear deformation effects.†

From Fig. 5.7b, the force equation of motion in the vertical direction has the form

$$\left[ Q(x, t) + \frac{\partial Q(x, t)}{\partial x}\, dx \right] - Q(x, t) + f(x, t)\, dx = m(x)\, dx\, \frac{\partial^2 y(x, t)}{\partial t^2} \quad (5.38)$$

On the other hand, ignoring the inertia torque associated with the rotation of the

† For more detailed discussion of these effects, see L. Meirovitch, *Analytical Methods in Vibrations*, sec. 5-2, The Macmillan Co., New York, 1967.

element, the moment equation of motion about the axis normal to $x$ and $y$ and passing through the center of the cross-sectional area is

$$\left[ M(x, t) + \frac{\partial M(x, t)}{\partial x} dx \right] - M(x, t) + \left[ Q(x, t) + \frac{\partial Q(x, t)}{\partial x} dx \right] dx$$

$$+ f(x, t) dx \frac{dx}{2} = 0 \quad (5.39)$$

Canceling appropriate terms and ignoring terms involving second powers in $dx$, we can write Eq. (5.39) in the simple form

$$\frac{\partial M(x, t)}{\partial x} + Q(x, t) = 0 \tag{5.40}$$

Moreover, canceling appropriate terms and considering (5.40), Eq. (5.38) reduces to

$$-\frac{\partial^2 M(x, t)}{\partial x^2} + f(x, t) = m(x) \frac{\partial^2 y(x, t)}{\partial t^2} \tag{5.41}$$

which must be satisfied over the domain $0 < x < L$.

Equation (5.41) relates the bending moment $M(x, t)$, the transverse force $f(x, t)$ and the bending displacement $y(x, t)$. Any elementary text on mechanics of materials, however, gives the relation between the bending moment and bending deformation in the form

$$M(x, t) = EI(x) \frac{\partial^2 y(x, t)}{\partial x^2} \tag{5.42}$$

Inserting Eq. (5.42) into (5.41), we obtain the differential equation for the flexural vibration of a bar

$$-\frac{\partial^2}{\partial x^2} \left[ EI(x) \frac{\partial^2 y(x, t)}{\partial x^2} \right] + f(x, t) = m(x) \frac{\partial^2 y(x, t)}{\partial t^2} \quad 0 < x < L \tag{5.43}$$

where we note that the equation contains spatial derivatives through fourth order.

To complete the formulation of the boundary-value problem, we must specify the boundary conditions. We list here the most common ones:

1. *Clamped end at $x = 0$.* The deflection and slope of the deflection curve are zero:

$$y(0, t) = 0 \qquad \frac{\partial y(x, t)}{\partial x}\bigg|_{x=0} = 0 \tag{5.44}$$

2. *Hinged end at $x = 0$.* The deflection and bending moment are zero:

$$y(0, t) = 0 \qquad EI(x) \frac{\partial^2 y(x, t)}{\partial x^2}\bigg|_{x=0} = 0 \tag{5.45}$$

Note that Eq. (5.42) was used in the second of conditions (5.45).

3. *Free end at x = 0.* The bending moment and shearing force are zero. Using Eqs. (5.40) and (5.42), the boundary conditions become

$$EI(x)\frac{\partial^2 y(x, t)}{\partial x^2}\bigg|_{x=0} = 0 \qquad \frac{\partial}{\partial x}\left[EI(x)\frac{\partial^2 y(x, t)}{\partial x^2}\right]\bigg|_{x=0} = 0 \qquad (5.46)$$

Analogous conditions can be written for the end $x = L$. Of course, there are less common boundary conditions, such as when the end is supported by springs, or when there is a concentrated mass at the end.

At this point a discussion of the character of the boundary conditions is in order. It is worth noting that boundary conditions (5.44) and the first of (5.45) are a result of the system geometry. For this reason they are called *geometric boundary conditions*. On the other hand, the second of boundary conditions (5.45) and both of (5.46) reflect the force and moment balance at the boundary; they are called *natural boundary conditions*. The significance of these definitions will become evident in Chap. 7, when approximate solutions of boundary-value problems are discussed.

Turning our attention to the corresponding eigenvalue problem, we first consider the free vibration characterized by $f(x, t) = 0$, in which case the solution of Eq. (5.43) becomes separable in space and time. Letting

$$y(x, t) = Y(x)F(t) \qquad (5.47)$$

and using the separation of variables method, as in Sec. 5.3, it can be shown that $F(t)$ is harmonic in this case also. This is no coincidence, however, as for all the conservative systems discussed here the time dependence is harmonic. Denoting the frequency of $F(t)$ by $\omega$, the eigenvalue problem formulation reduces to the differential equation

$$\frac{d^2}{dx^2}\left[EI(x)\frac{d^2 Y(x)}{dx^2}\right] = \omega^2 m(x)Y(x) \qquad 0 < x < L \qquad (5.48)$$

where the function $Y(x)$ must satisfy appropriate boundary conditions. Inserting Eq. (5.47) into (5.44) through (5.46), and eliminating the time dependence, we obtain boundary conditions similar in form to (5.44) through (5.46), with the exception that $y(x, t)$ is replaced by $Y(x)$ and partial derivatives with respect to $x$ by total derivatives with respect to $x$.

When the end is supported by a spring the time dependence can be eliminated quite easily in the same manner as above. On the other hand, a concentrated mass at the end has the effect of applying an inertia force at the end proportional to the acceleration of that end. Because the time dependence is harmonic with the frequency $\omega$, in this case the boundary condition involves the eigenvalue $\omega^2$.

## 5.6 NATURAL MODES OF A BAR IN BENDING VIBRATION

It should be clear by now that, to obtain the natural modes of a system, we must solve an eigenvalue problem. The eigenvalue problem associated with a bar in bending vibration, as derived in Sec. 5.5, consists of the differential equation

$$\frac{d^2}{dx^2}\left[EI(x)\frac{d^2 Y(x)}{dx^2}\right] = \omega^2 m(x) Y(x) \qquad 0 < x < L \qquad (5.49)$$

where $EI(x)$ is the flexural rigidity and $m(x)$ the mass per unit length at any point $x$. The solution $Y(x)$ is subject to given boundary conditions reflecting the manner in which the ends are supported. Several examples of boundary conditions can be obtained by replacing $y(x, t)$ by $Y(x)$ and partial derivatives by total derivatives with respect to $x$ in Eqs. (5.44)–(5.46). Equation (5.49) possesses coefficients depending on the spatial variable and has no general closed-form solution. Solutions can be obtained for certain special cases, most notably those in which the bar is uniform.

Let us consider the *uniform bar hinged at both ends* shown in Fig. 5.8, for which the differential equation (5.49) reduces to

$$\frac{d^4 Y(x)}{dx^4} - \beta^4 Y(x) = 0 \qquad \beta^4 = \frac{\omega^2 m}{EI} \qquad (5.50)$$

$$\omega = \beta^2$$

where $EI$ and $m$ are constant. The boundary conditions are obtained from Eqs. (5.45). Indeed, at the end $x = 0$, the boundary conditions are

$$Y(0) = 0 \qquad \frac{d^2 Y(x)}{dx^2}\bigg|_{x=0} = 0 \qquad Y''(0) = 0 \qquad (5.51)$$

whereas at the end $x = L$, the boundary conditions are

$$Y(L) = 0 \qquad \frac{d^2 Y(x)}{dx^2}\bigg|_{x=L} = 0 \qquad Y''(L) = 0 \qquad (5.52)$$

We note that the first boundary condition in both (5.51) and (5.52) is geometric and the second is natural.

The general solution of Eq. (5.50) can be easily verified to be

$$Y(x) = C_1 \sin \beta x + C_2 \cos \beta x + C_3 \sinh \beta x + C_4 \cosh \beta x \qquad (5.53)$$

where $C_i$ ($i = 1, 2, 3, 4$) are constants of integration. To evaluate three of these constants in terms of the fourth, as well as to derive the characteristic equation, we must use boundary conditions (5.51) and (5.52). Indeed, the first of boundary conditions (5.51) yields $C_2 + C_4 = 0$, whereas the second of (5.51) gives

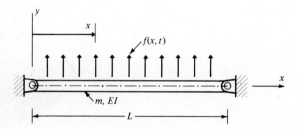

Figure 5.8

$-C_2 + C_4 = 0$, with the obvious conclusion that

$$C_2 = C_4 = 0 \tag{5.54}$$

Hence, solution (5.53) reduces to

$$Y(x) = C_1 \sin \beta x + C_3 \sinh \beta x \tag{5.55}$$

On the other hand, boundary conditions (5.52) lead to the two simultaneous equations

$$C_1 \sin \beta L + C_3 \sinh \beta L = 0$$
$$-C_1 \sin \beta L + C_3 \sinh \beta L = 0 \tag{5.56}$$

yielding

$$C_3 = 0 \tag{5.57}$$

and the characteristic equation

$$\sin \beta L = 0 \tag{5.58}$$

There are two other solutions of Eqs. (5.56), namely $C_1 = 0$, $\sinh \beta L = 0$ and $C_1 = C_3 = 0$, but they represent trivial solutions.

The solution of the characteristic equation is simply

$$\beta_r L = r\pi \qquad r = 1, 2, \ldots \tag{5.59}$$

yielding the natural frequencies

$$\omega_r = (r\pi)^2 \sqrt{\frac{EI}{mL^4}} \qquad r = 1, 2, \ldots \tag{5.60}$$

Moreover, recalling that $C_3 = 0$, using the values of $\beta_r$ ($r = 1, 2, \ldots$) given by Eq. (5.59) and normalizing according to $\int_0^L mY_r^2(x)\,dx = 1$ ($r = 1, 2, \ldots$), we obtain the normal modes

$$Y_r(x) = \sqrt{\frac{2}{mL}} \sin \frac{r\pi x}{L} \qquad r = 1, 2, \ldots \tag{5.61}$$

The first three modes are like those plotted in Fig. 5.3 but the frequencies are different. Note that the number of nodes is equal to the mode number minus 1.

Next let us consider the *clamped-free uniform bar* of Fig. 5.9. While the differential equation remains in the from (5.50), the boundary conditions at the clamped end, $x = 0$, are

$$Y(0) = 0 \qquad \frac{dY(x)}{dx}\bigg|_{x=0} = 0 \tag{5.62}$$

On the other hand, at the free end, $x = L$, the boundary conditions reduce to

$$\frac{d^2 Y(x)}{dx^2}\bigg|_{x=L} = 0 \qquad \frac{d^3 Y(x)}{dx^3}\bigg|_{x=L} = 0 \tag{5.63}$$

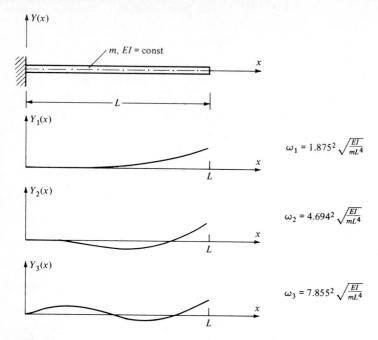

**Figure 5.9**

We note that boundary conditions (5.62) and (5.63) are geometric and natural, respectively.

The general solution remains in the form (5.53), but the constants $C_i$ ($i = 1, 2, 3, 4$) have different values. While the first of boundary conditions (5.62) leads once again to $C_2 + C_4 = 0$, the second of (5.62) yields $C_1 + C_3 = 0$, so that solution (5.53) takes the form

$$Y(x) = C_1(\sin \beta x - \sinh \beta x) + C_2(\cos \beta x - \cosh \beta x) \qquad (5.64)$$

Using boundary conditions (5.63), we arrive at the two simultaneous equations

$$C_1(\sin \beta L + \sinh \beta L) + C_2(\cos \beta L + \cosh \beta L) = 0 \qquad (5.65)$$

$$C_1(\cos \beta L + \cosh \beta L) - C_2(\sin \beta L - \sinh \beta L) = 0 \qquad (5.66)$$

Equation (5.66) can be solved for $C_2$ in terms of $C_1$ and the result inserted into (5.64) and (5.65) to yield

$$Y(x) = \frac{C_1}{\sin \beta L - \sinh \beta L} [(\sin \beta L - \sinh \beta L)(\sin \beta x - \sinh \beta x)$$

$$+ (\cos \beta L + \cosh \beta L)(\cos \beta x - \cosh \beta x)] \qquad (5.67)$$

and

$$C_1[(\sin \beta L + \sinh \beta L)(\sin \beta L - \sinh \beta L] + (\cos \beta L + \cosh \beta L)^2] = 0 \qquad (5.68)$$

Because for a nontrivial solution we must have $C_1 \neq 0$, the expression inside the brackets in (5.68) must be zero. After simplification, this leads to the characteristic equation

$$\cos \beta L \cosh \beta L = -1 \qquad (5.69)$$

The solution of Eq. (5.69) must be obtained numerically, yielding an infinite set of eigenvalues $\beta_r$ $(r = 1, 2, \ldots)$. Inserting these values into Eq. (5.67), we obtain the natural modes

$$Y_r(x) = A_r[(\sin \beta_r L - \sinh \beta_r L)(\sin \beta_r x - \sinh \beta_r x)$$

$$+ (\cos \beta_r L + \cosh \beta_r L)(\cos \beta_r x - \cosh \beta_r x)]$$

$$r = 1, 2, \ldots \quad (5.70)$$

where the notation $A_r = C_1/(\sin \beta_r L - \sinh \beta_r L)$ has been introduced for simplicity. The first three modes are plotted in Fig. 5.9, and we note once again that the mode $Y_r(x)$ has $r - 1$ nodes $(r = 1, 2, \ldots)$.

The natural frequencies and natural modes for a bar with a large variety of boundary conditions can be found in a report by D. Young and R. P. Felgar, Jr.†

Although the characteristic equation, Eq. (5.58) or Eq. (5.69), yields an infinity of characteristic values leading to the associated natural modes, we should recall that, because of the simple-beam theory limitations, the higher modes become increasingly inaccurate. This is so becaue the number of nodes increases with each mode, so that the distance between nodes decreases accordingly and the bar becomes progressively more "wrinkled." Hence, as the mode number increases, the rotation of a bar element can no longer be considered negligible compared with the translation, so that the simple-beam theory is not valid for the very high modes.

# 5.7 ORTHOGONALITY OF NATURAL MODES. EXPANSION THEOREM

In Sec. 5.3, it was mentioned that the eigenfunctions are orthogonal in a manner similar to the way in which the eigenvectors for discrete systems are. In fact, we shall prove the orthogonality property by first regarding the system as discrete and taking the limit, and then working directly with the continuous system.

Considering an $n$-degree-of-freedom discrete system, we have shown in Sec. 4.8 that two eigenvectors $\{u\}_r$ and $\{u\}_s$ corresponding to distinct eigenvalues $\omega_r^2$ and $\omega_s^2$ are orthogonal with respect to the mass matrix. Without loss of generality, we can assume that the mass matrix is diagonal, so that Eq. (4.80) can be written as the sum

$$\sum_{i=1}^{n} m_i u_{ir} u_{is} = 0 \qquad r \neq s \qquad (5.71)$$

† D. Young and R. P. Felgar, Jr., Tables of Characteristic Functions Representing Normal Modes of Vibration of a Beam, *The University of Texas Publication 4913*, July 1, 1949.

where $m_i$ is the mass in the position $x = x_i$, and $u_{ir}$ and $u_{is}$ are the displacements of $m_i$ in the modes $r$ and $s$, respectively. Following a pattern similar to that used in Sec. 5.2, we can increase the number $n$ of masses $m_i$ indefinitely, while reducing the size of the masses and the distance between any two masses, so that relations of the form

$$m_i = \rho(x_i)\,\Delta x_i \qquad i = 1, 2, \ldots, n \tag{5.72}$$

are preserved, where $\rho(x_i)$ is an equivalent mass per unit length at the point $x = x_i$. Inserting Eqs. (5.72) into (5.71), we obtain

$$\sum_{i=1}^{n} \rho(x_i)u_{ir}u_{is}\,\Delta x_i = 0 \qquad r \neq s \tag{5.73}$$

In the limit, as $\Delta x_i \to 0$, we can replace the indexed variable $x_i$ by the continuous independent variable $x$, so that the sum reduces to the integral

$$\int_0^L \rho(x)u_r(x)u_s(x)\,dx = 0 \qquad r \neq s \tag{5.74}$$

where $u_r(x)$ and $u_s(x)$ are the eigenfunctions obtained by letting the number of components of the eigenvectors $\{u\}_r$ and $\{u\}_s$ increase indefinitely. Equation (5.74) implies that the eigenfunctions $u_r(x)$ and $u_s(x)$ are orthogonal with respect to the mass density $\rho(x)$.

The orthogonality of eigenfunctions can be proved in a very general way, without the explicit knowledge of the eigenfunctions, by using operator notation.† However, because our text is more limited in scope, we would like to dispense with operator notation. Nevertheless, we can still use the idea of proving orthogonality for a given set of eigenfunctions without actually solving the eigenvalue problem. To this end, we consider the eigenvalue problem given by Eq. (5.48), subject to appropriate boundary conditions. Denoting two distinct solutions of the eigenvalue problem by $Y_r(x)$ and $Y_s(x)$, respectively, we can write

$$\frac{d^2}{dx^2}\left[ EI(x)\frac{d^2 Y_r(x)}{dx^2} \right] = \omega_r^2 m(x)Y_r(x) \qquad 0 < x < L \tag{5.75}$$

$$\frac{d^2}{dx^2}\left[ EI(x)\frac{d^2 Y_s(x)}{dx^2} \right] = \omega_s^2 m(x)Y_s(x) \qquad 0 < x < L \tag{5.76}$$

Next let us multiply Eq. (5.75) through by $Y_s(x)$, and integrate by parts over the domain $0 < x < L$, to obtain

$$\int_0^L Y_s(x)\frac{d^2}{dx^2}\left[ EI(x)\frac{d^2 Y_r(x)}{dx^2} \right] dx = \left\{ Y_s(x)\frac{d}{dx}\left[ EI(x)\frac{d^2 Y_r(x)}{dx^2} \right] \right\}\Bigg|_0^L$$

$$- \left[ \frac{dY_s(x)}{dx}EI(x)\frac{d^2 Y_r(x)}{dx^2} \right]\Bigg|_0^L + \int_0^L EI(x)\frac{d^2 Y_r(x)}{dx^2}\frac{d^2 Y_s(x)}{dx^2}\,dx$$

$$= \omega_r^2 \int_0^L m(x)Y_r(x)Y_s(x)\,dx \tag{5.77a}$$

† See L. Meirovitch, op. cit., sec. 5-5.

Multiplying Eq. (5.76) through by $Y_r(x)$, and performing a similar integration by parts, we arrive at

$$\int_0^L Y_r(x) \frac{d^2}{dx^2}\left[EI(x)\frac{d^2 Y_s(x)}{dx^2}\right] dx$$

$$= \left\{Y_r(x)\frac{d}{dx}\left[EI(x)\frac{d^2 Y_s(x)}{dx^2}\right]\right\}\Big|_0^L - \left[\frac{dY_r(x)}{dx}EI(x)\frac{d^2 Y_s(x)}{dx^2}\right]\Big|_0^L$$

$$+ \int_0^L EI(x)\frac{d^2 Y_r(x)}{dx^2}\frac{d^2 Y_s(x)}{dx^2} dx$$

$$= \omega_s^2 \int_0^L m(x)Y_r(x)Y_s(x)\, dx \qquad (5.77b)$$

Subtracting Eq. (5.77b) from (5.77a), we obtain

$$(\omega_r^2 - \omega_s^2)\int_0^L m(x)Y_r(x)Y_s(x)\, dx$$

$$= \left\{Y_s(x)\frac{d}{dx}\left[EI(x)\frac{d^2 Y_r(x)}{dx^2}\right]\right\}\Big|_0^L - \left[\frac{dY_s(x)}{dx}EI(x)\frac{d^2 Y_r(x)}{dx^2}\right]\Big|_0^L$$

$$- \left\{Y_r(x)\frac{d}{dx}\left[EI(x)\frac{d^2 Y_s(x)}{dx^2}\right]\right\}\Big|_0^L + \left[\frac{dY_r(x)}{dx}EI(x)\frac{d^2 Y_s(x)}{dx^2}\right]\Big|_0^L \qquad (5.78)$$

We shall consider only those systems for which the end conditions are such that the right side of (5.78) vanishes. Clearly, this is the case when the system has any combination of clamped, hinged, and free ends, as can be concluded from Sec. 5.5. It can be shown that the right side of (5.78) is zero also when the ends are supported by means of springs. Hence, Eq. (5.78) reduces to

$$(\omega_r^2 - \omega_s^2)\int_0^L m(x)Y_r(x)Y_s(x)\, dx = 0 \qquad (5.79)$$

But, according to our assumption $Y_r(x)$ and $Y_s(x)$ are eigenfunctions corresponding to distinct eigenvalues, $\omega_r^2 \neq \omega_s^2$ for $r \neq s$. It follows that

$$\int_0^L m(x)Y_r(x)Y_s(x)\, dx = 0 \qquad r \neq s \qquad (5.80)$$

so that the eigenfunctions $Y_r(x)$ and $Y_s(x)$ are orthogonal with respect to the mass density $m(x)$. We note the complete analogy with Eq. (5.74), where the latter was derived as a limiting case of a discrete system.

While the eigenvectors $\{u\}_r$ associated with a discrete system are also orthogonal with respect to the stiffness matrix, as stated by Eq. (4.81), the eigenfunctions are orthogonal with respect to the stiffness $EI(x)$ only in a certain sense. To explain the meaning of this statement, let us multiply Eq. (5.75) by $Y_s(x)$ and integrate over the length of the bar, so that

$$\int_0^L Y_s \frac{d^2}{dx^2}\left[EI(x)\frac{d^2 Y_r(x)}{dx^2}\right] dx = \omega_r^2 \int_0^L m(x)Y_s(x)Y_r(x)\, dx \qquad (5.81)$$

In view of Eq. (5.80), however, we can write

$$\int_0^L Y_s \frac{d^2}{dx^2}\left[EI(x)\frac{d^2 Y_r(x)}{dx^2}\right]dx = 0 \qquad r \neq s \tag{5.82}$$

so that the eigenfunctions are orthogonal with respect to the stiffness $EI(x)$ in the sense indicated by Eq. (5.82). Equation (5.82) can be shown to lead to a more convenient form. Indeed, integrating the equation by parts, we obtain

$$\int_0^L Y_s(x)\frac{d^2}{dx^2}\left[EI(x)\frac{d^2 Y_r(x)}{dx^2}\right]dx$$

$$= \left\{Y_s(x)\frac{d}{dx}\left[EI(x)\frac{d^2 Y_r(x)}{dx^2}\right]\right\}\bigg|_0^L - \left[\frac{dY_s(x)}{dx}EI(x)\frac{d^2 Y_r(x)}{dx^2}\right]\bigg|_0^L$$

$$+ \int_0^L EI(x)\frac{d^2 Y_r(x)}{dx^2}\frac{d^2 Y_s(x)}{dx^2}dx = 0 \qquad r \neq s \tag{5.83}$$

If the boundary conditions are as stipulated above, then we have

$$\int_0^L EI(x)\frac{d^2 Y_r(x)}{dx^2}\frac{d^2 Y_s(x)}{dx^2}dx = 0 \qquad r \neq s \tag{5.84}$$

so that the second derivatives of the eigenfunctions, but not the eigenfunctions themselves, are orthogonal with respect to the stiffness $EI(x)$. Note that the order of the derivatives involved in the orthogonality condition (5.84) is related to the order of the eigenvalue problem. Indeed, the order of the derivatives is equal to one-half the order of the eigenvalue problem. This can be explained easily by the fact that Eq. (5.84) is obtained from Eq. (5.82) through integrations by parts. Because the sum of the order of the highest derivatives of $Y_s$ and $Y_r$ in the two equations must be the same, and the order of the derivative of $Y_s$ in Eq. (5.82) is zero, it follows that the order of the derivatives of $Y_r$ and $Y_s$ in Eq. (5.84) is one-half the order of the highest derivative of $Y_r$ in Eq. (5.82), where the highest derivative determines the order of the eigenvalue problem.

When $r = s$ the integral in Eq. (5.80) is a positive quantity except in the case of the trivial solution, which presents no interest. Recalling that the eigenvalue problem is homogeneous, we can normalize the natural modes by writing

$$\int_0^L m(x)Y_r(x)Y_s(x)\,dx = \delta_{rs} \qquad r, s = 1, 2, \ldots \tag{5.85}$$

where $\delta_{rs}$ is the Kronecker delta. The natural modes satisfying Eqs. (5.85) are referred to as *normal modes*. It should be pointed out that normalization is not a unique process, and other definitions can be used. If the modes are normalized so that they satisfy Eq. (5.85), then upon integrating the left side of Eq. (5.81) and considering the boundary conditions leading to Eq. (5.84), it follows that

$$\int_0^L Y_s(x)\frac{d^2}{dx^2}\left[EI(x)\frac{d^2 Y_r(x)}{dx^2}\right]dx = \int_0^L EI(x)\frac{d^2 Y_r(x)}{dx^2}\frac{d^2 Y_s(x)}{dx^2}dx = \omega_r^2\delta_{rs}$$

$$r, s = 1, 2, \ldots \tag{5.86}$$

Although Eqs. (5.80) and (5.84) were derived using the eigenvalue problem for a bar in bending, the same reasoning can be used to derive similar formulas for other types of vibratory systems, such as strings in transverse vibration.

When one of the ends possesses a concentrated mass, formula (5.80) needs some modification. Assuming that at the end $x = L$ there is a concentrated mass $M$, Eq. (5.78) in conjunction with the proper boundary condition can be used to show that the corresponding orthogonality becomes

$$\int_0^L m(x) Y_r(x) Y_s(x) \, dx + M Y_r(L) Y_s(L) = 0 \qquad r \neq s \qquad (5.87)$$

A normalization scheme similar to that given by Eq. (5.85) can be used also in this case.

We observe that the integrals (5.80) and (5.84) are symmetric in the indices $r$ and $s$. This fact can be interpreted as being the counterpart for continuous systems of the fact that the matrices $[m]$ and $[k]$ are symmetric. Moreover, we observe that the integral (5.80) is always positive when $r = s$. This is the counterpart for continuous systems of the fact that the matrix $[m]$ is always positive definite. On the other hand, the integral (5.84) can be positive for $r = s$ or it can be zero without $Y_r(x)$ being identically zero. The counterpart for this is that the matrix $[k]$ can be positive definite or it can be positive semidefinite. Indeed, integral (5.84) is zero if $Y_r(x)$ is constant or a linear function of $x$, where the two configurations are recognized as the translational and rotational rigid-body modes, respectively. If the integral (5.84) is always positive for $r = s$ and if it becomes zero only if $Y_r(x)$ is identically zero, then the system is *positive definite*. On the other hand, if the system admits rigid-body modes, so that the integral (5.84) is generally positive for $r = s$ but can be zero without $Y_r(x)$ being identically zero, then the system is *positive semidefinite*. As might be expected, we conclude from Eq. (5.86) that the natural frequencies associated with the rigid-body modes are zero. Clearly, if both rigid-body modes are possible, then the zero natural frequency has multiplicity two. In this case the linear function of $x$, representing the rotational rigid-body mode, can be so chosen that the translational and rotational modes are orthogonal to one another. Moreover, by analogy with the approach used in Sec. 4.12 for discrete systems, the conservation of linear momentum and angular momentum can be invoked to demonstrate that the rigid-body modes are orthogonal to the elastic modes. Hence, all the modes are orthogonal to one another, regardless whether they are rigid-body or elastic modes. Of course, in all cases this is not ordinary orthogonality but orthogonality with respect to the mass density.

As for discrete systems, an expansion theorem exists for continuous systems, where the theorem is based on the orthogonality property. The *expansion theorem* can be stated in the form: *Any function $Y(x)$, satisfying the boundary conditions of the problem and such that $(d^2/dx^2)[EI(x)d^2 Y(x)/dx^2]$ is a continuous function, can be represented by the absolutely and uniformly convergent series of the system eigenfunctions*

$$Y(x) = \sum_{i=1}^{\infty} c_r Y_r(x) \qquad (5.88)$$

*where the constant coefficients $c_r$ are given by*

$$c_r = \int_0^L m(x)\,Y(x)\,Y_r(x)\,dx \qquad r = 1, 2, \ldots \tag{5.89}$$

If we recall that a periodic function can be represented by a Fourier series consisting of an infinite set of harmonic functions, the expansion theorem, Eqs. (5.88) and (5.89), can be regarded as a generalized Fourier series representation. In fact, in the special cases in which the eigenfunctions happen to be harmonic, the expansion theorem does reduce to a Fourier series representation.

Although we stated the expansion theorem in terms of the bending of a bar, the same theorem is applicable to an entire class of vibratory systems, including all the systems discussed in this chapter.†

## 5.8 RAYLEIGH'S QUOTIENT

In Sec. 4.13 we studied the properties of a certain scalar quantity called Rayleigh's quotient, defined in connection with discrete systems. As should be expected, a similar Rayleigh's quotient can be defined for continuous systems. The quotient can be defined in general form in terms of operator notation that makes it applicable to a large class of problems.‡ In this text, we introduce the concept by way of a specific example.

Let us consider the eigenvalue problem associated with a shaft clamped at the end $x = 0$ and free at the end $x = L$, as shown in Fig. 5.10, where $I(x)$ is the mass polar moment of inertia per unit length and $GJ(x)$ the torsional stiffness at point $x$. Denoting by $\theta(x, t) = \Theta(x)F(t)$ the angular displacement of the shaft, and recognizing that $F(t)$ is harmonic with frequency $\omega$, we can write the eigenvalue problem in the form of the differential equation

$$-\frac{d}{dx}\left[GJ(x)\frac{d\Theta(x)}{dx}\right] = \lambda I(x)\Theta(x) \qquad \lambda = \omega^2 \tag{5.90}$$

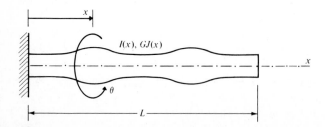

**Figure 5.10**

† See L. Meirovitch, op. cit., sec. 5-5.
‡ See L. Meirovitch, op. cit., sec. 5-14.

that must be satisfied over the domain $0 < x < L$, where $\Theta(x)$ is subject to the boundary conditions

$$\Theta(0) = 0 \qquad GJ(x)\frac{d\Theta(x)}{dx}\bigg|_{x=L} = 0 \tag{5.91}$$

Multiplying Eq. (5.90) by $\Theta(x)$, integrating over the domain $0 < x < L$ and considering boundary condition (5.91), we can write

$$\lambda = \omega^2 = R(\Theta) = \frac{-\displaystyle\int_0^L \Theta(x)(d/dx)\{GJ(x)[d\Theta(x)/dx]\}\,dx}{\displaystyle\int_0^L I(x)\Theta^2(x)\,dx}$$

$$= \frac{\displaystyle\int_0^L GJ(x)[d\Theta(x)/dx]^2\,dx}{\displaystyle\int_0^L I(x)\Theta^2(x)\,dx} \tag{5.92}$$

where $R(\Theta)$ is known as the *Rayleigh quotient* of the system. Note that $R(\Theta)$ is a functional, namely, a function of a function, and not a function of $\Theta$ in the ordinary sense.

If $\Theta(x)$ is an eigenfunction of the system, then Eq. (5.92) yields the associated eigenvalue. On the other hand, for a certain trial function $\Theta(x)$ satisfying all the boundary conditions of the problem but not the differential equation, Eq. (5.92) yields a scalar whose value depends on $\Theta(x)$. As was the case with discrete systems, it can be shown that Rayleigh's quotient has a stationary value when $\Theta$ is in the neighborhood of an eigenfunction. Indeed, by the expansion theorem, Eqs. (5.88) and (5.89), we can write the trial function $\Theta$ as a superposition of the system eigenfunctions $\Theta_i(x)$ $(i = 1, 2, \ldots)$ in the form

$$\Theta(x) = \sum_{i=1}^{\infty} c_i\Theta_i(x) \tag{5.93}$$

where we recall that the eigenfunctions $\Theta_i(x)$ are orthogonal. Let us assume that they are also normalized so as to satisfy

$$\int_0^L I(x)\Theta_i(x)\Theta_j(x)\,dx = \delta_{ij} \qquad i, j, = 1, 2, \ldots \tag{5.94}$$

where $\delta_{ij}$ is the Kronecker delta. Then, because the eigenfunctions must satisfy Eqs. (5.90) and (5.91), it follows that

$$-\int_0^L \Theta_i(x)\frac{d}{dx}\left[GJ(x)\frac{d\Theta_j(x)}{dx}\right]dx$$

$$= \int_0^L GJ(x)\frac{d\Theta_i(x)}{dx}\frac{d\Theta_j(x)}{dx}\,dx = \lambda_j\delta_{ij} \qquad i = 1, 2, \ldots \tag{5.95}$$

Introducing Eq. (5.93) into (5.92), and considering Eqs. (5.94) and (5.95), we can write

$$\lambda = \omega^2 = R(\Theta) = \frac{\displaystyle\sum_{i=1}^{\infty}\sum_{j=1}^{\infty} c_i c_j \int_0^L GJ(x)\frac{d\Theta_i(x)}{dx}\frac{d\Theta_j(x)}{dx}\,dx}{\displaystyle\sum_{i=1}^{\infty}\sum_{j=1}^{\infty} c_i c_j \int_0^L I(x)\Theta_i(x)\Theta_j(x)\,dx}$$

$$= \frac{\displaystyle\sum_{i=1}^{\infty} c_i^2 \lambda_i}{\displaystyle\sum_{i=1}^{\infty} c_i^2} \tag{5.96}$$

Following a procedure similar to that used for discrete systems, we let $\Theta(x)$ resemble the $r$th eigenfunction, so that

$$c_i = \epsilon_i c_r \qquad i = 1, 2, \ldots; i \neq r \tag{5.97}$$

where $\epsilon_i$ are small quantities, $\epsilon_i \ll 1$. Using the same approach as in Sec. 4.13, it is easy to show that

$$\lambda = R(\Theta) = \frac{\lambda_r + \displaystyle\sum_{i=1}^{\infty}(1-\delta_{ir})\lambda_i\epsilon_i^2}{1 + \displaystyle\sum_{i=1}^{\infty}(1-\delta_{ir})\epsilon_i^2} \cong \lambda_r + \sum_{i=1}^{\infty}(\lambda_i - \lambda_r)\epsilon_i^2 \tag{5.98}$$

Equation (5.98) indicates that if the trial function $\Theta$ differs from the eigenfunction $\Theta_r(x)$ by a small quantity of first order, then Rayleigh's quotient differs from the eigenvalue $\lambda_r$ by a small quantity of second order, with the implication that $R(\Theta)$ *has a stationary value in the neighborhood of* $\Theta_r$. Moreover, assuming that the eigenvalues $\lambda_i$ are such that $\lambda_1 < \lambda_2 < \ldots$, it is easy to see that

$$R(\Theta) \geqslant \lambda_1 \tag{5.99}$$

or, in words, *Rayleigh's quotient provides an upper bound for the lowest eigenvalue* $\lambda_1$.

Rayleigh's quotient can be used to provide an estimate of the first eigenvalue. As an example, let us consider the case in which the shaft of Fig. 5.10 has the parameters

$$I(x) = \frac{6}{5}I\left[1 - \frac{1}{2}\left(\frac{x}{L}\right)^2\right] \qquad GJ(x) = \frac{6}{5}GJ\left[1 - \frac{1}{2}\left(\frac{x}{L}\right)^2\right] \tag{5.100}$$

As the trial function we choose the first eigenfunction of the associated uniform shaft, namely, $\Theta(x) = \sin \pi x/2L$. First let us calculate

$$\int_0^L GJ(x)\left[\frac{d\Theta(x)}{dx}\right]^2 dx = \frac{3\pi^2}{10}\frac{GJ}{L^2}\int_0^L\left[1 - \frac{1}{2}\left(\frac{x}{L}\right)^2\right]\cos^2\frac{\pi x}{2L}\,dx$$

$$= \frac{1}{40}\frac{GJ}{L}(5\pi^2 + 6) \tag{5.101}$$

and

$$\int_0^L I(x)\Theta^2(x)\,dx = \frac{6}{5}I\int_0^L\left[1 - \frac{1}{2}\left(\frac{x}{L}\right)^2\right]\sin^2\frac{\pi x}{2L}\,dx = \frac{1}{10\pi^2}IL(5\pi^2 - 6) \tag{5.102}$$

from which it follows that

$$\lambda = \omega^2 = R(\Theta) = \frac{\displaystyle\int_0^L GJ(x)[d\Theta(x)/dx]^2\,dx}{\displaystyle\int_0^L I(x)\Theta^2(x)\,dx}$$

$$= \frac{\frac{1}{40}(GJ/L)(5\pi^2 + 6)}{(1/10\pi^2)IL(5\pi^2 - 6)} = 3.1504\,\frac{GJ}{IL^2} \tag{5.103}$$

Note that the estimated fundamental frequency is $\omega = 1.7749\sqrt{GJ/IL^2}$, which is higher than $\omega_1 = (\pi/2)\sqrt{GJ/IL^2}$, where the latter is the fundamental frequency of the associated uniform shaft. This is to be expected for two reasons: (1) Rayleigh's quotient yields higher estimated frequencies than the actual natural frequencies, and (2) the nonuniform shaft, having more mass toward the clamped end than toward the free end, tends to be stiffer and its natural frequencies higher than those of the uniform shaft. We shall return to this subject in Chap. 7 when we discuss approximate methods for solving eigenvalue problems.

## 5.9 RESPONSE OF SYSTEMS BY MODAL ANALYSIS

The response of a system to initial excitation, external excitation, or both initial and external excitation can be obtained conveniently by modal analysis. The method is based on the expansion theorem of Sec. 5.7 and regards the response as a superposition of the system eigenfunctions multiplied by corresponding time-dependent generalized coordinates, in a manner entirely analogous to that for discrete systems (Sec. 4.14). Of course, this necessitates first obtaining the solution of the system eigenvalue problem.

As an illustration of the method, let us consider a uniform bar in bending with both ends hinged (see Fig. 5.8). The bar is subjected to the external distributed force $f(x, t)$ and the initial conditions

$$y(x, 0) = y_0(x) \qquad \left.\frac{\partial y(x, t)}{\partial t}\right|_{t=0} = v_0(x) \tag{5.104}$$

From Sec. 5.5 we conclude that the boundary-value problem for a uniform bar reduces to

$$-EI\frac{\partial^4 y(x, t)}{\partial x^4} + f(x, t) = m\frac{\partial^2 y(x, t)}{dx^2} \qquad 0 < x < L \tag{5.105}$$

where the flexural stiffness $EI$ and mass per unit length $m$ are constant. Because both ends are hinged, the boundary conditions are

$$y(0, t) = 0 \qquad EI \left. \frac{\partial^2 y(x, t)}{\partial x^2} \right|_{x=0} = 0 \qquad (5.106)$$

$$y(L, t) = 0 \qquad EI \left. \frac{\partial^2 y(x, t)}{\partial x^2} \right|_{x=L} = 0 \qquad (5.107)$$

The eigenvalue problem associated with the system under consideration was solved in Sec. 5.6. Hence, from Sec. 5.6, we obtain the system natural frequencies

$$\omega_r = (r\pi)^2 \sqrt{\frac{EI}{mL^4}} \qquad r = 1, 2, \dots \qquad (5.108)$$

and the natural modes

$$Y_r(x) = \sqrt{\frac{2}{mL}} \sin \frac{r\pi x}{L} \qquad r = 1, 2, \dots \qquad (5.109)$$

where the modes are clearly orthogonal. Note that the modes have been normalized so as to satisfy Eqs. (5.85) and (5.86), or

$$\int_0^L m Y_r(x) Y_s(x) \, dx = \delta_{rs} \qquad r, s = 1, 2, \dots \qquad (5.110)$$

and

$$\int_0^L Y_s(x) EI \frac{d^4 Y_r(x)}{dx^4} \, dx = \omega_r^2 \delta_{rs} \qquad r, s = 1, 2, \dots \qquad (5.111)$$

According to modal analysis, we let the solution of Eq. (5.105) have the form

$$y(x, t) = \sum_{r=1}^{\infty} Y_r(x) q_r(t) \qquad (5.112)$$

so that, inserting (5.112) into (5.105), we arrive at

$$\sum_{r=1}^{\infty} \ddot{q}_r(t) m Y_r(x) + \sum_{r=1}^{\infty} q_r(t) EI \frac{d^4 Y_r(x)}{dx^4} = f(x, t) \qquad 0 < x < L \qquad (5.113)$$

Multiplying through by $Y_s(x)$, integrating over the domain, and considering Eqs. (5.110) and (5.111), we obtain the set of independent ordinary differential equations

$$\ddot{q}_r(t) + \omega_r^2 q_r(t) = Q_r(t) \qquad r = 1, 2, \dots \qquad (5.114)$$

where

$$Q_r(t) = \int_0^L f(x, t) Y_r(x) \, dx \qquad r = 1, 2, \dots \qquad (5.115)$$

are the generalized forces associated with the generalized coordinates $q_r(t)$.

Equations (5.114) resemble the equation of motion of an undamped single-degree-of-freedom system subjected to external excitation. As in Sec. 4.14, the response can be written in the general form

$$q_r(t) = \frac{1}{\omega_r} \int_0^t Q_r(\tau) \sin \omega_r(t - \tau) d\tau + q_{r0} \cos \omega_r t + \frac{\dot{q}_{r0}}{\omega_r} \sin \omega_r t \qquad (5.116)$$

where

$$q_{r0} = q_r(0) \qquad \dot{q}_{r0} = \dot{q}_r(0) \qquad (5.117)$$

are the initial generalized coordinates and velocities, respectively. The values of $q_{r0}$ and $\dot{q}_{r0}$ can be obtained by using the initial conditions (5.104) in conjunction with Eq. (5.112), as follows:

$$y(x, 0) = y_0(x) = \sum_{r=1}^{\infty} Y_r(x) q_r(0) = \sum_{r=1}^{\infty} Y_r(x) q_{r0} \qquad (5.118)$$

Multiplying through by $mY_s(x)$, integrating over the domain $0 < x < L$, and taking advantage of the orthogonality conditions (5.110), we obtain

$$q_{r0} = \int_0^L m y_0(x) Y_r(x) \, dx \qquad r = 1, 2, \ldots \qquad (5.119)$$

Analogously, we conclude that

$$\dot{q}_{r0} = \int_0^L m v_0(x) Y_r(x) \, dx \qquad r = 1, 2, \ldots \qquad (5.120)$$

The general response is obtained by inserting Eq. (5.116) into (5.112), with the result

$$y(x, t) = \sum_{r=1}^{\infty} Y_r(x) \left[ \frac{1}{\omega_r} \int_0^t Q_r(\tau) \sin \omega_r(t - \tau) \, d\tau + q_{r0} \cos \omega_r t + \frac{\dot{q}_{r0}}{\omega_r} \sin \omega_r t \right]$$
$$(5.121)$$

where $Y_r(x)$ is given by (5.109), $\omega_r$ by (5.108), $Q_r(t)$ by (5.115), $q_{r0}$ by (5.119) and $\dot{q}_{r0}$ by (5.120). Of course, before evaluating Eq. (5.121), we must know the distributed forcing function $f(x, t)$ and the initial conditions $y(x, 0) = y_0(x)$ and $\partial y(x, t)/\partial t|_{t=0} = v_0(x)$.

As a simple example of the use of formula (5.121), let us consider the case in which the initial conditions are zero and the distributed force has the form

$$f(x, t) = f_0 \alpha(t) \qquad (5.122)$$

where $f_0$ is a constant and $\alpha(t)$ the unit step function. Hence, inserting Eqs. (5.109) and (5.122) into Eq. (5.115), the generalized forces become

$$Q_r(t) = \int_0^L f_0 \alpha(t) Y_r(x) \, dx = f_0 \alpha(t) \sqrt{\frac{2}{mL}} \int_0^L \sin \frac{r\pi x}{L} \, dx$$

$$= f_0 \alpha(t) \sqrt{\frac{2}{mL}} \frac{L}{r\pi} (1 - \cos r\pi) \qquad r = 1, 2, 3, \ldots \qquad (5.123)$$

which reduce to

$$Q_r(t) = 2f_0 \sqrt{\frac{2}{mL}} \frac{L}{r\pi} u(t) \qquad r = 1, 3, 5, \ldots \tag{5.124}$$

implying that the generalized forces associated with the modes for which $r$ is an even number reduce to zero. This is to be expected because to even $r$ correspond modes antisymmetric with respect to the point $x = L/2$, which cannot be excited by virtue of the fact that the external excitation is uniform and hence symmetric. Inserting Eq. (5.124) into (5.121), with $q_{r0}$ and $\dot{q}_{r0}$ equal to zero, recalling the step response from Sec. 2.14 and making use of (5.108) and (5.109), we arrive finally at

$$y(x, t) = \frac{4f_0 L^4}{\pi^5 EI} \sum_{r=1}^{\infty} \frac{1}{(2r-1)^5} \sin \frac{(2r-1)\pi x}{L} \left[ 1 - \cos (2r-1)^2 \pi^2 \sqrt{\frac{EI}{mL^4}} t \right] \tag{5.125}$$

where, in replacing $r$ by $2r - 1$, we took into account the fact that only symmetric modes are excited. This enables us to sum up over all the integers $r$. It is easy to see that the first mode is by far the predominant one.

## 5.10 THE WAVE EQUATION

Referring to Sec. 5.2, the equation for the transverse displacement of a string in the absence of the distributed force $f(x, t)$ can be written in the form

$$\frac{\partial}{\partial x} \left[ T(x) \frac{\partial y(x, t)}{\partial x} \right] = \rho(x) \frac{\partial^2 y(x, t)}{\partial t^2} \tag{5.126}$$

where $T(x)$ is the tension in the string and $\rho(x)$ the mass per unit length at point $x$. For the moment, we leave the question of the length of the string open, so that we need not concern ourselves with boundary conditions. Moreover, we consider the case of a uniform string under constant tension, for which Eq. (5.126) reduces to

$$\frac{\partial^2 y(x, t)}{\partial x^2} = \frac{1}{c^2} \frac{\partial^2 y(x, t)}{\partial t^2} \qquad c = \sqrt{\frac{T}{\rho}} \tag{5.127}$$

Equation (5.127) is known as the *wave equation* and the constant $c$ as the *wave propagation velocity*. It can be easily verified that the solution of Eq. (5.127) has the general form

$$y(x, t) = F_1(x - ct) + F_2(x + ct) \tag{5.128}$$

where $F_1$ and $F_2$ are arbitrary functions of the arguments $x - ct$ and $x + ct$, respectively. The function $F_1(x - ct)$ represents a displacement wave traveling in the positive $x$ direction with the constant velocity $c$ without altering the shape of the wave or the wave profile. This profile is defined by the explicit form of the function $F_1$. Similarly, $F_2(x + ct)$ represents a displacement wave traveling in the negative $x$ direction with the velocity $c$. It follows that the most general type of motion of a string can be regarded as the superposition of two waves traveling in opposite directions.

A case of particular interest in vibrations is that of sinusoidal waves. Let us consider a sinusoidal wave of amplitude $A$ traveling in the positive $x$ direction. The wave can be expressed mathematically by

$$y(x, t) = A \sin \frac{2\pi}{\lambda} (x - ct) \tag{5.129}$$

where $\lambda$ is the *wavelength*, defined as the distance between two successive crests. Equation (5.129) can be rewritten in the form

$$y(x, t) = A \sin (2\pi kx - \omega t) \tag{5.130}$$

where

$$k = \frac{1}{\lambda} \tag{5.131}$$

is known as the *wave number*, representing the number of waves in a unit distance, and

$$\omega = c \frac{2\pi}{\lambda} \tag{5.132}$$

is the *frequency* of the wave. Moreover,

$$\tau = \frac{2\pi}{\omega} = \frac{\lambda}{c} \tag{5.133}$$

is the *period*, namely, the time necessary for a complete wave to pass through a given point.

Next let us consider a displacement consisting of two identical sinusoidal waves traveling in opposite directions. Hence, we have

$$y(x, t) = A \sin (2\pi kx - \omega t) + A \sin (2\pi kx + \omega t)$$

$$= 2A \sin 2\pi kx \cos \omega t \tag{5.134}$$

From Eq. (5.134) we conclude that in this special case the wave profile is no longer traveling, so that the two waves combine into a *stationary wave*, or *standing wave*, whose profile $2A \sin 2\pi kx$ oscillates about the equilibrium position with the frequency $\omega$. At the points for which $2kx$ has integer values the two traveling waves cancel each other, forming *nodes*, whereas at points for which $2kx$ is an odd multiple of $1/2$ the two waves reinforce each other, yielding the greatest amplitude. These latter points lie halfway between any two successive nodes and are called *loops*, or *antinodes*.

It may prove of interest to tie the above analysis to that of Sec. 5.3, where we discussed the transverse vibration of a string of length $L$ fixed at both ends. In Eq. (5.134) the frequency $\omega$ is arbitrary. As we well know, however, a string of finite length does not admit arbitrary frequencies but a denumerably infinite set of natural frequencies $\omega_r$ ($r = 1, 2, \ldots$). If the string considered above is fixed at the points $x = 0$ and $x = L$, then we must make sure that Eq. (5.134) has nodes at these points. Hence, we must have

$$2kL = r \qquad r = 1, 2, \ldots \tag{5.135}$$

Inserting the above values into (5.131), and considering (5.132), we obtain the natural frequencies

$$\omega_r = 2\pi k c = r\pi \frac{c}{L} = r\pi \sqrt{\frac{T}{\rho L^2}} \qquad r = 1, 2, \ldots \tag{5.136}$$

so that the normal-mode vibration of a string fixed at $x = 0$ and $x = L$ can be regarded as consisting of standing waves, where the wave profile oscillates about an equilibrium position with the natural frequency $\omega_r$. Note that Eq. (5.136) gives the same natural frequencies as in Example 5.1, Eq. (*i*).

Because of the analogy pointed out in Sec. 5.2, the above discussion is equally valid for rods in axial vibration or circular shafts in torsion as it is for strings in transverse vibration, provided the systems are uniform. In the case of bars in flexure, even when the bars are uniform, the motion does not satisfy the wave equation because the partial derivatives with respect to $x$ are of fourth order instead of second. Hence, wave motion in which the wave profile travels with constant velocity, without altering its shape, is not possible. Nevertheless, some type of wave motion in which the wave profile does alter its shape exists. A detailed discussion of this subject can be found in the text by L. Meirovitch.†

## 5.11 KINETIC AND POTENTIAL ENERGY FOR CONTINUOUS SYSTEMS

In Chap. 4, we derived general expressions for the kinetic and potential energy of discrete systems. In the case of linear systems these expressions possess quadratic forms involving the mass and stiffness coefficients. Similar expressions can be derived for continuous systems, except that no general expression in terms of the system parameters can be written for the potential energy, and different continuous systems possess different forms for the potential energy. To emphasize once again the parallel between discrete and continuous systems, we shall derive the kinetic and potential energy for a rod in longitudinal vibration by regarding it as a limiting case of a discrete system.

Let us consider the system of discrete masses $M_i$ ($i = 1, 2, \ldots, n$) shown in Fig. 5.11. The masses are connected by springs exhibiting linear behavior, where the springs' stiffnesses are denoted by $k_i$ ($i = 1, 2, \ldots, n$). The kinetic energy is simply

$$T(t) = \frac{1}{2} \sum_{i=1}^{n} M_i \left[ \frac{du_i(t)}{dt} \right]^2 \tag{5.137}$$

† Op. cit., sec. 8-7.

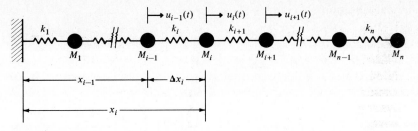

**Figure 5.11**

where $du_i(t)/dt$ is the velocity of mass $M_i$ measured relative to an inertial space. In the equilibrium configuration $M_i$ occupies the spatial position $x_i$. Introducing the notation $M_i = m_i \, \Delta x_i$, where $m_i$ can be regarded as a mass density at point $x_i$, letting $\Delta x_i \to 0$ while $x_1 \to 0$ and $x_n \to L$, and taking the limit, we can write

$$T(t) = \lim_{\Delta x_i \to 0} \frac{1}{2} \sum_{i=1}^{n} m_i \left[ \frac{du_i(t)}{dt} \right]^2 \Delta x_i = \frac{1}{2} \int_0^L m(x) \left[ \frac{\partial u(x, t)}{\partial t} \right]^2 dx \quad (5.138)$$

where the indexed position $x_i$ has been replaced by the continuous independent variable $x$ and $L$ represents the length of the equivalent rod in axial vibration. In the limiting process, the summation has been replaced by integration and total derivatives with respect to time by partial derivatives.

The potential energy requires a little more elaboration. Denoting by $F_i(t)$ the force across the spring $k_i$ and recalling that the system is linear, the potential energy can be written in the general form

$$V(t) = \frac{1}{2} \sum_{i=1}^{n} F_i(t) [u_i(t) - u_{i-1}(t)] \quad (5.139)$$

where $u_i - u_{i-1}$ represents the elongation of the spring $k_i$. Of course, $u_0$ must be set equal to zero since the left end of the spring $k_1$ is fixed. Because for a linear system the force is proportional to the elongation of the spring, $F_i(t) = k_i[u_i(t) - u_{i-1}(t)]$, Eq. (5.139) becomes

$$V(t) = \frac{1}{2} \sum_{i=1}^{n} k_i [u_i(t) - u_{i-1}(t)]^2 \quad (5.140)$$

At this point generality must be abandoned by specifying the stiffness in terms of an equivalent continuous element. Introducing the notation $k_i = EA_i/\Delta x_i$, $u_i(t) - u_{i-1}(t) = \Delta u_i(t)$, where $k_i$ is identified as an equivalent stiffness corresponding to a rod of longitudinal stiffness $EA_i$ and length $\Delta x_i$, Eq. (5.140) reduces to

$$V(t) = \frac{1}{2} \sum_{i=1}^{n} EA_i \left[ \frac{\Delta u_i(t)}{\Delta x_i} \right]^2 \Delta x_i \quad (5.141)$$

Now letting $\Delta x_i \to 0$ and taking the limit, we obtain the potential energy for a rod in longitudinal vibration

$$V(t) = \frac{1}{2} \int_0^L EA(x) \left[ \frac{\partial u(x, t)}{\partial x} \right]^2 dx \qquad (5.142)$$

Equation (5.142) can be derived directly from Eq. (5.139). Indeed, with the notation introduced above, Eq. (5.139) leads in the limit to

$$V(t) = \lim_{\Delta x_i \to 0} \frac{1}{2} \sum_{i=1}^n F_i(t) \frac{\Delta u_i(t)}{\Delta x_i} \Delta x_i = \frac{1}{2} \int_0^L F(x, t) \frac{\partial u(x, t)}{\partial x} dx \qquad (5.143)$$

Moreover, recalling that $F_i(t) = k_i[u_i(t) - u_{i-1}(t)] = EA_i \, \Delta u_i(t)/\Delta x_i$, we have

$$F(x, t) = \lim_{\Delta x_i \to 0} EA_i \frac{\Delta u_i(t)}{\Delta x_i} = EA(x) \frac{\partial u(x, t)}{\partial x} \qquad (5.144)$$

Introducing Eq. (5.144) into (5.143), we obtain Eq. (5.142).

Using the analogy of Sec. 5.2, expressions similar to (5.142) can be written for the potential energy of a string in transverse vibration and that of a shaft in torsion.

Of course, Eq. (5.142) can be derived by considering the system to be continuous from the beginning (see Prob. 5.19). Using such an approach it is not difficult to show that the potential energy of a bar in flexure is (see Prob. 5.20)

$$V(t) = \frac{1}{2} \int_0^L EI(x) \left[ \frac{\partial^2 y(x, t)}{\partial x^2} \right]^2 dx \qquad (5.145)$$

where $EI(x)$ is the flexural rigidity and $y(x, t)$ the transverse displacement of the bar.

## PROBLEMS

**5.1** Use the approach of Sec. 5.2 and derive the boundary-value problem for a thin rod in longitudinal vibration. The rod is fixed at $x = 0$ and is connected to a linear spring of stiffness $k$ at the end $x = L$, where the other end of the spring is anchored to a wall (see Fig. 5.12).

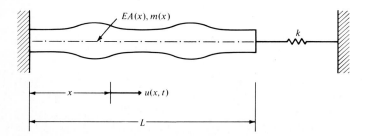

**Figure 5.12**

**5.2** Use the approach of Sec. 5.2 and derive the boundary-value problem for the torsional vibration of a circular shaft free at the end $x = 0$ and fixed at the end $x = L$.

**5.3** Derive the eigenvalue problems associated with the systems of Prob. 5.1 and 5.2.

**5.4** Consider a bar in flexural vibration and derive the boundary-value problem for the case in which the end $x = 0$ is fixed and a concentrated mass is attached at the end $x = L$. Derive the associated eigenvalue problem.

**5.5** Let the circular shaft of Prob. 5.2 be uniform, obtain a closed-form solution of the eigenvalue problem and plot the first three natural modes.

**5.6** Consider a uniform rod in axial vibration and solve the eigenvalue problem for the case in which both ends are free. Note that such a system is said to be *semidefinite* (see Sec. 5.7). Plot the first three modes and compare the results with those obtained in Example 4.9. Draw conclusions concerning the analogy between continuous and discrete systems.

**5.7** Let the rod of Prob. 5.1 be uniform, $EA(x) = EA = $ const, $m(x) = m = $ const, and solve the eigenvalue problem for the case $EA = 4kL$. Obtain the first three natural frequencies and modes and plot the modes.

**5.8** A uniform cable hangs from the ceiling with the lower end loose. Derive the eigenvalue problem for the lateral vibration of the cable and obtain a closed-form solution of the problem. (*Hint:* Measure the distance $x$ from the lower end and use a transformation of the independent variable to bring the differential equation defining the eigenvalue problem into the form of a Bessel equation.) Calculate the first three natural frequencies and plot the first three natural modes.

**5.9** A bar in flexural vibration is fixed at the end $x = 0$ and supported by means of a linear spring of stiffness $k$ at the end $x = L$ (see Fig. 5.13). Use the approach of Sec. 5.7 and prove the orthogonality of the natural modes.

**5.10** Consider the system of Prob. 5.4, use the approach of Sec. 5.7 and prove the orthogonality condition, Eq. (5.87).

**5.11** Derive an expression of Rayleigh's quotient for the system of Fig. 5.12 in terms of all the system parameters, including the spring stiffness $k$. (*Hint:* Begin with an expression based on the differential equation defining the eigenvalue problem and use integration by parts with due consideration of the boundary conditions.)

**5.12** Repeat Prob. 5.11 for the system of Fig. 5.13.

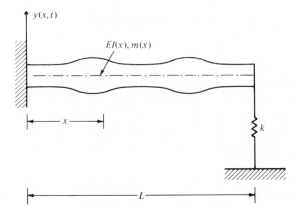

Figure 5.13

**5.13** Consider the system of Prob. 5.7 and estimate the first natural frequency by means of Rayleigh's quotient. Use as a trial function the first eigenfunction of the uniform rod fixed at $x = 0$ and free at $x = L$. Compare the result with the corresponding one obtained in Prob. 5.7. Note that, to account for the effect of the spring at $x = L$, it is necessary to use the expression of Rayleigh's quotient derived in Prob. 5.11.

**5.14** Let the bar of Prob. 5.4 be uniform. Moreover, let $M = 0.2mL$, where $M$ is the concentrated mass and $m$ the mass per unit length of bar, and use Rayleigh's quotient to obtain estimates of the first natural frequency. Repeat the problem for two different choices of trial functions. The trial functions must satisfy at least the geometric boundary conditions. Note that, to account for the concentrated mass at the end $x = L$, the mass distribution entering into the integral at the denominator of Rayleigh's quotient can be expressed in the form $m(x) = m + M\delta(x - L)$, where $\delta(x - L)$ is a spatial Dirac delta function.

**5.15** A uniform bar in flexure hinged at both ends is displaced initially according to $y(x, 0) = y_0(x) = Ax(1 - x/L)$, and then allowed to vibrate freely. Calculate the subsequent response.

**5.16** The uniform rod of Sec. 5.4 is subjected to the distributed forcing function $f(x, t) = f_0 \sin 6(EA/mL^2)^{1/2}t$. The initial conditions are zero. Calculate the system response, and draw conclusions as to the degree of participation of the natural modes.

**5.17** A uniform bar in flexure hinged at both ends is struck impulsively at $x = L/4$. Let the impulsive force be described by $f(x, t) = \hat{F}_0 \delta(x - L/4)\delta(t)$, where $\delta(x - L/4)$ is a spatial Dirac delta function, and calculate the response.

**5.18** The rod of Prob. 5.6 is subjected to a force in the form of a step function at the end $x = 0$. Let the force be described by $f(x, t) = F_0 \alpha(t)\delta(x)$, and calculate the response. (*Caution:* Do not forget to include the rigid-body mode in the solution.) Plot the response as a function of $x$ for two arbitrary values of time.

**5.19** Derive Eq. (5.142) by considering a continuous model.

**5.20** Derive Eq. (5.145) by considering a continuous model.

# ELEMENTS OF ANALYTICAL DYNAMICS

## 6.1 GENERAL DISCUSSION

Newton's laws were formulated for a single particle and can be extended to systems of particles and rigid bodies. In describing the motion, physical coordinates and forces are employed, quantities that can be represented by vectors. For this reason, this approach is often referred to as *vectorial mechanics*. Its main drawback is that it considers the individual components of a system separately, thus necessitating the calculation of interacting forces resulting from kinematical constraints. The calculation of these forces is quite often an added complication and, moreover, in many cases these forces are of no interest and must be eliminated from the equations of motion.

A different approach to mechanics, referred to as *analytical mechanics*, considers the system as a whole rather than its individual components, thus eliminating the need to calculate interacting forces. The appraoch is attributed to Leibnitz and Lagrange, and it formulates the problems of mechanics in terms of two scalar functions, the kinetic energy and the potential energy, and an infinitesimal expression, the virtual work associated with nonconservative forces. Analytical mechanics represents a broader point of view, as it formulates the problems of mechanics by means of generalized coordinates and generalized forces, which are not necessarily physical coordinates and forces, although in certain cases they can be. In this manner, the mathematical formulation is rendered independent of any special system of coordinates. Analytical mechanics relies heavily on the concept of virtual displacements, which led to the development of calculus of variations. For this reason, analytical mechanics is also referred to as the *variational approach to mechanics*.

In this chapter various concepts, such as work, energy, and virtual displacements, as well as the principle of virtual work and d'Alembert's principle, are introduced. These provide the groundwork for the real object of the chapter, namely, Lagrange's equations of motion. Finally, the differential equations governing the vibration of discrete linear systems are derived by means of Lagrange's equations.

## 6.2 WORK AND ENERGY

Let us consider a particle of mass $m$ moving along a curve $s$ under the action of the given force $\mathbf{F}$ (see Fig. 6.1), where the motion is unconstrained. The *increment of work* associated with the displacement of $m$ from position $\mathbf{r}$ to position $\mathbf{r} + d\mathbf{r}$ is defined as the dot product (scalar product) of the vectors $\mathbf{F}$ and $d\mathbf{r}$, or

$$\overline{dW} = \mathbf{F} \cdot d\mathbf{r} \tag{6.1}$$

where the overbar indicates that $\overline{dW}$ is not to be regarded as the true differential of a function $W$ but simply as an infinitesimal expression. We shall see shortly that only in special cases a function $W$ exists for which $\overline{dW} = dW$ is a true differential.

Newton's second law for the particle is simply $\mathbf{F} = m\ddot{\mathbf{r}}$, so that, recalling that $\ddot{\mathbf{r}} = d\dot{\mathbf{r}}/dt$ and $d\mathbf{r} = \dot{\mathbf{r}}\, dt$, Eq. (6.1) can be rewritten as

$$\overline{dW} = \mathbf{F} \cdot d\mathbf{r} = m\ddot{\mathbf{r}} \cdot d\mathbf{r} = m\dot{\mathbf{r}} \cdot d\dot{\mathbf{r}} = d(\tfrac{1}{2}m\dot{\mathbf{r}} \cdot \dot{\mathbf{r}}) = dT \tag{6.2}$$

In contrast to $\overline{dW}$, the right side of (6.2) does represent the true differential of a function, namely, the *kinetic energy* $T$ defined by

$$T = \tfrac{1}{2}m\dot{\mathbf{r}} \cdot \dot{\mathbf{r}} = \tfrac{1}{2}m\dot{r}^2 \tag{6.3}$$

where $\dot{r}$ is the magnitude of the velocity vector $\dot{\mathbf{r}}$. Note that $T$ is a scalar function. If the particle moves from position $\mathbf{r}_1$ to position $\mathbf{r}_2$ under the force $\mathbf{F}$, then the

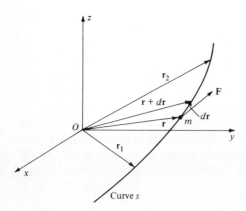

Figure 6.1

corresponding work is simply

$$\int_{\mathbf{r}_1}^{\mathbf{r}_2} \mathbf{F} \cdot d\mathbf{r} = \tfrac{1}{2} m \dot{\mathbf{r}}_2 \cdot \dot{\mathbf{r}}_2 - \tfrac{1}{2} m \dot{\mathbf{r}}_1 \cdot \dot{\mathbf{r}}_1 = T_2 - T_1 \qquad (6.4)$$

so that the work done raises the system kinetic energy from $T_1$ to $T_2$, where the subscripts are associated with positions $\mathbf{r}_1$ and $\mathbf{r}_2$, respectively.

In many physical problems the given force depends on the position alone, $\mathbf{F} = \mathbf{F}(\mathbf{r})$, and the quantity $\mathbf{F} \cdot d\mathbf{r}$ can be written in the form of a perfect differential. In such problems it is possible to introduce the definition

$$dW = \mathbf{F} \cdot d\mathbf{r} = -dV(\mathbf{r}) \qquad (6.5)$$

where $V(\mathbf{r})$ is a scalar function depending explicitly only on the position vector $\mathbf{r}$ and not on the velocity vector $\dot{\mathbf{r}}$ or the time $t$. The function $V$ is recognized as the system *potential energy*, and we notice that, unlike $\overline{dW}$ in Eq. (6.1), $dW$ in Eq. (6.5) is not merely an infinitesimal expression but the differential of the function $W = -V$, where $W$ is sometimes referred to as the *work function*. Concentrating on this particular case and combining Eqs. (6.2) and (6.5), we conclude that

$$d(T + V) = 0 \qquad (6.6)$$

But the sum of the kinetic and potential energy is the system total energy $E$, so that integrating Eq. (6.6) we obtain

$$T + V = E = \text{const} \qquad (6.7)$$

which states that, when Eq. (6.5) holds, the system total energy is constant. Equation (6.7) is known as the *principle of conservation of energy* and the corresponding force field $\mathbf{F}$ is said to be *conservative*.

Because $dW$ in Eq. (6.5) is a perfect differential, if we use cartesian coordinates, we obtain

$$\mathbf{F} \cdot d\mathbf{r} = -dV = -\left( \frac{\partial V}{\partial x} dx + \frac{\partial V}{\partial y} dy + \frac{\partial V}{\partial z} dz \right) = -\nabla V \cdot d\mathbf{r} \qquad (6.8)$$

where

$$\nabla = \frac{\partial}{\partial x} \mathbf{i} + \frac{\partial}{\partial y} \mathbf{j} + \frac{\partial}{\partial z} \mathbf{k} \qquad (6.9)$$

is an operator called *del* or *nabla*. Equation (6.8) implies that

$$\mathbf{F} = -\nabla V \qquad (6.10)$$

or, in words, for a conservative force field the force vector is the negative of the gradient of the potential energy. Equation (6.10) can be written in terms of the cartesian components

$$F_x = -\frac{\partial V}{\partial x} \qquad F_y = -\frac{\partial V}{\partial y} \qquad F_z = -\frac{\partial V}{\partial z} \qquad (6.11)$$

so that the components of the conservative force vector are derivable from a single scalar function, namely, the potential energy. Although we proved Eq. (6.10) by means of cartesian coordinates, the expression is valid also for curvilinear coordinates, in which case the specific form of $\nabla V$ depends on the type of coordinates used.

In general, the forces acting upon a particle can be divided into conservative and nonconservative, so that

$$\mathbf{F} = \mathbf{F}_c + \mathbf{F}_{nc} \tag{6.12}$$

where the meaning of the subscripts $c$ and $nc$ is self-evident. Recognizing that Eq. (6.5) is valid for conservative forces alone, considering Eq. (6.2) and taking the dot product of both sides of Eq. (6.12) with $d\mathbf{r}$, we obtain

$$d(T + V) = dE = \mathbf{F}_{nc} \cdot d\mathbf{r} \tag{6.13}$$

A division of Eq. (6.13) through by $dt$ leads to

$$\frac{d}{dt}(T + V) = \frac{dE}{dt} = \mathbf{F}_{nc} \cdot \dot{\mathbf{r}} \tag{6.14}$$

which states that the rate of work performed by the nonconservative force is equal to the rate of change of the system total energy.

## 6.3 THE PRINCIPLE OF VIRTUAL WORK

The principle of virtual work is basically a statement of the static equilibrium of a mechanical system and was formulated by Johann Bernoulli. To derive the principle it is necessary to introduce a new concept, namely, that of virtual displacements.

Let us concern ourselves with a system of $N$ particles moving in a three-dimensional space and define the *virtual displacements* $\delta x_1, \delta y_1, \delta z_1, \delta x_2, \ldots, \delta z_N$ as infinitesimal changes in the coordinates $x_1, y_1, z_1, x_2, \ldots, z_N$ that are consistent with the system constraints but are otherwise arbitrary. As an example, if the motion of a particle in the real situation is confined to a given smooth surface, then the virtual displacement must be parallel to that surface. The virtual displacements are not true displacements but small variations in the coordinates resulting from imagining the system in a slightly displaced position, a process that does not necessitate any corresponding change in time. Hence, the virtual displacements are assumed to take place contemporaneously. The symbol $\delta$ was introduced by Lagrange to emphasize the virtual character of the instantaneous variations, as opposed to the symbol $d$ which designates actual differentials of position coordinates taking place in the time interval $dt$, during which interval forces and constraints may change. The virtual displacements, being infinitesimal, obey the rules of differential calculus. Now, if the actual coordinates satisfy the constraint equation

$$g(x_1, y_1, z_1, x_2, y_2, z_2, x_3, \ldots, z_N, t) = c \tag{6.15}$$

then the virtual displacements must be such that

$$g(x_1 + \delta x_1, y_1 + \delta y_1, z_1 + \delta z_1, x_2 + \delta x_2, \ldots, z_N + \delta z_N, t) = c \qquad (6.16)$$

and we note that the time $t$ has not been varied. Expanding Eq. (6.16) in a Taylor series and retaining only the first-order terms in the virtual displacements, we can write

$$g(x_1, y_1, z_1, x_2, \ldots, z_N, t) + \sum_{i=1}^{N} \left( \frac{\partial g}{\partial x_i} \delta x_i + \frac{\partial g}{\partial y_i} \delta y_i + \frac{\partial g}{\partial z_i} \delta z_i \right) = c \qquad (6.17)$$

Considering Eq. (6.15), we conclude that for the virtual displacements $\delta x_1, \delta y_1, \delta z_1,$ $\delta x_2, \ldots, \delta z_N$ to be compatible with the system constraint they must satisfy the relation

$$\sum_{i=1}^{N} \left( \frac{\partial g}{\partial x_i} \delta x_i + \frac{\partial g}{\partial y_i} \delta y_i + \frac{\partial g}{\partial z_i} \delta z_i \right) = 0 \qquad (6.18)$$

which implies that only $3N - 1$ of the virtual displacements are arbitrary. In general the number of arbitrary virtual displacements coincides with the number of degrees of freedom of the system.

Let us assume that every one of the $N$ particles belonging to the system under consideration is acted upon by the resultant force

$$\mathbf{R}_i = \mathbf{F}_i + \mathbf{f}_i \qquad i = 1, 2, \ldots, N \qquad (6.19)$$

where $\mathbf{F}_i$ is the applied force and $\mathbf{f}_i$ the constraint force. Applied forces are of an external nature. Examples of applied forces are gravitational forces, aerodynamic lift and drag, magnetic forces, etc. On the other hand, constraint forces are of a reactive nature. The most common ones are the forces that confine the motion of a system to a given path or surface, or the internal forces in rigid bodies. An example of the latter are the forces in a dumbbell. If we regard the dumbbell as two particles connected by a rigid massless rod, the constraint forces are those forces that ensure that the distance between the particles does not change. For a system in equilibrium every particle must be at rest, so that the force on each particle must vanish, $\mathbf{R}_i = 0$, and the same can be said about the scalar product $\mathbf{R}_i \cdot \delta \mathbf{r}_i$, where $\delta \mathbf{r}_i = \delta x_i \mathbf{i} + \delta y_i \mathbf{j} + \delta z_i \mathbf{k}$ $(i = 1, 2, \ldots, N)$ is the virtual displacement vector of the $i$th particle. But $\mathbf{R}_i \cdot \delta \mathbf{r}_i$ represents the *virtual work* performed by the resultant force on the $i$th particle over the virtual displacement $\delta \mathbf{r}_i$. Summing up, it follows that the virtual work for the entire system must vanish, or

$$\overline{\delta W} = \sum_{i=1}^{N} \mathbf{R}_i \cdot \delta \mathbf{r}_i = 0 \qquad (6.20)$$

Introducing Eq. (6.19) into (6.20), we arrive at

$$\overline{\delta W} = \sum_{i=1}^{N} \mathbf{F}_i \cdot \delta \mathbf{r}_i + \sum_{i=1}^{N} \mathbf{f}_i \cdot \delta \mathbf{r}_i = 0 \qquad (6.21)$$

Moreover, we restrict ourselves to systems for which the total virtual work performed by the constraint forces is zero. This rules out friction forces such as those resulting from motion on a rough surface. It is clear that if the motion of a particle is confined to a smooth surface, then the constraint force is normal to the surface, whereas the virtual displacement is parallel to the surface, so that the virtual work is zero because the scalar product of two vectors normal to one another is zero. In the case of the dumbbell we observe that, whereas the virtual work done by the constraint force on each particle is not zero, the virtual work done by both constraint forces is zero. In view of the above, we can write

$$\sum_{i=1}^{N} \mathbf{f}_i \cdot \delta \mathbf{r}_i = 0 \qquad (6.22)$$

It follows from Eqs. (6.21) and (6.22) that

$$\overline{\delta W} = \sum_{i=1}^{N} \mathbf{F}_i \cdot \delta \mathbf{r}_i = 0 \qquad (6.23)$$

or *the work performed by the applied forces through infinitesimal virtual displacements compatible with the system constraints is zero.* This is the statement of the *principle of virtual work.* The principle can be used to calculate the position of static equilibrium of a system (see Example 6.1).

For a conservative system we can write

$$\overline{\delta W} = \sum_{i=1}^{N} \mathbf{F}_i \cdot \delta \mathbf{r}_i = -\delta V = -\sum_{i=1}^{N} \left( \frac{\partial V}{\partial x_i} \delta x_i + \frac{\partial V}{\partial y_i} \delta y_i + \frac{\partial V}{\partial z_i} \delta z_i \right) = 0 \quad (6.24)$$

If there are no constraints in the system, then all the virtual displacements $\delta x_i$, $\delta y_i$, $\delta z_i$ $(i = 1, 2, ..., N)$ are independent. Moreover, because they are arbitrary by definition, we can let them all be equal to zero except for one of them, which is different from zero. This implies that the quantity multiplying the virtual displacement different from zero must be zero itself. If we repeat the procedure $3N$ times every time letting a new virtual displacement be different from zero, we conclude that Eq. (6.24) is satisfied only if

$$F_{xi} = -\frac{\partial V}{\partial x_i} = 0 \qquad F_{yi} = -\frac{\partial V}{\partial y_i} = 0 \qquad F_{zi} = -\frac{\partial V}{\partial z_i} = 0 \qquad i = 1, 2, ..., N$$

$$(6.25)$$

or, for equilibrium, all the components of the applied forces must be equal to zero, as expected. However, in terms of the potential energy $V$, Eqs. (6.25) are precisely the conditions for $V$ to have a *stationary value.* A function is said to have a stationary value at a given point if the rate of change of the function with respect to every independent variable vanishes at that point. Special cases of stationary values are the extremal values of a function, namely, the maximum and the minimum. According to a theorem due to Lagrange, *an equilibrium point is stable if the potential energy has a minimum value at that point.*†

---

† See L. Meirovitch, *Methods of Analytical Dynamics*, p. 193, McGraw-Hill Book Co., New York, 1970.

**Example 6.1** Consider the system of Fig. 6.2 and calculate the angle $\theta$ corresponding to the position of static equilibrium by using the principle of virtual work. The spring exhibits linear behavior and when it is unstretched its length is $x_0$. The system is constrained as shown and the link, regarded as massless and rigid, is horizontal when the spring is unstretched.

First we calculate the position of the ends of the link for a given angle $\theta$. On geometrical grounds, we conclude that the position is defined by

$$x = L(1 - \cos \theta) \qquad y = L \sin \theta \qquad (a)$$

where $x$ is the elongation of the spring and $y$ the lowering of the weight $mg$. Because in this position there is a tensile force $kx$ in the spring, where the force is opposed in direction to the virtual displacement $\delta x$, the virtual work principle, Eq. (6.23), leads to

$$\overline{\delta W} = -kx \, \delta x + mg \, \delta y = 0 \qquad (b)$$

Note that, in writing Eq. (b), we regard the system as consisting of the link and weight alone, and view the spring force as external to the system. But, from Eqs. (a), we have

$$\delta x = L \sin \theta \, \delta \theta \qquad \delta y = L \cos \theta \, \delta \theta \qquad (c)$$

so that, introducing the first of Eqs. (a) and both Eqs. (c) into (b) and equating the coefficient of $\delta \theta$ to zero, we conclude that the angle $\theta$ corresponding to the equilibrium position can be calculated by solving the transcendental equation

$$(1 - \cos \theta) \tan \theta = \frac{mg}{kL} \qquad (d)$$

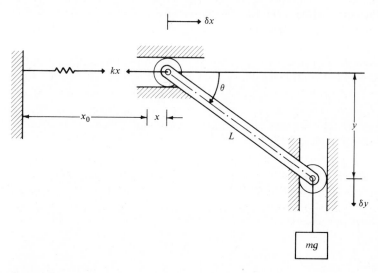

**Figure 6.2**

## 6.4 D'ALEMBERT'S PRINCIPLE

The principle of virtual work is concerned with the static equilibrium of systems. By itself it cannot be used to formulate problems in vibrations, which are basically problems of dynamics. However, we can extend the principle to dynamics, which can be done by a principle attributed to d'Alembert.

Referring to Newton's second law, Eq. (4.5), we can write

$$\mathbf{F}_i + \mathbf{f}_i - m_i\ddot{\mathbf{r}}_i = \mathbf{0} \qquad i = 1, 2, \ldots, N \tag{6.26}$$

where $-m_i\ddot{\mathbf{r}}_i$ can be regarded as an inertia force, which is simply the negative of the rate of change of the momentum vector $\mathbf{p}_i = m_i\dot{\mathbf{r}}_i$. Equation (6.26) is often referred to as d'Alembert's principle, and it permits us to regard problems of dynamics as if they were problems of statics. However, our interest in Eq. (6.26) can be traced to the fact that it enables us to extend the principle of virtual work to the dynamical case. Indeed, using Eq. (6.26), we can write the virtual work for the $i$th particle as

$$(\mathbf{F}_i + \mathbf{f}_i - m_i\ddot{\mathbf{r}}_i) \cdot \delta\mathbf{r}_i = 0 \qquad i = 1, 2, \ldots, N \tag{6.27}$$

Assuming virtual displacements $\delta\mathbf{r}_i$ compatible with the system constraints, we can sum over the entire system of particles and obtain

$$\sum_{i=1}^{N} (\mathbf{F}_i - m_i\ddot{\mathbf{r}}_i) \cdot \delta\mathbf{r}_i = 0 \tag{6.28}$$

where, according to Eq. (6.22), the virtual work associated with the constraint forces is zero. Equation (6.28) embodies both the principle of virtual work of statics and d'Alembert's principle, and is referred to as the *generalized principle of d'Alembert*.† The sum of the applied force and inertia force, $\mathbf{F}_i - m_i\ddot{\mathbf{r}}_i$, is sometimes called the *effective force*. Hence, *the virtual work performed by the effective forces through infinitesimal virtual displacements compatible with the system constraints is zero.*

Whereas d'Alembert's principle, Eq. (6.28), gives a complete formulation of the problems of mechanics, it is not very convenient for deriving the system equations of motion because the problems are formulated in terms of position coordinates, which may not all be independent. The principle, however, is useful in providing the transition to a formulation in terms of generalized coordinates that does not suffer from this drawback. In addition, this new formulation is extremely convenient, as it enables us to derive all the system differential equations of motion from two scalar functions, the kinetic energy and the potential energy, and an infinitesimal expression, the virtual work associated with the nonconservative forces. It eliminates the need for free-body diagrams or any knowledge of the constraint forces. The differential equations so derived are the celebrated Lagrange's equations.

† See L. Meirovitch, op. cit., p. 65.

## 6.5 LAGRANGE'S EQUATIONS OF MOTION

In Sec. 4.2 we pointed out that the physical coordinates $\mathbf{r}_i$ $(i = 1, 2, ..., N)$ are not always independent and that it is often desirable to describe the motion of the system by means of a set of independent generalized coordinates $q_k$ $(k = 1, 2, ..., n)$. To this end, we can use a coordinate transformation from $\mathbf{r}_i$ to $q_k$ in conjunction with the principle of d'Alembert, Eq. (6.28), to obtain a set of differential equations of motion in terms of the generalized coordinates $q_k$, where the equations are known as Lagrange's equations.

Assuming that the coordinates $\mathbf{r}_i$ do not depend explicitly on time and considering an $n$-degree-of-freedom system, we can write the coordinate transformation in the general form

$$\mathbf{r}_i = \mathbf{r}_i(q_i, q_2, ..., q_n) \qquad i = 1, 2, ..., N \tag{6.29}$$

The velocities $\dot{\mathbf{r}}_i$ are obtained by simply taking the total time derivative of Eqs. (6.29), leading to

$$\dot{\mathbf{r}}_i = \frac{\partial \mathbf{r}_i}{\partial q_1} \dot{q}_1 + \frac{\partial \mathbf{r}_i}{\partial q_2} \dot{q}_2 + \cdots + \frac{\partial \mathbf{r}_i}{\partial q_n} \dot{q}_n = \sum_{k=1}^{n} \frac{\partial \mathbf{r}_i}{\partial q_k} \dot{q}_k \qquad i = 1, 2, ..., N \tag{6.30}$$

Because the quantities $\partial \mathbf{r}_i / \partial q_k$ do not depend explicitly on the generalized velocities $\dot{q}_k$, Eqs. (6.30) yield

$$\frac{\partial \dot{\mathbf{r}}_i}{\partial \dot{q}_k} = \frac{\partial \mathbf{r}_i}{\partial q_k} \qquad i = 1, 2, ..., N; \ k = 1, 2, ..., n \tag{6.31}$$

Moreover, by analogy with Eqs. (6.30), we can write

$$\delta \mathbf{r}_i = \frac{\partial \mathbf{r}_i}{\partial q_1} \delta q_1 + \frac{\delta \mathbf{r}_i}{\partial q_2} \delta q_2 + \cdots + \frac{\partial \mathbf{r}_i}{\partial q_n} \delta q_n$$

$$= \sum_{k=1}^{n} \frac{\partial \mathbf{r}_i}{\partial q_k} \delta q_k \qquad k = 1, 2, ..., n \tag{6.32}$$

In view of Eqs. (6.32), the second term in Eq. (6.28) becomes

$$\sum_{i=1}^{N} m_i \ddot{\mathbf{r}}_i \cdot \delta \mathbf{r}_i = \sum_{i=1}^{N} m_i \ddot{\mathbf{r}}_i \cdot \sum_{k=1}^{n} \frac{\partial \mathbf{r}_i}{\partial q_k} \delta q_k = \sum_{k=1}^{n} \left( \sum_{i=1}^{N} m_i \ddot{\mathbf{r}}_i \cdot \frac{\partial \mathbf{r}_i}{\partial q_k} \right) \delta q_k \tag{6.33}$$

Concentrating on a typical term on the right side of (6.33), we observe that

$$m_i \ddot{\mathbf{r}}_i \cdot \frac{\partial \mathbf{r}_i}{\partial q_k} = \frac{d}{dt} \left( m_i \dot{\mathbf{r}}_i \cdot \frac{\partial \mathbf{r}_i}{\partial q_k} \right) - m_i \dot{\mathbf{r}}_i \cdot \frac{d}{dt} \left( \frac{\partial \mathbf{r}_i}{\partial q_k} \right) \tag{6.34}$$

Considering Eqs. (6.31) and assuming that the order of total derivatives with respect to time and partial derivatives with respect to $q_k$ is interchangeable, we can write Eq. (6.34) in the form

$$m_i\ddot{\mathbf{r}}_i \cdot \frac{\partial \mathbf{r}_i}{\partial q_k} = \frac{d}{dt}\left(m_i\dot{\mathbf{r}}_i \cdot \frac{\partial \mathbf{r}_i}{\partial q_k}\right) - m_i\dot{\mathbf{r}}_i \cdot \frac{\partial \dot{\mathbf{r}}_i}{\partial q_k}$$

$$= \left[\frac{d}{dt}\left(\frac{\partial}{\partial \dot{q}_k}\right) - \frac{\partial}{\partial q_k}\right](\tfrac{1}{2}m_i\dot{\mathbf{r}}_i \cdot \dot{\mathbf{r}}_i) \tag{6.35}$$

But the second term in parentheses on the right side of (6.35) is recognized as the kinetic energy of particle $i$ [see Eq. (6.3)]. Hence, insertion of (6.35) into (6.33) leads to

$$\sum_{i=1}^{N} m_i\ddot{\mathbf{r}}_i \cdot \delta\mathbf{r}_i = \sum_{k=1}^{n}\left\{\left[\frac{d}{dt}\left(\frac{\partial}{\partial \dot{q}_k}\right) - \frac{\partial}{\partial q_k}\right]\left(\sum_{i=1}^{N}\frac{1}{2}m_i\dot{\mathbf{r}}_i \cdot \dot{\mathbf{r}}_i\right)\right\}\delta q_k$$

$$= \sum_{k=1}^{n}\left[\frac{d}{dt}\left(\frac{\partial T}{\partial \dot{q}_k}\right) - \frac{\partial T}{\partial q_k}\right]\delta q_k \tag{6.36}$$

where, in view of transformation (6.30),

$$T = \frac{1}{2}\sum_{i=1}^{N} m_i\dot{\mathbf{r}}_i \cdot \dot{\mathbf{r}}_i = T(q_1, q_2, \dots, q_n, \dot{q}_1, \dot{q}_2, \dots, \dot{q}_n) \tag{6.37}$$

is the kinetic energy of the entire system.

It remains to write the forces $\mathbf{F}_i(\mathbf{r}_1, \mathbf{r}_2, \dots, \mathbf{r}_N, \dot{\mathbf{r}}_1, \dot{\mathbf{r}}_2, \dots, \dot{\mathbf{r}}_N, t)$ in terms of the generalized coordinates $q_k$ ($k = 1, 2, \dots, n$). This is done by using the virtual work expression, in conjunction with transformation (6.32), in the following manner:

$$\overline{\delta W} = \sum_{i=1}^{N} \mathbf{F}_i \cdot \delta\mathbf{r}_i = \sum_{i=1}^{N} \mathbf{F}_i \cdot \sum_{k=1}^{n}\frac{\partial \mathbf{r}_i}{\partial q_k}\delta q_k = \sum_{k=1}^{n}\left(\sum_{i=1}^{N}\mathbf{F}_i \cdot \frac{\partial \mathbf{r}_i}{\partial q_k}\right)\delta q_k \tag{6.38}$$

The virtual work, however, can be regarded as the product of $n$ generalized forces $Q_k$ acting over the virtual displacements $\delta q_k$, or

$$\overline{\delta W} = \sum_{k=1}^{n} Q_k\,\delta q_k \tag{6.39}$$

so that, comparing Eqs. (6.38) and (6.39), we conclude that the generalized forces have the form

$$Q_k = \sum_{i=1}^{N} \mathbf{F}_i \cdot \frac{\partial \mathbf{r}_i}{\partial q_k} \qquad k = 1, 2, \dots, n \tag{6.40}$$

In actual situations the generalized forces are derived by identifying physically a set of generalized coordinates and writing the virtual work in the form (6.39), rather than by using formula (6.40) (see Example 6.2). We note that the generalized forces are not necessarily forces. They can be moments or any other quantities such that the product $Q_k\delta q_k$ has units of work.

If the forces acting upon the system can be divided into conservative forces, which are derivable from the potential energy $V = V(q_1, q_2, \dots, q_n)$, and nonconservative forces, which are not, then the first term in Eq. (6.28) becomes

$$\sum_{i=1}^{N} \mathbf{F}_i \cdot \delta \mathbf{r}_i = \overline{\delta W} = \delta W_c + \overline{\delta W}_{nc} = -\delta V + \sum_{k=1}^{n} Q_{knc} \, \delta q_k$$

$$= -\left( \frac{\partial V}{\partial q_1} \delta q_1 + \frac{\partial V}{\partial q_2} \delta q_2 + \cdots + \frac{\partial V}{\partial q_n} \delta q_n \right) + \sum_{k=1}^{n} Q_{knc} \, \delta q_k$$

$$= -\sum_{k=1}^{n} \left( \frac{\partial V}{\partial q_k} - Q_{knc} \right) \delta q_k \tag{6.41}$$

where $Q_{knc}$ $(k = 1, 2, \ldots, n)$ are nonconservative generalized forces. Introducing Eqs. (6.36) and (6.41) into Eq. (6.28), we obtain

$$-\sum_{k=1}^{n} \left[ \frac{d}{dt} \left( \frac{\partial T}{\partial \dot{q}_k} \right) - \frac{\partial T}{\partial q_k} + \frac{\partial V}{\partial q_k} - Q_{knc} \right] \delta q_k = 0 \tag{6.42}$$

However, by definition, the generalized virtual displacements $\delta q_k$ are both arbitrary and independent. Hence, letting $\delta q_k = 0$ $(k = 1, 2, \ldots, n;\ k \neq j)$ and $\delta q_j \neq 0$, we conclude that Eq. (6.42) can be satisfied if and only if the coefficient of $\delta q_j$ is zero. The procedure can be repeated $n$ times for $j = 1, 2, \ldots, n$. Moreover, with the understanding that $Q_j$ represents nonconservative forces, we can drop the subscript $nc$ in $Q_{jnc}$ and arrive at the set of equations

$$\frac{d}{dt} \left( \frac{\partial T}{\partial \dot{q}_j} \right) - \frac{\partial T}{\partial q_j} + \frac{\partial V}{\partial q_j} = Q_j \qquad j = 1, 2, \ldots, n \tag{6.43}$$

which are the famous *Lagrange's equations of motion*. In general, the potential energy does not depend on the generalized velocities $\dot{q}_j$ $(j = 1, 2, \ldots, n)$. In view of this, we can introduce the *Lagrangian* defined by

$$L = T - V \tag{6.44}$$

and reduce Eqs. (6.43) to the more compact form

$$\frac{d}{dt} \left( \frac{\partial L}{\partial \dot{q}_j} \right) - \frac{\partial L}{\partial q_j} = Q_j \qquad j = 1, 2, \ldots, n \tag{6.45}$$

Of the nonconservative forces, there are some that deserve special consideration, namely, those due to viscous damping. If the damping forces are proportional to the generalized velocities, it is possible to devise a function, known as *Rayleigh's dissipation function*,† in the form

$$\mathcal{F} = \frac{1}{2} \sum_{r=1}^{n} \sum_{s=1}^{n} c_{rs} \dot{q}_r \dot{q}_s \tag{6.46}$$

where the constant coefficients $c_{rs}$ are symmetric in $r$ and $s$. This enables us to derive viscous damping forces in a manner analogous to that for conservative

† See L. Meirovitch, op. cit., sec. 2-12.

forces. In particular, viscous damping forces can be derived from Rayleigh's dissipation function by means of the formula

$$Q_j = -\frac{\partial \mathscr{F}}{\partial \dot{q}_j} \qquad j = 1, 2, \ldots, n \tag{6.47}$$

Assuming that the nonconservative forces $Q_j$ in Eqs. (6.45) can be divided into dissipative forces and forces impressed upon the system by external factors, we can rewrite Lagrange's equations (6.45) as follows:

$$\frac{d}{dt}\left(\frac{\partial L}{\partial \dot{q}_j}\right) - \frac{\partial L}{\partial q_j} + \frac{\partial \mathscr{F}}{\partial \dot{q}_j} = Q_j \qquad j = 1, 2, \ldots, n \tag{6.48}$$

where this time the symbol $Q_j$ is understood to designate only impressed forces.

Of course, in many problems there are no nonconservative forces involved, in which cases $\partial \mathscr{F}/\partial \dot{q}_j = 0$ and $Q_j = 0$ ($j = 1, 2, \ldots, n$). Hence, Lagrange's equations for conservative systems are simply

$$\frac{d}{dt}\left(\frac{\partial L}{\partial \dot{q}_j}\right) - \frac{\partial L}{\partial q_j} = 0 \qquad j = 1, 2, \ldots, n \tag{6.49}$$

The Lagrangian approach is very efficient for deriving the system equations of motion, especially when the number of degrees of freedom is large. All the differential equations of motion are derived from two scalar functions, namely, the kinetic energy $T$ and the potential energy $V$, as well as the virtual work $\overline{\delta W}_{nc}$ associated with the nonconservative forces. The equations apply to linear as well as nonlinear systems. Although it appears that the identification of the generalized coordinates and generalized forces is a major stumbling block in using this approach, this is actually not the case; in most physical systems considered in this text this aspect presents no particular difficulty. A distinct feature of the Lagrangian approach is that it obviates the computation of constraint forces.

**Example 6.2** Consider the double pendulum shown in Fig. 6.3a and derive the equations of motion by means of (a) the Newtonian approach and (b) the Lagrangian approach. Discuss the difference between the two sets of equations and show how the difference can be reconciled.

The Newtonian approach requires accelerations and the Lagrangian approach requires velocities. It will prove convenient to use tangential and normal components, as shown in Fig. 6.3b; the corresponding unit vectors are depicted in Fig. 6.3c. The velocities of the masses $m_1$ and $m_2$ are

$$\mathbf{v}_1 = v_{t1}\mathbf{u}_{t1} = L_1\dot{\theta}_1\mathbf{u}_{t1}$$
$$\mathbf{v}_1 = v_{t2}\mathbf{u}_{t2} + v_{n2}\mathbf{u}_{n2} = \mathbf{v}_1 + L_2\dot{\theta}_2\mathbf{u}_{t2} = L_1\dot{\theta}_1\mathbf{u}_{t1} + L_2\dot{\theta}_2\mathbf{u}_{t2} \tag{a}$$

The velocity $\mathbf{v}_2$ contains the unit vectors $\mathbf{u}_{t1}$ and $\mathbf{u}_{t2}$. To express $\mathbf{v}_2$ in terms of unit vectors associated with $m_2$, we turn to Fig. 6.3c and write the relations

$$\mathbf{u}_{t1} = \mathbf{u}_{t2}\cos(\theta_2 - \theta_1) - \mathbf{u}_{n2}\sin(\theta_2 - \theta_1)$$
$$\mathbf{u}_{n1} = \mathbf{u}_{t2}\sin(\theta_2 - \theta_1) + \mathbf{u}_{n2}\cos(\theta_2 - \theta_1) \tag{b}$$

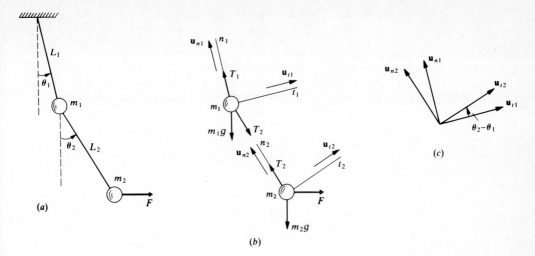

**Figure 6.3**

Hence, $v_2$ can be rewritten as

$$\mathbf{v}_2 = L_1\dot{\theta}_1[\mathbf{u}_{t2}\cos(\theta_2 - \theta_1) - \mathbf{u}_{n2}\sin(\theta_2 - \theta_1)] + L_2\dot{\theta}_2\mathbf{u}_{t2}$$

$$= [L_1\dot{\theta}_1\cos(\theta_2 - \theta_1) + L_2\dot{\theta}_2]\mathbf{u}_{t2} - L_1\dot{\theta}_1\sin(\theta_2 - \theta_1)\mathbf{u}_{n2} \qquad (c)$$

so that

$$v_{t2} = L_1\dot{\theta}_1\cos(\theta_2 - \theta_1) + L_2\dot{\theta}_2, \quad v_{n2} = -L_1\dot{\theta}_1\sin(\theta_2 - \theta_1) \qquad (d)$$

Similarly, the accelerations of $m_1$ and $m_2$ are

$$\mathbf{a}_1 = a_{t1}\mathbf{u}_{t1} + a_{n1}\mathbf{u}_{n1} = L_1\ddot{\theta}_1\mathbf{u}_{t1} + L_1\dot{\theta}_1^2\mathbf{u}_{n1}$$

$$\mathbf{a}_2 = a_{t2}\mathbf{u}_{t2} + a_{n2}\mathbf{u}_{n2} = \mathbf{a}_1 + L_2\ddot{\theta}_2\mathbf{u}_{t2} + L_2\dot{\theta}_2^2\mathbf{u}_{n2} \qquad (e)$$

$$= L_1\ddot{\theta}_1\mathbf{u}_{t1} + L_1\dot{\theta}_1^2\mathbf{u}_{n1} + L_2\ddot{\theta}_2\mathbf{u}_{t2} + L_2\dot{\theta}_2^2\mathbf{u}_{n2}$$

Using Eqs. (b), we can write

$$\mathbf{a}_2 = L_1\ddot{\theta}_1[\mathbf{u}_{t2}\cos(\theta_1 - \theta_1) - \mathbf{u}_{n2}\sin(\theta_2 - \theta_1)]$$

$$+ L_1\dot{\theta}_1^2[\mathbf{u}_{t2}\sin(\theta_2 - \theta_1) + \mathbf{u}_{n2}\cos(\theta_2 - \theta_1)]$$

$$+ L_2\ddot{\theta}_2\mathbf{u}_{t2} + L_2\dot{\theta}_2^2\mathbf{u}_{n2}$$

$$= [L_1\ddot{\theta}_1\cos(\theta_2 - \theta_1) + L_1\dot{\theta}_1^2\sin(\theta_2 - \theta_1) + L_2\ddot{\theta}_2]\mathbf{u}_{t2}$$

$$+ [-L_1\ddot{\theta}_1\sin(\theta_2 - \theta_1) + L_1\dot{\theta}_1^2\cos(\theta_2 - \theta_1) + L_2\dot{\theta}_2^2]\mathbf{u}_{n2} \qquad (f)$$

so that

$$a_{t2} = L_1\ddot{\theta}_1\cos(\theta_2 - \theta_1) + L_1\dot{\theta}_1^2\sin(\theta_2 - \theta_1) + L_2\ddot{\theta}_2$$

$$a_{n2} = -L_1\ddot{\theta}_1\sin(\theta_2 - \theta_1) + L_1\dot{\theta}_1^2\cos(\theta_2 - \theta_1) + L_2\dot{\theta}_2^2 \qquad (g)$$

Newton's equations of motion have the general form

$$\sum F_{t1} = m_1 a_{t1} \qquad \sum F_{n1} = m_1 a_{n1} \qquad \sum F_{t2} = m_2 a_{t2} \qquad \sum F_{n2} = m_2 a_{n2}$$

$$(h)$$

so that, using the free-body diagrams of Fig. 6.3b in conjunction with the acceleration components given by the first of Eqs. (e) and Eqs. (g), we obtain Newton's equations of motion in the explicit form

$$T_2 \sin (\theta_2 - \theta_1) - m_1 g \sin \theta_1 = m_1 L_1 \ddot{\theta}_1$$

$$T_1 - T_2 \cos (\theta_2 - \theta_1) - m_1 g \cos \theta_1 = m_1 L_1 \dot{\theta}_1^2$$

$$F \cos \theta_2 - m_2 g \sin \theta_2$$

$$= m_2 [L_1 \ddot{\theta}_1 \cos (\theta_2 - \theta_1) + L_1 \dot{\theta}_1^2 \sin (\theta_2 - \theta_1) + L_2 \ddot{\theta}_2]$$

$$-F \sin \theta_2 + T_2 - m_2 g \cos \theta_2$$

$$= m_2 [-L_1 \ddot{\theta}_1 \sin (\theta_2 - \theta_1) + L_1 \dot{\theta}_1^2 \cos (\theta_2 - \theta_1) + L_2 \dot{\theta}_2^2]$$

$$(i)$$

Using as generalized coordinates the angular displacements, $q_1 = \theta_1$, $q_2 = \theta_2$, Lagrange's equations of motion can be written in the general form

$$\frac{d}{dt} \left( \frac{\partial L}{\partial \dot{\theta}_1} \right) - \frac{\partial L}{\partial \theta_1} = \Theta_1 \qquad \frac{d}{dt} \left( \frac{\partial L}{\partial \dot{\theta}_2} \right) - \frac{\partial L}{\partial \theta_2} = \Theta_2 \qquad (j)$$

where $L$ is the Lagrangian and $\Theta_1$ and $\Theta_2$ are generalized forces. Hence, to derive explicit Lagrange's equations of motion, we must first derive expressions for the kinetic energy, potential energy and virtual work. From the first of Eqs. (a) and from Eqs. (d), we can write the kinetic energy

$$T = \tfrac{1}{2} m_1 (v_{t1}^2 + v_{n1}^2) + \tfrac{1}{2} m_2 (v_{t2}^2 + v_{n2}^2)$$

$$= \tfrac{1}{2} m_1 (L_1 \dot{\theta}_1)^2 + \tfrac{1}{2} m_2 \{ [L_1 \dot{\theta}_1 \cos (\theta_2 - \theta_1) + L_2 \dot{\theta}_2]^2$$

$$+ [-L_1 \dot{\theta}_1 \sin (\theta_2 - \theta_1)]^2 \}$$

$$= \tfrac{1}{2} \{ m_1 L_1^2 \dot{\theta}_1^2 + m_2 [L_1^2 \dot{\theta}_1^2 + 2L_1 L_2 \dot{\theta}_1 \dot{\theta}_2 \cos (\theta_2 - \theta_1) + L_2^2 \dot{\theta}_2^2] \}$$

$$= \tfrac{1}{2} [(m_1 + m_2) L_1^2 \dot{\theta}_1^2 + 2m_2 L_1 L_2 \dot{\theta}_1 \dot{\theta}_2 \cos (\theta_2 - \theta_1) + m_2 L_2^2 \dot{\theta}_2^2] \qquad (k)$$

The potential energy is due to gravitational forces alone and has the form

$$V = m_1 g L_1 (1 - \cos \theta_1) + m_2 g [L_1 (1 - \cos \theta_1) + L_2 (1 - \cos \theta_2)]$$

$$= (m_1 + m_2) g L_1 (1 - \cos \theta_1) + m_2 g L_2 (1 - \cos \theta_2) \qquad (l)$$

Hence, the Lagrangian has the expression

$$L = T - V = \tfrac{1}{2} [(m_1 + m_2) L_1^2 \dot{\theta}_1^2 + 2m_2 L_1 L_2 \dot{\theta}_1 \dot{\theta}_2 \cos (\theta_2 - \theta_1) + m_2 L_2^2 \dot{\theta}_2^2]$$

$$- (m_1 + m_2) g L_1 (1 - \cos \theta_1) - m_2 g L_2 (1 - \cos \theta_2) \qquad (m)$$

Before we write the virtual work expression due to the external force $F$, we observe that the displacement component in the same direction as $F$ is

$$x_2 = L_1 \sin \theta_1 + L_2 \sin \theta_2 \qquad (n)$$

so that the virtual work is

$$\overline{\delta W} = F \delta x_2 = F \delta (L_1 \sin \theta_1 + L_2 \sin \theta_2)$$
$$= FL_1 \cos \theta_1 \, \delta\theta_1 + FL_2 \cos \theta_2 \, \delta\theta_2 \qquad (o)$$

Next, let us calculate the derivatives

$$\frac{\partial L}{\partial \dot{\theta}_1} = (m_1 + m_2)L_1^2 \dot{\theta}_1 + m_2 L_1 L_2 \dot{\theta}_2 \cos (\theta_2 - \theta_1)$$

$$\frac{d}{dt}\left(\frac{\partial L}{\partial \dot{\theta}_1}\right) = (m_1 + m_2)L_1^2 \ddot{\theta}_1$$
$$+ m_2 L_1 L_2 [\ddot{\theta}_2 \cos (\theta_2 - \theta_1) - \dot{\theta}_2(\dot{\theta}_2 - \dot{\theta}_1) \sin (\theta_2 - \theta_1)]$$

$$\frac{\partial L}{\partial \theta_1} = m_2 L_1 L_2 \dot{\theta}_1 \dot{\theta}_2 \sin (\theta_2 - \theta_1) - (m_1 + m_2)gL_1 \sin \theta_1$$

$$\frac{\partial L}{\partial \dot{\theta}_2} = m_2 L_1 L_2 \dot{\theta}_1 \cos (\theta_2 - \theta_1) + m_2 L_2^2 \dot{\theta}_2 \qquad (p)$$

$$\frac{d}{dt}\left(\frac{\partial L}{\partial \dot{\theta}_2}\right) = m_2 L_1 L_2 [\ddot{\theta}_1 \cos (\theta_2 - \theta_1) - \dot{\theta}_1(\dot{\theta}_2 - \dot{\theta}_1) \sin (\theta_2 - \theta_1)]$$
$$+ m_2 L_2^2 \ddot{\theta}_2$$

$$\frac{\partial L}{\partial \theta_2} = -m_2 L_1 L_2 \dot{\theta}_1 \dot{\theta}_2 \sin (\theta_2 - \theta_1) - m_2 g L_2 \sin \theta_2$$

But, the virtual work can be expressed in terms of generalized forces and virtual generalized displacements as follows:

$$\overline{\delta W} = \Theta_1 \, \delta\theta_1 + \Theta_2 \, \delta\theta_2 \qquad (q)$$

so that, comparing Eqs. ($o$) and ($q$), we conclude that the generalized forces are simply

$$\Theta_1 = FL_1 \cos \theta_1 \qquad \Theta_2 = FL_2 \cos \theta_2 \qquad (r)$$

and we observe that the generalized forces are really torques. Finally, inserting Eqs. ($p$) and ($r$) into Eqs. ($j$), we obtain Lagrange's equations of motion in the explicit form

$$(m_1 + m_2)L_1^2 \ddot{\theta}_1 + m_2 L_1 L_2 [\ddot{\theta}_2 \cos (\theta_2 - \theta_1) - \dot{\theta}_2^2 \sin (\theta_2 - \theta_1)]$$
$$+ (m_1 + m_2)gL_1 \sin \theta_1 = FL_1 \cos \theta_1$$

$$(s)$$

$$m_2 L_1 L_2 [\ddot{\theta}_1 \cos (\theta_2 - \theta_1) + \dot{\theta}_1^2 \sin (\theta_2 - \theta_1)]$$
$$+ m_2 L_2^2 \ddot{\theta}_2 + m_2 g L_2 \sin \theta_2 = FL_2 \cos \theta_2$$

We observe that there are four Newton's equations of motion. The unknowns are $\theta_1$, $\theta_2$, $T_1$, and $T_2$, so that the tensile forces $T_1$ and $T_2$ in the strings play the role of unknowns, supplementing the angular displacements $\theta_1$ and $\theta_2$. By contrast, there are only two Lagrange's equations, as forces internal to the system, such as $T_1$ and $T_2$, do not appear. Note that Newton's equations are force equations, whereas Lagrange's equations are moment equations. Quite often the interest lies only in the motion of the system and the tension in the strings is irrelevant. In such cases, it is possible to eliminate $T_1$ and $T_2$ and produce two equations in terms of $\theta_1$ and $\theta_2$ alone. Indeed, Lagrange's equations can be obtained from Newton's equations if the constraint forces are eliminated. In this particular case it is only necessary to eliminate $T_2$, as the equation containing $T_1$ can be ignored. It is easy to verify that the first of Lagrange's equations can be obtained by multiplying the first of Newton's equations by $L_1$, the third by $L_1 \cos (\theta_2 - \theta_1)$, the fourth by $-L_1 \sin (\theta_2 - \theta_1)$ and summing the resulting equations. On the other hand, the second of Lagrange's equations is simply the third of Newton's equations multiplied by $L_2$.

## 6.6 LAGRANGE'S EQUATIONS OF MOTION FOR LINEAR SYSTEMS

The interest lies in the motion of a multi-degree-of-freedom system in the neighborhood of an equilibrium position. Without loss of generality, we assume that the equilibrium position is given by the trivial solution $q_1 = q_2 = \cdots = q_n = 0$. Moreover, we assume that the generalized displacements from the equilibrium position are sufficiently small that the linear force-displacement and force-velocity relations hold, so that the generalized coordinates and their time derivatives appear in the differential equations of motion at most to the first power. This represents, in essence, the so-called *small-motions assumption*, leading to a linear system of equations.

In this section, we derive the differential equations of motion of a multi-degree-of-freedom linear system by means of the Lagrangian approach, and in Example 6.3 we apply the equations to a simple three-degree-of-freedom system. To this end, we must obtain first the kinetic energy, the potential energy and Rayleigh's dissipation function for linear systems. Because we do not admit powers larger than one in the differential equations of motion, the coefficients $\partial \mathbf{r}_i / \partial q_k$ in Eqs. (6.30) must be constant and not functions of the generalized coordinates. Inserting transformation (6.30) into Eq. (6.37), the system kinetic energy becomes

$$
T = \frac{1}{2} \sum_{i=1}^{N} m_i \dot{\mathbf{r}}_i \cdot \dot{\mathbf{r}}_i = \frac{1}{2} \sum_{i=1}^{N} m_i \left( \sum_{r=1}^{n} \frac{\partial \mathbf{r}_i}{\partial q_r} \dot{q}_r \right) \cdot \left( \sum_{s=1}^{n} \frac{\partial \mathbf{r}_i}{\partial q_s} \dot{q}_s \right)
$$

$$
= \frac{1}{2} \sum_{r=1}^{n} \sum_{s=1}^{n} \left( \sum_{i=1}^{N} m_i \frac{\partial \mathbf{r}_i}{\partial q_r} \cdot \frac{\partial \mathbf{r}_i}{\partial q_s} \right) \dot{q}_r \dot{q}_s = \frac{1}{2} \sum_{r=1}^{n} \sum_{s=1}^{n} m_{rs} \dot{q}_r \dot{q}_s \qquad (6.50)
$$

where

$$m_{rs} = \sum_{i=1}^{N} m_i \frac{\partial \mathbf{r}_i}{\partial q_r} \cdot \frac{\partial \mathbf{r}_i}{\partial q_s} = m_{sr} \qquad r, s = 1, 2, \ldots, n \qquad (6.51)$$

are constant *mass coefficients*, or *inertia coefficients*, symmetric in $r$ and $s$. Note that we replaced the dummy index $k$ in transformation (6.30) by $r$ and $s$, in turn, to permit cross products to appear in Eq. (6.50), as they should.

The potential energy did not appear in Sec. 6.5 in an explicit form, but in the general form $V = V(q_1, q_2, \ldots, q_n)$, where $V$ is generally a nonlinear function of the generalized coordinates $q_k$ and it depends on the reference position chosen. Because the potential energy is defined within an arbitrary additive constant, without loss of generality, we can choose the reference position to coincide with the trivial equilibrium position $q_1 = q_2 = \cdots = q_n = 0$. Under these circumstances, the Taylor series expansion of $V$ about the equilibrium point is

$$V(q_1, q_2, \ldots, q_n)$$

$$= \frac{\partial V}{\partial q_1} q_1 + \frac{\partial V}{\partial q_2} q_2 + \cdots + \frac{\partial V}{\partial q_n} q_n$$

$$+ \frac{1}{2} \left( \frac{\partial^2 V}{\partial q_1^2} q_1^2 + \frac{\partial^2 V}{\partial q_2^2} q_2^2 + \cdots + \frac{\partial^2 V}{\partial q_n^2} q_n^2 + 2 \frac{\partial^2 V}{\partial q_1 \partial q_2} q_1 q_2 \right.$$

$$\left. + 2 \frac{\partial^2 V}{\partial q_1 \partial q_3} q_1 q_3 + \cdots + 2 \frac{\partial^2 V}{\partial q_{n-1} \partial q_n} q_{n-1} q_n \right) + \cdots \qquad (6.52)$$

where all the partial derivatives of $V$ in (6.52) are evaluated at the equilibrium point $q_k = 0$ ($k = 1, 2, \ldots, n$), and hence are constant. By analogy with Eqs. (6.25), however, $\partial V / \partial q_k = 0$ ($k = 1, 2, \ldots, n$) at an equilibrium point, with the implication that the generalized conservative forces reduce to zero at an equilibrium. Moreover, because of the small-motions assumption, terms of order higher than 2 are to be discarded, so that Eq. (6.52) reduces to

$$V = \frac{1}{2} \sum_{r=1}^{n} \sum_{s=1}^{n} \frac{\partial^2 V}{\partial q_r \partial q_s} q_r q_s = \frac{1}{2} \sum_{r=1}^{n} \sum_{s=1}^{n} k_{rs} q_r q_s \qquad (6.53)$$

where

$$k_{rs} = \frac{\partial^2 V}{\partial q_r \partial q_s} = \frac{\partial^2 V}{\partial q_s \partial q_r} = k_{sr} \qquad r, s = 1, 2, \ldots, n \qquad (6.54)$$

are constant symmetric coefficients, which can be identified as the *stiffness coefficients*.

Both the kinetic energy, Eq. (6.50), and the potential energy, Eq. (6.53), as well as Rayleigh's dissipation function, Eq. (6.46), are in a form generally known as quadratic, the first and third in the generalized velocities and the second in the generalized coordinates. Properties of quadratic forms of this type were studied in Sec. 4.5.

Let us derive now the differential equations of motion for the system by means of the Lagrangian approach, where the system is subject to dissipative forces of the Rayleigh type as well as to externally impressed forces. Because the potential energy does not depend on generalized velocities, we can differentiate Eq. (6.50) to obtain

$$
\frac{\partial L}{\partial \dot{q}_j} = \frac{\partial T}{\partial \dot{q}_j} = \frac{1}{2} \sum_{r=1}^{n} \sum_{s=1}^{n} m_{rs} \left( \frac{\partial \dot{q}_r}{\partial \dot{q}_j} \dot{q}_s + \dot{q}_r \frac{\partial \dot{q}_s}{\partial \dot{q}_j} \right)
$$

$$
= \frac{1}{2} \sum_{r=1}^{n} \sum_{s=1}^{n} m_{rs} (\dot{q}_s \delta_{rj} + \dot{q}_r \delta_{sj})
$$

$$
= \frac{1}{2} \sum_{s=1}^{n} m_{js} \dot{q}_s + \frac{1}{2} \sum_{r=1}^{n} m_{rj} \dot{q}_r = \sum_{s=1}^{n} m_{js} \dot{q}_s \qquad j = 1, 2, \ldots, n \qquad (6.55)
$$

where $\delta_{rj}$ is the Kronecker delta, which is equal to zero for $r \neq s$ and equal to one for $r = s$. Moreover, use has been made of the symmetry of the mass coefficients and of the fact that $r$ and $s$ are dummy indices. By analogy, we have from Eq. (6.53)

$$
\frac{\partial L}{\partial q_j} = \frac{\partial V}{\partial q_j} = \sum_{s=1}^{n} k_{js} q_s \qquad j = 1, 2, \ldots, n \qquad (6.56)
$$

and from Eq. (6.46)

$$
\frac{\partial \mathscr{F}}{\partial \dot{q}_j} = \sum_{s=1}^{n} c_{js} \dot{q}_s \qquad j = 1, 2, \ldots, n \qquad (6.57)
$$

Introducing Eqs. (6.55), (6.56), and (6.57) into (6.48), we obtain Lagrange's equations of motion for a general linear system

$$
\sum_{s=1}^{n} [m_{js} \ddot{q}_s(t) + c_{js} \dot{q}_s(t) + k_{js} q_s(t)] = Q_j(t) \qquad j = 1, 2, \ldots, n \qquad (6.58)
$$

where, as mentioned in Sec. 6.5, the quantities $Q_j(t)$ represent externally impressed forces. Equations (6.58) constitute a set of $n$ simultaneous second-order differential equations in the generalized coordinates $q_s(t)$ ($s = 1, 2, \ldots, n$) that are completely identical to the equations of motion, Eqs. (4.11), derived in Sec. 4.3 by means of Newton's second law. As in Sec. 4.3, the equations can be written in the matrix form

$$
[m]\{\ddot{q}(t)\} + [c]\{\dot{q}(t)\} + [k]\{q(t)\} = \{Q(t)\} \qquad (6.59)
$$

where the symmetric matrices

$$
[m] = [m]^T \qquad [c] = [c]^T \qquad [k] = [k]^T \qquad (6.60)
$$

are the inertia, damping, and stiffness matrices, respectively, and $\{q(t)\}$ and $\{Q(t)\}$ are the column matrices of the generalized coordinates $q_s(t)$ and generalized impressed forces $Q_s(t)$. Note that the matrices $[m]$, $[c]$, and $[k]$ are the matrices of the coefficients of the quadratic forms $T$, $\mathscr{F}$, and $V$, respectively.

**Example 6.3** Consider the three-degree-of-freedom system of Example 4.1 and derive the system differential equations of motion by the Lagrangian approach.

As in Example 4.1, the generalized coordinates $q_1(t)$, $q_2(t)$, and $q_3(t)$ represent the horizontal translation of masses $m_1$, $m_2$, and $m_3$, respectively, and $Q_1(t)$, $Q_2(t)$, and $Q_3(t)$ are the associated generalized externally applied forces. To derive Lagrange's equations, Eqs. (6.48), it is necessary to calculate the Lagrangian and Rayleigh's dissipation function. The generalized coordinates are the displacements of the associated masses, so that the kinetic energy has the simple expression

$$T = \tfrac{1}{2}(m_1\dot{q}_1^2 + m_2\dot{q}_2^2 + m_3\dot{q}_3^2) \tag{a}$$

which is free of cross products. Because the elongations of the springs $k_1$, $k_2$, and $k_3$ are $q_1$, $q_2 - q_1$, and $q_3 - q_2$, respectively, the potential energy has the form

$$V = \tfrac{1}{2}[k_1 q_1^2 + k_2(q_2 - q_1)^2 + k_3(q_3 - q_2)^2]$$
$$= \tfrac{1}{2}[(k_1 + k_2)q_1^2 + (k_2 + k_3)q_2^2 + k_3 q_3^2 - 2k_2 q_1 q_2 - 2k_3 q_2 q_3] \tag{b}$$

By analogy, Rayleigh's dissipation function can be written directly as

$$\mathscr{F} = \tfrac{1}{2}[(c_1 + c_2)\dot{q}_1^2 + (c_2 + c_3)\dot{q}_2^2 + c_3\dot{q}_3^2 - 2c_2\dot{q}_1\dot{q}_2 - 2c_3\dot{q}_2\dot{q}_3] \tag{c}$$

To derive Lagrange's equations of motion, we recall that $L = T - V$, take the appropriate derivatives in Eqs. (a), (b), and (c) and insert the results into Eqs. (6.48). This is not necessary, however, because all these operations were already performed before writing the compact matrix form of the equations of motion, Eq. (6.59). Matrices $[m]$, $[c]$, and $[k]$ entering into Eq. (6.59) are simply the matrices of the coefficients of the quadratic forms (a), (c), and (b), respectively. Hence, the equations of motion are fully defined by the matrices

$$[m] = \begin{bmatrix} m_1 & 0 & 0 \\ 0 & m_2 & 0 \\ 0 & 0 & m_3 \end{bmatrix} \tag{d}$$

$$[c] = \begin{bmatrix} c_1 + c_2 & -c_2 & 0 \\ -c_2 & c_2 + c_3 & -c_3 \\ 0 & -c_3 & c_3 \end{bmatrix} \tag{e}$$

$$[k] = \begin{bmatrix} k_1 + k_2 & -k_2 & 0 \\ & k_2 + k_3 & -k_3 \\ 0 & -k_3 & k_3 \end{bmatrix} \tag{f}$$

provided the generalized externally applied forces $Q_j(t)$ ($j = 1, 2, 3$) are given. The results are identical to those obtained in Example 4.1 by using Newton's second law.

# PROBLEMS

**6.1** The system of Fig. 6.4 consists of a uniform rigid link of mass $m$ and two linear springs of stiffnesses $k_1$ and $k_2$, respectively. When the springs are unstretched the link is horizontal. Use the principle of virtual work and calculate the angle $\theta$ corresponding to the position of static equilibrium.

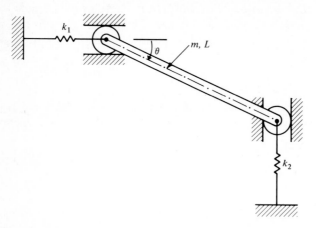

**Figure 6.4**

**6.2** Two masses $m_1 = 0.5m$ and $m_2 = m$ are suspended on a massless string, as shown in Fig. 6.5. The tension $T$ in the string is constant, and remains unchanged during the motion of the masses. Assume small displacements and use the principle of virtual work to calculate the equilibrium configuration of the system.

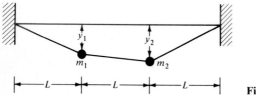

**Figure 6.5**

**6.3** Consider the system of Fig. 6.5 and use d'Alembert's principle to derive the system equations of motion.

**6.4** Derive the three Newton's equations of motion and the single Lagrange equation for the system of Problem 6.1, discuss differences, and show how Newton's equations can be reduced to Lagrange's equation.

**6.5** Derive Newton's and Lagrange's equations of motion for the system shown in Fig. 4.18, discuss differences, and show how Newton's equations can be reduced to Lagrange's equations.

**6.6** Repeat Prob. 6.5 for the triple pendulum shown in Fig. 4.19.

**6.7** The upper end of a pendulum is attached to a linear spring of stiffness $k$, where the spring is constrained so as to move in the vertical direction (Fig. 6.6). Derive the Lagrange equations of motion for the system by using $y$ and $\theta$ as generalized coordinates, where $y$ is measured from the equilibrium position.

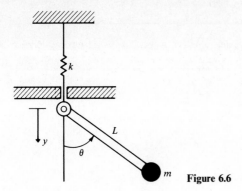

**Figure 6.6**

**6.8** Derive the Lagrange equations of motion for the system of Fig. 4.15.
**6.9** Derive the Lagrange equations of motion for the system of Fig. 6.5.
**6.10** Derive the Lagrange equations of motion for the system of Fig. 4.17.

# SEVEN

# CONTINUOUS SYSTEMS.
# APPROXIMATE METHODS

## 7.1 GENERAL CONSIDERATIONS

This chapter is of particular importance to the practicing engineer because it presents various methods of treating eigenvalue problems for which exact solutions do not exist, or are not feasible. It should be pointed out that the vast majority of continuous systems lead to eigenvalue problems that do not lend themselves to closed-form solutions, owing to nonuniform mass or stiffness distributions. Hence, quite often it is necessary to seek approximate solutions of the eigenvalue problem. In view of this, one would be tempted to conclude that the study in Chap. 5 of exact solutions for uniformly distributed systems was a wasted effort. This would be a premature conclusion, however, because such solutions can be very helpful in obtaining approximate solutions for nonuniformly distributed systems.

The approximate methods considered here consist of schemes for the discretization of continuous systems, that is to say, procedures for replacing a continuous system by an equivalent discrete one. The discretization methods can be divided into two major classes, the first representing the solution as a finite series consisting of space-dependent functions multiplied by time-dependent generalized coordinates, and the second lumping the masses at discrete points of the otherwise continuous system. In the first method the space-dependent functions do not satisfy the differential equation, but must satisfy all or some of the boundary conditions, depending on the type of formulation. It is here that the exact solutions of Chap. 5 prove helpful, as the functions in question can often be chosen as the eigenfunctions of an associated uniform system. The lumped methods are advised when the system nonuniformity is pronounced. In this chapter several approximate methods of both classes are presented.

There is another discretization method that belongs rightfully in this chapter, namely, the finite element method. However, because of the special place it occupies in mechanics and because of the wealth of information it involves, the finite element method is treated separately in the next chapter.

## 7.2 RAYLEIGH'S ENERGY METHOD

Rayleigh's method is a procedure designed to estimate the fundamental frequency of a system without solving the associated eigenvalue problem. The method is based on *Rayleigh's principle*, which can be stated in the form: *The estimated frequency of vibration of a conservative system, oscillating about the equilibrium position, has a stationary value in the neighborhood of a natural mode.* This stationary value is a minimum in the neighborhood of the fundamental mode, a fact demonstrated in Secs. 4.13 and 5.8 by means of Rayleigh's quotient for discrete and continuous systems, respectively. Hence, we should expect the Rayleigh's method to be applicable to both types of mathematical models. Clearly, Rayleigh's principle is merely an enunciation of results derived in Secs. 4.13 and 5.8 by means of Rayleigh's quotient.

To emphasize once again the analogy between discrete and continuous systems, we shall begin the discussion of Rayleigh's energy method by first considering a discrete system and then extending the results to continuous systems.

In Chaps. 4 and 5 we showed that natural modes execute harmonic motion. Hence, considering an $n$-degree-of-freedom conservative system of the type shown in Fig. 4.7 and denoting by $q_i(t)$ the generalized displacement of a typical mass $m_i$, as measured relative to an inertial space, we can write

$$q_i(t) = u_i f(t) \qquad i = 1, 2, \ldots, n \tag{7.1}$$

where $u_i$ is a constant amplitude and $f(t)$ a harmonic function of time. Equations (7.1) can be written in the compact matrix form

$$\{q(t)\} = f(t)\{u\} \tag{7.2}$$

and, because $\{u\}$ is constant, it follows that

$$\{\dot{q}(t)\} = \dot{f}(t)\{u\} \tag{7.3}$$

Introducing Eq. (7.3) into (4.42) (with $\{\dot{u}\}$ replaced by $\{\dot{q}\}$), we obtain the kinetic energy

$$T(t) = \tfrac{1}{2}\{\dot{q}(t)\}^T[m]\{\dot{q}(t)\} = \tfrac{1}{2}\dot{f}^2(t)\{u\}^T[m]\{u\} \tag{7.4}$$

Note that the vector $\{q(t)\}$ here plays the role of the time-dependent vector $\{u\}$ of Sec. 4.5, whereas the vector $\{u\}$ of the present section is constant. Similarly, inserting Eq. (7.2) into (4.38) (with $\{u\}$ replaced by $\{q\}$), we arrive at the potential energy

$$V(t) = \tfrac{1}{2}\{q(t)\}^T[k]\{q(t)\} = \tfrac{1}{2}f^2(t)\{u\}^T[k]\{u\} \tag{7.5}$$

Denoting the harmonic function of time by $f(t) = \cos(\omega t - \phi)$, it follows that $\dot{f}(t) = -\omega \sin(\omega t - \phi)$, so that Eqs. (7.4) and (7.5) become

$$T(t) = \tfrac{1}{2}\{u\}^T[m]\{u\}\omega^2 \sin^2(\omega t - \phi) \tag{7.6}$$

and

$$V(t) = \tfrac{1}{2}\{u\}^T[k]\{u\} \cos^2(\omega t - \phi) \tag{7.7}$$

respectively. From Eqs. (7.6) and (7.7) we conclude that when $\cos(\omega t - \phi) = 0$ the potential energy is equal to zero, with the implication that the system passes through the equilibrium position. At the same time $\sin(\omega t - \phi) = \pm 1$, so that when $\cos(\omega t - \phi) = 0$ the kinetic energy attains its maximum value. Similarly, when $\cos(\omega t - \phi) = \pm 1$ and $\sin(\omega t - \phi) = 0$ the potential energy attains its maximum value and the kinetic energy is zero. But for a conservative system the total energy is constant, from which it follows that

$$E = T_{\max} + 0 = 0 + V_{\max} \tag{7.8}$$

or

$$T_{\max} = V_{\max} \tag{7.9}$$

Introducing the notation

$$T^* = \tfrac{1}{2}\{u\}^T[m]\{u\} \tag{7.10}$$

where $T^*$ is known as the *reference kinetic energy*, we have

$$T_{\max} = \tfrac{1}{2}\{u\}^T[m]\{u\}\omega^2 = T^*\omega^2 \tag{7.11}$$

In addition

$$V_{\max} = \tfrac{1}{2}\{u\}^T[k]\{u\} \tag{7.12}$$

Inserting Eqs. (7.11) and (7.12) into (7.9), we arrive at

$$\omega^2 = R(\{u\}) = \frac{V_{\max}}{T^*} = \frac{\{u\}^T[k]\{u\}}{\{u\}^T[m]\{u\}} \tag{7.13}$$

where, by comparing Eq. (7.13) to (4.146), we conclude that (7.13) represents Rayleigh's quotient. Although we derived Eq. (7.13) on the basis of energy considerations, there is no difference between the Rayleigh's quotient derived here and the one derived in Sec. 4.13. Hence, the conclusion reached in Sec. 4.13, namely, that Rayleigh's quotient can be used to obtain an estimate for the lowest eigenvalue $\omega_1^2$, where $\omega_1$ is the system fundamental frequency, remains valid. Because Rayleigh's quotient has a minimum value in the neighborhood of the first mode, to obtain an estimate for $\omega_1^2$, we must insert into Eq. (7.13) a trial vector $\{u\}$ resembling as closely as possible the first eigenvector $\{u\}_1$ of the system. The closeness of the estimate to the actual value $\omega_1^2$ depends on how close the trial vector $\{u\}$ is to $\{u\}_1$, which depends in turn on the skill and experience of the analyst. The use of Rayleigh's quotient to estimate the lowest natural frequency is known as *Rayleigh's energy method*.

Actually Rayleigh's method is more useful for continuous systems than for discrete systems, because for a large number of continuous systems, such as those involving nonuniform mass or stiffness distribution, a closed-form solution of the eigenvalue problem is generally not possible. We note that when Eq. (7.13) is expressed in terms of the maximum potential energy and reference kinetic energy, the equation is equally valid for continuous systems. Of course, in contrast to discrete systems, for continuous systems it is no longer possible to write general expressions for the kinetic and potential energy (especially for the latter), but specific expressions can be written for a particular system considered. As an example, let us consider the torsional vibration of a nonuniform shaft of circular cross section. Denoting by $\theta(x, t)$ the angular displacement of the shaft, and using the analogy pointed out in Sec. 5.2 together with Eq. (5.144) it is possible to verify that the torque at point $x$ has the expression

$$M(x, t) = GJ(x)\frac{\partial\theta(x, t)}{\partial x} \tag{7.14}$$

where $GJ(x)$ is the torsional rigidity. Moreover, the angle of twist corresponding to a differential element of shaft of length $dx$ is $[\partial\theta(x, t)/\partial x]\, dx$. Considering a linear system, for which the angle of twist is proportional to the torque, the potential energy for a shaft of length $L$ (whose ends are not supported by torsional springs capable of storing potential energy) can be written as

$$V(t) = \frac{1}{2}\int_0^L GJ(x)\left[\frac{\partial\theta(x, t)}{\partial x}\right]^2 dx \tag{7.15}$$

which has the same structure as the potential energy of a rod in longitudinal vibration, Eq. (5.142). On the other hand, if $I(x)$ is the mass polar moment of inertia per unit length of shaft, then the kinetic energy is simply

$$T(t) = \frac{1}{2}\int_0^L I(x)\left[\frac{\partial\theta(x, t)}{\partial t}\right]^2 dx \tag{7.16}$$

Note the analogy with Eq. (5.138). Considering again a conservative system, and assuming that the angular displacement $\theta(x, t)$ is separable in space and time

$$\theta(x, t) = \Theta(x)F(t) \tag{7.17}$$

where $F(t)$ is harmonic, $F(t) = \cos(\omega t - \phi)$, we are led to Rayleigh's quotient

$$\omega^2 = R(\Theta) = \frac{V_{max}}{T^*} = \frac{\displaystyle\int_0^L GJ(x)[d\Theta(x)/dx]^2\, dx}{\displaystyle\int_0^L I(x)\Theta^2(x)\, dx} \tag{7.18}$$

Hence, for any trial function $\Theta(x)$ resembling the fundamental mode $\Theta_1(x)$ to a reasonable degree, Eq. (7.18) yields an estimate for the first eigenvalue $\omega_1^2$.

Considering the tapered shaft fixed at $x = 0$ and free at $x = L$ investigated in Sec. 5.8, where

$$I(x) = \frac{6}{5} I \left[ 1 - \frac{1}{2} \left( \frac{x}{L} \right)^2 \right] \qquad GJ(x) = \frac{6}{5} GJ \left[ 1 - \frac{1}{2} \left( \frac{x}{L} \right)^2 \right] \qquad (7.19)$$

and assuming the trial function $\Theta(x) = \sin \pi x/2L$, we obtain

$$\omega^2 = \frac{\pi^2 GJ \displaystyle\int_0^L [1 - \frac{1}{2}(x/L)^2] \cos^2 (\pi x/2L) \, dx}{4IL^2 \displaystyle\int_0^L [1 - \frac{1}{2}(x/L)^2] \sin^2 (\pi x/2L) \, dx}$$

$$= \frac{\pi^2 GJ(L/12\pi^2)(5\pi^2 + 6)}{4IL^2(L/12\pi^2)(5\pi^2 - 6)} = 3.1504 \frac{GJ}{IL^2} \qquad (7.20$$

As is to be expected, the result is exactly the same as that obtained in Sec. 5.8, Eq. (5.103).

Rayleigh's method is concerned only with a crude approximation for the system fundamental frequency. It should be noted that the estimates obtained by Rayleigh's method are at least as high as the actual fundamental frequency. To obtain more accurate estimates it is advisable to use more refined methods, such as the Rayleigh-Ritz method described in Sec. 7.3. The idea behind the Rayleigh-Ritz method is to lower the estimate, thus approaching the true natural frequencies from above.

## 7.3 THE RAYLEIGH-RITZ METHOD.
## THE INCLUSION PRINCIPLE

Rayleigh's energy method is generally used when one is interested in a quick (but not particularly accurate) estimate of the fundamental frequency of a continuous system for which a solution of the eigenvalue problem cannot be readily obtained. It is based on the fact that Rayleigh's quotient has a minimum in the neighborhood of the lowest natural modes of vibration. Of particular interest here is the fact that Rayleigh's quotient provides an upper bound for the first eigenvalue $\lambda_1$ (see Sec. 5.8).

$$R(u) \geqslant \lambda_1 \qquad (7.21)$$

where $\lambda_1$ is related to the fundamental frequency $\omega_1$, and $u$ is a trial function satisfying all the boundary conditions of the problem but not the differential equation (otherwise $u$ would be an eigenfunction). It is desirable that the function $u$ resemble as closely as possible the first natural mode. In fact, the closer the function $u$ resembles the first mode, the closer the estimate is to the first eigenvalue. *The Rayleigh-Ritz method is simply a procedure for lowering the estimate of $\lambda_1$, by producing a trial function $u$ reasonably close to the first natural mode.* However, the method is not concerned with the first eigenvalue alone, as it furnishes estimates for a finite number of higher eigenvalues.

We recall from Sec. 5.8 that the expression for Rayleigh's quotient depends on the system considered. Moreover, the trial functions used in the quotient must satisfy all the boundary conditions of the problem. Before proceeding with the Rayleigh-Ritz method, however, it will prove beneficial to derive an expression for Rayleigh's quotient that is valid for a large class of continuous systems, a class which includes all the systems discussed in this text. In addition, in using this general expression, some relaxation of the number of boundary conditions to be satisfied by the trial functions is achieved. Using the analogy with discrete systems, in Sec. 7.2 we wrote a general expression for Rayleigh's quotient in terms of the maximum potential energy and the reference kinetic energy. In this section we arrive at the same general expression by beginning with an arbitrary continuous system.

Let us consider the longitudinal vibration of a thin rod having the end $x = 0$ fixed and the end $x = L$ attached to a spring of stiffness $k$ (see Fig. 7.1). The eigenvalue problem is defined by the differential equation

$$-\frac{d}{dx}\left[EA(x)\frac{du(x)}{dx}\right] = \omega^2 m(x)u(x) \qquad 0 < c < L \qquad (7.22)$$

and the boundary conditions

$$u(0) = 0 \qquad EA(x)\frac{du(x)}{dx}\bigg|_{x=L} = -ku(L) \qquad (7.23)$$

Using the analogy with the shaft in torsion, we conclude from Eq. (5.92) that one of the forms of Rayleigh's quotient is

$$\lambda = \omega^2 = R(u) = \frac{-\displaystyle\int_0^L u(x)(d/dx)\{EA(x)[du(x)/dx]\}\,dx}{\displaystyle\int_0^L m(x)u^2(x)\,dx} \qquad (7.24)$$

The value of Rayleigh's quotient depends on the trial function $u(x)$, so that a

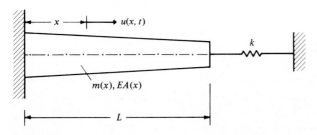

**Figure 7.1**

closer examination of the nature of the trial functions is warranted. In the first place, we observe from Eq. (7.24) that the numerator involves the stiffness term on the left side of Eq. (7.22). Hence, the trial functions must be differentiable as many times as the order of the system. We also observe that the boundary conditions, Eqs. (7.23), do not appear explicitly in Eq. (7.24). As a result, to ensure that the characteristics of the system are taken into consideration, the trial functions must satisfy the boundary conditions of the problem. *Functions that are differentiable as many times as the order of the system and satisfy all the boundary conditions* of the problem are referred to as *comparison functions*. Hence, *the trial functions entering into the Rayleigh's quotient expression given by Eq. (7.24) must belong to the class of comparison functions*. Clearly, the class of comparison functions is appreciably larger than the class of eigenfunctions, as the class of eigenfunctions represents only a small subset of the class of comparison functions.

The first of boundary conditions (7.23) is geometric and its physical significance is obvious. On the other hand, the second of boundary conditions (7.23) is natural and it expresses the force balance at the end $x = L$. Satisfaction of natural boundary conditions can be troublesome at times, so that a way of circumventing this requirement is desirable. This involves using a form for Rayleigh's quotient different from that given by Eq. (7.24), as shown in the sequel.

Integrating the numerator of Eq. (7.24) by parts, and considering boundary conditions (7.23), we obtain

$$-\int_0^L u(x) \frac{d}{dx}\left[ EA(x) \frac{du(x)}{dx} \right] dx$$

$$= -u(x)EA(x) \frac{du(x)}{dx}\bigg|_0^L + \int_0^L EA(x) \left[ \frac{du(x)}{dx} \right]^2 dx$$

$$= ku^2(L) + \int_0^L EA(x) \left[ \frac{du(x)}{dx} \right]^2 dx \qquad (7.25)$$

In view of the discussion of Sec. 7.2, the right side of Eq. (7.25) can be identified as twice the maximum potential energy:

$$-\int_0^L u(x) \frac{d}{dx}\left[ EA(x) \frac{du(x)}{dx} \right] dx = ku^2(L) + \int_0^L EA(x) \left[ \frac{du(x)}{dx} \right]^2 dx = 2V_{max}$$

$$(7.26)$$

On the other hand, the denominator in Eq. (7.24) is recognized as twice the system reference kinetic energy:

$$\int_0^L m(x)u^2(x)\, dx = 2T^* \qquad (7.27)$$

Hence, inserting Eqs. (7.26) and (7.27) into Eq. (7.24), we obtain the general expression for Rayleigh's quotient

$$\lambda = \omega^2 = R(u) = \frac{V_{\text{max}}}{T^*} \tag{7.28}$$

which is entirely analogous to Eq. (7.13) for discrete systems.

Equation (7.28) is valid for any continuous system and for any type of boundary conditions, provided they can be accounted for in $V_{\text{max}}$ and $T^*$. As another illustration of the way in which boundary conditions are accounted for in Eq. (7.28), let us consider the case in which the spring $k$ at the end $x = L$ is replaced by the rigid mass $M$ (see Fig. 7.2). Whereas the differential equation remains in the form (7.22), the boundary conditions in this case can be shown to be

$$u(0) = 0 \qquad EA(x) \frac{du(x)}{dx}\bigg|_{x=L} = \omega^2 M u(L) \tag{7.29}$$

Once again integrating the numerator in Eq. (7.24) by parts and considering boundary conditions (7.29), we obtain

$$-\int_0^L u(x) \frac{d}{dx}\left[ EA(x) \frac{du(x)}{dx} \right] dx$$

$$= -u(x) EA(x) \frac{du(x)}{dx}\bigg|_0^L + \int_0^L EA(x) \left[\frac{du(x)}{dx}\right]^2 dx$$

$$= -\omega^2 M u^2(L) + \int_0^L EA(x) \left[\frac{du(x)}{dx}\right]^2 dx \tag{7.30}$$

Inserting Eq. (7.30) into (7.24), we can write

$$\omega^2 = \frac{-\displaystyle\int_0^L u(x)(d/dx)\{EA(x)[du(x)/dx]\}\, dx}{\displaystyle\int_0^L m(x)u^2(x)\, dx}$$

$$= \frac{-\omega^2 M u^2(L) + \displaystyle\int_0^L EA(x)[du(x)/dx]^2\, dx}{\displaystyle\int_0^L m(x)u^2(x)\, dx} \tag{7.31}$$

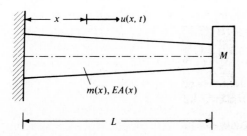

**Figure 7.2**

where we omitted $R(u)$ from Eq. (7.24) on purpose. Equation (7.31) can be solved for $\omega^2$, with the result

$$\omega^2 = \frac{\displaystyle\int_0^L EA(x)[du(x)/dx]^2 \, dx}{\displaystyle\int_0^L m(x)u^2(x) \, dx + Mu^2(L)} = \frac{V_{max}}{T^*} \tag{7.32}$$

where the numerator is once again recognized as $2V_{max}$ and the denominator as $2T^*$. Hence, if we redefine $R(u)$ to be equal to $\omega^2$ in Eq. (7.32), we obtain once again the general expression (7.28). Note that Eq. (7.32) can be obtained, perhaps in a more direct way, by regarding the system of Fig. 7.2 as being free at the end $x = L+$, where $L+$ is the point immediately to the right of $M$, and having the mass distribution

$$m_1(x) = m(x) + M\delta(x - L) \tag{7.33}$$

where $\delta(x - L)$ is a spatial Dirac delta function, defined as

$$\delta(x - L) = 0 \qquad x \neq L$$

$$\int_0^{L+} \delta(x - L) \, dx = 1 \tag{7.34}$$

Note that $M\delta(x - L)$ has units of mass/length.

At this point, let us return to the examination of the nature of the trial functions, particularly as it concerns the use of the form (7.28) of Rayleigh's quotient. Equation (7.28) was obtained from Eq. (7.24) through an integration by parts. As a result, the numerator involves derivatives of the trial functions of one-half the order of the system, as can be concluded from Eq. (7.26). Moreover, the effect of any elastic supports, such as the spring at the end $x = L$ in Fig. 7.1, is taken into account automatically in $V_{max}$. Similarly, the effect of any lumped masses at boundaries, such as the lumped mass at the end $x = L$ in Fig. 7.2, is taken into account automatically in $T^*$. The only characteristic not reflected in the form (7.28) of Rayleigh's quotient is the satisfaction of geometric boundary conditions. It follows that, in using the Rayleigh's quotient in the form (7.28), the system characteristics are taken into account by trial *functions that are differentiable half as many times as the order of the system and satisfy only the geometric boundary conditions* of the problem. We refer to such functions as *admissible functions*. Hence, *the trial functions entering into Rayleigh's quotient expression given by Eq.* (7.28) *must belong to the class of admissible functions.* This is very significant, as the class of admissible functions is much more abundant than the class of comparison functions. Future uses of Rayleigh's quotient will be confined to the form (7.28) in conjunction with admissible functions. Of course, comparison functions can always be used, as they are by definition admissible, but this is not necessary.

According to the Rayleigh-Ritz method, an approximate solution of the eigenvalue problem associated with an arbitrary continuous system can be constructed in the form of the linear combination

$$u(x) = \sum_{i=1}^{n} a_i \phi_i(x) \tag{7.35}$$

where $a_i$ are coefficients to be determined and $\phi_i(x)$ are trial functions, which are known functions of the spatial coordinate $x$ prescribed by the analyst. In seeking an approximate solution of the eigenvalue problem, we have the choice between attempting to satisfy the differential equation (7.22), in conjunction with the use of trial functions $\phi_i(x)$ in the form of comparison functions, and solving the mathematically equivalent problem of rendering the value of Rayleigh's quotient stationary, where Rayleigh's quotient is in the form (7.24). Rendering Rayleigh's quotient stationary is our preferred choice. However, instead of working with the form (7.24) of Rayleigh's quotient in conjunction with comparison functions, we shall use the form (7.28) in conjunction with admissible functions. The set of admissible functions $\phi_i(x)$ is referred to as a *generating set*. The coefficients $a_i$ are determined so as to bring about a close resemblance between the function $u(x)$ and the natural modes. Mathematically this is equivalent to seeking those values of $a_i$ for which Rayleigh's quotient is rendered stationary.

For the sake of this development, let us express Rayleigh's quotient in the form

$$\lambda = \omega^2 = R(u) = \frac{V_{\max}}{T^*} = \frac{N(u)}{D(u)} = \frac{N(a_1, a_2, \ldots, a_n)}{D(a_1, a_2, \ldots, a_n)} \tag{7.36}$$

where $N$ and $D$ denote the numerator and denominator of the quotient, respectively. We note that, by virtue of the fact that the admissible functions $\phi_i(x)$ are given, the integrations over the spatial domain involved in $V_{\max}$ and $T^*$ can actually be carried out, thus eliminating the dependence of the quotient on the spatial variable $x$ and leaving $N$ and $D$ as mere quadratic forms in the undetermined coefficients $a_i$ $(i = 1, 2, \ldots, n)$. Then, the values of the coefficients are determined so as to render Rayleigh's quotient stationary. The quotient has a stationary value if its variation vanishes, or

$$\delta R = \frac{\partial R}{\partial a_1} \delta a_1 + \frac{\partial R}{\partial a_2} \delta a_2 \cdots + \frac{\partial R}{\partial a_n} \delta a_n = 0 \tag{7.37}$$

But, because the coefficients $a_i$ $(i = 1, 2, \ldots, n)$ are independent, Eq. (7.37) can be satisfied only if the quantity multiplying every $\delta a_r$ $(r = 1, 2, \ldots, n)$ is equal to zero independently. Hence, the necessary conditions for the stationarity of the quotient are

$$\frac{\partial R}{\partial a_r} = \frac{D(\partial N/\partial a_r) - N(\partial D/\partial a_r)}{D^2} = 0 \qquad r = 1, 2, \ldots, n \tag{7.38}$$

Denoting the value of $\lambda$ associated with the stationary value of Rayleigh's quotient by $\Lambda$, and considering Eq. (7.36), Eqs. (7.38) become

$$\frac{\partial N}{\partial a_r} - \Lambda \frac{\partial D}{\partial a_r} = 0 \qquad r = 1, 2, \ldots, n \tag{7.39}$$

Moreover, introducing the notation

$$N = \sum_{i=1}^{n} \sum_{j=1}^{n} k_{ij} a_i a_j \qquad D = \sum_{i=1}^{n} \sum_{j=1}^{n} m_{ij} a_i a_j \qquad (7.40)$$

where the constant coefficients $k_{ij}$ and $m_{ij}$ are symmetric, $k_{ij} = k_{ji}$, $m_{ij} = m_{ji}$ ($i, j = 1, 2, \ldots, n$), we can write

$$\frac{\partial N}{\partial a_r} = \sum_{i=1}^{n} \sum_{j=1}^{n} k_{ij} \left( \frac{\partial a_i}{\partial a_r} a_j + a_i \frac{\partial a_j}{\partial a_r} \right)$$

$$= \sum_{i=1}^{n} \sum_{j=1}^{n} k_{ij} (\delta_{ir} a_j + \delta_{jr} a_i)$$

$$= \sum_{j=1}^{n} k_{rj} a_j + \sum_{i=1}^{n} k_{ir} a_i = 2 \sum_{j=1}^{n} k_{rj} a_j \qquad r = 1, 2, \ldots, n) \qquad (7.41)$$

where $\delta_{ir}$ and $\delta_{jr}$ are Kronecker deltas. Note that in the second sum we took into consideration that $k_{ir} = k_{ri}$ and, moreover, we replaced the dummy index $i$ by $j$. In a similar fashion, we obtain

$$\frac{\partial D}{\partial a_r} = 2 \sum_{k=1}^{n} m_{rj} a_j \qquad r = 1, 2, \ldots, n \qquad (7.42)$$

Inserting Eqs. (7.41) and (7.42) into (7.39), we arrive at the homogeneous set of algebraic equations

$$\sum_{j=1}^{n} (k_{rj} - \Lambda m_{rj}) a_j = 0 \qquad r = 1, 2, \ldots, n \qquad (7.43)$$

where $a_j$ are the unknowns and $\Lambda$ is a parameter. Equations (7.43), known as *Galerkin's equations*, are recognized as representing the eigenvalue problem associated with an $n$-degree-of-freedom discrete system. They can be written in the matrix form

$$[k]\{a\} = \Lambda[m]\{a\} \qquad (7.44)$$

where $[k]$ and $[m]$ are $n \times n$ constant symmetric matrices, referred to as the stiffness and mass matrix, respectively.

Before we delve into the meaning of the solution of the eigenvalue problem (7.44), let us calculate the coefficients $k_{ij}$ and $m_{ij}$ for the system of Fig. 7.1 as an illustration. Considering solution (7.35), and recalling Eq. (7.26), the numerator of Rayleigh's quotient becomes

$$N = 2V_{\max} = ku^2(L) + \int_0^L EA(x) \left[ \frac{du(x)}{dx} \right]^2 dx$$

$$= k \left[ \sum_{i=1}^{n} a_i \phi_i(L) \sum_{j=1}^{n} a_j \phi_j(L) \right]$$

$$+ \int_0^L EA(x) \left[ \sum_{i=1}^{n} a_i \frac{d\phi_i(x)}{dx} \right] \left[ \sum_{j=1}^{n} a_j \frac{d\phi_j(x)}{dx} \right] dx$$

$$= \sum_{i=1}^{n} \sum_{j=1}^{n} a_i a_j \left[ k\phi_i(L)\phi_j(L) + \int_0^L EA(x) \frac{d\phi_i(x)}{dx} \frac{d\phi_j(x)}{dx} dx \right] \quad (7.45)$$

with the obvious conclusion that the coefficients $k_{ij}$ have the form

$$k_{ij} = k\phi_i(L)\phi_j(L) + \int_0^L EA(x) \frac{d\phi_i(x)}{dx} \frac{d\phi_j(x)}{dx} dx \quad i,j = 1, 2, \ldots, n \quad (7.46)$$

and it is clear that the coefficients $k_{ij}$ are symmetric. Moreover, the denominator of Rayleigh's quotient is simply

$$D = 2T^* = \int_0^L m(x) \left[ \sum_{i=1}^{n} a_i \phi_i(x) \right] \left[ \sum_{j=1}^{n} a_j \phi_j(x) \right] dx$$

$$= \sum_{i=1}^{n} \sum_{j=1}^{n} a_i a_j \int_0^L m(x)\phi_i(x)\phi_j(x)\, dx \quad (7.47)$$

so that

$$m_{ij} = \int_0^L m(x)\phi_i(x)\phi_j(x)\, dx \quad i,j = a, 2, \ldots, n \quad (7.48)$$

As expected, the coefficients $m_{ij}$ are also symmetric.

Next, let us examine how the solution of the eigenvalue problem (7.44) is related to the approximate solution generated by the Rayleigh-Ritz method. The solution of Eq. (7.44) yields $n$ eigenvalues $\Lambda_r$ and associated eigenvectors $\{a\}_r$ ($r = 1, 2, \ldots, n$) (see Sec. 4.7). The computed eigenvalues $\Lambda_r$ ($r = 1, 2, \ldots, n$) represent estimates of the first $n$ actual eigenvalues $\lambda_r$ of the continuous system. Moreover, inserting the eigenvectors $\{a\}_r$ into Eq. (7.35), we obtain the estimated eigenfunctions

$$u_r(x) = \sum_{i=1}^{n} a_{ir}\phi_i(x) \quad r = 1, 2, \ldots, n \quad (7.49)$$

where $a_{ir}$ is the $i$th component of the vector $\{a\}_r$. It is not difficult to show that the eigenfunctions $u_r(x)$ are orthogonal with respect to the distributed mass $m(x)$ of the continuous system because the eigenvectors $\{a\}_r$ are orthogonal with respect to the mass matrix $[m]$ (see Prob. 7.4).

The Rayleigh-Ritz method calls for the use of a sequence of approximations obtained by letting $r = 1, 2, 3, \ldots$ in the series given by Eq. (7.35), solving the eigenvalue problem (7.44) and observing the improvement in the computed eigenvalues. The process is stopped when a desired number of eigenvalues reach sufficient accuracy, i.e., when the addition of terms to the series does not produce meaningful improvement in these eigenvalues. Note that in general the number of terms in the series must be significantly larger, perhaps by a factor of 2, than the number of accurate eigenvalues desired.

The question remains as to how the computed eigenvalues and eigenvectors relate to the actual ones. For convenience, we let the actual and computed eigenvalues be ordered so as to satisfy $\lambda_1 \leqslant \lambda_2 \leqslant \cdots$ and $\Lambda_1 \leqslant \Lambda_2 \leqslant \cdots \leqslant \Lambda_n$, respectively. Then, assuming that the admissible functions $\phi_1(x), \phi_2(x), \ldots, \phi_n(x)$

are from a complete set, which implies that the difference between the approximate solution and the actual solution can be made as small as desired by simply increasing $n$ in Eq. (7.35), we conclude that the computed solution of the eigenvalue problem must approach the actual solution as $n \to \infty$. In using only $n$ functions $\phi_i(x)$ in series (7.35), instead of an infinite number of functions, we essentially reduce a continuous system with an infinite number of degrees of freedom to a discrete one with $n$ degrees of freedom. This discretization and truncation is tantamount to the statement that the higher-order terms in the generating set are ignored, so that the constraints

$$a_{n+1} = a_{n+2} = \cdots = 0 \tag{7.50}$$

are imposed on the system. Because constraints tend to increase the system stiffness, the computed eigenvalues tend to be higher than the actual eigenvalues, or

$$\Lambda_r \geqslant \lambda_r \qquad r = 1, 2, \ldots, n \tag{7.51}$$

Inequalities (7.51) can be demonstrated in a more rigorous manner.† How well the computed eigenvalues approximate the actual eigenvalues depends on the choice of admissible functions and their number, but the lower eigenvalues tend to be better approximations than the higher eigenvalues. As the number of terms in series (7.35) increases the errors between the computed and actual eigenvalues tend to decrease, or at least not to increase, with the most significant improvement occurring in the higher eigenvalues. This is true because there is less room for improvement in the lower eigenvalues. Because the admissible functions $\phi_i(x)$ belong to a complete set, the errors should vanish as $n \to \infty$. To corroborate this statement, let us denote the mass and stiffness matrices corresponding to $n$ terms in series (7.35) by $[m]^{(n)}$ and $[k]^{(n)}$, respectively. Similarly, we denote the corresponding computed eigenvalues by $\Lambda_r^{(n)}$ ($r = 1, 2, \ldots, n$). Then, if we add one term to series (7.35), for a total number of $n + 1$ terms, we obtain the mass and stiffness matrices $[m]^{(n+1)}$ and $[k]^{(n+1)}$, respectively, yielding the computed eigenvalues $\Lambda_r^{(n+1)}$ ($r = 1, 2, \ldots, n + 1$). The eigenvalues can be arranged so as to satisfy $\Lambda_1^{(n+1)} \leqslant \Lambda_2^{(n+1)} \leqslant \cdots \leqslant \Lambda_{n+1}^{(n+1)}$. The mass and stiffness matrices possess the embedding property defined as

$$[m]^{(n+1)} = \begin{bmatrix} [m]^{(n)} & \times \\ & \times \\ \times & \times & \times \end{bmatrix} \qquad [k]^{(n+1)} = \begin{bmatrix} [k]^{(n)} & \times \\ & \times \\ \times & \times & \times \end{bmatrix} \tag{7.52}$$

which means that the matrices $[m]^{(n+1)}$ and $[k]^{(n+1)}$ are obtained by adding one row and one column to matrices $[m]^{(n)}$ and $[k]^{(n)}$, respectively. The embedding property displayed by Eqs. (7.52) can be used to prove that eigenvalues computed by means of the Rayleigh-Ritz method satisfy the inequalities

$$\Lambda_1^{(n+1)} \leqslant \Lambda_1^{(n)} \leqslant \Lambda_2^{(n+1)} \leqslant \Lambda_2^{(n)} \leqslant \cdots \leqslant \Lambda_n^{(n)} \leqslant \Lambda_{n+1}^{(n+1)} \tag{7.53}$$

which are known as the *inclusion principle*.‡ If in addition we consider inequalities

† See L. Meirovitch, *Computational Methods in Structural Dynamics*, sec. 8.2, Sijthoff & Noordhoff, The Netherlands, 1980.

‡ See L. Meirovitch and H. Baruh, "On the Inclusion Principle for the Hierarchical Finite Element Method," *International Journal for Numerical Methods in Engineering*, vol. 19, pp. 281–291, 1983.

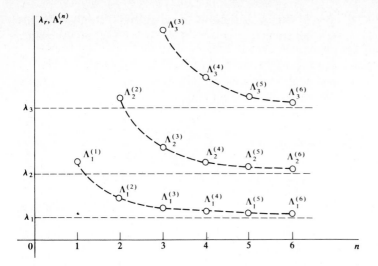

**Figure 7.3**

(7.51), then we conclude that *as n increases, the computed eigenvalues approach the actual eigenvalues asymptotically and from above.* Hence, we can write

$$\lim_{n \to \infty} \Lambda_r^{(n)} = \lambda_r \qquad r = 1, 2, \ldots, n \tag{7.54}$$

The above results are illustrated in Fig. 7.3. Unfortunately, there is no parallel analysis for the computed eigenfunctions.

One question asked frequently is what constitutes a good set of admissible functions. Clearly, the functions should be linearly independent and should form a complete set. In this regard, we mention power series, trigonometric functions, Bessel functions, etc. Quite often the eigenfunctions of a simpler but related system can serve as a good set of admissible functions, as demonstrated in Example 7.1.

**Example 7.1** Consider the longitudinal vibration of a nonuniform thin rod fixed at $x = 0$ and free at $x = L$, and obtain estimates of the lowest eigenvalues by the Rayleigh-Ritz method. The stiffness and mass distributions are

$$EA(x) = \frac{6}{5} EA \left[ 1 - \frac{1}{2} \left( \frac{x}{L} \right)^2 \right] \qquad m(x) = \frac{6}{5} m \left[ 1 - \frac{1}{2} \left( \frac{x}{L} \right)^2 \right] \tag{a}$$

The coefficients $k_{ij}$ and $m_{ij}$ are given by Eqs. (7.46) and (7.48). except that $k$ must be set equal to zero in Eqs. (7.46). As a generating set we can use the eigenfunctions corresponding to a uniform rod clamped at $x = 0$ and free at $x = L$, namely,

$$\phi_i(x) = \sin (2i - 1) \frac{\pi x}{2L} \qquad i = 1, 2, \ldots, n \tag{b}$$

and note that the trial functions $\phi_i(x)$ are actually comparison functions for the system at hand. This is perfectly all right as comparison functions belong to the class of admissible functions. Letting $k = 0$ in Eqs. (7.46), we can write

$$k_{ij} = \frac{6}{5} EA \frac{(2i-1)\pi}{2L} \frac{(2j-1)\pi}{2L} \int_0^L \left[ 1 - \frac{1}{2}\left(\frac{x}{L}\right)^2 \right]$$

$$\times \cos \frac{(2i-1)\pi x}{2L} \cos \frac{(2j-1)\pi x}{2L} dx \qquad i, j = 1, 2, \ldots, n \quad (c)$$

Similarly, Eqs. (7.48) yield simply

$$m_{ij} = \frac{6}{5} m \int_0^L \left[ 1 - \frac{1}{2}\left(\frac{x}{L}\right)^2 \right] \sin (2i-1) \frac{\pi x}{2L} \sin (2j-1) \frac{\pi x}{2L} dx$$

$$i, j = 1, 2, \ldots, n \quad (d)$$

As a very crude approximation, we let $n = 1$ in series (7.35), as well as in Eqs. (c) and (d), leading to the coefficients

$$k_{11} = \frac{EA}{40L} (5\pi^2 + 6) \qquad m_{11} = \frac{1}{10\pi^2} mL(5\pi^2 - 6) \qquad (e)$$

The eigenvalue problem (7.44) reduces to the single equation $k_{11}a_1 = \Lambda_1 m_{11}a_1$, yielding the eigenvalue

$$\Lambda_1 = \frac{k_{11}}{m_{11}} = \frac{(EA/40L)(5\pi^2 + 6)}{(1/10\pi^2)mL(5\pi^2 - 6)} = 3.1504 \frac{EA}{mL^2} \qquad (f)$$

which is a first approximation for $\omega_1^2$. This is precisely the value given by Eq. (7.20), obtained by Rayleigh's energy method. This should surprise no one, because by using only one term in series (7.35) the Rayleigh-Ritz method reduces essentially to Rayleigh's energy method.

A better approximation for $\omega_1^2$ and a first approximation for $\omega_2^2$ can be obtained by letting $n = 2$ in series (7.35), in which case Eqs. (c) and (d) yield the matrices

$$[k] = \frac{EA}{40L} \begin{bmatrix} 5\pi^2 + 6 & \frac{27}{2} \\ \frac{27}{2} & 45\pi^2 + 6 \end{bmatrix} \qquad [m] = \frac{mL}{10\pi^2} \begin{bmatrix} 5\pi^2 - 6 & \frac{15}{2} \\ \frac{15}{2} & 5\pi^2 - \frac{2}{3} \end{bmatrix} \qquad (g)$$

Inserting matrices (g) into Eq. (7.44), and solving the eigenvalue problem, we obtain the result

$$\Lambda_1 = 3.1482 \frac{EA}{mL^2} \qquad \{a\}_1 = \begin{Bmatrix} 0.9999 \\ -0.0101 \end{Bmatrix}$$

$$\Lambda_2 = 23.2840 \frac{EA}{mL^2} \qquad \{a\}_2 = \begin{Bmatrix} -0.1598 \\ 0.9871 \end{Bmatrix}$$

$$(h)$$

Comparing $\Lambda_1$ from Eqs. (f) and (h), it is clear that the latter provides a better estimate for $\omega_1^2$ while $\Lambda_2$ provides a first estimate for $\omega_2^2$. Moreover, introducing $\{a\}_1$ and $\{a\}_2$ into Eqs. (7.49), we obtain the first two estimated eigenfunctions

$$u_1(x) = 0.9999 \sin \frac{\pi x}{2L} - 0.0101 \sin \frac{3\pi x}{2L}$$

$$u_2(x) = -0.1598 \sin \frac{\pi x}{2L} + 0.9871 \sin \frac{3\pi x}{2L}$$

(i)

The eigenfunctions are plotted in Fig. 7.4a.

To develop a better appreciation for the effect of the number of terms in series (7.35) on the results, let us consider the case in which $n = 3$. From Eqs. (c) and (d), we obtain the matrices

$$[k] = \frac{EA}{40L} \begin{bmatrix} 5\pi^2 + 6 & \frac{27}{2} & -\frac{25}{6} \\ \frac{27}{2} & 45\pi^2 + 6 & \frac{675}{8} \\ -\frac{25}{6} & \frac{675}{8} & 125\pi^2 + 6 \end{bmatrix}$$

$$[m] = \frac{mL}{10\pi^2} \begin{bmatrix} 5\pi^2 - 6 & \frac{15}{2} & -\frac{13}{6} \\ \frac{15}{2} & 5\pi^2 - \frac{2}{3} & \frac{51}{8} \\ -\frac{13}{6} & \frac{51}{8} & 5\pi^2 - \frac{6}{25} \end{bmatrix}$$

(j)

so that, solving the eigenvalue problem (7.35), we arrive at

$$\Lambda_1 = 3.1480 \frac{EA}{mL^2} \qquad \{a\}_1 = \begin{Bmatrix} 0.9999 \\ -0.0105 \\ 0.0019 \end{Bmatrix}$$

$$\Lambda_2 = 23.2532 \frac{EA}{mL^2} \qquad \{a\}_2 = \begin{Bmatrix} -0.1610 \\ 0.9866 \\ -0.0275 \end{Bmatrix}$$

(k)

$$\Lambda_3 = 62.9118 \frac{EA}{mL^2} \qquad \{a\}_3 = \begin{Bmatrix} 0.0674 \\ -0.1131 \\ 0.9913 \end{Bmatrix}$$

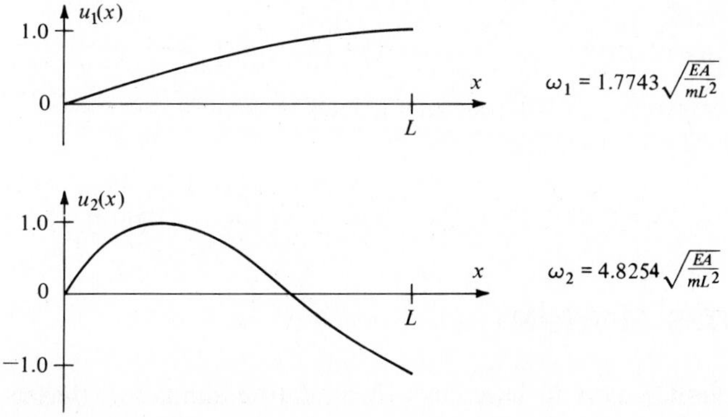

$$\omega_1 = 1.7743 \sqrt{\frac{EA}{mL^2}}$$

$$\omega_2 = 4.8254 \sqrt{\frac{EA}{mL^2}}$$

Figure 7.4a

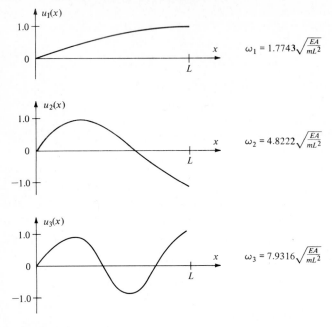

**Figure 7.4b**

It is clear from the first of Eqs. (k) that using $n = 3$ in series (7.35) leads to a better estimate for $\omega_1^2$ and $\omega_2^2$ while providing a first estimate for $\omega_3^2$. In addition, inserting the eigenvectors $\{a\}_r$ ($r = 1, 2, 3$) into (7.49), we obtain the estimated eigenfunctions

$$u_1(x) = 0.9999 \sin \frac{\pi x}{2L} - 0.0105 \sin \frac{3\pi x}{2L} + 0.0019 \sin \frac{5\pi x}{2L}$$

$$u_2(x) = -0.1610 \sin \frac{\pi x}{2L} + 0.9866 \sin \frac{3\pi x}{2L} - 0.0275 \sin \frac{5\pi x}{2L} \qquad (l)$$

$$u_3(x) = 0.0674 \sin \frac{\pi x}{2L} - 0.1131 \sin \frac{3\pi x}{2L} + 0.9913 \sin \frac{5\pi x}{2L}$$

The eigenfunctions are plotted in Fig. 7.4b.

Clearly, the computed eigenvalues satisfy the inclusion principle. Note that the superscripts were omitted from the notation of the eigenvalues.

## 7.4 ASSUMED-MODES METHOD

Although the assumed-modes method leads to a formulation similar to that of the Rayleigh-Ritz method, its discussion may prove rewarding because it will most likely improve the understanding of discretization by means of a series solution.

The method assumes a solution of the boundary-value problem associated with a conservative continuous system in the form

$$y(x, t) = \sum_{i=1}^{n} \phi_i(x) q_i(t) \tag{7.55}$$

where $\phi_i(x)$ are trial functions and $q_i(t)$ generalized coordinates, and uses this solution in conjunction with Lagrange's equations to obtain an approximate formulation of the equations of motion. It essentially regards a continuous system as an $n$-degree-of-freedom system in a manner similar to the Rayleigh-Ritz method.

The kinetic and potential energy of a continuous system have integral expressions depending on partial derivatives of $y(x, t)$ with respect to $t$ and $x$, respectively. Using the series solution (7.55), and performing the corresponding integration with respect to $x$, the kinetic energy can be written in the familiar form

$$T(t) = \frac{1}{2} \sum_{i=1}^{n} \sum_{j=1}^{n} m_{ij} \dot{q}_i(t) \dot{q}_j(t) \tag{7.56}$$

where $m_{ij}$ are constant symmetric mass coefficients depending on the mass distribution of the system and the trial functions $\phi_i(x)$ chosen. In a similar fashion, the potential energy can be written as

$$V(t) = \frac{1}{2} \sum_{i=1}^{n} \sum_{j=1}^{n} k_{ij} q_i(t) q_j(t) \tag{7.57}$$

where $k_{ij}$ are constant symmetric stiffness coefficients depending on the stiffness distribution and the functions $\phi_i(x)$. The coefficients $k_{ij}$ contain derivatives of $\phi_i(x)$ of orders half as large as the order of the differential equation of the continuous system under consideration. The natural boundary conditions are of no particular concern here because they are automatically accounted for in the kinetic and potential energy. Hence $\phi_i(x)$ need be admissible functions only. Note that comparison functions can always be used, as they are a subset of the set of admissible functions.

Because normal-mode vibration is by definition associated with conservative systems, we consider Lagrange's equations for such systems, namely,

$$\frac{d}{dt} \left( \frac{\partial T}{\partial \dot{q}_r} \right) - \frac{\partial T}{\partial q_r} + \frac{\partial V}{\partial q_r} = 0 \qquad r = 1, 2, \ldots, n \tag{7.58}$$

But $T$ does not depend on the coordinates $q_i(t)$ and $V$ does not depend on the velocities $\dot{q}_i(t)$, so that, inserting Eqs. (7.56) and (7.57) into Eqs. (7.58), we obtain the equations of motion

$$\sum_{j=1}^{n} m_{rj} \ddot{q}_j(t) + \sum_{j=1}^{n} k_{rj} q_j(t) = 0 \qquad r = 1, 2, \ldots, n \tag{7.59}$$

which can be written in the matrix form

$$[m]\{\ddot{q}(t)\} + [k]\{q(t)\} = \{0\} \tag{7.60}$$

Moreover, recognizing that for normal-mode vibration the time dependence of $\{q(t)\}$ is harmonic, or

$$\{q(t)\} = \{a\} \cos(\omega t - \phi) \tag{7.61}$$

where $\{a\}$ is a constant vector, Eq. (7.60) yields the eigenvalue problem

$$[k]\{a\} = \Lambda[m]\{a\} \qquad \Lambda = \omega^2 \tag{7.62}$$

which is the same as that obtained by the Rayleigh-Ritz method, Eq. (7.44). Its solution yields $n$ eigenvalues $\Lambda_r$, related to the estimated natural frequencies $\omega_r$, and $n$ associated eigenvectors $\{a\}_r$ $(r = 1, 2, \ldots, n)$, where the latter lead to the estimated eigenfunctions

$$y_r(x) = \sum_{i=1}^{n} a_{ir}\phi_i(x) \qquad r = 1, 2, \ldots, n \tag{7.63}$$

where $a_{ir}$ is the $i$th component of the vector $\{a\}_r$.

To illustrate the procedure, let us consider a bar in flexure with one end clamped and with a concentrated mass attached to the other end, as shown in Fig. 7.5. The kinetic energy of the system is

$$\begin{aligned}
T(t) &= \frac{1}{2} \int_0^L m(x) \left[\frac{\partial y(x, t)}{\partial t}\right]^2 dx + \frac{1}{2} M \left[\frac{\partial y(L, t)}{\partial t}\right]^2 \\
&= \frac{1}{2} \int_0^L m(x) \left[\sum_{i=1}^n \phi_i(x)\dot{q}_i(t)\right]\left[\sum_{j=1}^n \phi_j(x)\dot{q}_j(t)\right] dx \\
&\quad + \frac{1}{2} M \left[\sum_{j=1}^n \phi_i(L)\dot{q}_i(t)\right]\left[\sum_{j=1}^n \phi_j(L)\dot{q}_j(t)\right] \\
&= \frac{1}{2} \sum_{i=1}^n \sum_{j=1}^n \dot{q}_i(t)\dot{q}_j(t)\left[\int_0^L m(x)\phi_i(x)\phi_j(x)\, dx + M\phi_i(L)\phi_j(L)\right] \quad (7.64)
\end{aligned}$$

from which we conclude that the mass coefficients have the form

$$m_{ij} = \int_0^L m(x)\phi_i(x)\phi_j(x)\, dx + M\phi_i(L)\phi_j(L) \qquad i, j = 1, 2, \ldots, n \tag{7.65}$$

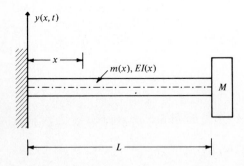

**Figure 7.5**

On the other hand, the potential energy can be written as

$$V(t) = \frac{1}{2} \int_0^L EI(x) \left[ \frac{\partial^2 y(x, t)}{\partial x^2} \right]^2 dx$$

$$= \frac{1}{2} \int_0^L EI(x) \left[ \sum_{i=1}^n \frac{d^2 \phi_i(x)}{dx^2} q_i(t) \right] \left[ \sum_{j=1}^n \frac{d^2 \phi_j(x)}{dx^2} q_j(t) \right] dx$$

$$= \frac{1}{2} \sum_{i=1}^n \sum_{j=1}^n q_i(t) q_j(t) \left[ \int_0^L EI(x) \frac{d^2 \phi_i(x)}{dx^2} \frac{d^2 \phi_j(x)}{dx^2} dx \right] \tag{7.66}$$

so that the stiffness coefficients are

$$k_{ij} = \int_0^L EI(x) \frac{d^2 \phi_i(x)}{dx^2} \frac{d^2 \phi_j(x)}{dx^2} dx \qquad i, j = 1, 2, \ldots, n \tag{7.67}$$

and note that $k_{ij}$ contains derivatives of $\phi_i(x)$ of second order, which is consistent with the fact that the differential equation of a bar in flexure is of order four. Clearly, the coefficients $m_{ij}$ and $k_{ij}$ are symmetric.

## 7.5 SYMMETRIC AND ANTISYMMETRIC MODES

When the system possess symmetric mass and stiffness properties and, in addition, the boundary conditions are symmetrical, the solution of the eigenvalue problem consists of eigenfunctions of two types, namely, symmetric and antisymmetric with respect to the symmetry center (see, for example, Sec. 5.4). While this fact is not particularly significant when a closed-form solution of the eigenvalue problem can be readily obtained, it has important implications when an approximate solution of the eigenvalue problem is sought. In this latter case it is advantageous to assume a series solution consisting of both symmetric and antisymmetric admissible functions, because in doing so the eigenvalue problem can be separated into two eigenvalue problems of correspondingly smaller order, one for the symmetric and the other for the antisymmetric modes. From a computational point of view, the solution of two eigenvalue problems of smaller order requires less effort than a single eigenvalue problem of correspondingly larger order.

Although the concepts are as valid for two- and three-dimensional systems as they are for one-dimensional systems, to illustrate the procedure let us consider the assumed-modes method and formulate the problem of a shaft in torsion clamped at both ends, as shown in Fig. 7.6. The system inertia and stiffness properties are symmetric, as indicated by the expressions

$$I(x) = I(-x) \qquad GJ(x) = GJ(-x) \tag{7.68}$$

where we note that $x$ is measured from the middle of the shaft, which coincides with the symmetry center. Because both boundary conditions are geometric, admissible functions are also comparison functions, so that in this case there is no difference between the two classes of functions. Denoting the angular displacement of the

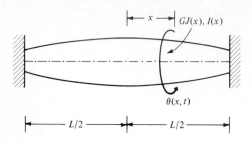

Figure 7.6

shaft at any point by $\theta(x, t)$, the system kinetic energy can be written in the form

$$T(t) = \frac{1}{2} \int_{-L/2}^{L/2} I(x) \left[ \frac{\partial \theta(x, t)}{\partial t} \right]^2 dx \qquad (7.69)$$

whereas the potential energy has the expression

$$V(t) = \frac{1}{2} \int_{-L/2}^{L/2} GJ(x) \left[ \frac{\partial \theta(x, t)}{\partial x} \right]^2 dx \qquad (7.70)$$

Letting the displacement $\theta(x, t)$ be represented by the series

$$\theta(x, t) = \sum_{i=1}^{n} \phi_i(x) q_i(t) \qquad (7.71)$$

where $\phi_i(x)$ are admissible functions and $q_i(t)$ generalized coordinates, and following the pattern of Sec. 7.4, the corresponding eigenvalue problem is

$$[k]\{a\} = \Lambda[I]\{a\} \qquad (7.72)$$

where the stiffness coefficients are given by

$$k_{ij} = \int_{-L/2}^{L/2} GJ(x) \frac{d\phi_i(x)}{dx} \frac{d\phi_j(x)}{dx} dx \qquad i, j = 1, 2, \ldots, n \qquad (7.73)$$

and the inertia coefficients by

$$I_{ij} = \int_{-L/2}^{L/2} I(x) \phi_i(x) \phi_j(x) dx \qquad i, j = 1, 2, \ldots, n \qquad (7.74)$$

Now let us assume that $r$ admissible functions are symmetric and $n - r$ are antisymmetric. Mathematically this can be expressed by

$$\phi_i(x) = \phi_i(-x) \qquad 1 \leqslant i \leqslant r$$
$$\phi_i(x) = -\phi_i(-x) \qquad r < i \leqslant n \qquad (7.75a)$$

It also follows that

$$\frac{d\phi_i(x)}{dx} = -\frac{d\phi_i(-x)}{dx} \qquad 1 \leqslant i \leqslant r$$

$$\frac{d\phi_i(x)}{dx} = \frac{d\phi_i(-x)}{dx} \qquad r < i \leqslant n \qquad (7.75b)$$

Inserting Eqs. (7.75) into (7.73) and (7.74), respectively, it is not difficult to show that if both $\phi_i$ and $\phi_j$ are symmetric, or if both are antisymmetric, then

$$k_{ij} = 2 \int_0^{L/2} GJ(x) \frac{d\phi_i(x)}{dx} \frac{d\phi_j(x)}{dx} dx$$

$$I_{ij} = 2 \int_0^{L/2} I(x)\phi_i(x)\phi_j(x) dx \qquad (7.76)$$

On the other hand, if $\phi_i$ is symmetric and $\phi_j$ antisymmetric, or if $\phi_i$ is antisymmetric and $\phi_j$ symmetric, then

$$k_{ij} = 0 \qquad I_{ij} = 0 \qquad (7.77)$$

Denoting quantities pertaining to symmetric and antisymmetric modes by the subscripts $s$ and $a$, respectively, Eq. (7.72) can be written in terms of partitioned matrices as follows:

$$\begin{bmatrix} [k]_s & [0] \\ \hline [0] & [k]_a \end{bmatrix} \begin{Bmatrix} \{a\}_s \\ \{a\}_a \end{Bmatrix} = \Lambda \begin{bmatrix} [I]_s & [0] \\ \hline [0] & [I]_a \end{bmatrix} \begin{Bmatrix} \{a\}_s \\ \{a\}_q \end{Bmatrix} \qquad (7.78)$$

which can be separated into

$$[k]_s\{a\}_s = \Lambda[I]_s\{a\}_s \qquad (7.79a)$$

and

$$[k]_a\{a\}_a = \Lambda[I]_a\{a\}_a \qquad (7.79b)$$

where the independent eigenvalue problems (7.79) are of order $r$ and $n - r$, respectively. Note that matrices $[k]_s$, $[I]_s$, $[k]_a$, and $[I]_a$ are symmetric. Because the complexity of solving an eigenvalue problem increases at a much faster rate than its order, the solution of the two eigenvalue problems (7.79) is less laborious than that of a single eigenvalue problem of order $n$.

As a specific example, let us consider the case in which

$$I(x) = \frac{12}{11} I \left[ 1 - \left(\frac{x}{L}\right)^2 \right] \qquad GJ(x) = \frac{12}{11} GJ \left[ 1 - \left(\frac{x}{L}\right)^2 \right] \qquad (7.80)$$

As admissible functions, we choose the eigenfunctions of a uniform shaft clamped at both ends. These consist of the symmetric modes

$$\phi_i(x) \cos (2i - 1) \frac{\pi x}{2L} \qquad i = 1, 2, \ldots, r \qquad (7.81a)$$

and the antisymmetric modes

$$\phi_i(x) = \sin \frac{i\pi x}{L} \qquad i = r + 1, r + 2, \ldots, n \qquad (7.81b)$$

where we let $2r = n$. Inserting Eqs. (7.80) and (7.81) into (7.73) and (7.74), it can be easily verified that the eigenvalue problem (7.72) does indeed separate into two eigenvalue problems of order $r$, one for the symmetric and the other for the antisymmetric modes (see Prob. 7.9).

# 7.6 RESPONSE OF SYSTEMS BY THE ASSUMED-MODES METHOD

In Sec. 7.4 we pointed out that the assumed-modes method yields the same eigenvalue problem as the Rayleigh-Ritz method, provided that in the latter Rayleigh's quotient is expressed in terms of energies. The assumed-modes method proves particularly convenient in deriving the response of a system to external forces or initial excitation. Of course, its main advantage is that we can derive general expressions for the response in terms of only admissible functions.

Let us assume the response of a continuous system in the form of the series

$$y(x, t) = \sum_{i=1}^{n} \phi_i(x) q_i(t) \tag{7.82}$$

where $\phi_i(x)$ are admissible functions and $q_i(t)$ generalized coordinates. We have shown in Sec. 7.4 that the system kinetic energy can be written as

$$T(t) = \frac{1}{2} \sum_{i=1}^{n} \sum_{j=1}^{n} m_{ij} \dot{q}_i(t) \dot{q}_j(t) \tag{7.83}$$

in which the constant symmetric mass coefficients $m_{ij}$ depend on the continuous system mass properties and the functions $\phi_i(x)$. The potential energy has the expression

$$V(t) = \frac{1}{2} \sum_{i=1}^{n} \sum_{j=1}^{n} k_{ij} q_i(t) q_j(t) \tag{7.84}$$

where the constant symmetric stiffness coefficients $k_{ij}$ depend on the continuous system stiffness properties and derivatives of $\phi_i(x)$.

External forces are generally regarded as nonconservative, so that Lagrange's equations of motion have the form

$$\frac{d}{dt}\left(\frac{\partial T}{\partial \dot{q}_r}\right) - \frac{\partial T}{\partial q_r} + \frac{\partial V}{\partial q_r} = Q_r(t) \qquad r = 1, 2, \ldots, n \tag{7.85}$$

where $Q_r(t)$ $(r = 1, 2, \ldots, n)$ are generalized nonconservative forces. These generalized forces can be expressed in terms of the actual forces and the admissible functions by means of the virtual work performed by these forces. We shall assume that the distributed external forces can be written in the form $f(x, t) + F_j(t)\delta(x - x_j)$, where $f(x, t)$ represents a distributed force and $F_j(t)$ $(j = 1, 2, \ldots, l)$ are $l$ concentrated forces acting at the points $x = x_j$. Note that $\delta(x - x_j)$ represents a spatial Dirac delta function given by

$$\delta(x - x_j) = 0 \qquad x \neq x_j$$
$$\int_0^L \delta(x - x_j)\, dx = 1 \tag{7.86}$$

so that $F_j(t)\delta(x - x_j)$ has units of distributed force. In view of (7.86), if we extend definition (6.38) to continuous systems and recall Eq. (6.39), then we can use Eq.

(7.82) and write the virtual work as follows:

$$\overline{\delta W}(t) = \int_0^L [f(x, t) + F_j(t)\delta(x - x_j)]\delta y(x, t)\, dx$$

$$= \int_0^L [f(x, t) + F_j(t)\delta(x - x_j)] \sum_{r=1}^{n} \phi_r(x)\delta q_r(t)\, dx$$

$$= \sum_{r=1}^{n} \left[ \int_0^L f(x, t)\phi_r(x)\, dx + \sum_{j=1}^{l} F_j(t)\phi_r(x_j) \right] \delta q_r(t)$$

$$= \sum_{r=1}^{n} Q_r(t)\delta q_r(t) \tag{7.87}$$

with the obvious conclusion that the generalized forces have the expressions

$$Q_r(r) = \int_0^L f(x, t)\phi_r(x)\, dx + \sum_{j=1}^{l} F_j(t)\phi_r(x_j) \qquad r = 1, 2, \ldots, n \tag{7.88}$$

Inserting Eqs. (7.83) and (7.84) into (7.85), Lagrange's equations of motion become

$$\sum_{j=1}^{n} m_{rj}\ddot{q}_j(t) + \sum_{j=1}^{n} k_{rj}q_j(t) = Q_r(t) \qquad r = 1, 2, \ldots, n \tag{7.89}$$

which can be written in the matrix form

$$[m]\{\ddot{q}(t)\} + [k]\{q(t)\} = \{Q(t)\} \tag{7.90}$$

Equation (7.90) is identical in form to that of an $n$-degree-of-freedom discrete system, and the response is the same as that given in Sec. 4.14.

As an illustration, let us consider the nonuniform rod in longitudinal vibration studied in Example 7.1, but with the end $x = L$ attached to a spring $k$ instead of being free. The mass and stiffness distributions of the rod are

$$m(x) = \frac{6}{5} m \left[ 1 - \frac{1}{2}\left(\frac{x}{L}\right)^2 \right] \qquad EA(x) = \frac{6}{5} EA \left[ 1 - \frac{1}{2}\left(\frac{x}{L}\right)^2 \right] \tag{7.91}$$

As admissible functions, we use the eigenfunctions of the corresponding uniform rod clamped at $x = 0$ and free at $x = L$, namely,

$$\phi_i(x) = \sin(2i - 1)\frac{\pi x}{2L} \qquad i = 1, 2, \ldots, n \tag{7.92}$$

The kinetic energy has the general form

$$T(t) = \frac{1}{2} \int_0^L m(x) \left[ \frac{\partial y(x, t)}{\partial t} \right]^2 dx$$

$$= \frac{1}{2} \int_0^L m(x) \left[ \sum_{i=1}^{n} \phi_i(x)\dot{q}_i(t) \right]\left[ \sum_{j=1}^{n} \phi_j(x)\dot{q}_j(t) \right] dx$$

$$= \frac{1}{2} \sum_{i=1}^{n} \sum_{j=1}^{n} \dot{q}_i(t)\dot{q}_j(t) \int_0^L m(x)\phi_i(x)\phi_j(x)\, dx \tag{7.93}$$

so that the mass coefficients are

$$m_{ij} = \int_0^L m(x)\phi_i(x)\phi_j(x)\,dx = \frac{6}{5}m\int_0^L\left[1 - \frac{1}{2}\left(\frac{x}{L}\right)^2\right]$$

$$\times \sin(2i-1)\frac{\pi x}{2L}\sin(2j-1)\frac{\pi x}{2L}\,dx \qquad i,j = 1,2,\ldots,n \quad (7.94)$$

On the other hand, the potential energy has the expression

$$V(t) = \frac{1}{2}\int_0^L EA(x)\left[\frac{\partial y(x,t)}{\partial x}\right]^2 dx + \tfrac{1}{2}ky^2(L,t)$$

$$= \frac{1}{2}\int_0^L EA(x)\left[\sum_{i=1}^n \frac{d\phi_i(x)}{dx}q_i(t)\right]\left[\sum_{j=1}^n \frac{d\phi_j(x)}{dx}q_j(t)\right]dx$$

$$+ \frac{1}{2}k\left[\sum_{i=1}^n \phi_i(L)q_i(t)\right]\left[\sum_{j=1}^n \phi_j(L)q_j(t)\right]$$

$$= \frac{1}{2}\sum_{i=1}^n\sum_{j=1}^n q_i(t)q_j(t)\left[\int_0^L EA(x)\frac{d\phi_i(x)}{dx}\frac{d\phi_j(x)}{dx}dx + k\phi_i(L)\phi_j(L)\right] \quad (7.95)$$

from which it follows that the stiffness coefficients are

$$k_{ij} = \int_0^L EA(x)\frac{d\phi_i(x)}{dx}\frac{d\phi_j(x)}{dx}dx + k\phi_i(L)\phi_j(L)$$

$$= \frac{6}{5}EA\frac{(2i-1)\pi}{2L}\frac{(2j-1)\pi}{2L}\int_0^L\left[1 - \frac{1}{2}\left(\frac{x}{L}\right)^2\right]$$

$$\times \cos(2i-1)\frac{\pi x}{2L}\cos(2j-1)\frac{\pi x}{2L}dx + k(-1)^{i+j} \qquad i,j = 1,2,\ldots,n \quad (7.96)$$

Assuming that the rod is subjected to the external force

$$f(x,t) = f_0 u(t) \qquad\qquad\qquad (7.97)$$

where $f_0$ is a constant and $u(t)$ the unit step function, the generalized forces become

$$f_r(t) = \int_0^L f(x,t)\phi_r(x)\,dx = f_0 u(t)\int_0^L \sin(2r-1)\frac{\pi x}{2L}dx$$

$$= \frac{2f_0 L}{(2r-1)\pi}u(t) \qquad r = 1,2,\ldots,n \qquad (7.98)$$

Equations (7.94), (7.96) and (7.98) define the equations of motion, Eq. (7.90), completely.

## 7.7 HOLZER'S METHOD FOR TORSIONAL VIBRATION

In Secs. 7.3 through 7.5 we examined discretization schemes for continuous systems. All these schemes had one thing in common, namely, they all regarded a continuous system as an $n$-degree-of-freedom discrete system by representing the displacement of the system by a finite series consisting of $n$ terms. This approach is suitable for cases in which the system nonuniformity is not particularly pronounced. In the case of systems with pronounced nonuniformity, or with a relatively large number of concentrated masses, other approaches are advised. We refer to these approaches as lumped-parameter methods.

A lumped-parameter method for the torsional vibration of shafts was developed by Holzer and extended to the flexural vibration of bars by Myklestad. According to Holzer's method, the system is regarded as consisting of $n$ *lumped rigid masses* concentrated at $n$ points called *stations*. The segments of shaft between the lumped masses, assumed to be massless and of uniform stiffness are referred to as *fields*. In replacing a continuous system by a discrete one, the system differential equation of motion and the load-deformation relation are replaced by corresponding finite difference equations, so that there is one finite difference equation relating the angular displacements and torques on both sides of a station and another equation relating the angular displacements and torques on both sides of a field. In essence, this is a step-by-step, or chain, method. An identical approach can be used for the transverse vibration of strings or longitudinal vibration of rods with lumped masses.

In Sec. 7.2 we pointed out that the relation between the angular displacement $\theta(x, t)$ and torque $M(x, t)$ is

$$\frac{\partial \theta(x, t)}{\partial x} = \frac{M(x, t)}{GJ(x)} \tag{7.99}$$

whereas, using the analogy of Sec. 5.2, the differential equation for the free vibration of a shaft in torsion can be written in the form

$$\frac{\partial M(x, t)}{\partial x} = I(x) \frac{\partial^2 \theta(x, t)}{\partial t^2} \tag{7.100}$$

where use has been made of Eq. (7.99). Because free vibration is harmonic, we can write

$$\theta(x, t) = \Theta(x) \cos(\omega t - \phi) \qquad M(x, t) = M(x) \cos(\omega t - \phi) \tag{7.101}$$

where $\omega$ is the frequency of oscillation, eliminate the time dependence and replace Eqs. (7.99) and (7.100) by

$$\frac{d\Theta(x)}{dx} = \frac{M(x)}{GJ(x)} \tag{7.102}$$

and

$$\frac{dM(x)}{dx} = -\omega^2 I(x)\Theta(x) \tag{7.103}$$

respectively. Equations (7.102) and (7.103) form the basis for the finite difference approach.

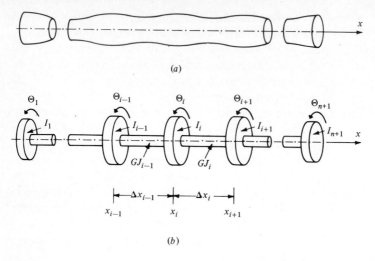

(a)

(b)

**Figure 7.7**

Next, let us consider the nonuniform shaft of Fig. 7.7$a$ and represent it by $n + 1$ rigid disks connected by $n$ massless circular shafts of uniform stiffness, as shown in Fig. 7.7$b$. The disks possess mass polar moments of inertia.

$$I_i = \tfrac{1}{2}I(x_i)(\Delta x_{i-1} + \Delta x_i) \cong I(x_i)\,\Delta x_i \qquad i = 2.\,3, \ldots, n$$

$$I_1 = \tfrac{1}{2}I(x_1)\,\Delta x_1 \qquad I_{n+1} = \tfrac{1}{2}I(x_{n+1})\,\Delta x_n \tag{7.104}$$

where the increments $\Delta x_i$ are sufficiently small that the above approximation can be justified. Moreover, we use the notation

$$GJ_i = GJ(x_i + \tfrac{1}{2}\Delta x_i) \qquad i = 1, 2, \ldots, n \tag{7.105}$$

Figure 7.8 shows free-body diagrams for station and field $i$. The superscripts $R$ and $L$ refer to the *right and left sides of a station*, respectively. In keeping with this notation, we observe that the left and right sides of field $i$ use the notation corresponding to the right side of station $i$ and the left side of station $i + 1$, respectively.

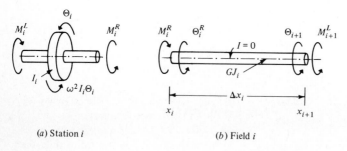

(a) Station $i$          (b) Field $i$

**Figure 7.8**

At this point we wish to invoke Eqs. (7.102) and (7.103) and write expressions relating the angular displacements and torques on both sides of station $i$ and field $i$. Because the disks are rigid, the displacements on both sides of station $i$ are the same,

$$\Theta_i^R = \Theta_i^L = \Theta_i \qquad (7.106)$$

On the other hand, Eq. (7.103) in incremental form becomes

$$\Delta M(x_i) = -\tfrac{1}{2}\omega^2 I(x_i)\Theta(x_i)(\Delta x_{i-1} + \Delta x_i) \cong -\omega^2 I(x_i)\Theta(x_i)\,\Delta x_i \qquad (7.107)$$

so that, using Eqs. (7.104) and (7.106), Eq. (7.107) leads to

$$M_i^R = M_i^L - \omega^2 I_i \Theta_i^L \qquad (7.108)$$

Because the segment of shaft associated with field $i$ is assumed to be massless (and hence possesses no mass moment of inertia), Eq. (7.103) yields directly

$$M_{i+1}^L = M_i^R \qquad (7.109)$$

whereas Eq. (7.102) in incremental form, as applied to field $i$, can be written as

$$\Delta\Theta(x_i + \tfrac{1}{2}\Delta x_i) = M(x_i + \tfrac{1}{2}\Delta x_i)\,\frac{\Delta x_i}{GJ(x_i + \tfrac{1}{2}\Delta x_i)} \cong \tfrac{1}{2}(M_{i+1}^L + M_i^R)\frac{\Delta x_i}{GJ_i} \qquad (7.110)$$

Using Eq. (7.109), Eq. (7.110) reduces to

$$\Theta_{i+1}^L = \Theta_i^R + a_i M_i^R \qquad (7.111)$$

where

$$a_i = \frac{\Delta x_i}{GJ_i} \qquad (7.112)$$

represents a torsional flexibility influence coefficient that can be interpreted as the angular displacement of disk $i + 1$ due to a unit moment $M_{i+1}^L = M_i^R = 1$ at station $i + 1$, where disk $i$ is prevented from rotating.

We note that Eqs. (7.106) and (7.108) give the angular displacement and torque on the right side of station $i$ in terms of the analogous quantities on the left side. The equations can be written in the matrix form

$$\begin{Bmatrix} \Theta_i^R \\ M_i^R \end{Bmatrix} = \begin{bmatrix} 1 & 0 \\ -\omega^2 I_i & 1 \end{bmatrix} \begin{Bmatrix} \Theta_i^L \\ M_i^L \end{Bmatrix} \qquad (7.113)$$

Letting

$$\begin{Bmatrix} \Theta_i^R \\ M_i^R \end{Bmatrix} = \begin{Bmatrix} \Theta \\ M \end{Bmatrix}_i^R \qquad \begin{Bmatrix} \Theta_i^L \\ M_i^L \end{Bmatrix} = \begin{Bmatrix} \Theta \\ M \end{Bmatrix}_i^L \qquad (7.114)$$

be the *state vectors* consisting of the angular displacements and torques on the right and left sides of station $i$, and introducing the *station transfer matrix*

$$[T_S]_i = \begin{bmatrix} 1 & 0 \\ -\omega^2 I_i & 1 \end{bmatrix} \qquad (7.115)$$

relating these two state vectors, Eq. (7.113) can be written in the compact form

$$\left\{\begin{matrix}\Theta\\M\end{matrix}\right\}_i^R = [T_S]_i \left\{\begin{matrix}\Theta\\M\end{matrix}\right\}_i^L \tag{7.116}$$

In a similar way, Eqs. (7.109) and (7.111) can be written as

$$\left\{\begin{matrix}\Theta\\M\end{matrix}\right\}_{i+1}^L = [T_F]_i \left\{\begin{matrix}\Theta\\M\end{matrix}\right\}_i^R \tag{7.117}$$

where

$$[T_F]_i = \begin{bmatrix} 1 & a_i \\ 0 & 1 \end{bmatrix} \tag{7.118}$$

is referred to as a *field transfer matrix*. Inserting Eq. (7.116) into (7.117), we obtain

$$\left\{\begin{matrix}\Theta\\M\end{matrix}\right\}_{i+1}^L = [T]_i \left\{\begin{matrix}\Theta\\M\end{matrix}\right\}_i^L \tag{7.119}$$

in which

$$[T]_i = [T_F]_i[T_S]_i \tag{7.120}$$

represents the *transfer matrix* relating the state vector on the left side of station $i + 1$ to that on the left side of station $i$.

Beginning with the first disk, $i = 1$, it is not difficult to show that

$$\left\{\begin{matrix}\Theta\\M\end{matrix}\right\}_{i+1}^L = [T]_i[T]_{i-1}\cdots[T]_2[T]_1 \left\{\begin{matrix}\Theta\\M\end{matrix}\right\}_1^L \qquad i = 1, 2, \ldots, n \tag{7.121}$$

Moreover, from Fig. 7.7b we conclude that

$$\left\{\begin{matrix}\Theta\\M\end{matrix}\right\}_{n+1}^R = [T] \left\{\begin{matrix}\Theta\\M\end{matrix}\right\}_1^L \tag{7.122}$$

where

$$[T] = [T_S]_{n+1}[T]_n[T]_{n-1}\cdots[T]_2[T]_1 \tag{7.123}$$

is known as the *overall transfer matrix*, relating the state vector on the left side of station 1 to that on the right side of station $n + 1$. In the sequel, we show that the frequency equation for any type of boundary conditions can be derived from Eq. (7.122).

Equation (7.122) can be written in the explicit form

$$\Theta_{n+1}^R = T_{11}\Theta_1^L + T_{12}M_1^L \qquad M_{n+1}^R = T_{21}\Theta_1^L + T_{22}M_1^L \tag{7.124}$$

where the elements $T_{ij}$ $(i, j = 1, 2)$ of the overall transfer matrix $[T]$ represent polynomials in $\omega^2$, because the station transfer matrices depend on $\omega^2$. The system frequency equation can be obtained by setting one of these elements, or a combination of these elements, equal to zero, depending on the end conditions. We examine several cases:

1. *Free-free shaft.* In the absence of torques at the ends, the boundary conditions are

$$M_1^L = 0 \qquad M_{n+1}^R = 0 \tag{7.125}$$

Inserting Eqs. (7.125) into the second of Eqs. (7.124), and recognizing that for a free end $\Theta_1^L \neq 0$, we must have

$$T_{21} = 0 \tag{7.126}$$

which is identified as the *frequency equation*, in this case an algebraic equation of degree $n + 1$ in $\omega^2$. It turns out, however, that $\omega^2$ can be factored out in the polynomial $T_{21}$, so that $\omega^2 = 0$ is one root of the frequency equation. The fact that one root is zero is to be expected, because the system is unconstrained and hence semidefinite (see Sec. 4.12).

2. *Clamped-free shaft.* Because at the left end the displacement is zero and at the right end the torque is zero, the boundary conditions are

$$\Theta_1^L = 0 \qquad M_{n+1}^R = 0 \tag{7.127}$$

From the second of Eqs. (7.124), we conclude that the frequency equation for this case is

$$T_{22} = 0 \tag{7.128}$$

which is of degree $n$ in $\omega^2$.

3. *Clamped-clamped shaft.* In this case the boundary conditions are simply

$$\Theta_1^L = 0 \qquad \Theta_{n=1}^R = 0 \tag{7.129}$$

leading to the frequency equation

$$T_{12} = 0 \tag{7.130}$$

which is of degree $n - 1$ in $\omega^2$. This comes as no surprise, because when both ends are clamped we really have a system with only $n - 1$ degrees of freedom.

Another possible boundary condition is that in which one of the ends is elastically supported. For example, if *the left end is clamped and the right end is supported by means of a torsional spring* of stiffness $k$, then the boundary conditions are

$$\Theta_1^L = 0 \qquad M_{n+1}^R = -k\Theta_{n+1}^R \tag{7.131}$$

Inserting Eqs. (7.131) into (7.124), it is not difficult to show that the frequency equation for this case is

$$T_{22} + kT_{12} = 0 \tag{7.132}$$

The solution of the frequency equation can be obtained by a root-finding technique. This task is facilitated by the fact that the roots $\omega_r^2$ are known to be real and positive. After the natural frequencies $\omega_r$ have been obtained, they can be inserted into the transfer matrices in Eq. (7.121), which enables us to plot the natural modes $\{\Theta\}_r$. As a by-product, we can also plot the torques $\{M\}_r$ corresponding to the natural modes.

The approach described above can be used also for the bending vibration of bars. The procedure is known as Myklestad's method and the basic difference is that in the case of bending the state vector is four-dimensional, consisting of the displacement, slope, being moment and shearing force. We do not pursue this subject here and the interested reader is referred to the text by Meirovitch.†

## 7.8 LUMPED-PARAMETER METHOD EMPLOYING INFLUENCE COEFFICIENTS

The lumped-parameter method employing influence coefficients is very simple conceptually. The continuous system is merely divided into segments, the mass associated with these segments lumped into discrete masses and the eigenvalue problem derived by regarding the system as discrete. In a way it resembles Holzer's method, with the exception that it makes no assumptions concerning the stiffness properties. Because of this, the system stiffness properties are simulated better than in Holzer's method. On the other hand, it necessitates the calculation of flexibility influence coefficients, which may prove a difficult task in many cases.

Although the method is applicable to many kinds of continuous systems, including two- and three-dimensional ones, for comparison purposes we consider a shaft clamped at the end $x = 0$ and free at the end $x = L$, as shown in Fig. 7.9a. The shaft has a circular cross-sectional area but nonuniform mass polar moment of inertia per unit length $I(x)$ and torsional rigidity $GJ(x)$. Figure 7.9b shows the lumped model consisting of $n$ disks of moments of inertia $I_i$ located at distances $x = x_i$ from the left end. The torsional flexibility influence coefficient $a_{ij}$ is defined as the angular displacement of disk $i$ due to a unit torque, $M_j = 1$, applied at station $j$. This is really the same definition as that given in Sec. 4.4 for discrete systems, except that in calculating the coefficients the system is regarded here as possessing continuous stiffness distribution, as it does in fact. It follows that the angular displacement $\Theta_i$ of disk $i$ due to arbitrary torques $M_j$ $(j = 1, 2, \ldots n)$ is

$$\Theta_i = \sum_{j=1}^{n} a_{ij} M_j \qquad i = 1, 2, \ldots, n \qquad (7.133)$$

For free vibration, however, there are no external torques present and the only torques are inertial, or

$$M_j = \omega^2 I_j \Theta_j \qquad j = 1, 2, \ldots, n \qquad (7.134)$$

(see Fig. 7.8a). Inserting Eqs. (7.134) into (7.133), we obtain

$$\Theta_i = \omega^2 \sum_{j=1}^{n} a_{ij} I_j \Theta_j \qquad i = 1, 2, \ldots, n \qquad (7.135)$$

† *Analytical Methods in Vibrations*, sec. 6.12, The Macmillan Co., New York, 1967.

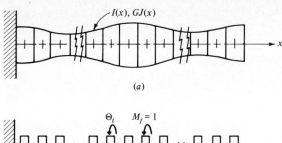

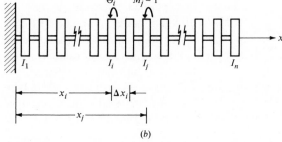

**Figure 7.9**

which is recognized as the eigenvalue problem for an $n$-degree-of-freedom system. In matrix form, Eqs. (7.135) become

$$\{\Theta\} = \lambda[a][I]\{\Theta\} \qquad \lambda = \omega^2 \tag{7.136}$$

where $[a]$ is the familiar flexibility matrix and $[I]$ the inertia matrix, in this case a diagonal matrix. The eigenvalue problem (7.136) can be solved by one of the methods described in Chap. 4.

As an example, let us consider a shaft with the inertia and stiffness distributions

$$I(x) = \frac{6}{5}I\left[1 - \frac{1}{2}\left(\frac{x}{L}\right)^2\right] \qquad GJ(x) = \frac{6}{5}GJ\left[1 - \frac{1}{2}\left(\frac{x}{L}\right)^2\right] \tag{7.137}$$

Dividing the shaft into $n$ equal increments, we have $\Delta x_i = \Delta x = L/n$. With every one of these increments we associate a disk of mass polar moment of inertia

$$I_i = \int_{(i-1)(L/n)}^{i(L/n)} I(x)\,dx = \frac{6}{5}I \int_{(i-1)(L/n)}^{i(L/n)} \left[1 - \frac{1}{2}\left(\frac{x}{L}\right)^2\right] dx$$

$$= \frac{IL}{5n^3}(6n^2 - 3i^2 + 3i - 1) \qquad i = 1, 2, \ldots, n \tag{7.138}$$

The disks are assumed to be located at their inertia centers. Hence, the positions of these disks are defined by

$$x_i = \frac{1}{I_i} \int_{(i-1)(L/n)}^{i(L/n)} xI(x)\,dx = \frac{6I}{5I_i} \int_{(i-1)(L/n)}^{i(L/n)} x\left[1 - \frac{1}{2}\left(\frac{x}{L}\right)^2\right] dx$$

$$= \frac{3IL^2}{20I_in^4}[4n^2(2i - 1) - 4i^3 + 6i^2 - 4i + 1] \qquad i = 1, 2, \ldots, n \tag{7.139}$$

Moreover, from mechanics of materials, the influence coefficients can be shown to have the form

$$a_{ij} = a_{ji} = \int_0^{x_i} \frac{dx}{\frac{6}{5}GJ[1 - \frac{1}{2}(x/L)^2]} = \frac{5L}{6\sqrt{2GJ}} \log \frac{\sqrt{2L} + x_i}{\sqrt{2L} - x_i}$$

$$i, j = 1, 2, \ldots, n \quad (7.140)$$

where $x_i \leqslant x_j$. Equation (7.140) can be easily explained by noticing that the displacement at point $x_i$ is the same regardless at which point $x_j$ the unit torque is acting, as long as $x_j \geqslant x_i$. For $x_i > x_j$ we have $a_{ij} = a_{jj}$. Perhaps a better appreciation for the evaluation of flexibility influence coefficients for cases in which the stiffness is distributed can be gained by considering influence functions.†

The eigenvalue problem, corresponding to the data given by Eqs. (7.138) to (7.140), was solved for the case $n = 10$. The first three natural modes, together with the corresponding natural frequencies, are displayed in Fig. 7.10.

## PROBLEMS

**7.1** Consider the cable of Prob. 5.8 and use the trial function $Y(x) = 1 - (x/L)^2$ in conjunction with Rayleigh's energy method to obtain an estimate of the fundamental frequency of the system.

**7.2** Consider the shaft in torsion of Sec. 7.2, and use the trial function $\Theta(x) = x/L - \frac{1}{3}(x/L)^3$ in conjunction with Rayleigh's energy method to obtain an estimate of the fundamental frequency of the system.

**7.3** Repeat Prob. 7.2 by using the trial function $\Theta(x) = a_1[(x/L) - \frac{1}{3}(x/L)^3] + a_2[(x/L)^3 - \frac{3}{5}(x/L)^5]$, where $a_1$ and $a_2$ are undetermined constants. In this case the estimated natural frequency depends on ratio $a_2/a_1$. Determine the value of $a_2/a_1$ so as to render the estimated natural frequency a minimum.

**7.4** Prove the orthogonality with respect to mass of the estimated eigenfunctions, Eqs. (7.49), where the orthogonality conditions are given by

$$\int_0^L m(x)u_r(x)u_s(x) \, dx = 0 \qquad r \neq s$$

The eigenvectors $\{a\}_r$ and $\{a\}_s$ can be assumed to be orthogonal with respect to the mass matrix $[m]$, where the elements of $[m]$ are given by Eq. (7.48).

**7.5** Solve Prob. 7.1 by the Rayleigh-Ritz method by assuming the trial functions $Y(x) = a_1[1 - (x/L)^2] + a_2[1 - (x/L)^3]$ and $Y(x) = a_1[1 - (x/L)^2] + a_2[1 - (x/L)^3] + a_3[1 - (x/L)^4]$, in sequence. Compare the results obtained here with those of Prob. 7.1 and draw conclusions.

**7.6** Consider the system of Fig. 7.1, let $EA(x) = EA = $ const, $m(x) = m = $ const and $k = EA/4L$, and solve the eigenvalue problem by the Rayleigh-Ritz method using an approximate solution in the form $u(x) = \sum_{i=1}^n a_i \sin (2i - 1)\pi x/2L$ for the three cases $n = 1, 2, 3$. Plot the computed natural modes. Compare the results with those obtained in Prob. 5.7.

**7.7** Consider the system of Fig. 7.5, Let $EI(x) = EI = $ const, $m(x) = m = $ const and $M = 0.2mL$, and solve the eigenvalue problem by the Rayleigh-Ritz method using an approximate solution in the form $u(x) = \sum_{i=1}^n a_i(x/L)^{i+1}$ for the two cases $n = 2, 3$. Plot the computed natural modes.

**7.8** Consider the system of Fig. 5.13 and derive the equations of motion of the associated discrete system by the assumed-modes method. Give general expressions for the coefficients $m_{ij}$ and $k_{ij}$ ($i, j = 1, 2, \ldots, n$).

† See, for example, L. Meirovitch, op. cit., sec. 3–2, The Macmillan Co., New York, 1967.

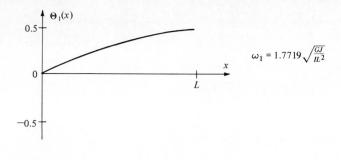

$$\omega_1 = 1.7719 \sqrt{\frac{GJ}{IL^2}}$$

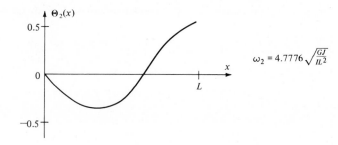

$$\omega_2 = 4.7776 \sqrt{\frac{GJ}{IL^2}}$$

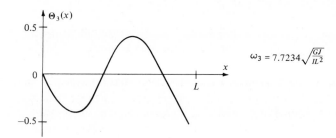

$$\omega_3 = 7.7234 \sqrt{\frac{GJ}{IL^2}}$$

**Figure 7.10**

**7.9** Complete the formulation of the problem of Sec. 7.5 by deriving the matrices $[m]$ and $[k]$ in explicit form. Solve the eigenvalue problem by using two symmetric and two antisymmetric admissible functions.

**7.10** Let the system of Prob. 7.6 be acted upon by an impulsive uniformly distributed force $f(x, t) = \hat{f}_0 \delta(t)$ $(0 < x < L)$ acting in the axial direction. Derive the system response by modal analysis (Sec. 7.6) for the case $n = 3$. What can be said about the mode participation in the total response?

**7.11** Let the system of Prob. 7.7 be acted upon by a concentrated transverse force whose time-dependence is in the form of a step function. The force is applied to the mass $M$, and can be expressed mathematically by $f(x, t) = F_0 \delta(x - L) \alpha(t)$. Derive the system response by modal analysis (Sec. 7.6) for the case $n = 3$ and discuss the mode participation in the total response.

**7.12** Solve Prob. 7.9 by means of Holzer's method by dividing the shaft into six lumped disks (six degrees of freedom). Plot the first three natural modes.

**7.13** Solve Prob. 7.12 by the lumped-parameter method employing influence coefficients (Sec. 7.8). Compare the results obtained here with those of Prob. 7.12.

# EIGHT

## THE FINITE ELEMENT METHOD

### 8.1 GENERAL CONSIDERATIONS

The increasing complexity of structures and sophistication of digital computers have been instrumental in the development of new methods of analysis, particularly of the finite element method. The idea behind the finite element method is to provide a formulation which can exploit digital computer automation for the analysis of irregular systems. To this end, the method regards a complex structure as an assemblage of finite elements, where every such element is part of a continuous structural member. By requiring that the displacements be compatible and the internal forces in balance at certain points shared by several elements, where the points are known as nodes, the entire structure is compelled to act as one entity.

Although the finite element method considers continuous individual elements, it is in essence a discretization procedure, as it expresses the displacement at any point of the continuous element in terms of a finite number of displacements at the nodal points multiplied by given interpolation functions. To illustrate the idea, we refer to the one-dimensional system shown in Fig. 8.1. The system is divided into a finite number $N$ of elements of width $h$, where $Nh = L$, and the motion of the system is defined in terms of the *nodal displacements* $u_j(t)$ $(j = 1, 2, ..., N)$. The advantage of the finite element method over any other method is that the equations of motion for the system can be derived by first deriving the equations of motion for a typical finite element and then assembling the individual elements' equations of motion. The motion at any point inside the element is obtained by means of interpolation, where the interpolation functions are generally low-degree polynomials and they are the same for every element.

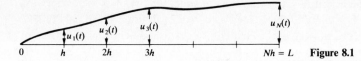

The finite element method, as practiced today, began as a method of structural analysis, being related to the direct stiffness method. This direct approach may be satisfactory for static problems, but encounters difficulties in handling dynamic problems, such as in vibrations of continuous media. Such problems are treated better by a variational approach. In fact, the finite element method can be regarded as a special case of the Rayleigh-Ritz method, although since its inception the method has acquired a life of its own, going well beyond the original structural applications.

The purpose of this chapter is to present some of the basic ideas involved in the use of the finite element method for vibration problems rather than an exhaustive treatment of the subject. Consistent with the scope of this text, we shall be concerned only with one-dimensional elements, although the concepts and developments presented are quite general and can be readily applied to two- and three-dimensional elements.

## 8.2 DERIVATION OF THE ELEMENT STIFFNESS MATRIX BY THE DIRECT APPROACH

As pointed out in Sec. 4.4, the stiffness matrix relates a displacement vector to a force vector. The entries in the stiffness matrix can be identified as the stiffness influence coefficients, which represent a strictly static concept. In this section, we adopt a similar approach by deriving the element stiffness matrix as the matrix relating the nodal displacement vector to the nodal force vector. To this end, we consider a rod in axial vibration and derive the stiffness matrix for a typical element, such as that shown in Fig. 8.2. We carry out the task in two steps. In the first step we derive an expression for the axial displacement of an arbitrary point inside the element in terms of the nodal displacements and in the second step we use this expression to relate the nodal displacements to the nodal forces. Although

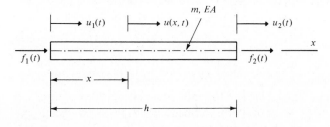

**Figure 8.2**

in vibrations all these quantities are functions of time, for the purpose of deriving the stiffness matrix, they can all be regarded as constant in time. The axial stiffness can be assumed to be constant over the element, so that the differential equation for the axial displacement $u(x)$ is

$$EA \frac{d^2 u(x)}{dx^2} = 0 \qquad 0 < x < h \tag{8.1}$$

Integrating Eq. (8.1) twice, we obtain

$$u(x) = c_1 x + c_2 \tag{8.2}$$

where $c_1$ and $c_2$ are constants of integration. But, from Fig. 8.2, at $x = 0$ the axial displacement $u(x)$ is equal to the nodal displacement $u_1$ and at $x = h$ it is equal to the nodal displacement $u_2$. Hence, using Eq. (8.2), we can write

$$u(0) = u_1 = c_2 \qquad u(h) = c_1 h + c_2 = u_2 \tag{8.3}$$

Equations (8.3) have the solution

$$c_1 = \frac{u_2 - u_1}{h} \qquad c_2 = u_1 \tag{8.4}$$

so that, inserting the constants $c_1$ and $c_2$ just obtained into Eq. (8.2), we obtain the expression for the axial displacement

$$u(x) = \left(1 - \frac{x}{h}\right) u_1 + \frac{x}{h} u_2 \tag{8.5}$$

The displacement $u(x)$ is related to the nodal forces through the boundary conditions

$$EA \frac{du(x)}{dx}\bigg|_{x=0} = -f_1 \qquad EA \frac{du(x)}{dx}\bigg|_{x=h} = f_2 \tag{8.6}$$

so that, using Eq. (8.5), we have

$$EA \frac{u_2 - u_1}{h} = -f_1 \qquad EA \frac{u_2 - u_1}{h} = f_2 \tag{8.7}$$

Equations (8.7) can be written in the matrix form

$$[k]\{u\} = \{f\} \tag{8.8}$$

where

$$\{u\} = \begin{Bmatrix} u_1 \\ u_2 \end{Bmatrix} \qquad \{f\} = \begin{Bmatrix} f_1 \\ f_2 \end{Bmatrix} \tag{8.9}$$

are the nodal displacement vector and nodal force vector, respectively, and

$$[k] = \frac{EA}{h} \begin{bmatrix} 1 & -1 \\ -1 & 1 \end{bmatrix} \tag{8.10}$$

is the desired element stiffness matrix.

Another case of particular interest is the bar in bending vibration. We propose to use the same approach as above to derive the corresponding element stiffness matrix. To this end, we consider the element shown in Fig. 8.3. For uniform bending stiffness, the differential equation for the displacement $w(x)$ is

$$EI \frac{d^4 w(x)}{dx^4} = 0 \qquad 0 < x < h \tag{8.11}$$

Integrating Eq. (8.11) four times, we have

$$w(x) = \tfrac{1}{6} c_1 x^3 + \tfrac{1}{2} c_2 x^2 + c_3 x + c_4 \tag{8.12}$$

where $c_i$ ($i = 1, 2, 3, 4$) are constants of integration. To determine these constants, we refer to Fig. 8.3 and write

$$w(0) = w_1 \qquad \frac{dw(x)}{dx}\bigg|_{x=0} = \theta_1 \qquad w(h) = w_2 \qquad \frac{dw(x)}{dx}\bigg|_{x=h} = \theta_2 \tag{8.13}$$

where $w_1$ and $w_2$ are nodal displacements and $\theta_1$ and $\theta_2$ are nodal rotations, or nodal angular displacements. Introducing Eq. (8.12) into Eqs. (8.13), we obtain

$$w(0) = c_4 = w_1 \qquad \frac{dw(x)}{dx}\bigg|_{x=0} = c_3 = \theta_1$$

$$w(h) = \tfrac{1}{6} c_1 h^3 + \tfrac{1}{2} c_2 h^2 + c_3 h + c_4 = w_2 \tag{8.14}$$

$$\frac{dw(x)}{dx}\bigg|_{x=h} = \tfrac{1}{2} c_1 h^2 + c_2 h + c_3 = \theta_2$$

which have the solution

$$c_1 = \frac{6}{h^3} (2w_1 + h\theta_1 - 2w_2 + h\theta_2) \qquad c_2 = \frac{2}{h^2} (-3w_1 - 2h\theta_1 + 3w_2 - h\theta_2)$$

$$\tag{8.15}$$

$$c_3 = \theta_1 \qquad c_4 = w_1$$

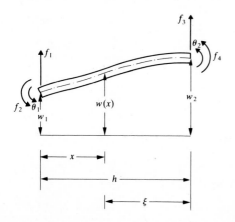

**Figure 8.3**

Hence, introducing Eqs. (8.15) into Eq. (8.12), we obtain the expression for the bending displacement

$$w(x) = \left[1 - 3\left(\frac{x}{h}\right)^2 + 2\left(\frac{x}{h}\right)^3\right]w_1 + \left[\frac{x}{h} - 2\left(\frac{x}{h}\right)^2 + \left(\frac{x}{h}\right)^3\right]h\theta_1$$

$$+ \left[3\left(\frac{x}{h}\right)^2 - 2\left(\frac{x}{h}\right)^3\right]w_2 + \left[-\left(\frac{x}{h}\right)^2 + \left(\frac{x}{h}\right)^3\right]h\theta_2 \qquad (8.16)$$

The bending displacement is related to the nodal forces $f_1, f_2, f_3,$ and $f_4$ as follows:

$$EI\frac{d^3w(x)}{dx^3}\bigg|_{x=0} = f_1 \qquad\qquad EI\frac{d^2w(x)}{dx^2} = f_2$$

$$EI\frac{d^3w(x)}{dx^3}\bigg|_{x=h} = -f_3 \qquad EI\frac{d^2w(x)}{dx^2}\bigg|_{x=h} = f_4 \qquad (8.17)$$

which yield

$$f_1 = \frac{EI}{h^3}(12w_1 + 6h\theta_1 - 12w_2 + 6h\theta_2)$$

$$f_2 = \frac{EI}{h^2}(6w_1 + 4h\theta_1 - 6w_2 + 2h\theta_2)$$

$$f_3 = \frac{EI}{h^3}(-12w_1 - 6h\theta_1 + 12w_2 - 6h\theta_2) \qquad (8.18)$$

$$f_4 = \frac{EI}{h^2}(6w_1 + 2h\theta_1 - 6w_2 + 4h\theta_2)$$

Equations (8.18) have the matrix form

$$[k]\{w\} = \{f\} \qquad (8.19)$$

where

$$\{w\} = \begin{Bmatrix} w_1 \\ h\theta_1 \\ w_2 \\ h\theta_2 \end{Bmatrix} \qquad \{f\} = \begin{Bmatrix} f_1 \\ f_2/h \\ f_3 \\ f_4/h \end{Bmatrix} \qquad (8.20)$$

are the nodal displacement vector and nodal force vector, respectively, and

$$[k] = \frac{EI}{h^3} \begin{bmatrix} 12 & 6 & -12 & 6 \\ 6 & 4 & -6 & 2 \\ -12 & -6 & 12 & -6 \\ 6 & 2 & -6 & 4 \end{bmatrix} \qquad (8.21)$$

is the element stiffness matrix.

Equation (8.5) can be written in the form

$$u(x) = L_1(x)u_1 + L_2(x)u_2 \qquad (8.22)$$

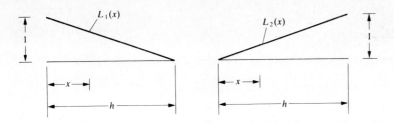

**Figure 8.4**

where

$$L_1(x) = 1 - \frac{x}{h} \qquad L_2 = \frac{x}{h} \tag{8.23}$$

are known as *shape functions*, or *interpolation functions*. They are plotted in Fig. 8.4. The term interpolation functions can be easily justified, as the functions $L_1(x)$ and $L_2(x)$ permit us to determine the displacement at any distance $x$ from the left end through an interpolation between the nodal displacements $u_1$ and $u_2$. Similarly, Eq. (8.16) can be expressed as

$$w(x) = L_1(x)w_1 + L_2(x)h\theta_1 + L_3(x)w_2 + L_4(x)h\theta_2 \tag{8.24}$$

where the interpolation functions

$$L_1(x) = 1 - 3\left(\frac{x}{h}\right)^2 + 2\left(\frac{x}{h}\right)^3 \qquad L_2(x) = \frac{x}{h} - 2\left(\frac{x}{h}\right)^2 + \left(\frac{x}{h}\right)^3$$

$$L_3(x) = 3\left(\frac{x}{h}\right)^2 - 2\left(\frac{x}{h}\right)^2 \qquad L_4(x) = -\left(\frac{x}{h}\right)^2 + \left(\frac{x}{h}\right)^3 \tag{8.25}$$

are known as *Hermite cubics*. They are plotted in Fig. 8.5.

The interpolation functions and the stiffness matrices were derived on the basis of the static deformation pattern under nodal forces. It turns out that the interpolation functions are not unique and other choices are possible. The interpolation functions derived here, however, represent the lowest-degree polynomials that can be used for second-order and fourth-order problems. This subject is discussed in more detail later in this chapter.

## 8.3 ELEMENT EQUATIONS OF MOTION. A CONSISTENT APPROACH

In the finite element method, the equations of motion for a structure are obtained by deriving first the element equations of motion and then assembling the equations for all the elements. In this section, we derive the element equations of motion, leaving the assembly process for a later section. The element stiffness matrices derived in Sec. 8.2 for elements in axial deformation and in bending,

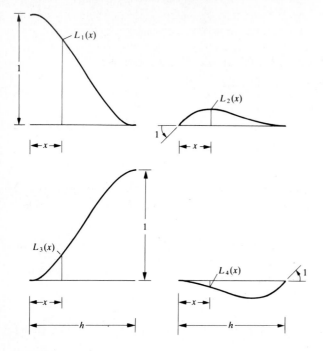

**Figure 8.5**

respectively, are the same matrices entering into the element equations of motion. The task of deriving the element mass matrix and the element nodal forces resulting from external loads remains.

The simplest way of generating the element mass matrix is by lumping the mass at the nodes, which is the way it was done in the early days of the finite element method. This approach has several drawbacks, however. In the first place, lumping is an arbitrary process, so that some control over the error involved in discretization is lost, which is true of all lumping methods. Perhaps more serious is the fact that lumping can lead to singular mass matrices, which is at odds with the fact that mass matrices are positive definite by definition. Singular mass matrices arise in bending problems because lumped masses are generally regarded as point masses, so that the mass coefficients corresponding to the rotational coordinates are zero. Of course, one could assign mass moments of inertia to the lumped masses, but this would make the lumping process even more arbitrary. In this section, we derive mass matrices by an approach precluding the occurrence of singularities. In fact, the mass matrices are derived by the same approach as the stiffness matrices, so that the mass matrices are consistent with the stiffness matrices. In view of this, mass matrices derived by lumping are known as *inconsistent mass matrices*. In this text, we use only consistent mass matrices.

A most satisfactory derivation of the element equations of motion can be effected by means of a variational approach. As can be concluded from Chap. 6, this amounts to deriving Lagrange's equations of motion for the element. This reduces, in turn, to writing the kinetic energy, the potential energy, and the virtual work expressions in terms of the nodal coordinates. We propose to derive here element equations of motion for the second-order and fourth-order systems considered in Sec. 8.2. The element equations of motion can be regarded as having been derived as soon as the mass matrix, the stiffness matrix, and the force vector for the element have been derived.

Let us assume that the axial displacement of the second-order system depicted in Fig. 8.2 can be written in the form

$$u(x, t) = L_1(x)u_1(t) + L_2(x)u_2(t) = \{L(x)\}^T\{u(t)\} \tag{8.26}$$

where $\{L(x)\}$ is a two-dimensional vector of interpolation functions and $\{u(t)\}$ is the corresponding vector of nodal displacements. Using Eq. (8.26), the kinetic energy for the element is simply

$$
\begin{aligned}
T(t) &= \frac{1}{2} \int_0^h m(x) \left[ \frac{\partial u(x, t)}{\partial t} \right]^2 dx \\
&= \frac{1}{2} \int_0^h m(x)\{\dot{u}(t)\}^T\{L(x)\}\{L(x)\}^T\{\dot{u}(t)\} \, dx \\
&= \tfrac{1}{2}\{\dot{u}(t)\}^T[m]\{\dot{u}(t)\}
\end{aligned}
\tag{8.27}
$$

where

$$[m] = \int_0^h m(x)\{L(x)\}\{L(x)\}^T \, dx \tag{8.28}$$

is the $2 \times 2$ *element mass matrix*. Similarly, the potential energy is

$$
\begin{aligned}
V(t) &= \frac{1}{2} \int_0^h EA(x) \left[ \frac{\partial u(x, t)}{\partial x} \right]^2 dx \\
&= \frac{1}{2} \int_0^h EA(x)\{u(t)\}^T\{L'(x)\}\{L'(x)\}^T\{u(t)\} \, dx \\
&= \tfrac{1}{2}\{u(t)\}^T[k]\{u(t)\}
\end{aligned}
\tag{8.29}
$$

where

$$[k] = \int_0^h EA(x)\{L'(x)\}\{L'(x)\}^T \, dx \tag{8.30}$$

is the $2 \times 2$ *element stiffness matrix*, in which primes indicate differentiations with respect to $x$. To derive the nodal force vector, we turn to the virtual work expression. Assuming that the element is subjected to the distributed axial nonconservative force $f(x, t)$ and considering Eq. (8.26), we can write

$$
\begin{aligned}
\overline{\delta W}(t) &= \int_0^h f(x, t) \, \delta u(x, t) \, dx = \int_0^h f(x, t)\{L(x)\}^T\{\delta u(t)\} \, dx \\
&= \{f(t)\}^T\{\delta u(t)\}
\end{aligned}
\tag{8.31}
$$

where

$$\{f(t)\} = \int_0^h f(x, t)\{L(x)\}\, dx \tag{8.32}$$

is the *nodal nonconservative force vector*. Note that concentrated forces can be included in the distributed force $f(x, t)$ by means of spatial Dirac delta functions.

In the case of the fourth-order system shown in Fig. 8.3, the bending displacement can be written as

$$w(x, t) = \{L(x)\}^T\{x(t)\} \tag{8.33}$$

where $\{L(x)\}$ is now a four-dimensional vector of interpolation functions and $\{w(t)\}$ is a four-dimensional vector of nodal displacements. The element kinetic energy has the form

$$T(t) = \frac{1}{2}\int_0^h m(x)\left[\frac{\partial w(x, t)}{\partial t}\right]^2 dx = \tfrac{1}{2}\{\dot{w}(t)\}^T[m]\{\dot{w}(t)\} \tag{8.34}$$

where the $4 \times 4$ element mass matrix is

$$[m] = \int_0^h m(x)\{L(x)\}\{L(x)\}^T dx \tag{8.35}$$

Similarly, the element potential energy can be written as

$$V(t) = \frac{1}{2}\int_0^h EI(x)\left[\frac{\partial^2 w(x, t)}{\partial x^2}\right]^2 dx = \tfrac{1}{2}\{w(t)\}^T[k]\{w(t)\} \tag{8.36}$$

where the $4 \times 4$ element stiffness matrix has the expression

$$[k] = \int_0^h EI(x)\{L''(x)\}\{L''(x)\}^T dx \tag{8.37}$$

in which the notation is obvious. The nodal force vector has the same general form as that given by Eq. (8.32), except that now it is a four-dimensional vector including forces and moments.

**Example 8.1** Consider an element in axial vibration, such as that shown in Fig. 8.2, use the interpolation functions given by Eqs. (8.23) and calculate the element mass and stiffness matrices, as well as the nodal force vector for the distributed load $f(x, t) = a + bx$. Assume that the mass and the axial stiffness are constant over the element.

Inserting Eqs. (8.23) into Eq. (8.28), we obtain the element mass matrix

$$[m] = m\int_0^h \left\{\begin{array}{c} 1 - \dfrac{x}{h} \\[2mm] \dfrac{x}{h} \end{array}\right\} \left\{\begin{array}{c} 1 - \dfrac{x}{h} \\[2mm] \dfrac{x}{h} \end{array}\right\}^T dx$$

$$= m \int_0^h \begin{bmatrix} \left(1 - \dfrac{x}{h}\right)^2 & \left(1 - \dfrac{x}{h}\right)\dfrac{x}{h} \\ \left(1 - \dfrac{x}{h}\right)\dfrac{x}{h} & \left(\dfrac{x}{h}\right)^2 \end{bmatrix} dx = \frac{mh}{6}\begin{bmatrix} 2 & 1 \\ 1 & 2 \end{bmatrix} \qquad (a)$$

For the stiffness matrix, we need $\{L'(x)\}$. From Eqs. (8.23), we can write

$$\{L'(x)\} = \frac{d}{dx}\{L(x)\} = \frac{d}{dx}\begin{Bmatrix} 1 - \dfrac{x}{h} \\ \dfrac{x}{h} \end{Bmatrix} = \frac{1}{h}\begin{Bmatrix} -1 \\ 1 \end{Bmatrix} \qquad (b)$$

Hence, introducing Eq. (b) into Eq. (8.30), we obtain the element stiffness matrix

$$[k] = \frac{EA}{h^2}\int_0^h \begin{Bmatrix} -1 \\ 1 \end{Bmatrix}\begin{Bmatrix} -1 \\ 1 \end{Bmatrix}^T dx = \frac{EA}{h}\begin{bmatrix} 1 & -1 \\ -1 & 1 \end{bmatrix} \qquad (c)$$

which is the same matrix as that given by Eq. (8.10).

Finally, to calculate the modal force vector, we use Eq. (8.32) and write

$$\{f(t)\} = \int_0^h (a + bx)\begin{Bmatrix} 1 - \dfrac{x}{h} \\ \dfrac{x}{h} \end{Bmatrix} dx = \int_0^h \begin{Bmatrix} a + \left(b - \dfrac{a}{h}\right)x - \dfrac{b}{h}x^2 \\ \dfrac{a}{h}x + \dfrac{b}{h}x^2 \end{Bmatrix} dx$$

$$= \begin{Bmatrix} \frac{1}{2}ah + \frac{1}{6}bh^2 \\ \frac{1}{2}ah + \frac{1}{3}bh^2 \end{Bmatrix} \qquad (d)$$

**Example 8.2** Consider an element in bending vibration, such as that shown in Fig. 8.3, use the interpolation functions given by Eqs. (8.25) and calculate the element mass and stiffness matrices, as well as the nodal force vector corresponding to the concentrated force $P(t)$ applied at $x = h/3$. Assume that the mass and bending stiffness are constant over the element.

Introducing Eqs. (8.25) into Eq. (8.28), we obtain the element mass matrix

$$[m] = m \int_0^h \begin{Bmatrix} 1 - 3\left(\dfrac{x}{h}\right)^2 + 2\left(\dfrac{x}{h}\right)^3 \\ \dfrac{x}{h} - 2\left(\dfrac{x}{h}\right)^2 + \left(\dfrac{x}{h}\right)^3 \\ 3\left(\dfrac{x}{h}\right)^2 - 2\left(\dfrac{x}{h}\right)^3 \\ -\left(\dfrac{x}{h}\right)^2 + \left(\dfrac{x}{h}\right)^3 \end{Bmatrix} \begin{Bmatrix} 1 - 3\left(\dfrac{x}{h}\right)^2 + 2\left(\dfrac{x}{h}\right)^3 \\ \dfrac{x}{h} - 2\left(\dfrac{x}{h}\right)^2 + \left(\dfrac{x}{h}\right)^3 \\ 3\left(\dfrac{x}{h}\right)^2 - 2\left(\dfrac{x}{h}\right)^3 \\ -\left(\dfrac{x}{h}\right)^2 + \left(\dfrac{x}{h}\right)^3 \end{Bmatrix}^T dx \qquad (a)$$

which has the entries

$$m_{11} = m \int_0^h \left[ 1 - 3\left(\frac{x}{h}\right)^2 + 2\left(\frac{x}{h}\right)^3 \right]^2 dx = \frac{13}{35} mh$$

$$m_{12} = m_{21} = m \int_0^h \left[ 1 - 3\left(\frac{x}{h}\right)^2 + 2\left(\frac{x}{h}\right)^3 \right]\left[ \frac{x}{h} - 2\left(\frac{x}{h}\right)^2 + \left(\frac{x}{h}\right)^3 \right] dx$$

$$= \frac{11}{210} mh$$

$$m_{13} = m_{31} = m \int_0^h \left[ 1 - 3\left(\frac{x}{h}\right)^2 + 2\left(\frac{x}{h}\right)^3 \right]\left[ 3\left(\frac{x}{h}\right)^2 - 2\left(\frac{x}{h}\right)^3 \right] dx = \frac{9}{70} mh$$

$$m_{14} = m_{41} = m \int_0^h \left[ 1 - 3\left(\frac{x}{h}\right)^2 + 2\left(\frac{x}{h}\right)^3 \right]\left[ -\left(\frac{x}{h}\right)^2 + \left(\frac{x}{h}\right)^3 \right] dx$$

$$= -\frac{13}{420} mh$$

$$m_{22} = m \int_0^h \left[ \frac{x}{h} - 2\left(\frac{x}{h}\right)^2 + \left(\frac{x}{h}\right)^3 \right]^2 dx = \frac{1}{105} mh \qquad (b)$$

$$m_{23} = m_{32} = m \int_0^h \left[ \frac{x}{h} - 2\left(\frac{x}{h}\right)^2 + \left(\frac{x}{h}\right)^3 \right]\left[ 3\left(\frac{x}{h}\right)^2 - 2\left(\frac{x}{h}\right)^3 \right] dx = \frac{13}{420} mh$$

$$m_{24} = m_{42} = m \int_0^h \left[ \frac{x}{h} - 2\left(\frac{x}{h}\right)^2 + \left(\frac{x}{h}\right)^3 \right]\left[ -\left(\frac{x}{h}\right)^2 + \left(\frac{x}{h}\right)^3 \right] dx$$

$$= -\frac{1}{140} mh$$

$$m_{33} = m \int_0^h \left[ 3\left(\frac{x}{h}\right)^2 - 2\left(\frac{x}{h}\right)^3 \right]^2 dx = \frac{13}{35} mh$$

$$m_{34} = m_{43} = m \int_0^h \left[ 3\left(\frac{x}{h}\right)^2 - 2\left(\frac{x}{h}\right)^3 \right]\left[ -\left(\frac{x}{h}\right)^2 + \left(\frac{x}{h}\right)^3 \right] dx = -\frac{11}{210} mh$$

$$m_{44} = m \int_0^h \left[ -\left(\frac{x}{h}\right)^2 + \left(\frac{x}{h}\right)^3 \right]^2 dx = \frac{1}{105} mh$$

Hence, the element mass matrix is

$$[m] = \frac{mh}{420} \begin{bmatrix} 156 & 22 & 54 & -13 \\ 22 & 4 & 13 & -3 \\ 54 & 13 & 156 & -22 \\ -13 & -3 & -22 & 4 \end{bmatrix} \qquad (c)$$

Before we compute the stiffness matrix, we use Eqs. (8.25) and write

$$\{L''(x)\} = \frac{d^2}{dx^2}\{L(x)\} = \frac{2}{h^2}\begin{Bmatrix} -3 + 6\dfrac{x}{h} \\[2mm] -2 + 3\dfrac{x}{h} \\[2mm] 3 - 6\dfrac{x}{h} \\[2mm] -1 + 3\dfrac{x}{h} \end{Bmatrix} \qquad (d)$$

Introducing Eq. (d) into Eq. (8.37), we obtain the element stiffness matrix

$$[k] = \frac{4EI}{h^4}\int_0^h \begin{Bmatrix} -3 + 6\dfrac{x}{h} \\[2mm] -2 + 3\dfrac{x}{h} \\[2mm] 3 - 6\dfrac{x}{h} \\[2mm] -1 + 3\dfrac{x}{h} \end{Bmatrix}\begin{Bmatrix} -3 + 6\dfrac{x}{h} \\[2mm] -2 + 3\dfrac{x}{h} \\[2mm] 3 - 6\dfrac{x}{h} \\[2mm] -1 + 3\dfrac{x}{h} \end{Bmatrix}^T dx \qquad (e)$$

which has the entries

$$k_{11} = \frac{4EI}{h^4}\int_0^h \left(-3 + 6\frac{x}{h}\right)^2 dx = \frac{12EI}{h^3}$$

$$k_{12} = k_{21} = \frac{4EI}{h^4}\int_0^h \left(-3 + 6\frac{x}{h}\right)\left(-2 + 3\frac{x}{h}\right) dx = \frac{6EI}{h^3}$$

$$k_{13} = k_{31} = \frac{4EI}{h^4}\int_0^h \left(-3 + 6\frac{x}{h}\right)\left(3 - 6\frac{x}{h}\right) = -\frac{12EI}{h^3}$$

$$k_{14} = k_{41} = \frac{4EI}{h^4}\int_0^h \left(-3 + 6\frac{x}{h}\right)\left(-1 + 3\frac{x}{h}\right) dx = \frac{6EI}{h^3}$$

$$k_{22} = \frac{4EI}{h^4}\int_0^h \left(-2 + 3\frac{x}{h}\right)^2 dx = \frac{4EI}{h^3}$$

$$k_{23} = k_{32} = \frac{4EI}{h^4}\int_0^h \left(-2 + 3\frac{x}{h}\right)\left(3 - 6\frac{x}{h}\right) dx = -\frac{6EI}{h^3}$$  $\qquad (f)$

$$k_{24} = k_{42} = \frac{4EI}{h^4}\int_0^h \left(-2 + 3\frac{x}{h}\right)\left(-1 + 3\frac{x}{h}\right) dx = \frac{2EI}{h^3}$$

$$k_{33} = \frac{4EI}{h^3}\int_0^h \left(3 - 6\frac{x}{h}\right)^2 dx = \frac{12EI}{h^3}$$

$$k_{34} = k_{43} = \frac{4EI}{h^4} \int_0^h \left(3 - 6\frac{x}{h}\right)\left(-1 + 3\frac{x}{h}\right) dx = -\frac{6EI}{h^3}$$

$$k_{44} = \frac{4EI}{k^4} \int_0^h \left(-1 + 3\frac{x}{h}\right)^2 dx = \frac{4EI}{h^3}$$

Hence, the element stiffness matrix is

$$[k] = \frac{EI}{h^3}\begin{bmatrix} 12 & 6 & -12 & 6 \\ 6 & 4 & -6 & 2 \\ -12 & -6 & 12 & -6 \\ 6 & 2 & -6 & 4 \end{bmatrix} \tag{g}$$

To compute the nodal force vector, we first express the concentrated force in the distributed form

$$f(x, t) = P(t)\delta\left(x - \frac{h}{3}\right) \tag{h}$$

where $\delta(x - h/3)$ is a spatial Dirac delta function. Introducing Eq. (h) into Eq. (8.32), we obtain

$$\{f(t)\} = P(t) \int_0^h \delta\left(x - \frac{h}{3}\right) \left\{ \begin{array}{c} 1 - 3\left(\dfrac{x}{h}\right)^2 + 2\left(\dfrac{x}{h}\right)^3 \\[2mm] \dfrac{x}{h} - 2\left(\dfrac{x}{h}\right)^2 + \left(\dfrac{x}{h}\right)^3 \\[2mm] 3\left(\dfrac{x}{h}\right)^2 - 2\left(\dfrac{x}{h}\right)^3 \\[2mm] -\left(\dfrac{x}{h}\right)^2 + \left(\dfrac{x}{n}\right)^3 \end{array} \right\} dx$$

$$= P(t) \left\{ \begin{array}{c} 1 - 3(\tfrac{1}{3})^2 + 2(\tfrac{1}{3})^3 \\[1mm] \tfrac{1}{3} - 2(\tfrac{1}{3})^2 + (\tfrac{1}{3})^3 \\[1mm] 3(\tfrac{1}{3})^2 - 2(\tfrac{1}{3})^3 \\[1mm] -(\tfrac{1}{3})^2 + (\tfrac{1}{3})^3 \end{array} \right\} = \frac{1}{27} P(t) \left\{ \begin{array}{c} 20 \\ 4 \\ 7 \\ -2 \end{array} \right\} \tag{i}$$

## 8.4 REFERENCE SYSTEMS

We recall that, according to the finite element method, the dynamical system is regarded as an assemblage of individual discrete elements. The displacement components at the joints of any individual element are chosen in a direction that depends on the nature of the element considered. For example, in the case of a slender bar with the ends denoted by *a* and *b*, it is convenient to choose the displacement components of any one end so that one component is in the axial direction *x* and the other two in orthogonal transverse directions *y* and *z* (see Fig. 8.6). The displacement components of the ends *a* and *b* along these axes are denoted by $u_1$, $u_2$, $u_3$ and $u_4$, $u_5$, $u_6$, respectively. But the individual elements are generally parts of structural members. In turn, the structural members can be parts of a more complex structure, such as a truss. Although ordinarily a structural member is divided into a given number of finite elements, for the sake of this discussion we assume that the individual members are modeled by a single finite element each. Then, because the individual elements have different orientations in space, it becomes obvious immediately that expressing the displacements in a coordinate system particular to every such element, where such a system is often referred to as a *local coordinate system*, can create difficulties in matching the displacements at a given node. For this reason it is advisable to work with displacement components in a single set of coordinates, while retaining the advantages of identifying the displacement components of any one element with the directions most convenient for that particular element. Specifically, we wish to choose a *global reference system* $\bar{x}$, $\bar{y}$, $\bar{z}$ and denote displacement components along these directions at *a* by $\bar{u}_1$, $\bar{u}_2$, $\bar{u}_3$ and at *b* by $\bar{u}_4$, $\bar{u}_5$, $\bar{u}_6$, respectively. Then a simple coordinate transformation can resolve the displacement components along the local coordinates *x*, *y*, *z* peculiar to the element in question into components along the global reference system $\bar{x}$, $\bar{y}$, $\bar{z}$. To this end, we introduce the matrix of direction cosines

$$[l] = \begin{bmatrix} l_{x\bar{x}} & l_{x\bar{y}} & l_{x\bar{z}} \\ l_{y\bar{x}} & l_{y\bar{y}} & l_{y\bar{z}} \\ l_{z\bar{x}} & l_{z\bar{y}} & l_{z\bar{z}} \end{bmatrix} \tag{8.38}$$

where $l_{x\bar{x}}$ represents the cosine of the angle between axes *x* and $\bar{x}$, etc. This enables

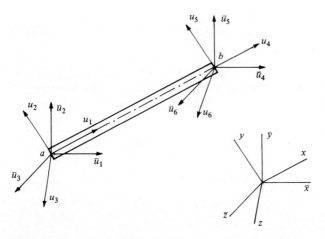

**Figure 8.6**

us to write the coordinate transformation

$$\begin{Bmatrix} x \\ y \\ z \end{Bmatrix} = [l] \begin{Bmatrix} \bar{x} \\ \bar{y} \\ \bar{z} \end{Bmatrix} \tag{8.39}$$

in which $[l]$ plays the role of a transformation matrix. The same coordinate transformation applies to displacement components, so that

$$\begin{Bmatrix} u_1 \\ u_2 \\ u_3 \end{Bmatrix} = [l] \begin{Bmatrix} \bar{u}_1 \\ \bar{u}_2 \\ \bar{u}_3 \end{Bmatrix} \qquad \begin{Bmatrix} u_4 \\ u_5 \\ u_6 \end{Bmatrix} = [l] \begin{Bmatrix} \bar{u}_4 \\ \bar{u}_5 \\ \bar{u}_6 \end{Bmatrix} \tag{8.40}$$

Equations (8.40) can be combined so as to apply to the entire element by writing simply

$$\{u\} = [L]\{\bar{u}\} \tag{8.41}$$

where $\{u\}$ and $\{\bar{u}\}$ are column matrices with elements $u_i$ $(i = 1, 2, \ldots, 6)$ and $\bar{u}_j$ $(j = 1, 2, \ldots, 6)$, respectively, and the transformation matrix $[L]$ is defined by

$$[L] = \begin{bmatrix} [l] & \vdots & [0] \\ \hline [0] & \vdots & [l] \end{bmatrix} \tag{8.42}$$

Clearly, there are different matrices $[L]$ for different elements, unless some of the elements have the same orientation in space, i.e., the local coordinates are parallel. It should be noted that matrices $[L]$ are orthonormal, $[L]^{-1} = [L]^T$, because $[l]$ represents a transformation between two orthogonal systems of axes.

In the special case of planar structures, all local systems have one axis parallel to one axis of the global system. Figure 8.7 shows the case in which axis $z$ of an arbitrary local system is parallel to axis $\bar{z}$ of the global system. In this case, the matrix of direction cosines can be verified to be

$$[l] = \begin{bmatrix} \cos \alpha & -\sin \alpha & 0 \\ \sin \alpha & \cos \alpha & 0 \\ 0 & 0 & 1 \end{bmatrix} \tag{8.43}$$

For future reference, it will prove useful to express the inertia and stiffness matrices, as well as the vector of the nodal forces, in terms of the global reference system $\bar{x}, \bar{y}, \bar{z}$ instead of the local system $x, y, z$. To this end, we recall from Sec. 8.3

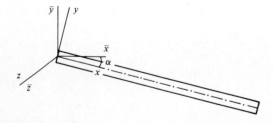

**Figure 8.7**

that the kinetic energy and potential energy can be written in the form of the triple matrix products

$$T = \tfrac{1}{2}\{\dot{u}\}^T[m]\{\dot{u}\} \tag{8.44}$$

and

$$V = \tfrac{1}{2}\{u\}^T[k]\{u\} \tag{8.45}$$

whereas the virtual work has the expression

$$\overline{\delta W} = \{\delta u\}^T\{f\} \tag{8.46}$$

But, if the local and global components of displacements are related by Eq. (8.41), then the local and global components of velocities are related by

$$\{\dot{u}\} = [L]\{\dot{\bar{u}}\} \tag{8.47}$$

and the corresponding virtual displacements are related by

$$\{\delta u\} = [L]\{\delta \bar{u}\} \tag{8.48}$$

Hence, inserting Eq. (8.47) into (8.44), we obtain

$$T = \tfrac{1}{2}\{\dot{\bar{u}}\}^T[L]^T[m][L]\{\dot{\bar{u}}\} = \tfrac{1}{2}\{\dot{\bar{u}}\}^T[\bar{m}]\{\dot{\bar{u}}\} \tag{8.49}$$

where

$$[\bar{m}] = [L]^T[m][L] \tag{8.50}$$

is the inertia matrix of the element in terms of the global coordinates $\bar{x}$, $\bar{y}$, $\bar{z}$. Moreover, using Eq. (8.41), we can write the potential energy as

$$V = \tfrac{1}{2}\{\bar{u}\}^T[L]^T[k][L]\{\bar{u}\} = \tfrac{1}{2}\{\bar{u}\}^T[\bar{k}]\{\bar{u}\} \tag{8.51}$$

where

$$[\bar{k}] = [L]^T[k][L] \tag{8.52}$$

is the stiffness matrix of the element in terms of the global coordinates. Note that $[\bar{m}]$ and $[\bar{k}]$ are symmetric because $[m]$ and $[k]$ are symmetric. In addition, inserting Eq. (8.48) into the virtual work, Eq. (8.46), we obtain

$$\overline{\delta W} = \{\delta \bar{u}\}^T[L]^T\{f\} = \{\delta \bar{u}\}^T\{\bar{f}\} \tag{8.53}$$

where

$$\{\bar{f}\} = [L]^T\{f\} \tag{8.54}$$

is recognized as the vector of the nodal forces in terms of global components.

The above expressions can be used to write the equations of motion for a single element in terms of the global reference system. This step can be skipped, however, as the interest lies in the equations of motion of the complete structure in terms of global coordinates, not merely in those of a single element.

**Example 8.3** The truss depicted in Fig. 8.8 consists of seven members, each modeled by a single finite element. Derive the inertia matrices $[\bar{m}]_i$ and the

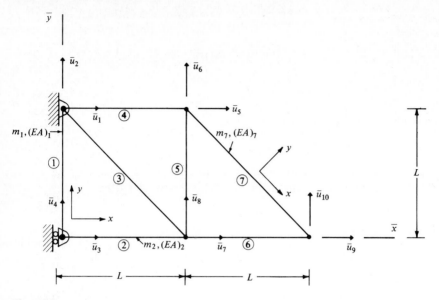

**Figure 8.8**

stiffness matrices $[\bar{k}]_i$ $(i = 1, 2, \ldots, 7)$ for the elements in terms of the global coordinates $\bar{x}, \bar{y}$ shown.

Choosing the local coordinates $x, y$ for every element as shown in Fig. 8.8, the matrices of direction cosines are simply

$$[l]_1 = [l]_2 = [l]_4 = [l]_5 = [l]_6 = \begin{bmatrix} 1 & 0 \\ 0 & 1 \end{bmatrix} \tag{a}$$

$$[l]_3 = [l]_7 = \frac{\sqrt{2}}{2} \begin{bmatrix} 1 & -1 \\ 1 & 1 \end{bmatrix}$$

from which it follows that

$$[L]_1 = [L]_2 = [L]_4 = [L]_5 = [L]_6 = \begin{bmatrix} 1 & 0 & 0 & 0 \\ 0 & 1 & 0 & 0 \\ 0 & 0 & 1 & 0 \\ 0 & 0 & 0 & 1 \end{bmatrix} \tag{b}$$

$$[L]_3 = [L]_7 = \frac{\sqrt{2}}{2} \begin{bmatrix} 1 & -1 & 0 & 0 \\ 1 & 1 & 0 & 0 \\ 0 & 0 & 1 & -1 \\ 0 & 0 & 1 & 1 \end{bmatrix}$$

The only elastic deformations experienced by a given element are ordinarily assumed to be in the axial direction. Hence, because bending is ignored, the element undergoes only rigid-body transverse displacements. It turns out that both the axial and the transverse displacement of a certain element can be expressed in terms of linear interpolation functions, using Eq. (*a*) of Example 8.1, so that the inertia matrices in terms of local coordinates can be shown to be

$$[m]_i = \frac{m_i L}{6} \begin{bmatrix} 2 & 0 & 1 & 0 \\ 0 & 2 & 0 & 1 \\ 1 & 0 & 2 & 0 \\ 0 & 1 & 0 & 2 \end{bmatrix} \qquad i = 1, 2, 4, 5, 6$$

$$[m]_i = \frac{\sqrt{2}m_i L}{6} \begin{bmatrix} 2 & 0 & 1 & 0 \\ 0 & 2 & 0 & 1 \\ 1 & 0 & 2 & 0 \\ 0 & 1 & 0 & 2 \end{bmatrix} \qquad i = 3, 7$$

(*c*)

Introducing Eqs. (*b*) and (*c*) into Eq. (8.50), we obtain

$$[\bar{m}]_i = [m]_i \qquad\qquad i = 1, 2, 4, 5, 6$$

(*d*)

$$[\bar{m}]_i = \frac{m_i L}{6\sqrt{2}} \begin{bmatrix} 2 & -2 & 1 & -1 \\ -2 & 2 & -1 & 1 \\ 1 & -1 & 2 & -2 \\ -1 & 1 & -2 & 2 \end{bmatrix} \qquad i = 3, 7$$

Similarly, using Eq. (*c*) of Example 8.1, the stiffness matrices in terms of local coordinates are

$$[k]_i = \frac{(EA)_i}{L} \begin{bmatrix} 0 & 0 & 0 & 0 \\ 0 & 1 & 0 & -1 \\ 0 & 0 & 0 & 0 \\ 0 & -1 & 0 & 1 \end{bmatrix} \qquad i = 1, 5$$

$$[k]_i = \frac{(EA)_i}{L} \begin{bmatrix} 1 & 0 & -1 & 0 \\ 0 & 0 & 0 & 0 \\ -1 & 0 & 1 & 0 \\ 0 & 0 & 0 & 0 \end{bmatrix} \qquad i = 2, 4, 6$$

(*e*)

$$[k]_i = \frac{(EA)_i}{\sqrt{2}L} \begin{bmatrix} 1 & 0 & -1 & 0 \\ 0 & 0 & 0 & 0 \\ -1 & 0 & 1 & 0 \\ 0 & 0 & 0 & 0 \end{bmatrix} \qquad i = 3, 7$$

so that, inserting Eqs. (*b*) and (*e*) into Eq. (8.52), we obtain

$$[\bar{k}]_i = [k]_i \qquad\qquad i = 1, 2, 4, 5, 6$$

(*f*)

$$[\bar{k}]_i = \frac{(EA)_i}{2\sqrt{2}L} \begin{bmatrix} 1 & -1 & -1 & 1 \\ -1 & 1 & 1 & -1 \\ -1 & 1 & 1 & -1 \\ 1 & -1 & -1 & 1 \end{bmatrix} \qquad i = 3, 7$$

## 8.5 THE EQUATIONS OF MOTION FOR THE COMPLETE SYSTEM. THE ASSEMBLING PROCESS

In Sec. 8.3 we derived the equations of motion for a single element in terms of local coordinates, and in Sec. 8.4 we showed how local coordinates can be related to global coordinates. The question remains as to how to extend the results obtained for individual elements to the complete structure.

The essence of the finite element method is to regard the continuous structure as an assemblage of individual elements. For this assemblage of individual elements to represent the structure adequately, there must be geometric compatibility at the element nodes, i.e., the displacements at the nodes shared by several elements must be the same for every such element. Moreover, the corresponding nodal forces must be statically equivalent to the applied forces. Note that the displacements include rotations and the forces include torques.

Let us assume that the complete system consists of $E$ elements and identify quantities pertaining to individual elements by the subscript $e$ $(e = 1, 2, ..., E)$. Then, considering a typical element, we denote the nodal displacement vector by $\{\bar{u}\}_e$, the nodal force vector by $\{\bar{f}\}_e$, the mass matrix by $[\bar{m}]_e$, and the stiffness matrix by $[\bar{k}]_e$, where all quantities pertain to the element $e$ and are expressed in terms of global coordinates. Next, we assume that the system has a total of $N$ nodal displacements $\bar{u}_j$ $(j = 1, 2, ..., N)$ and denote the $N$-dimensional vector of nodal displacements for the complete system by $\{\bar{U}\}$. To carry out the assembling process, we introduce the *extended element nodal displacement vector* $\{\bar{U}\}_e$, which is obtained by adding to the vector $\{\bar{u}\}_e$ as many zero components as to make the dimension of the vector $\{\bar{U}\}_e$ equal to $N$. In a similar fashion, we define the $N$-dimensional *extended element nodal force vector* $\{\bar{F}\}_e$, as well as the $N \times N$ *extended element mass matrix* $[\bar{M}]_e$ and *extended element stiffness matrix* $[\bar{K}]_e$ obtained from the corresponding element quantities by adding as many zero entries as necessary.

The equations of motion for the complete system can be obtained by an assembling process that amounts to expressing the kinetic energy, the potential energy and the virtual work in terms of contributions from the individual elements. Hence, the kinetic energy can be written in the form

$$T(t) = \frac{1}{2} \sum_{e=1}^{E} \{\dot{\bar{u}}\}_e^T [\bar{m}]_e \{\dot{\bar{u}}\}_e = \frac{1}{2} \sum_{e=1}^{E} \{\dot{\bar{U}}\}_e^T [\bar{M}]_e \{\dot{\bar{U}}\}_e$$

$$= \frac{1}{2} \{\dot{\bar{U}}\}^T [\bar{M}] \{\dot{\bar{U}}\} \tag{8.55}$$

where

$$[\bar{M}] = \sum_{e=1}^{E} [\bar{M}]_e \tag{8.56}$$

is the symmetric *mass matrix for the complete system*, which is obtained by a simple addition of the extended element mass matrices. Similarly, the potential energy can be written as

$$V(t) = \frac{1}{2} \sum_{e=1}^{E} \{\bar{u}\}_e^T [k]_e \{\bar{u}\}_e = \frac{1}{2} \sum_{e=1}^{E} \{\bar{U}\}_e^T [\bar{K}]_e \{\bar{U}\}_e$$

$$= \frac{1}{2} \{\bar{U}\}^T [\bar{K}] \{\bar{U}\} \tag{8.57}$$

where

$$[\bar{K}] = \sum_{e=1}^{E} [\bar{K}]_e \tag{8.58}$$

is the symmetric *stiffness matrix for the complete system.* Moreover, the virtual work can be expressed in the form

$$\overline{\delta W} = \sum_{e=1}^{E} \{\bar{f}\}_e^T \{\delta \bar{u}\}_e = \sum_{e=1}^{E} \{\bar{F}\}_e^T \{\delta \bar{U}\}_e = \{\bar{F}\}^T \{\delta \bar{U}\} \tag{8.59}$$

where

$$\{\bar{F}\} = \sum_{e=1}^{E} \{\bar{F}\}_e \tag{8.60}$$

is the *vector of nodal nonconservative forces for the complete system.* Using the formulation of Sec. 6.6, in conjunction with Eqs. (8.55), (8.57), and (8.59), we obtain Lagrange's *equations of motion for the complete structure* in the matrix form

$$[\bar{M}]\{\ddot{\bar{U}}\} + [\bar{K}]\{\bar{U}\} = \{\bar{F}\} \tag{8.61}$$

The vector $\{\bar{F}\}$ in Eq. (8.61) represents the vector of nonconservative nodal forces. If the system possesses viscous damping of the Rayleigh type, then the equations of motion become

$$[\bar{M}]\{\ddot{\bar{U}}\} + [\bar{C}]\{\dot{\bar{U}}\} + [\bar{K}]\{\bar{U}\} = \{\bar{F}\} \tag{8.62}$$

where now the vector $\{\bar{F}\}$ of nonconservative forces excludes viscous damping forces, and the symmetric *damping matrix for the complete structure*, $[\bar{C}]$, can be obtained by analogy with $[\bar{M}]$ or $[\bar{K}]$ in the form

$$[\bar{C}] = \sum_{e=1}^{E} [\bar{C}]_e \tag{8.63}$$

where $[\bar{C}]_e$ is a symmetric *extended element matrix of damping coefficients* associated with the element $e$.

The preceding discussion regards every element as possessing free nodes, i.e., nodes which can undergo displacements as if they were unrestrained. The implication is that the complete structure is unrestrained and capable of rigid-body motion, so that the matrix $[\bar{K}]$ is singular (see Sec. 4.12). Many structures, however, are supported so as to prevent rigid-body motion, which is reflected in the geometric boundary conditions. Other structures, such as indeterminate structures, are supported in such a manner that the displacements are zero at a number of

points exceeding the number required to prevent rigid-body motion (see Fig. 8.21).
A simple way of treating the problem in which $[\bar{K}]$ is singular and the structure is
supported so that a given number of joint displacements are zero is to eliminate
from the matrices $[\bar{M}]$, $[\bar{C}]$, $[\bar{K}]$, and $\{\bar{F}\}$ the corresponding number of rows and
columns. To illustrate the procedure, let us denote by $\{\bar{U}\}_0 = \{0\}$ the null vector
corresponding to zero displacements and by $\{\bar{U}\}_1$ the vector in $\{\bar{U}\}$ consisting of
the remaining elements, so that $\{\bar{U}\}$ can be partitioned as follows:

$$\{\bar{U}\} = \left\{ \frac{\{\bar{U}\}_0}{\{\bar{U}\}_1} \right\} = \left\{ \frac{\{0\}}{\{\bar{U}\}_1} \right\} \tag{8.64}$$

and analogous expressions exist for the velocity vector $\{\dot{\bar{U}}\}$ and the acceleration
vector $\{\ddot{\bar{U}}\}$. In a similar manner, we partition the matrices $[\bar{M}]$, $[\bar{C}]$, $[\bar{K}]$, and $\{\bar{F}\}$
by writing

$$[\bar{M}] = \left[ \begin{array}{c|c} [\bar{M}]_{00} & [\bar{M}]_{01} \\ \hline [\bar{M}]_{10} & [\bar{M}]_{11} \end{array} \right] \qquad [\bar{C}] = \left[ \begin{array}{c|c} [\bar{C}]_{00} & [\bar{C}]_{01} \\ \hline [\bar{C}]_{10} & [\bar{C}]_{11} \end{array} \right]$$

$$[\bar{K}] = \left[ \begin{array}{c|c} [\bar{K}]_{00} & [\bar{K}]_{01} \\ \hline [\bar{K}]_{10} & [\bar{K}]_{11} \end{array} \right] \qquad \{\bar{F}\} = \left[ \begin{array}{c} \{\bar{F}\}_0 \\ \hline \{\bar{F}\}_1 \end{array} \right] \tag{8.65}$$

where the dimensions of the submatrices of $[\bar{M}]$, $[\bar{C}]$, $[\bar{K}]$, and $\{\bar{F}\}$ are such that
the matrix products resulting from inserting Eqs. (8.64) and (8.65) into Eq. (8.62)
are defined. Clearly, $\{\bar{F}\}_0$ and $\{\bar{F}\}_1$ must have the same dimensions as $\{\bar{U}\}_0 = \{0\}$
and $\{\bar{U}\}_1$, respectively. It follows immediately that $[\bar{M}]_{00}$, $[\bar{C}]_{00}$, and $[\bar{K}]_{00}$ are
square matrices having the same dimensions as the dimension of $\{\bar{U}\}_0$ and $[M]_{11}$,
$[C]_{11}$, and $[K]_{11}$ are square matrices having the same dimensions as the
dimensions of $\{U\}_1$. Inserting Eqs. (8.64) and (8.65) into Eq. (8.62), we obtain the
two matrix equations

$$[\bar{M}]_{11}\{\ddot{\bar{U}}\}_1 + [\bar{C}]_{11}\{\dot{\bar{U}}\}_1 + [\bar{K}]_{11}\{\bar{U}\}_1 = \{\bar{F}\}_1 \tag{8.66}$$

and

$$[\bar{M}]_{01}\{\ddot{\bar{U}}\}_1 + [\bar{C}]_{01}\{\dot{\bar{U}}\}_1 + [\bar{K}]_{01}\{\bar{U}\}_1 = \{\bar{F}\}_0 \tag{8.67}$$

Equations (8.66) and (8.67) yield the system response as well as the reactions
for any given external excitation. Indeed, Eq. (8.66) can be solved for the nonzero
joint displacement vector $\{\bar{U}\}_1$ for any given initial excitation and external
excitation $\{\bar{F}\}_1$. On the other hand, inserting the solution $\{U\}_1$ into Eq. (8.67), we
obtain the vector $\{\bar{F}\}_0$, where $\{\bar{F}\}_0$ represents the forces associated with the null
submatrix of $\{\bar{U}\}$ and can be identified as the dynamic reaction forces due to the
motion $\{\bar{U}\}_1$. To these we must add the share of the external load originally
allocated to the points corresponding to $\{\bar{U}\}_0$ (see Example 8.4). If the reactions
present no interest, Eq. (8.67) can be ignored.

**Example 8.4** Consider the uniform circular shaft of Fig. 8.9 and derive the
equations of motion. The shaft is subjected to a uniformly distributed torque
$f(x, t) = f(t)$, and is fixed at the end $x = 0$ and free at the end $x = L$. Give a
general expression for the reaction at the fixed end.

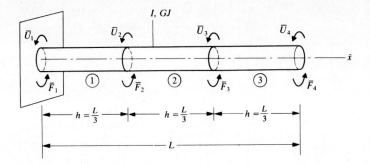

**Figure 8.9**

For simplicity, we divide the shaft into three elements of equal length $h = L/3$. The problem being one-dimensional, the orientation of the local and global coordinates is the same. Hence, from Example 8.1, we can write directly the mass and stiffness matrices for the elements in the form

$$[\bar{m}]_e = [m]_e = \frac{Ih}{6}\begin{bmatrix} 2 & 1 \\ 1 & 2 \end{bmatrix} \qquad e = 1, 2, 3 \qquad (a)$$

and

$$[\bar{k}]_e = [k]_2 = \frac{GJ}{h}\begin{bmatrix} 1 & -1 \\ -1 & 1 \end{bmatrix} \qquad e = 1, 2, 3 \qquad (b)$$

so that the extended element mass matrices are

$$[\bar{M}]_1 = \frac{Ih}{6}\begin{bmatrix} 2 & 1 & 0 & 0 \\ 1 & 2 & 0 & 0 \\ 0 & 0 & 0 & 0 \\ 0 & 0 & 0 & 0 \end{bmatrix} \qquad [\bar{M}]_2 = \frac{Ih}{6}\begin{bmatrix} 0 & 0 & 0 & 0 \\ 0 & 2 & 1 & 0 \\ 0 & 1 & 2 & 0 \\ 0 & 0 & 0 & 0 \end{bmatrix}$$

$$(c)$$

$$[\bar{M}]_3 = \frac{Ih}{6}\begin{bmatrix} 0 & 0 & 0 & 0 \\ 0 & 0 & 0 & 0 \\ 0 & 0 & 2 & 1 \\ 0 & 0 & 1 & 2 \end{bmatrix}$$

and the extended element stiffness matrices are

$$[\bar{K}]_1 = \frac{GJ}{h}\begin{bmatrix} 1 & -1 & 0 & 0 \\ -1 & 1 & 0 & 0 \\ 0 & 0 & 0 & 0 \\ 0 & 0 & 0 & 0 \end{bmatrix} \qquad [\bar{K}]_2 = \frac{GJ}{h}\begin{bmatrix} 0 & 0 & 0 & 0 \\ 0 & 1 & -1 & 0 \\ 0 & -1 & 1 & 0 \\ 0 & 0 & 0 & 0 \end{bmatrix}$$

$$(d)$$

$$[\bar{K}]_3 = \frac{GJ}{h}\begin{bmatrix} 0 & 0 & 0 & 0 \\ 0 & 0 & 0 & 0 \\ 0 & 0 & 1 & -1 \\ 0 & 0 & -1 & 1 \end{bmatrix}$$

Inserting Eqs. (c) into Eq. (8.56), we obtain the mass matrix for the complete shaft in the form

$$[\bar{M}] = \sum_{e=1}^{3} [\bar{M}]_e = \frac{Ih}{6} \begin{bmatrix} 2 & 1 & 0 & 0 \\ 1 & 4 & 1 & 0 \\ 0 & 1 & 4 & 1 \\ 0 & 0 & 1 & 2 \end{bmatrix} \tag{e}$$

Moreover, introducing Eqs. (d) into Eq. (8.58), we obtain the stiffness matrix for the complete shaft as follows:

$$[\bar{K}] = \sum_{e=1}^{3} [\bar{K}]_e = \frac{GJ}{h} \begin{bmatrix} 1 & -1 & 0 & 0 \\ -1 & 2 & -1 & 0 \\ 0 & -1 & 2 & -1 \\ 0 & 0 & -1 & 1 \end{bmatrix} \tag{f}$$

Next we must calculate the force vector. Using Eq. (8.32), in conjunction with Eqs. (8.23), we obtain

$$\bar{f}_{1e}(t) = f_{1e}(t) = \int_0^h f(t)\left(1 - \frac{x}{h}\right) dx = \tfrac{1}{2}f(t)h$$

$$\bar{f}_{2e}(t) = f_{2e}(t) = \int_0^h f(t)\frac{x}{h}\, dx = \tfrac{1}{2}f(t)h$$

$$\qquad e = 1, 2, 3 \tag{g}$$

Hence, the extended element nodal force vectors are

$$\{\bar{F}\}_1 = \tfrac{1}{2}f(t)h \begin{Bmatrix} 1 \\ 1 \\ 0 \\ 0 \end{Bmatrix} \qquad \{\bar{F}\}_2 = \tfrac{1}{2}f(t)h \begin{Bmatrix} 0 \\ 1 \\ 1 \\ 0 \end{Bmatrix} \qquad \{\bar{F}\}_3 = \tfrac{1}{2}f(t)h \begin{Bmatrix} 0 \\ 0 \\ 1 \\ 1 \end{Bmatrix} \tag{h}$$

so that the vector of nodal forces for the complete shaft is

$$\{\bar{F}\} = \tfrac{1}{2}f(t)h \begin{Bmatrix} 1 \\ 2 \\ 2 \\ 1 \end{Bmatrix} \tag{i}$$

The system equations of motion are obtained by inserting Eqs. (e), (f), and (i) into Eq. (8.61).

It is easy to see, by inspection, that matrix $[\bar{K}]$ is singular. This can be verified by considering the determinant of $[\bar{K}]$ and simply adding the second, third, and fourth row to the first, which results in a row with all its elements equal to zero. Hence, the determinant of $[\bar{K}]$ is zero, which implies that $[\bar{K}]$ is singular. This can be easily explained by the fact that matrix $[\bar{K}]$ was derived on the basis of three free-free elements, so that such a system admits rigid-body rotation. But the shaft is clamped at the left end, so that we must have $\bar{U}_1 = 0$. Hence, the discretized system possess only three degrees of freedom. Following the procedure outlined earlier, we can partition matrices (e), (f), and (i) according to the dashed lines and write the equations of motion for the nonzero coordinates

$$\frac{IL}{18}\begin{bmatrix} 4 & 1 & 0 \\ 1 & 4 & 1 \\ 0 & 1 & 2 \end{bmatrix}\begin{Bmatrix} \ddot{\bar{U}}_2 \\ \ddot{\bar{U}}_3 \\ \ddot{\bar{U}}_4 \end{Bmatrix} + \frac{3GJ}{L}\begin{bmatrix} 2 & -1 & 0 \\ -1 & 2 & -1 \\ 0 & -1 & 1 \end{bmatrix}\begin{Bmatrix} \bar{U}_2 \\ \bar{U}_3 \\ \bar{U}_4 \end{Bmatrix} = \tfrac{1}{6}f(t)L\begin{Bmatrix} 2 \\ 2 \\ 1 \end{Bmatrix}$$

$$(j)$$

The dynamic reactive torque is obtained from the equation corresponding to Eq. (8.67). This, together with the original part allocated to the point $x = 0$, yields

$$\bar{F}_0 = \tfrac{1}{6}f(t)L + \frac{IL}{18}\ddot{\bar{U}}_2 - \frac{3GJ}{L}\bar{U}_2 \qquad (k)$$

**Example 8.5** Consider the uniform bar in bending clamped at $x = 0$ and free at $x = L$, as shown in Fig. 8.10, and derive the equations for the free vibration of the system.

Dividing the bar into two equal elements of length $h = L/2$, and recognizing once again that the problem is one-dimensional, we can use Eqs. (c) and (g) of Example 8.2 and write the element mass matrices

$$[\bar{m}]_e = [m]_e = \frac{mh}{420}\begin{bmatrix} 156 & 22 & 54 & -13 \\ 22 & 4 & 13 & -3 \\ 54 & 13 & 156 & -22 \\ -13 & -3 & -22 & 4 \end{bmatrix} \qquad e = 1, 2 \qquad (a)$$

and the element stiffness matrices

$$[\bar{k}]_e = [k]_e = \frac{EI}{h^3}\begin{bmatrix} 12 & 6 & -12 & 6 \\ 6 & 4 & -6 & 2 \\ -12 & -6 & 12 & -6 \\ 6 & 2 & -6 & 4 \end{bmatrix} \qquad e = 1, 2 \qquad (b)$$

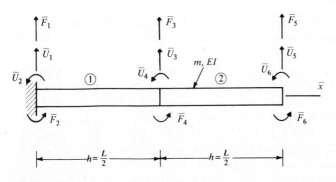

**Figure 8.10**

Hence, the extended element mass matrices are

$$[\bar{M}]_1 = \frac{mh}{420} \begin{bmatrix} 156 & 22 & 54 & -13 & 0 & 0 \\ 22 & 4 & 13 & -3 & 0 & 0 \\ 54 & 13 & 156 & -22 & 0 & 0 \\ -13 & -3 & -22 & 4 & 0 & 0 \\ 0 & 0 & 0 & 0 & 0 & 0 \\ 0 & 0 & 0 & 0 & 0 & 0 \end{bmatrix}$$

(c)

$$[\bar{M}]_2 = \frac{mh}{420} \begin{bmatrix} 0 & 0 & 0 & 0 & 0 & 0 \\ 0 & 0 & 0 & 0 & 0 & 0 \\ 0 & 0 & 156 & 22 & 54 & -13 \\ 0 & 0 & 22 & 4 & 13 & -3 \\ 0 & 0 & 54 & 13 & 156 & -22 \\ 0 & 0 & -13 & -3 & -22 & 4 \end{bmatrix}$$

and the extended element stiffness matrices are

$$[\bar{K}]_1 = \frac{EI}{h3} \begin{bmatrix} 12 & 6 & -12 & 6 & 0 & 0 \\ 6 & 4 & -6 & 2 & 0 & 0 \\ -12 & -6 & 12 & -6 & 0 & 0 \\ 6 & 2 & -6 & 4 & 0 & 0 \\ 0 & 0 & 0 & 0 & 0 & 0 \\ 0 & 0 & 0 & 0 & 0 & 0 \end{bmatrix}$$

(d)

$$[\bar{K}]_2 = \frac{EI}{h^3} \begin{bmatrix} 0 & 0 & 0 & 0 & 0 & 0 \\ 0 & 0 & 0 & 0 & 0 & 0 \\ 0 & 0 & 12 & 6 & -12 & 6 \\ 0 & 0 & 6 & 4 & -6 & 2 \\ 0 & 0 & -12 & -6 & 12 & -6 \\ 0 & 0 & 6 & 2 & -6 & 4 \end{bmatrix}$$

so that, using Eqs. (8.56) and (8.58), we obtain the mass matrix for the complete structure in the form

$$[\bar{M}] = \sum_{e=1}^{2} [\bar{M}]_e = \frac{mh}{420} \begin{bmatrix} 156 & 22 & 54 & -13 & 0 & 0 \\ 22 & 4 & 13 & -3 & 0 & 0 \\ 54 & 13 & 312 & 0 & 54 & -13 \\ -13 & -3 & 0 & 8 & 13 & -3 \\ 0 & 0 & 54 & 13 & 156 & -22 \\ 0 & 0 & -13 & -3 & -22 & 4 \end{bmatrix}$$

(e)

and the corresponding stiffness matrix in the form

$$[\bar{K}] = \sum_{e=1}^{2} [\bar{K}]_e = \frac{EI}{h^3} \begin{bmatrix} 12 & 6 & -12 & 6 & 0 & 0 \\ 6 & 4 & -6 & 2 & 0 & 0 \\ -12 & -6 & 24 & 0 & -12 & 6 \\ 6 & 2 & 0 & 8 & -6 & 2 \\ 0 & 0 & -12 & -6 & 12 & -6 \\ 0 & 0 & 6 & 2 & -6 & 4 \end{bmatrix} \quad (f)$$

Because the bar is clamped at $x = 0$, the translational and rotational displacements must be zero, $\bar{U}_1 = 0$ and $\bar{U}_2 = 0$. Hence, deleting the first and second rows and columns in $[\bar{M}]$ and $[\bar{K}]$, we can write the equations of motion

$$\frac{mL}{840} \begin{bmatrix} 312 & 0 & 54 & -13 \\ 0 & 8 & 13 & -3 \\ 54 & 13 & 156 & -22 \\ -13 & -3 & -22 & 4 \end{bmatrix} \begin{Bmatrix} \ddot{\bar{U}}_3 \\ h\ddot{\bar{U}}_4 \\ \ddot{\bar{U}}_5 \\ h\ddot{\bar{U}}_6 \end{Bmatrix}$$

$$+ \frac{16EI}{L^3} \begin{bmatrix} 12 & 0 & -6 & 3 \\ 0 & 4 & -3 & 1 \\ -6 & -3 & 6 & -3 \\ 3 & 1 & -3 & 2 \end{bmatrix} \begin{Bmatrix} \bar{U}_3 \\ h\bar{U}_4 \\ \bar{U}_5 \\ h\bar{U}_6 \end{Bmatrix} = \begin{Bmatrix} 0 \\ 0 \\ 0 \\ 0 \end{Bmatrix} \quad (g)$$

Moreover, using Eq. (8.67), the dynamic reactions can be obtained from

$$\frac{mL}{840} \begin{bmatrix} 54 & -13 & 0 & 0 \\ 13 & -3 & 0 & 0 \end{bmatrix} \begin{Bmatrix} \ddot{\bar{U}}_3 \\ h\ddot{\bar{U}}_4 \\ \ddot{\bar{U}}_5 \\ h\ddot{\bar{U}}_6 \end{Bmatrix}$$

$$+ \frac{16EI}{L^3} \begin{bmatrix} -6 & 3 & 0 & 0 \\ -3 & 1 & 0 & 0 \end{bmatrix} \begin{Bmatrix} \bar{U}_3 \\ h\bar{U}_4 \\ \bar{U}_5 \\ h\bar{U}_6 \end{Bmatrix} = \begin{Bmatrix} \bar{F}_1 \\ \bar{F}_2 \end{Bmatrix} \quad (h)$$

**Example 8.6** Consider the truss of Example 8.3 and derive the equations for the free vibration for the complete structure. Use the notation and the system properties indicated in Fig. 8.11.

The element mass matrices $[\bar{m}]_e$ and element stiffness matrices $[\bar{k}]_2$ ($e = 1, 2, \ldots, 7$) were calculated in Example 8.3. Hence, using Eqs. (d) of

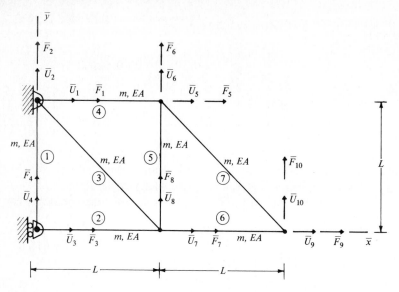

**Figure 8.11**

Example 8.3, we can write the extended element mass matrices

$$[\bar{M}]_1 = \frac{mL}{6} \begin{bmatrix} 2 & 0 & 1 & 0 & 0 & 0 & 0 & 0 & 0 & 0 \\ 0 & 2 & 0 & 1 & 0 & 0 & 0 & 0 & 0 & 0 \\ 1 & 0 & 2 & 0 & 0 & 0 & 0 & 0 & 0 & 0 \\ 0 & 1 & 0 & 2 & 0 & 0 & 0 & 0 & 0 & 0 \\ 0 & 0 & 0 & 0 & 0 & 0 & 0 & 0 & 0 & 0 \\ 0 & 0 & 0 & 0 & 0 & 0 & 0 & 0 & 0 & 0 \\ 0 & 0 & 0 & 0 & 0 & 0 & 0 & 0 & 0 & 0 \\ 0 & 0 & 0 & 0 & 0 & 0 & 0 & 0 & 0 & 0 \\ 0 & 0 & 0 & 0 & 0 & 0 & 0 & 0 & 0 & 0 \\ 0 & 0 & 0 & 0 & 0 & 0 & 0 & 0 & 0 & 0 \end{bmatrix}$$

............................................................................................ (a)

$$[\bar{M}]_7 = \frac{mL}{6\sqrt{2}} \begin{bmatrix} 0 & 0 & 0 & 0 & 0 & 0 & 0 & 0 & 0 & 0 \\ 0 & 0 & 0 & 0 & 0 & 0 & 0 & 0 & 0 & 0 \\ 0 & 0 & 0 & 0 & 0 & 0 & 0 & 0 & 0 & 0 \\ 0 & 0 & 0 & 0 & 0 & 0 & 0 & 0 & 0 & 0 \\ 0 & 0 & 0 & 0 & 2 & -2 & 0 & 0 & 1 & -1 \\ 0 & 0 & 0 & 0 & -2 & 2 & 0 & 0 & -1 & 1 \\ 0 & 0 & 0 & 0 & 0 & 0 & 0 & 0 & 0 & 0 \\ 0 & 0 & 0 & 0 & 0 & 0 & 0 & 0 & 0 & 0 \\ 0 & 0 & 0 & 0 & 1 & -1 & 0 & 0 & 2 & -2 \\ 0 & 0 & 0 & 0 & -1 & 1 & 0 & 0 & -2 & 2 \end{bmatrix}$$

Moreover, using Eqs. (*e*) of Example 8.3, we obtain the extended element stiffness matrices

$$[\bar{K}]_1 = \frac{EA}{L}\begin{bmatrix} 0 & 0 & 0 & 0 & 0 & 0 & 0 & 0 & 0 & 0 \\ 0 & 1 & 0 & -1 & 0 & 0 & 0 & 0 & 0 & 0 \\ 0 & 0 & 0 & 0 & 0 & 0 & 0 & 0 & 0 & 0 \\ 0 & -1 & 0 & 1 & 0 & 0 & 0 & 0 & 0 & 0 \\ 0 & 0 & 0 & 0 & 0 & 0 & 0 & 0 & 0 & 0 \\ 0 & 0 & 0 & 0 & 0 & 0 & 0 & 0 & 0 & 0 \\ 0 & 0 & 0 & 0 & 0 & 0 & 0 & 0 & 0 & 0 \\ 0 & 0 & 0 & 0 & 0 & 0 & 0 & 0 & 0 & 0 \\ 0 & 0 & 0 & 0 & 0 & 0 & 0 & 0 & 0 & 0 \\ 0 & 0 & 0 & 0 & 0 & 0 & 0 & 0 & 0 & 0 \end{bmatrix}$$

$$\cdots\cdots\cdots\cdots\cdots\cdots\cdots\cdots\cdots\cdots\cdots\cdots \quad (b)$$

$$[\bar{K}]_7 = \frac{EA}{2\sqrt{2}L}\begin{bmatrix} 0 & 0 & 0 & 0 & 0 & 0 & 0 & 0 & 0 & 0 \\ 0 & 0 & 0 & 0 & 0 & 0 & 0 & 0 & 0 & 0 \\ 0 & 0 & 0 & 0 & 0 & 0 & 0 & 0 & 0 & 0 \\ 0 & 0 & 0 & 0 & 0 & 0 & 0 & 0 & 0 & 0 \\ 0 & 0 & 0 & 0 & 1 & -1 & 0 & 0 & -1 & 1 \\ 0 & 0 & 0 & 0 & -1 & 1 & 0 & 0 & 1 & -1 \\ 0 & 0 & 0 & 0 & 0 & 0 & 0 & 0 & 0 & 0 \\ 0 & 0 & 0 & 0 & 0 & 0 & 0 & 0 & 0 & 0 \\ 0 & 0 & 0 & 0 & -1 & 1 & 0 & 0 & 1 & -1 \\ 0 & 0 & 0 & 0 & 1 & -1 & 0 & 0 & -1 & 1 \end{bmatrix}$$

The system mass matrix is obtained by introducing Eqs. (*a*) into Eq. (8.56). The result is

$$[\bar{M}] = \sum_{e=1}^{7} [\bar{M}]_e$$

$$= \frac{mL}{6\sqrt{2}}\begin{bmatrix} 4+4\sqrt{2} & 0 & \sqrt{2} & 0 & \sqrt{2} & 0 & 2 & 0 & 0 & 0 \\ 0 & 4+4\sqrt{2} & 0 & \sqrt{2} & 0 & \sqrt{2} & 0 & 2 & 0 & 0 \\ \sqrt{2} & 0 & 4\sqrt{2} & 0 & 0 & 0 & \sqrt{2} & 0 & 0 & 0 \\ 0 & \sqrt{2} & 0 & 4\sqrt{2} & 0 & 0 & 0 & \sqrt{2} & 0 & 0 \\ \sqrt{2} & 0 & 0 & 0 & 4+4\sqrt{2} & 0 & \sqrt{2} & 0 & 2 & 0 \\ 0 & \sqrt{2} & 0 & 0 & 0 & 4+4\sqrt{2} & 0 & \sqrt{2} & 0 & 2 \\ 2 & 0 & \sqrt{2} & 0 & \sqrt{2} & 0 & 4+6\sqrt{2} & 0 & \sqrt{2} & 0 \\ 0 & 2 & 0 & \sqrt{2} & 0 & \sqrt{2} & 0 & 4+6\sqrt{2} & 0 & \sqrt{2} \\ 0 & 0 & 0 & 0 & 2 & 0 & \sqrt{2} & 0 & 4+2\sqrt{2} & 0 \\ 0 & 0 & 0 & 0 & 0 & 2 & 0 & \sqrt{2} & 0 & 4+2\sqrt{2} \end{bmatrix}$$

$$(c)$$

Moreover, inserting Eqs. (b) into Eq. (8.58), we obtain the system stiffness matrix

$$[\bar{K}] = \sum_{e=1}^{7} [\bar{K}]_e$$

$$= \frac{EA}{2\sqrt{2}L}
\begin{bmatrix}
1+2\sqrt{2} & -1 & 0 & 0 & -2\sqrt{2} & 0 & -1 & 1 & 0 & 0 \\
-1 & 1+2\sqrt{2} & 0 & -2\sqrt{2} & 0 & 0 & 1 & -1 & 0 & 0 \\
0 & 0 & 2\sqrt{2} & 0 & 0 & 0 & -2\sqrt{2} & 0 & 0 & 0 \\
0 & -2\sqrt{2} & 0 & 2\sqrt{2} & 0 & 0 & 0 & 0 & 0 & 0 \\
-2\sqrt{2} & 0 & 0 & 0 & 1+2\sqrt{2} & -1 & 0 & 0 & -1 & 1 \\
0 & 0 & 0 & 0 & -1 & 1+2\sqrt{2} & 0 & -2\sqrt{2} & 1 & -1 \\
-1 & 1 & -2\sqrt{2} & 0 & 0 & 0 & 1+4\sqrt{2} & -1 & -2\sqrt{2} & 0 \\
1 & -1 & 0 & 0 & 0 & -2\sqrt{2} & -1 & 1+2\sqrt{2} & 0 & 0 \\
0 & 0 & 0 & 0 & -1 & 1 & -2\sqrt{2} & 0 & 1+2\sqrt{2} \\
0 & 0 & 0 & 0 & 1 & -1 & 0 & 0 & -1 & 1
\end{bmatrix}$$

$$(d)$$

From Fig. 8.11, we see that the truss is supported in such a way that $\bar{U}_1 = \bar{U}_2 = \bar{U}_3 = 0$, so that we can write the equations of motion

$$\frac{mL}{6\sqrt{2}}
\begin{bmatrix}
4\sqrt{2} & 0 & 0 & 0 & \sqrt{2} & 0 & 0 \\
0 & 4+4\sqrt{2} & 0 & \sqrt{2} & 0 & 2 & 0 \\
0 & 0 & 4+4\sqrt{2} & 0 & \sqrt{2} & 0 & 2 \\
0 & \sqrt{2} & 0 & 4+6\sqrt{2} & 0 & \sqrt{2} & 0 \\
\sqrt{2} & 0 & \sqrt{2} & 0 & 4+6\sqrt{2} & 0 & \sqrt{2} \\
0 & 2 & 0 & \sqrt{2} & 0 & 4+2\sqrt{2} & 0 \\
0 & 0 & 2 & 0 & \sqrt{2} & 0 & 4+2\sqrt{2}
\end{bmatrix}
\begin{Bmatrix}
\ddot{U}_4 \\ \ddot{U}_5 \\ \ddot{U}_6 \\ \ddot{U}_7 \\ \ddot{U}_8 \\ \ddot{U}_9 \\ \ddot{U}_{10}
\end{Bmatrix}$$

$$+ \frac{EA}{2\sqrt{2}L}
\begin{bmatrix}
2\sqrt{2} & 0 & 0 & 0 & 0 & 0 & 0 \\
0 & 1+2\sqrt{2} & -1 & 0 & 0 & -1 & 1 \\
0 & -1 & 1+2\sqrt{2} & 0 & -2\sqrt{2} & 1 & -1 \\
0 & 0 & 0 & 1+4\sqrt{2} & -1 & -2\sqrt{2} & 0 \\
0 & 0 & -2\sqrt{2} & -1 & 1+2\sqrt{2} & 0 & 0 \\
0 & -1 & 1 & -2\sqrt{2} & 0 & 1+2\sqrt{2} & -1 \\
0 & 1 & -1 & 0 & 0 & -1 & 1
\end{bmatrix}
\begin{Bmatrix}
\bar{U}_4 \\ \bar{U}_5 \\ \bar{U}_6 \\ \bar{U}_7 \\ \bar{U}_8 \\ \bar{U}_9 \\ \bar{U}_{10}
\end{Bmatrix}
=
\begin{Bmatrix}
0 \\ 0 \\ 0 \\ 0 \\ 0 \\ 0 \\ 0
\end{Bmatrix} \quad (e)$$

The calculation of the reactions is left as an exercise to the reader.

## 8.6 THE EIGENVALUE PROBLEM.
## THE FINITE ELEMENT METHOD
## AS A RAYLEIGH-RITZ METHOD

Let us consider the undamped free vibration problem associated with a continuous system discretized by the finite element method. As shown in Chap. 4, undamped free vibration executes harmonic oscillation at the system natural frequencies. To compute the natural frequencies and natural modes, we must solve the eigenvalue problem

$$[K]\{U\} = \Lambda[M]\{U\} \qquad \Lambda = \omega^2 \qquad (8.68)$$

where $[K]$ and $[M]$ are $n \times n$ stiffness and mass matrices, respectively, and where $\{U\}$ is an $n$-dimensional vector of modal coordinates, obtained by imposing the geometric boundary conditions on the $N$-dimensional free-free vector. Note that $[K]$, $[M]$, and $\{U\}$ corresponds to $[\bar{K}]_{11}$, $[\bar{M}]_{11}$, and $\{\bar{U}\}_1$ of Sec. 8.5, respectively. The $n \times n$ stiffness matrix $[K]$ and mass matrix $[M]$ depend on the stiffness and mass distribution, as well as on the interpolation functions.

To gain some insight into the nature of the eigenvalue problem derived by the finite element method, let us take a closer look at the interpolation functions. To this end, we consider the finite element approximation of a second-order system by means of linear interpolation functions, as shown in Figs. 8.12a and 8.12b. Figure 8.12a shows the entire system and Fig. 8.12b shows a typical finite element. Inside that element, the displacement has the expression

$$u(x) = \frac{jh - x}{h} u_{j-1} + \frac{x - (j-1)h}{h} u_j \qquad j = 1, 2, \ldots, n; \; (j-1)h < x < jh \quad (8.69)$$

and we note that $u_0 = 0$. It will prove convenient to introduce the notation

$$\frac{jh - x}{h} = j - \frac{x}{h} = \xi \qquad \frac{x - (j-1)h}{h} = \frac{x}{h} - (j-1) = 1 - \xi \qquad (8.70)$$

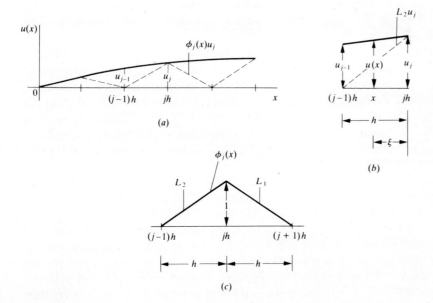

(a)

(b)

(c)

**Figure 8.12**

where $\xi$ can be interpreted as a *nondimensional local coordinate*. In view of Eqs. (8.70), Eq. (8.69) can be rewritten as

$$u(x) = L_1 u_{j-1} + L_2 u_j \qquad j = 1, 2, \ldots, n \tag{8.71}$$

where

$$L_1 = \xi \qquad L_2 = 1 - \xi \tag{8.72}$$

are recognized as the linear interpolation functions encountered earlier in this chapter.

A slightly different perspective can be obtained by combining two interpolation functions $L_1$ and $L_2$ sharing the node $j$, giving rise to a so-called *roof function* $\phi_j$ (Fig. 8.12c). We observe that the functions $\phi_j$ ($j = 1, 2, \ldots, n$) are defined over subdomains of length $2h$ of the domain $L$ of the system, they are differentiable once and they satisfy the geometric boundary condition. In general, the differentiability of the interpolation functions is consistent with the definition of the potential energy. Hence, the functions $\phi_j$ are simply *admissible functions* for the system. Moreover, the displacement can be rewritten in the form

$$u(x) = \sum_{j=1}^{n} \phi_j(x) u_j \tag{8.73}$$

which represents a linear combination of admissible functions multiplying the nodal coordinates. It follows that *the finite element method can be regarded as a Rayleigh-Ritz method*. In view of this, it will prove convenient to refer to the version of the Rayleigh-Ritz method introduced in Chap. 7 as the *classical Rayleigh-Ritz method*, as opposed to the finite element version of the Rayleigh-Ritz method. To be sure, significant differences exist between the finite element method and the classical Rayleigh-Ritz method. In the sequel, we wish to explore these differences in detail.

In the classical Rayleigh-Ritz method, the admissible functions are *global*, in the sense that they are defined over the entire domain of the system, and they tend to be complicated and hard to work with. The latter is particularly true when integrations of the admissible functions are involved. The admissible functions are all different, although they may belong to the same set of functions, such as trigonometric functions, Bessel functions, etc. Each of these sets of functions satisfies a given orthogonality relation, but this relation is in general not the one specified by the problem. The use of global admissible functions makes the use of the classical Rayleigh-Ritz method more suitable for systems with nearly uniform mass and stiffness distributions. The computation of the mass and stiffness matrices tends to be involved and tailored to the particular problem of interest at the moment. On the other hand, these matrices tend to be of relatively low order. The coefficients of the series are generally abstract in nature, and they merely represent the contribution of a particular admissible function to the displacement profile. Improvement in the accuracy of the computed solution of the eigenvalue problem is brought about by an increase in the number of terms in the series. This requires the computation of additional entries in the mass and stiffness matrices, *leaving the entries computed earlier unaffected.* Finally, because the admissible functions are generally from a complete set, *convergence to the actual solution is guaranteed.*

In the finite element method, the admissible functions are *local*, in the sense that they are defined over small subdomains of the system, and they tend to be very simple and easy to work with. In fact, for the most part they are low-degree polynomials, quite often satisfying the minimum differentiability requirements. The admissible functions are all the same for every element and they are nearly orthogonal. Indeed, from Fig. 8.12a, we conclude that $\phi_j$ and $\phi_{j+1}$ overlap over a segment of width $h$, whereas $\phi_j$ and $\phi_{j+2}$ do not overlap at all, so that $\phi_j$ and $\phi_{j+2}$ are orthogonal, no matter what the mass and stiffness distributions are. As a result, the mass and stiffness matrices tend to be banded. Moreover, their computation tends to be very easy and readily adapted to automation, as it merely consists of assembling element matrices. Because the finite element method uses local admissible functions, the method is better able to handle systems with abrupt variations in the mass and stiffness distributions. However, the finite element method tends to lead to high-order mass and stiffness matrices. The coefficients of the series are nodal coordinates and they have a great deal of physical content, as they represent displacements and slopes at the nodal points. To improve the accuracy of the computed solution of the eigenvalue problem, the width $h$ of the elements must be reduced. This requires the computation of entirely new mass and stiffness matrices. Although the number of admissible functions can be increased so as to produce a solution as accurate as desired, the local interpolation functions do not fall within the definition of a complete set, so that monotonic convergence cannot always be guaranteed.

Later in this chapter we present an approach combining the advantages of both the classical Rayleigh-Ritz method and the finite element method.

## 8.7 HIGHER-DEGREE INTERPOLATION FUNCTIONS. INTERNAL NODES

Earlier in this chapter we established that linear interpolation functions can be used for a finite element approximation of second-order systems, such as strings in transverse vibration, rods in axial vibration, and shafts in torsional vibration. Then, in Sec. 8.6, we indicated that the linear interpolation functions can be regarded as admissible functions (in a Rayleigh-Ritz sense) satisfying the minimum differentiability requirements. The question arises naturally whether some other low-degree polynomials can be used as interpolation functions. In the sequel, we propose to address this question.

Let us explore the possibility of approximating the displacement of a second-order system by a quadratic function of the form

$$u(x) = a + bx + cx^2 \tag{8.74}$$

where $a$, $b$, and $c$ are constant coefficients. In trying to determine the value of the coefficients in terms of the nodal coordinates, according to the pattern established in Sec. 8.2, we encounter the problem of determining three coefficients in terms of only two nodal coordinates. To circumvent this problem, we must add another node, which must be an internal node. For simplicity, we take the internal node at $x = h/2$, as shown in Fig. 8.13. Then, we determine the coefficients $a$, $b$, and $c$ by solving the equations

$$u(0) = u_1 = a$$

$$u\left(\frac{h}{2}\right) = u_2 = a + \frac{h}{2}b + \frac{h^2}{4}c \qquad (8.75)$$

$$u(h) = u_3 = a + hb + h^2c$$

with the result

$$a = u_1 \qquad b = \frac{1}{h}(-3u_1 + 4u_2 - u_3) \qquad c = \frac{2}{h^2}(u_1 - 2u_2 + u_3) \qquad (8.76)$$

Inserting Eqs. (8.76) into Eq. (8.74), we can write

$$u(x) = L_1(x)u_1 + L_2(x)u_2 + L_3(x)u_3 \qquad (8.77)$$

where

$$L_1(x) = 1 - 3\frac{x}{h} + 2\left(\frac{x}{h}\right)^2 \qquad L_2(x) = 4\frac{x}{h}\left(1 - \frac{x}{h}\right) \qquad L_3(x) = \frac{x}{L}\left(2\frac{x}{h} - 1\right)$$

$$(8.78)$$

are the desired quadratic interpolation functions. They can be expressed in terms of a nondimensional coordinate $\xi$ by introducing the transformation

$$1 - \frac{x}{h} = \xi \qquad \frac{x}{h} = 1 - \xi \qquad (8.79)$$

into Eqs. (8.78), which results in

$$L_1(\xi) = \xi(2\xi - 1) \qquad L_2(\xi) = 4\xi(1 - \xi) \qquad L_3(\xi) = 1 - 3\xi + 2\xi^2 \quad (8.80)$$

The quadratic interpolation functions given by Eqs. (8.80) are plotted in Fig. 8.14.

Using the same approach, we can derive cubic interpolation functions, which requires two internal nodes. This task is left as an exercise to the reader (see Prob. 8.13).

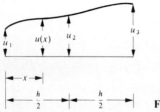

**Figure 8.13**

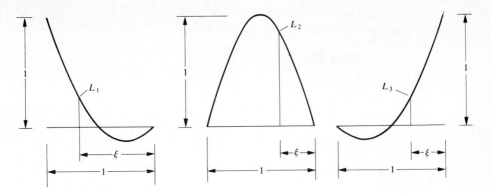

**Figure 8.14**

**Example 8.7** Derive the element mass and stiffness matrices for an element in axial vibration in terms of quadratic interpolation functions. The mass density $m$ and stiffness $EA$ can be assumed to be constant.

The element mass and stiffness matrices are still given by Eqs. (8.28) and (8.30), respectively, except that now the vector $\{L(x)\}$ has the three components given by Eqs. (8.78). The evaluation of the matrices can be simplified somewhat by working with the nondimensional coordinate $\xi$ instead of $x$. To this end, we recognize from Eqs. (8.79) that

$$dx = -h\,d\xi \qquad \frac{d}{dx} = \frac{d}{d\xi}\frac{d\xi}{dx} = -\frac{1}{h}\frac{d}{d\xi} \tag{a}$$

Moreover, when $x = h$, $\xi = 0$, and when $x = 0$, $\xi = 1$. Hence, inserting Eqs. (8.80) into Eq. (8.28), considering the first of Eqs. (a) and adjusting the integral limits, we obtain

$$[m] = m(-h)\int_{1}^{0} \left\{ \begin{array}{c} \xi(2\xi - 1) \\ 4\xi(1 - \xi) \\ 1 - 3\xi + 2\xi^2 \end{array} \right\} \left\{ \begin{array}{c} \xi(2\xi - 1) \\ 4\xi(1 - \xi) \\ 1 - 3\xi + 2\xi^2 \end{array} \right\}^T d\xi$$

$$= mh\int_{0}^{1} \left[ \begin{array}{ccc} \xi^2(2\xi - 1)^2 & 4\xi^2(2\xi - 1)(1 - \xi) & \xi(2\xi - 1)(1 - 3\xi + 2\xi^2) \\ & 16\xi^2(1 - \xi)^2 & 4\xi(1 - \xi)(1 - 3\xi + 2\xi^2) \\ \text{symm} & & (1 - 3\xi + 2\xi^2)^2 \end{array} \right] d\xi$$

$$= \frac{mh}{30} \left[ \begin{array}{ccc} 4 & 2 & -1 \\ 2 & 16 & 2 \\ -1 & 2 & 4 \end{array} \right] \tag{b}$$

Before evaluating the element stiffness matrix, we write

$$\{L'(x)\} = -\frac{1}{h}\frac{d}{d\xi}\{L(\xi)\} = -\frac{1}{h} \left\{ \begin{array}{c} 4\xi - 1 \\ 4(1 - 2\xi) \\ -3 + 4\xi \end{array} \right\} \tag{c}$$

so that

$$[k] = \frac{EA(-h)}{h^2} \int_1^0 \begin{Bmatrix} 4\xi - 1 \\ 4(1 - 2\xi) \\ -3 + 4\xi \end{Bmatrix} \begin{Bmatrix} 4\xi - 1 \\ 4(1 - 2\xi) \\ -3 + 4\xi \end{Bmatrix}^T d\xi$$

$$= \frac{EA}{h} \int_0^1 \begin{bmatrix} (4\xi - 1)^2 & 4(4\xi - 1)(1 - 2\xi) & (4\xi - 1)(-3 + 4\xi) \\ & 16(1 - 2\xi)^2 & 4(1 - 2\xi)(-3 + 4\xi) \\ \text{symm} & & (-3 + 4\xi)^2 \end{bmatrix} d\xi$$

$$= \frac{EA}{3h} \begin{bmatrix} 7 & -8 & 1 \\ -8 & 16 & -8 \\ 1 & -8 & 7 \end{bmatrix} \tag{d}$$

**Example 8.8**  Consider a uniform, free-free rod in axial vibration and derive the eigenvalue problem in two ways: (1) by using four finite elements in conjunction with linear interpolation functions and (2) by using two elements in conjunction with quadratic interpolation functions. Note that in each case there are five nodal coordinates. Solve the two eigenvalue problems, plot the modes and draw conclusions as to accuracy.

In the first case $h = L/4$, so that from Example 8.1 the element mass and stiffness matrices are

$$[m]_e = \frac{mL}{24} \begin{bmatrix} 2 & 1 \\ 1 & 2 \end{bmatrix} \qquad [k]_e = \frac{4EA}{L} \begin{bmatrix} 1 & -1 \\ -1 & 1 \end{bmatrix} \tag{a}$$

Hence, using the assembling technique described in Sec. 8.5, in the case of the linear interpolation functions the mass and stiffness matrices for the complete system are

$$[M] = \frac{mL}{24} \begin{bmatrix} 2 & 1 & 0 & 0 & 0 \\ 1 & 4 & 1 & 0 & 0 \\ 0 & 1 & 4 & 1 & 0 \\ 0 & 0 & 1 & 4 & 1 \\ 0 & 0 & 0 & 1 & 2 \end{bmatrix}$$

$$[K] = \frac{4EA}{L} \begin{bmatrix} 1 & -1 & 0 & 0 & 0 \\ -1 & 2 & -1 & 0 & 0 \\ 0 & -1 & 2 & -1 & 0 \\ 0 & 0 & -1 & 2 & -1 \\ 0 & 0 & 0 & -1 & 1 \end{bmatrix} \tag{b}$$

In the second case $h = L/2$, so that from Example 8.7 the element mass and stiffness matrices are

$$[m]_e = \frac{mL}{60} \begin{bmatrix} 4 & 2 & -1 \\ 2 & 16 & 2 \\ -1 & 2 & 4 \end{bmatrix} \qquad [k]_e = \frac{2EA}{3L} \begin{bmatrix} 7 & -8 & 1 \\ -8 & 16 & -8 \\ 1 & -8 & 7 \end{bmatrix} \tag{c}$$

Hence, once again referring to the assembling technique discussed in Sec. 8.5, in the case of quadratic interpolation functions we obtain the mass and stiffness matrices for the complete system

$$[M] = \frac{mL}{60} \begin{bmatrix} 4 & 2 & -1 & 0 & 0 \\ 2 & 16 & 2 & 0 & 0 \\ -1 & 2 & 8 & 2 & -1 \\ 0 & 0 & 2 & 16 & 2 \\ 0 & 0 & -1 & 2 & 4 \end{bmatrix}$$

$$[K] = \frac{2EA}{3L} \begin{bmatrix} 7 & -8 & 1 & 0 & 0 \\ -8 & 16 & -8 & 0 & 0 \\ 1 & -8 & 14 & -8 & 1 \\ 0 & 0 & -8 & 16 & -8 \\ 0 & 0 & 1 & -8 & 7 \end{bmatrix}$$

$$(d)$$

The eigenvalue problem based on linear interpolation functions is defined by the mass and stiffness matrices given by Eqs. (b) and has the solution

$$\Lambda_1 = 0$$

$$\{U\}_1 = \frac{1}{\sqrt{mL}} [1 \quad 1 \quad 1 \quad 1 \quad 1]^T$$

$$\Lambda_2 = 10.3866 \frac{EA}{mL^2}$$

$$\{U\}_2 = \frac{1}{\sqrt{mL}} [1.4888 \quad 1.0527 \quad 0 \quad -1.0527 \quad -1.4888]^T$$

$$\Lambda_3 = 48.0000 \frac{EA}{mL^2}$$

$$(e)$$

$$\{U\}_3 = \frac{1}{\sqrt{mL}} [1.7321 \quad 0 \quad -1.7321 \quad 0 \quad 1.7321]^T$$

$$\Lambda_4 = 126.7562 \frac{EA}{mL^2}$$

$$\{U\}_4 = \frac{1}{\sqrt{mL}} [-2.1542 \quad 1.5233 \quad 0 \quad -1.5233 \quad 2.1542]^T$$

$$\Lambda_5 = 192.0000 \frac{EA}{mL^2}$$

$$\{U\}_5 = \frac{1}{\sqrt{mL}} [1.7321 \quad -1.7321 \quad 1.7321 \quad -1.7321 \quad 1.7321]^T$$

The above eigenvectors can be used in conjunction with the linear interpolation functions given by Eqs. (8.72) to generate the approximate modes $u_r(x)$ $(r = 1, 2, \ldots, 5)$. The modes are plotted in Fig. 8.15.

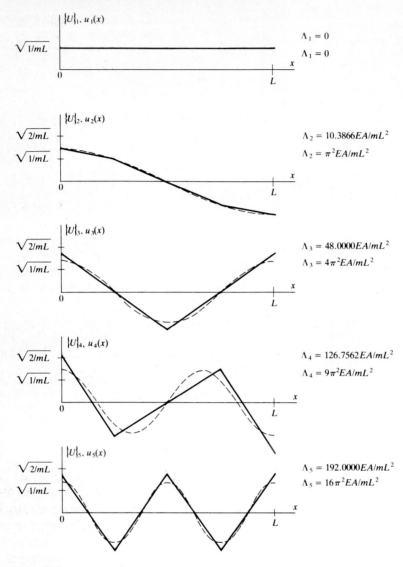

**Figure 8.15**

The eigenvalue problem based on quadratic interpolation functions is defined by the mass and stiffness matrices given by Eqs. (d) and has the solution

$$\Lambda_1 = 0$$

$$\{U\}_1 = \frac{1}{\sqrt{mL}} [1 \quad 1 \quad 1 \quad 1 \quad 1]^T$$

$$\Lambda_2 = 9.9438 \frac{EA}{mL^2}$$

$$\{U\}_2 = \frac{1}{\sqrt{mL}} [1.4228 \quad 1.0056 \quad 0 \quad -1.0056 \quad -1.4228]^T$$

$$\Lambda_3 = 48.0000 \frac{EA}{mL^2}$$

$$\{U\}_3 = \frac{1}{\sqrt{mL}} [1.7321 \quad 0 \quad -1.7321 \quad 0 \quad 1.7321]^T$$

$$\Lambda_4 = 128.7228 \frac{EA}{mL^2}$$

$$\{U\}_4 = \frac{1}{\sqrt{mL}} [-2.4445 \quad 0.9944 \quad 0 \quad -0.9944 \quad 2.4445]^T$$

$$\Lambda_5 = 240.0000 \frac{EA}{mL^2}$$

$$\{U\}_5 = \frac{1}{\sqrt{mL}} [2.2361 \quad -1.1180 \quad 2.2361 \quad -1.1180 \quad 2.2361]^T$$

(f)

The eigenvectors in (f) together with the quadratic interpolation functions given by Eqs. (8.80) yield the approximate modes shown in Fig. 8.16.

The problem considered here can actually be solved in closed form. The eigenvalues and eigenfunctions are

$$\Lambda_1 = 0 \qquad\qquad u_1(x) = \frac{1}{\sqrt{mL}}$$

(g)

$$\Lambda_r = (r-1)^2 \pi^2 \frac{EA}{mL^2} \qquad u_r(x) = \frac{2}{\sqrt{mL}} \cos \frac{(r-1)\pi x}{L} \qquad r = 2, 3, 4, 5$$

The exact eigenfunctions are plotted in Figs. 8.15 and 8.16 in dashed lines. As can be concluded from Eqs. (e) and (f) and Figs. 8.15 and 8.16, only the second eigenvalue in each case is close to the actual value and only the second mode

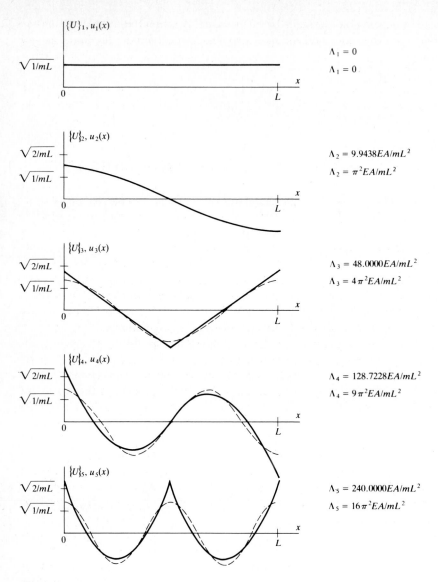

**Figure 8.16**

based on the quadratic interpolation functions resembles the actual corresponding eigenfunction, the rigid-body mode excluded. It is typical of finite element discretization that less than half the eigenvalues retain any degree of accuracy.

Although the quadratic interpolation functions lead to somewhat better results, it is clear that the solution is in need of substantial improvement, which requires a larger number of finite elements.

## 8.8 THE HIERARCHICAL FINITE ELEMENT METHOD

As pointed out in Sec. 8.6, the finite element method can be regarded as a special case of the Rayleigh-Ritz method, with the main difference between the two lying in the choice of admissible functions used in the series representation of the solution. In the classical Rayleigh-Ritz method the admissible functions are global functions, i.e., functions defined over the entire domain of the system. On the other hand, in the finite element method the admissible functions are local functions, i.e., functions defined over smaller subdomains, where these subdomains extend over a few elements, and are zero everywhere else. The local admissible functions are ordinarily very simple functions, such as low-degree polynomials.

The accuracy of the solution of the eigenvalue problem derived by the Rayleigh-Ritz method can be improved by simply increasing the number of admissible functions in the series. On the other hand, in the finite element method the accuracy is improved by refining the mesh, which amounts to increasing the number of elements. This in turn implies decreasing the width $h$ of the finite elements. For this reason, this procedure is known as the *h-version* of the finite element method. The procedure is characterized by the fact that the degree $p$ of the polynomials used in the approximation is a fixed, generally low number.

Another way of improving the accuracy of the finite element approximation is to keep $h$ constant and to increase the number of polynomials over the elements, which implies increasing the degree $p$ of the polynomials. This approach is known as the *p-version* of the finite element method. Because in the $p$-version accuracy is improved by increasing the number of admissible functions in the approximation, this version is similar to the classical Rayleigh-Ritz method. Of course, differences remain as in the classical Rayleigh-Ritz method the admissible functions used are global functions, whereas in the $p$-version of the finite element method they are local functions. This gives the $p$-version greater versatility. Moreover, the rate of convergence of the $p$-version can be higher than that of the classical Rayleigh-Ritz method or that of the $h$-version. In the $p$-version of the finite element method it is possible to choose from a variety of sets of polynomials, provided the sets are complete. Particularly desirable polynomials are the so-called *hierarchical* ones, which have the property that the set of functions corresponding to a polynomial approximation of order $p$ constitutes a subset of the set of functions corresponding to the approximation of order $p + 1$. This version is referred to as the *hierarchical finite element method* and is characterized by the fact that the mass and stiffness matrices possess the embedding property indicated by Eqs. (7.52), so that the inclusion principle holds true.† As an illustration, in the case of bending vibration the ordinarily used polynomials are the Hermite cubics

$$L_1 = 3\xi^2 - 2\xi^3 \qquad L_2 = \xi^2 - \xi^3$$
$$L_3 = 1 - 3\xi^2 + 2\xi^3 \qquad L_4 = -\xi + 2\xi^2 - \xi^3 \tag{8.81}$$

which can be obtained from Eqs. (8.25) by letting $1 - x/h = \xi$. A suitable set of hierarchical functions are the polynomials

$$L_5 = \xi^2(1 - \xi)^2$$

$$L_6 = \xi^2(1 - \xi)^2(1 - 2\xi)$$

$$L_7 = \xi^2(1 - \xi)^2(1 - 3\xi)(2 - 3\xi) \tag{8.82}$$

. . . . . . . . . . . . . . . . . . . . . . . . . . . . . . . . .

$$L_{4+i} = \xi^2(1 - \xi)^2 \prod_{j=2}^{i} (j - 1 - i\xi)$$

Note that all hierarchical functions have zero amplitudes and slopes at the nodes $\xi = 0$ and $\xi = 1$. As a result, when one hierarchical function is added, the order of the element mass and stiffness matrices thus obtained is increased by one and the original element mass and stiffness matrices are embedded in these new element mass and stiffness matrices. Hence, when one hierarchical function is added to a single element of an existing approximation, the order of the mass and stiffness matrices for the complete system increases by one and the old matrices are embedded in the new matrices. It follows that the inclusion principle is valid.

To gain some feel for the type of results that can be expected from the hierarchical finite element, let us refer to a numerical example presented in the paper by Meirovitch and Baruh cited earlier in this section. The example is concerned with the numerical solution of the eigenvalue problem for a uniform cantilever beam. The numerical results, obtained by both the hierarchical and the *h*-version of the finite element method, are summerized in Table 8.1. The first and sixth columns were obtained by the *h*-version of the finite element method and second through fifth columns by the hierarchical finite element method. Only the first five eigenvalues are listed. The eigenvalues in column one are obtained by using four elements in conjunction with Hermite cubics; the eigenvalue problem is of order eight. The eigenvalues in column two are obtained by adding $L_5$, $L_5$ and $L_6$, and $L_5$, $L_6$ and $L_7$ to one element. Similarly, columns three, four and five are obtained by adding the same hierarchical functions to two, three and all four elements. Finally, column six gives the eigenvalues obtained by the *h*-version using six, eight and ten elements. Note that in the case of ten elements the order of the eigenvalue problem is twenty, which is the same as that in which three hierarchical functions are added to all four elements. As we move from left to right and

---

† See L. Meirovitch and H. Baruh, "On the Inclusion Principle for the Hierarchical Finite Element Method," *International Journal for Numerical Methods in Engineering*, vol. 19, pp. 281–291, 1983.

**Table 8.1**

| Hermite cubics only | Four elements | | | | Six elements |
|---|---|---|---|---|---|
| | Hermite cubics and one polynomial on 4 | Hermite cubics and one polynomial on 3 and 4 | Hermite cubics and one polynomial on 2, 3, and 4 | Hermite cubics and one polynomial on all | Hermite cubics only |
| 0.14065 | 0.14065 | 0.14064 | 0.14064 | 0.14064 | 0.14064 |
| 0.88241 | 0.88221 | 0.88190 | 0.88142 | 0.88140 | 0.88160 |
| 2.48700 | 2.48687 | 2.48008 | 2.47145 | 2.46875 | 2.47240 |
| 4.90631 | 4.89846 | 4.89464 | 4.89295 | 4.85038 | 4.86724 |
| 9.12550 | 8.92140 | 8.53673 | 8.26909 | 8.01927 | 8.11453 |

| | Four elements | | | | Eight elements |
|---|---|---|---|---|---|
| | Hermite cubics and two polynomials on 4 | Hermite cubics and two polynomials on 3 and 4 | Hermite cubics and two polynomials on 2, 3, and 4 | Hermite cubics and two polynomials on all | Hermite cubics only |
| | 0.14064 | 0.14064 | 0.14064 | 0.14064 | 0.14064 |
| | 0.88220 | 0.88188 | 0.88141 | 0.88138 | 0.88145 |
| | 2.48633 | 2.47948 | 2.47074 | 2.46790 | 2.46991 |
| | 4.89359 | 4.88554 | 4.87950 | 4.83619 | 4.84691 |
| | 8.90123 | 8.51833 | 8.24920 | 7.99920 | 8.04064 |

| | Four elements | | | | Ten elements |
|---|---|---|---|---|---|
| | Hermite cubics and three polynomials on 4 | Hermite cubics and three polynomials on 3 and 4 | Hermite cubics and three polynomials on 2, 3, and 4 | Hermite cubics and three polynomials on all | Hermite cubics only |
| | 0.14064 | 0.14064 | 0.14064 | 0.14064 | 0.14064 |
| | 0.88220 | 0.88188 | 0.88141 | 0.88138 | 0.88141 |
| | 2.48633 | 2.47947 | 2.47073 | 2.46789 | 2.46852 |
| | 4.89351 | 4.88545 | 4.87941 | 4.83609 | 4.84068 |
| | 8.89793 | 8.51363 | 8.24405 | 7.99442 | 8.01453 |

from top to bottom, the eigenvalues in the first five columns improve in all three cases. The eigenvalues in the sixth column, resulting from eigenvalue problems equal in order to those in the fifth column, are never lower than those in the fifth column, thus demonstrating the effectiveness of the hierarchical finite element method. Note that, in moving from left to right, the results in the first five columns verify the validity of the inclusion principle for the hierarchical finite element method.

## 8.9 THE INCLUSION PRINCIPLE REVISITED

The inclusion principle applies to eigenvalue problems derived by the classical Rayleigh-Ritz method and by the hierarchical finite element method. The reason for this is that in both methods the embedding property is preserved for the mass and stiffness matrices, which implies that higher-order approximations can be obtained by adding a single admissible function to the series representing the solution and, moreover, the entries in the original mass and stiffness matrices remain unaffected. The question remains as to whether the inclusion principle applies to the $h$-version of the finite element method as well. It turns out that the principle applies in certain special cases, but it does not apply in general. We recall from Sec. 7.3 that the inclusion principle ensures the convergence of the Rayleigh-Ritz method.

In attempting to explore the validity of the inclusion principle in the case of the $h$-version of the finite element method, a direct approach is not very useful because of difficulties in demonstrating whether or not the embedding property holds true. Hence, our strategy is to explore the circumstances under which the $h$-version of the finite element method is equivalent to the hierarchical finite element method. To this end, we consider second-order and fourth-order systems separately.

Figure 8.17a shows an element for a second-order system together with linear interpolation functions. In Sec. 8.6 we demonstrated that the displacement can be expressed in terms of the nondimensional coordinate $\xi$ as follows:

$$u(\xi) = L_1(\xi)u_1 + L_2(\xi)u_2 \tag{8.83}$$

where

$$L_1(\xi) = \xi \qquad L_2(\xi) = 1 - \xi \tag{8.84}$$

are the linear interpolation functions. In the $h$-version of the finite element method, we subdivide the element into two elements, as shown in Fig. 8.17b. Then the displacement can be expressed as

$$u(x) = \begin{cases} L_1\left(\dfrac{\xi - h_2}{h_1}\right)u_1 + L_2\left(\dfrac{\xi - h_2}{h_1}\right)u_3 & h_2 \leqslant \xi \leqslant 1 \\ \\ L_2\left(\dfrac{\xi}{h_2}\right)u_2 + L_1\left(\dfrac{\xi}{h_2}\right)u_3 & 0 \leqslant \xi \leqslant h_2 \end{cases} \tag{8.85}$$

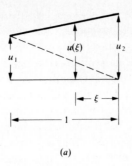

*(a)*

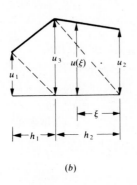

*(b)*

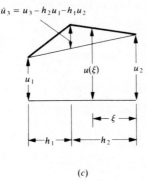

*(c)*

**Figure 8.17**

On the other hand, in the hierarchical finite element method, we simply add an extra function in the form of a triangle with the height $\bar{u}_3$, as shown in Fig. 8.17c. In this case, the displacement has the form

$$u(\xi) = L_1(\xi)u_1 + L_2(\xi)u_2 + L_3(\xi)\bar{u}_3 \qquad (8.86)$$

Considering

$$u_3 = u(h_2) = L_1(h_2)u_1 + L_2(h_2)u_2 + \bar{u}_3 \qquad h_1 + h_2 = 1$$

$$L_3(h_2) = 1 \qquad (8.87)$$

we conclude that the representations in Figs. 8.17b and 8.17c are identical, provided

$$L_3(\xi) = \begin{cases} L_1\left(\dfrac{\xi}{h_2}\right) & 0 \leqslant \xi \leqslant h_2 \\[2ex] L_2\left(\dfrac{\xi - h_2}{h_1}\right) & h_2 \leqslant \xi \leqslant 1 \end{cases} \qquad (8.88)$$

and we note that $L_3(0) = L_3(1) = 0$, so that the term $L_3(\xi)\bar{u}_3$ does not affect any element other than the one under consideration. The above proof of equivalence of the *h*-version of the finite element method and the hierarchical finite element method permits us to state the following

**Theorem** *The inclusion principle is valid for second-order systems, provided linear interpolation functions are used as admissible functions.*

The situation is entirely different in the case of fourth-order systems. It can be shown† that the *h*-version of the finite element method is equivalent to the hierarchical finite element method, provided Hermite cubics are used as admissible functions. However, even in this case the inclusion principle is not valid for fourth-order systems. The reason for this is that in fourth-order systems the equivalence is predicated on the *addition of two hierarchical functions* and not one, as required by the inclusion principle. In the reference cited above, Meirovitch and Silverberg advance two bracketing theorems for fourth-order systems to replace the classical inclusion principle. They read as follows:

**Bracketing theorem 1** *If the order of the approximation in the h-version of the finite element method is increased by subdividing one element into two, the two sets of computed eigenvalues satisfy the chains of inequalities*

$$\Lambda_1^{(n+2)} \leqslant \Lambda_1^{(n)} \leqslant \Lambda_3^{(n+2)} \leqslant \Lambda_3^{(n)} \leqslant \Lambda_5^{(n+2)} \leqslant \cdots \leqslant \Lambda_{n-1}^{(n+2)} \leqslant \Lambda_{n-1}^{(n)} \leqslant \Lambda_{n+1}^{(n+2)}$$

$$(8.89a)$$

$$\Lambda_2^{(n+2)} \leqslant \Lambda_2^{(n)} \leqslant \Lambda_4^{(n+2)} \leqslant \Lambda_4^{(n)} \leqslant \Lambda_6^{(n+2)} \leqslant \cdots \leqslant \Lambda_n^{(n+2)} \leqslant \Lambda_n^{(n)} \leqslant \Lambda_{n+2}^{(n+2)}$$

$$(8.89b)$$

**Bracketing theorem 2** *Any two adjacent eigenvalues of the lower-order approximation bracket none, one or two eigenvalues of the higher-order approximation.*

The above reference also contains a numerical example verifying the two bracketing theorems. The theorem for second-order systems and the two theorems for fourth-order systems are sufficient to ensure convergence of the *h*-version of the finite element method for the two special cases considered.

# PROBLEMS

**8.1** Show that matrices (8.10) and (8.21) can be obtained by using the definition of Sec. 4.4 for the stiffness influence coefficients.

**8.2** Consider the truss shown in Fig. 8.18, assume that each truss member can be modeled by a single finite element, let $L = \frac{3}{4}H$ and derive the equations of motion for the elements in terms of the local coordinates. Let $m_i = m$ $(i = 1, 2, \ldots, 5)$ and $(EA)_1 = (EA)_4 = EA$, $(EA)_2 = (EA)_3 = (EA)_5 = \frac{3}{2}EA$. Rewrite the equations in terms of the global coordinates $\bar{x}$, $\bar{y}$ shown.

**8.3** Repeat Prob. 8.2 but model each truss member by two finite elements.

---

† See L. Meirovitch and L. M. Silverberg, "Two Braketing Theorems Characterizing the Eigensolution for the *h*-Version of the Finite Element Method," *International Journal for Numerical Methods in Engineering*, vol. 19, pp. 1691–1704, 1983.

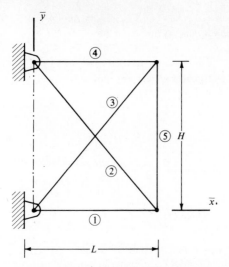

Figure 8.18

**8.4** Consider the frame of Fig. 8.19 and derive the equations of motion for the elements in terms of local coordinates by considering both axial and bending displacements. Rewrite the equations in terms of the global reference system shown in the figure.

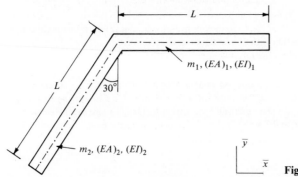

Figure 8.19

**8.5** Derive the equations of motion for the truss of Prob. 8.2 and write the equations for the dynamic reactions.

**8.6** Derive the equations of motion for the truss of Prob. 8.3 and write the equations for the dynamic reactions.

**8.7** Let the frame of Fig. 8.19 be clamped at both ends and derive the equations of motion for the system for the case $m_1 = m_2 = m$ and $(EI)_1 = (EI)_2 = EI$, $(EA)_1 = (EA)_2 = EA$. Model each member by two finite elements.

**8.8** Use the results derived in Example 8.5 and write the equations of motion for the uniform bar hinged at $x = 0$ and free at $x = L$ (see Fig. 8.20).

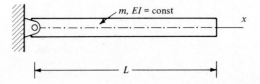

Figure 8.20

**8.9** Solve the eigenvalue problem for the truss of Prob. 8.5.

**8.10** Solve the eigenvalue problem for the frame of Prob. 8.7 for a radius of gyration $r = 0.020L$. Sketch the modes.

**8.11** Solve the eigenvalue problem for the bar of Prob. 8.8. Sketch the modes.

**8.12** Consider the system of Fig. 8.21, use the results of Example 8.5 and write down the equations of motion. Solve the corresponding eigenvalue problem and sketch the modes.

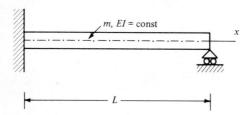

$$m, EI = \text{const}$$

$$x$$

$$L$$

**Figure 8.21**

**8.13** Use the approach of Sec. 8.7 to generate cubic interpolation functions for second-order systems. Then, derive the element mass and stiffness matrices for an element in torsional vibration in terms of the cubic interpolation functions. The polar mass moment of inertia $I$ per unit length and the torsional stiffness $GJ$ can be assumed to be constant.

**8.14** Solve the problem of Example 8.8 by using twice the number of finite elements in each case. Compare with the results obtained in Example 8.8 and draw conclusions.

**8.15** Consider a uniform shaft in torsion fixed at $x = 0$ and free at $x = L$ and derive the eigenvalue problem in three ways: (1) by using six finite elements in conjunction with linear interpolation functions, (2) by using three elements in conjunction with quadratic interpolation functions, and (3) by using two elements in conjunction with cubic interpolation functions. Solve the three eigenvalue problems, plot the modes and draw conclusion as to accuracy.

**8.16** Solve the eigenvalue problem for a uniform rod in axial vibration clamped at $x = 0$ and free at $x = L$. Model the system by the finite element method using four elements in conjunction with linear interpolation functions. Then, add the hierarchical functions $L_3(\xi) = \xi(1 - \xi)$ to (a) element 1, (b) elements 1 and 2, (c) elements 1, 2, and 3, and (d) all four elements, solve the eigenvalue problem for all four cases, and verify the inclusion principle.

**8.17** The horizontal member of the frame of Prob. 8.10 is subjected to the uniform vertical load $f(x, t) = f_0 \sin \omega t$. Derive the system reponse.

**8.18** The system of Prob. 8.11 is subjected to the impulsive force $f(x, t) = f_0 \delta(x - L)\delta(t)$ applied vertically at the end $x = L$. Derive the system response and write an expression for the dynamic reaction.

# NINE

# NONLINEAR SYSTEMS. GEOMETRIC THEORY

## 9.1 INTRODUCTION

The question as to what constitutes a linear system cannot be answered unequivocally without specifying the range over which the system is expected to operate, because the relation between the excitation and response of certain components of the system can depend on that range. For example, it was pointed out in Sec. 1.2 that when a spring is stretched or compressed, a tensile or compressive restoring force, respectively, arises. Over a given range the force-deformation relation tends to be linear, but beyond that the restoring force increases at a higher rate than the deformation for a "hardening spring," and at a lower rate for a "softening spring." Hence, a simple mass-spring oscillator must be regarded as a linear system if it operates within the linear range of the spring and as a nonlinear one if it operates beyond the linear range. Similarly, a simple pendulum must be regarded as a linear system if the amplitude $\theta$ remains sufficiently small that $\sin \theta$ can be assumed to be equal to $\theta$ itself, but must be regarded as a nonlinear system for larger amplitudes invalidating this assumption.

The study of nonlinear systems is considerably more complicated than that of linear systems, which can be attributed to the fact that the superposition principle, whereby the responses of a system to different excitations can be added linearly, is not valid for nonlinear systems. As a result, the treatment of nonlinear systems often requires entirely different methods of attack. It should be recognized at the outset that the theory of nonlinear differential equations is not nearly as well developed as that of linear differential equations, and, in fact, it relies quite heavily on approximations based upon the linear theory. Indeed, under certain circumstances, it is possible to use methods of the linear theory in the study of nonlinear systems by examining the motion in the neighborhood of known motions, a process referred to as linearization. To be sure, caution must be exercised in using this approach, as will be demonstrated later.

There are two basic approaches to nonlinear systems, namely, qualitative and quantitative. The qualitative approach is concerned with the general stability characteristics of a system in the neighborhood of a known solution, rather than with the explicit time history of the motion. On the other hand, the quantitative approach is concerned with just these time histories. Such solutions can be obtained by so-called perturbation methods or by numerical integration. This chapter is devoted to the study of qualitative methods, Chap. 10 is devoted to perturbation methods, and Sec. 12.7 to numerical integration. In our study of the qualitative behavior we shall adopt a geometric approach in describing the motion characteristics.

## 9.2 FUNDAMENTAL CONCEPTS IN STABILITY

Let us concern ourselves with an $n$-degree-of-freedom system described by the differential equations

$$\ddot{q}_i(t) = f_i(q_1, q_2, \ldots, q_n, \dot{q}_1, \dot{q}_2, \ldots, \dot{q}_n, t) \qquad i = 1, 2, \ldots, n \qquad (9.1)$$

where $f_i$ are nonlinear functions of the generalized coordinates $q_i(t)$, generalized velocities $\dot{q}_i(t)$, and time $t$. Physically, the functions $f_i$ represent generalized forces per unit mass. They are not restricted to elastically restoring forces or viscous damping forces, as were almost all the forces encountered in Chaps. 1 through 8, but are to be regarded as of a more general nature.

The solution of Eqs. (9.1) depends on the initial conditions $q_i(0)$, $\dot{q}_i(0)$ $(i = 1, 2, \ldots, n)$, and can be given a geometric interpretation by imagining an $n$-dimensional cartesian space defined by the variables $q_i$ and known as the *configuration space*. For a given value of time, the solutions $q_i(t)$ can be represented by an $n$-dimensional vector in that space, with the tip of the vector defining a point $P$ called the *representative point*. With time, point $P$ traces a curve, or a path, in the configuration space showing how the solution of the system varies with time, although the time may appear only implicitly. As a simple illustration, we can envision the planar motion of an artillery shell, where the motion is given by a curve in the $xy$ plane, with the time $t$ playing the role of a parameter. Different paths are obtained corresponding to different initial conditions; and in certain situations some of these paths may intersect, which implies that to the same position correspond different velocities, and hence different slopes. Because this geometric description does not have a unique slope for the trajectory at any given point, we wish to consider a different space which does not suffer from this drawback.

Equations (9.1) constitute a system of $n$ second-order Lagrangian differential equations of motion in the variables $q_i(t)$ $(i = 1, 2, \ldots, n)$. It is possible to use a set of auxiliary variables in the form of the generalized momenta defined by $p_i = \partial L / \partial \dot{q}_i$, where $L$ is the Lagrangian, and convert Eqs. (9.1) into a system of $2n$

first-order Hamiltonian differential equations of motion in the variables $q_i(t)$, $p_i(t)$ ($i = 1, 2, \ldots, n$) (see the text by Meirovitch†). Then we can describe the solution of the dynamical system in the $2n$-dimensional space defined by $q_i$ and $p_i$ and called the *phase space*. But generalized momenta are related linearly to generalized velocities. Hence, an alternative phase space is that defined by $q_i$ and $\dot{q}_i$. We shall use the latter definition, and, to this end, introduce the notation

$$
\begin{array}{ll}
q_i = x_i & \dot{q}_i = x_{n+i} \\
x_{n+i} = X_i & f_i = X_{n+i}
\end{array} \qquad i = 1, 2, \ldots, n \tag{9.2}
$$

so that Eqs. (9.1), together with the definition of the auxiliary variables $\dot{q}_i$, yield the $2n$ first-order differential equations

$$
\dot{x}_i(t) = X_i(x_1, x_2, \ldots, x_{2n}, t) \qquad i = 1, 2, \ldots, 2n \tag{9.3}
$$

The quantities $x_i$ and $X_i$ can be regarded as the components of $2n$-dimensional vectors **x** and **X**, which can be represented by the column matrices $\{x\}$ and $\{X\}$, respectively. The vector **x**$(t)$ defines the state of the system uniquely for any time $t$ and is often referred to as the *state vector*; analogously, the space defined by **x** is also known as the *state space*. Using matrix notation, Eqs. (9.3) can be written in the compact form

$$
\{\dot{x}\} = \{X\} \tag{9.4}
$$

For a certain set of initial conditions $x_i(0) = \alpha_i$ ($i = 1, 2, \ldots, 2n$), where $\alpha_i$ are given constants, the set of Eqs. (9.3), or Eq. (9.4), has the unique solution

$$
x_i(t) = \phi_i(\alpha_1, \alpha_2, \ldots, \alpha_{2n}, t) \qquad i = 1, 2, \ldots, 2n \tag{9.5}
$$

For different sets of initial conditions $\alpha_i$, Eqs. (9.5) yield different solutions that can be represented in the phase space by corresponding paths. The totality of paths, representing all possible solutions, is referred to as the *phase portrait*. The phase portrait has an orderly appearance, with all trajectories having unique slopes at any point, so that no two paths intersect, except at certain points to be discussed shortly.

If the time $t$ is regarded as an additional coordinate, then it is possible to introduce a $(2n + 1)$-dimensional space defined by $x_1, x_2, \ldots, x_{2n}, t$ and known as the *motion space*. The motion in that space can be visualized as a fluid flow, with the fluid velocity at any point $(\mathbf{x}, t)$ defined uniquely by the vector **X**. The integral curves (9.5) in the motion space corresponding to various sets of initial conditions $\alpha_i$ are called *characteristics*.

When none of the functions $X_i$ ($i = 1, 2, \ldots, 2n$) depends explicitly on the time $t$, the system is said to be *autonomous*. If at least one of the functions $X_i$ contains the time explicitly, the system is *nonautonomous*. In the autonomous case the fluid flow analogy implies that the flow is steady. More important, however, is the fact that when the system is autonomous the characteristic curves in the motion space can be projected onto the phase space, where the projected paths are called *trajectories* and represent the system motion without regard to time. This is another way of

---

† L. Meirovitch, *Methods of Analytical Dynamics*, sec. 2.13, McGraw-Hill Book Co., New York, 1970.

saying that the time can be eliminated from the problem formulation, so that its role is reduced to that of a parameter. The trajectories corresponding to $t \geqslant 0$ are called *positive half-trajectories* and those corresponding to $t \leqslant 0$ are *negative half-trajectories*. We shall concern ourselves primarily with positive half-trajectories.

A point for which $\{X\}^T\{X\} = \sum_{i=1}^{2n} X_i^2 > 0$ is referred to as an *ordinary point*, or *regular point*. On the other hand, a point for which $\{X\} = \{0\}$ is called a *singular point*, or an *equilibrium point*. Recognizing that the vector $\{x\}$ consists of both displacements and velocities, and that at a point for which $\{X\}$ is zero $\{\dot{x}\}$ vanishes, we conclude that the velocities and accelerations are zero at a singular point, which explains why such a point is called an equilibrium point. If in a given neighborhood there is only one equilibrium point, then the point in question is said to be an *isolated equilibrium point*. In this text we are concerned only with isolated equilibrium points. Because at an equilibrium point $\{\dot{x}\} = \{0\}$, with the implication that the solution must be constant at that point, another definition of an equilibrium point is a set of constants $\alpha_i$ satisfying the equations

$$\phi_i(\alpha_1, \alpha_2, \dots, \alpha_{2n}, t) = \alpha_i \qquad i = 1, 2, \dots, 2n \tag{9.6}$$

It should be pointed out that, because at an equilibrium point the velocities and accelerations are zero, from a mathematical point of view a particle moving along a trajectory can approach an equilibrium point on the trajectory only for $t \to \pm\infty$. In practice it can approach the equilibrium point for reasonably large values of time, positive or negative.

Next let us consider a given solution $x_i = \phi_i$ $(i = 1, 2, \dots, 2n)$ of Eqs. (9.3), and refer to it as the *unperturbed solution*. The interest lies in the motion $x_i(t)$ in the neighborhood of $\phi_i(t)$, where $x_i(t)$ is called the *perturbed motion. There are two classes of unperturbed solutions that are of particular interest, namely, constant solutions and periodic solutions. The first class corresponds to an equilibrium point and the second to a closed trajectory.* We shall discuss both cases.

In the special case in which $\phi_i(t) \equiv 0$ $(i = 1, 2, \dots, n)$ the unperturbed solution is referred to as the *null*, or *trivial*, *solution*. In this case the equilibrium point coincides with the origin of the phase space. In the general case, however, we can introduce the *perturbations* $y_i(t)$ from the given solution $\phi_i(t)$ in the form

$$y_i(t) = x_i(t) - \phi_i(t) \qquad i = 1, 2, \dots, 2n \tag{9.7}$$

Inserting Eqs. (9.7) into (9.3), we can write

$$\dot{y}_i(t) + \dot{\phi}_i(t) = X_i(y_1 + \phi_1, y_2 + \phi_2, \dots, y_{2n} + \phi_{2n}, t) \qquad i = 1, 2, \dots, 2n \tag{9.8}$$

Because $\phi_i(t)$ are solutions of Eqs. (9.3), they must satisfy

$$\dot{\phi}(t) = X_i(\phi_1, \phi_2, \dots, \phi_{2n}, t) \qquad i = 1, 2, \dots, 2n \tag{9.9}$$

so that, introducing the notation

$$Y_i(y_1, y_2, \dots, y_{2n}, t) = X_i(y_1 + \phi_1, y_2 + \phi_2, \dots, y_{2n} + \phi_{2n}, t)$$

$$- X_i(\phi_1, \phi_2, \dots, \phi_{2n}, t) \qquad i = 1, 2, \dots, 2n \tag{9.10}$$

Eqs. (9.8) can be written as

$$\dot{y}_i(t) = Y_i(y_1, y_2, \ldots, y_{2n}, t) \qquad i = 1, 2, \ldots, 2n \qquad (9.11)$$

where Eqs. (9.11) are referred to as the *differential equations of the perturbed motion*. From Eqs. (9.10), however, we observe that if $y_i \equiv 0$ $(i = 1, 2, \ldots, 2n)$, then $Y_i(0, 0, \ldots 0, t)$ reduce to zero for all $i$. Hence, if we imagine a phase space defined by $y_i$, then we conclude that the origin of that space is an equilibrium point.

When $\phi_i$ are equal to a set of constants, say $\alpha_i$, the origin of the phase space can be made to coincide with the equilibrium point $x_i = \phi_i$ by a coordinate transformation representing simple translation, so that once again the unperturbed motion is the trivial solution.

Of particular interest in mechanics is the problem of stability of motion of dynamical systems when they are perturbed from an equilibrium state. Before stability can be defined more precisely, it is necessary to introduce a quantity that can serve as a measure of the amplitude of motion (in a general sense) at any time $t$. In view of the preceding discussion, if we assume that the origin of the phase space defined by $x_i$ $(i = 1, 2, \ldots, 2n)$ coincides with an equilibrium point, then the problem reduces to the stability of the trivial solution. In this case a measure of the amplitude of motion is simply the distance from the origin to any point on the integral curve $\mathbf{x}(t)$. A measure of this distance is provided by the *Euclidean norm*, or *Euclidean length*, of the vector $\mathbf{x}$, defined by $\|\mathbf{x}\| = (\{x\}^T\{x\})^{1/2} = (\Sigma_{i=1}^{2n} x_i^2)^{1/2}$. Then a sphere of radius $r$ with the center at the origin of the phase space can be written simply as $\|\mathbf{x}\| = r$, and the domain enclosed by that sphere as $\|\mathbf{x}\| < r$. There are many definitions of stability. We give here only the most frequently used ones.

Assuming that the origin is an equilibrium point, the definitions due to Liapunov can be stated as follows:

1. The trivial solution is *stable in the sense of Liapunov* if for any arbitrary positive quantity $\epsilon$ there exists a positive quantity $\delta$ such that the satisfaction of the inequality

$$\|\mathbf{x}_0\| < \delta \qquad (9.12)$$

   implies the satisfaction of the inequality

$$\|\mathbf{x}(t)\| < \epsilon \qquad 0 \leqslant t < \infty \qquad (9.13)$$

   where $\mathbf{x}_0 = \mathbf{x}(0)$.

2. The trivial solution is *asymptotically stable* if it is Liapunov stable and in addition

$$\lim_{t \to \infty} \|\mathbf{x}(t)\| = 0 \qquad (9.14)$$

3. The trivial solution is *unstable* if it is not stable.

Geometrically, the trivial solution is stable if any motion initiated inside the sphere $\|\mathbf{x}\| = \delta$ remains inside the sphere $\|\mathbf{x}\| = \epsilon$ for all times. If the motion approaches the origin as $t \to \infty$, the trivial solution is asymptotically stable, and if it reaches the boundary of the sphere $\|x\| = \epsilon$ in finite time, it is unstable. The three possibilities are illustrated in Fig. 9.1a.

The preceding definitions are concerned with the stability of the trivial solution and preclude other types of equilibrium that must be considered stable, namely, equilibrium motions associated with periodic phenomena. In this case, the unperturbed solutions $\phi_i(t)$ $(i = 1, 2, \ldots, 2n)$ are periodic functions of time represented by closed trajectories in the phase space. Denoting a given closed trajectory by $C$, stability must be interpreted in terms of the behavior of every trajectory in the neighborhood of $C$. In particular, if every trajectory in the neighborhood of $C$ remains in the neighborhood of $C$, then the unperturbed motion is said to be *orbitally stable*. If the trajectories approach $C$ as $t \to \infty$, the unperturbed motion is said to be *asymptotically orbitally stable* (see Fig. 9.1b). On the other hand, if there are trajectories that tend to leave the neighborhood of $C$ (or approach $C$ as $t \to -\infty$), the unperturbed motion is *orbitally unstable*. Orbital stability is also referred to as *stability in the sense of Poincaré*, and is associated with closed trajectories generally known as *limit cycles* (see Sec. 9.6). It should be pointed out that most of the theory concerning limit cycles is confined to second-order systems.

If the vector $x(t)$ representing an integral curve of system (9.4) is such that $\|x(t)\| \leqslant r$ for some $r$, then the integral curve is said to be *bounded*. *Stability in the sense of Lagrange* requires only that the solution be bounded.

When a nonlinear system can be approximated by a linearized one, a stable system is referred to as *infinitesimally stable* (see Sec. 9.3).

**Example 9.1** As an illustration of the geometric description of motion, let us consider a simple pendulum. The differential equation of motion of the simple pendulum can be shown to have the form

$$\ddot{\theta} + \omega^2 \sin \theta = 0 \qquad \omega^2 = \frac{g}{L} \tag{a}$$

where $g$ is the acceleration due to gravity and $L$ the length of the pendulum. To

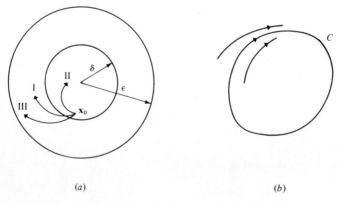

(a)          (b)

**Figure 9.1**

use the analogy with Eqs. (9.2), we introduce the notation

$$\theta = x_1 \qquad\qquad \dot\theta = x_2$$
$$x_2 = X_1 \qquad -\omega^2 \sin x_1 = X_2 \qquad\qquad (b)$$

so that the second-order differential equation (a) can be written in the form of the two first-order differential equations

$$\dot x_1 = X_1 = x_2 \qquad \dot x_2 = X_2 = -\omega^2 \sin x_1 \qquad\qquad (c)$$

From Eqs. (c), we conclude that the system has equilibrium points at

$$x_1 = \pm j\pi \qquad j = 0, 1, 2, \dots \qquad x_2 = 0 \qquad\qquad (d)$$

so that the origin is one of the equilibrium points. Because the right side of Eqs. (c) does not depend explicitly on time, the system is autonomous.

To obtain the trajectories of the system, we eliminate the time by dividing the second of Eqs. (c) by the first, with the result

$$\frac{dx_2}{dx_1} = -\frac{\omega^2 \sin x_1}{x_2} \qquad\qquad (e)$$

Equation (e) can be rearranged in the form

$$x_2 \, dx_2 = -\omega^2 \sin x_1 \, dx_1 \qquad\qquad (f)$$

which yields the integral

$$\tfrac{1}{2}x_2^2 + \omega^2(1 - \cos x_1) = E = \text{const} \qquad\qquad (g)$$

where $E$ is a constant proportional to the system total energy, and its value depends on the initial conditions. Equation (g) represents the equation of the trajectories. This being a second-order system, the phase space reduces to the phase plane defined by $x_1$ and $x_2$. By varying the value of $E$ we can obtain the phase portrait. Figure 9.2 shows a phase portrait limited to three trajectories that typify the various possible motions of the pendulum, as explained below.

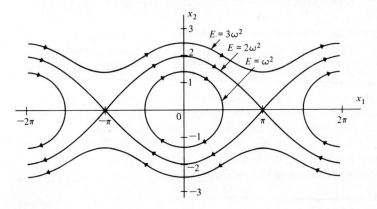

**Figure 9.2**

Figure 9.2 will now be used to interpret the motion in a qualitative way. We notice that for $E < 2\omega^2$ we obtain closed trajectories, so that the motion repeats itself. The implication is that for $E < 2\omega^2$ the motion is periodic but not necessarily harmonic; it is harmonic only for small amplitudes. In fact, *for relatively large amplitudes the period depends on the amplitude*, while it is a known fact that the period of a harmonic oscillator is $T = 2\pi/\omega = 2\pi\sqrt{L/g}$, a constant independent of amplitude. It should be pointed out that the system reduces to a harmonic oscillator only when $\sin x_1$ can be replaced by $x_1$. Hence, in general, periodic motion takes place for values of $x_1$ such that $-\pi < x_{1\,max} < \pi$, where $x_{1\,max}$ denotes the maximum angular displacement of the pendulum. From the first of Eqs. (c) and solution (g), we can write the period in the form

$$T = 4 \int_0^{x_{1\,max}} \frac{dx_1}{\{2[E - \omega^2(1 - \cos x_1)]\}^{1/2}} \qquad (h)$$

By letting $x_1 = 2z$, $x_{1\,max} = 2z_{max}$, Eq. (h) reduces to

$$T = \frac{8}{\sqrt{2E}} \int_0^{z_{max}} \frac{dz}{(1 - k^2 \sin^2 z)^{1/2}}, \qquad k^2 = \frac{2\omega^2}{E} \qquad (i)$$

which represents an elliptic integral of the first kind, whose value can be obtained from tables.† Equation (i) clearly shows that $T$ depends on $E$, which in turn controls the maximum angular displacement $x_{1\,max} = 2z_{max}$ through Eq. (g).

For $E > 2\omega^2$ the trajectories are open and the motion of the pendulum is nonuniformly rotary, with the pendulum going over the top. The highest velocity is obtained for $x_1 = \pm 2j\pi$ ($j = 0, 1, 2, \ldots$) and the lowest for $x_1 = \pm(2j + 1)\pi$ ($j = 0, 1, 2, \ldots$). As pointed out earlier, for $E < 2\omega^2$ the trajectories are closed and the motion periodic. For $E \to 0$ the trajectories become ellipses with the centers at $x_1 = \pm 2j\pi$ ($j = 0, 1, 2, \ldots$), $x_2 = 0$ and the motion becomes harmonic. For $E = 0$ the trajectories reduce to the equilibrium points $x_1 = \pm 2j\pi$ ($j = 0, 1, 2, \ldots$), $x_2 = 0$, with the implication that there is no motion for $E = 0$.

The trajectories corresponding to $E = 2\omega^2$, intersecting at the equilibrium points $x_1 = \pm(2j + 1)\pi$ ($j = 0, 1, 2, \ldots$), $x_2 = 0$, separate the two types of motion, namely, oscillatory and rotary, for which reason these trajectories are called *separatrices*.

We recognize that for $E \leqslant 2\omega^2$ we obtain the equilibrium points given by (d), corresponding to positions for which the pendulum is aligned with the vertical. Specifically, for $E < 2\omega^2$ we obtain the equilibrium points $x_1 = \pm 2j\pi$ ($j = 0, 1, 2, \ldots$), $x_2 = 0$, with the pendulum pointing downward, and for $E = 2\omega^2$ we obtain the equilibrium points $x_1 = \pm(2j + 1)\pi$ ($j = 0, 1, 2, \ldots$), $x_2 = 0$, with the pendulum pointing upward. Although mathematically we

† See, for example, B. O. Peirce and R. M. Foster, *A Short Table of Integrals*, 4th ed., p. 134, Ginn and Company, Boston, 1957.

obtain different equilibrium points for different values of the integer $j$, physically there are only two equilibrium points, namely, $x_1 = 0$, $x_2 = 0$ and $x_1 = \pi$, $x_2 = 0$. A more general discussion of trajectories and equilibrium points is presented in Sec. 9.3.

It should be pointed out that *in all systems for which the velocities are introduced as auxiliary variables, the equilibrium points are located on the axes designating coordinates.* This can be easily verified by examining Fig. 9.2, where all equilibrium points are on the $x_1$ axis.

## 9.3 SINGLE-DEGREE-OF-FREEDOM AUTONOMOUS SYSTEMS. PHASE PLANE PLOTS

The usefulness of the geometric theory of nonlinear systems is limited largely to low-order autonomous systems, although some of the concepts can be extended to higher-order systems. The geometric theory is particularly useful for second-order systems, because for such systems the phase space reduces to a phase plane, permitting two-dimensional trajectory plots. This fact was already established in Example 9.1, but in this section we propose to expand on and generalize many of the ideas presented there. The foundation for the geometric theory of nonlinear systems was laid to a large extent by Poincaré.

Let us consider a single-degree-of-freedom autonomous system described by the two first-order differential equations

$$\dot{x}_1 = X_1(x_1, x_2) \qquad \dot{x}_2 = X_2(x_1, x_2) \qquad (9.15)$$

where $X_1$ and $X_2$ are generally nonlinear functions of the state variables $x_1$ and $x_2$, possessing first-order partial derivatives with respect to these variables. Because the system is autonomous, which is reflected in the fact that the right side of Eqs. (9.15) does not contain the time explicitly, the time dependence can be eliminated altogether by dividing the second of Eqs. (9.15) by the first, with the result

$$\frac{dx_2}{dx_1} = \frac{X_2(x_1, x_2)}{X_1(x_1, x_2)} \qquad X_1(x_1, x_2) \neq 0 \qquad (9.16)$$

where Eq. (9.16) gives the tangent to the trajectories at any point in the phase plane without reference to time, with the exception of points at which $X_1$ and $X_2$ are zero simultaneously, which are by definition equilibrium points. Hence, Eq. (9.16) determines the tangent to the trajectories uniquely at any ordinary point of the phase plane, but not at equilibrium points. To obtain the direction of motion along a given trajectory in the phase plane, we must refer back to Eqs. (9.15).

It was pointed out in Sec. 9.2 that two integral curves never intersect, except perhaps at equilibrium points. It follows that *through any regular point of the phase plane there passes at most one trajectory*, so that *two trajectories have no ordinary point in common.*

A problem of particular interest is the nature of motion in the neighborhood of an equilibrium point. Denoting the coordinates of an equilibrium point by $x_1 = \alpha_1$,

$x_2 = \alpha_2$, where $\alpha_1$ and $\alpha_2$ are constants, it follows that these values must satisfy the algebraic equations

$$X_1(\alpha_1, \alpha_2) = 0 \qquad X_2(\alpha_1, \alpha_2) = 0 \tag{9.17}$$

Because $X_1$ and $X_2$ are generally nonlinear, there can be more than one solution of Eqs. (9.17), a fact that can be verified from Example 9.1. We are concerned with one particular case, namely, that in which the equilibrium point coincides with the origin of the phase plane, $\alpha_1 = \alpha_2 = 0$. There is no loss of generality in this, because the origin can be always translated by means of a coordinate transformation so as to cause it to coincide with an equilibrium point. Expanding a Taylor's series for $X_1$ and $X_2$ in the neighborhood of the origin, we can write Eqs. (9.15) in the form

$$\dot{x}_1 = a_{11}x_1 + a_{12}x_2 + \epsilon_1(x_1, x_2) \qquad \dot{x}_2 = a_{21}x_1 + a_{22}x_2 + \epsilon_2(x_1, x_2) \tag{9.18}$$

where the coefficients $a_{ij}$ have the expressions

$$a_{ij} = \left. \frac{\partial X_i}{\partial x_j} \right|_{x_j = 0} \qquad i, j = 1, 2 \tag{9.19}$$

which explains why the functions $X_i$ ($i = 1, 2$) must possess first-order partial derivatives with respect to $x_1$ and $x_2$. The functions $\epsilon_1$ and $\epsilon_2$ are nonlinear, which implies that they are at least of degree 2 in $x_1$ and $x_2$. Introducing the matrix notation

$$\{x\} = \begin{Bmatrix} x_1 \\ x_2 \end{Bmatrix} \qquad \{\epsilon\} = \begin{Bmatrix} \epsilon_1 \\ \epsilon_2 \end{Bmatrix} \qquad [a] = \begin{bmatrix} a_{11} & a_{12} \\ a_{21} & a_{22} \end{bmatrix} \tag{9.20}$$

Eqs. (9.18) can be written in the compact form

$$\{\dot{x}\} = [a]\{x\} + \{\epsilon\} \tag{9.21}$$

The differential equations represented by (9.21) are referred to as the *complete nonlinear equations* of the system. Assuming that the functions $\epsilon_1$ and $\epsilon_2$ are negligibly small in the neighborhood of the origin, it is reasonable to expect that Eq. (9.21) can be approximated by

$$\{\dot{x}\} = [a]\{x\} \tag{9.22}$$

where the equations represented by (9.22) are called the *linearized equations*. An analysis based on the linearized equations, Eq. (9.22), instead of the complete nonlinear equations, Eq. (9.21), is referred to as an *infinitesimal analysis*. The infinitesimal analysis can generally be expected to yield reliable information concerning the nature of motion in the neighborhood of the origin. There are cases, however, when the linearized equations do not provide conclusive information concerning the behavior of the complete nonlinear system. These cases are discussed later.

The behavior of the system in the neighborhood of the origin depends on the eigenvalues of the matrix $[a]$. To show this, let the solution of Eq. (9.22) have the form

$$\{x(t)\} = e^{\lambda t}\{x_0\} \tag{9.23}$$

where $\{x_0\}$ is a constant column matrix. Inserting solution (9.23) into Eq. (9.22), and dividing through by $e^{\lambda t}$, we obtain the eigenvalue problem

$$\lambda\{x_0\} = [a]\{x_0\} \tag{9.24}$$

leading to the characteristic equation

$$\det([a] - \lambda[1]) = 0 \tag{9.25}$$

Equation (9.25) has two solutions, $\lambda_1$ and $\lambda_2$, which are recognized as the eigenvalues of the matrix $[a]$. The type of motion obtained depends on the nature of the roots $\lambda_1$ and $\lambda_2$ of the characteristic equation. We note that for Eq. (9.25) to have nonzero roots, we must have $\det[a] \neq 0$, or the *matrix $[a]$ must be nonsingular*.

The solution of Eq. (9.22) is conveniently discussed by introducing the linear transformation

$$\{x(t)\} = [b]\{u(t)\} \tag{9.26}$$

where $[b]$ is a constant nonsingular matrix. Introducing Eq. (9.26) into Eq. (9.22) and premultiplying the result by $[b]^{-1}$, we obtain

$$\{\dot{u}\} = [c]\{u\} \tag{9.27}$$

where

$$[c] = [b]^{-1}[a][b] \tag{9.28}$$

Equation (9.28) represents a *similarity transformation*, and matrices $[c]$ and $[a]$ are said to be *similar*. Systems (9.22) and (9.27) have the same dynamic characteristics, because matrices $[a]$ and $[c]$ possess the same eigenvalues. This can be proved easily by recalling that the determinant of a product of matrices is equal to the product of the determinants of the matrices in question. Moreover, recognizing that $\det[b]^{-1} = (\det[b])^{-1}$, we obtain

$$\det[c] = \det([b]^{-1}[a][b]) = \det[b]^{-1} \det[a] \det[b] = \det[a] \tag{9.29}$$

*Because $[a]$ and $[c]$ possess the same determinant, they must possess the same eigenvalues*. The object of this analysis is to find a transformation matrix $[b]$ such that $[c]$ reduces to a simple form, diagonal if possible or at least triangular. The simplest possible form of $[c]$ for a given system is known as the *Jordan canonical form*; its diagonal elements are the system eigenvalues. An examination of the various possible Jordan forms provides the desired information concerning the nature of motion in the neighborhood of the trivial solution.

There are basically three distinct Jordan forms possible, depending on the eigenvalues $\lambda_1$ and $\lambda_2$, although one of them represents a special case seldom encountered in practice. We wish to distinguish the following cases:

1. *The eigenvalues $\lambda_1$ and $\lambda_2$ are real and distinct*, in which case the Jordan form is diagonal,

$$[c] = \begin{bmatrix} \lambda_1 & 0 \\ 0 & \lambda_2 \end{bmatrix} \tag{9.30}$$

Inserting Eq. (9.30) into (9.27), we obtain

$$\dot{u}_1 = \lambda_1 u_1 \qquad \dot{u}_2 = \lambda_2 u_2 \qquad (9.31)$$

which have the solutions

$$u_1 = u_{10} e^{\lambda_1 t} \qquad u_2 = u_{20} e^{\lambda_2 t} \qquad (9.32)$$

where $u_{10}$ and $u_{20}$ are the initial values of $u_1$ and $u_2$, respectively. The type of motion depends on whether $\lambda_1$ and $\lambda_2$ are of the same sign or of opposite signs.

If the roots $\lambda_1$ and $\lambda_2$ are of the same sign, the equilibrium point is called a *node.* Figure 9.3a shows the phase portrait corresponding to the case $\lambda_2 < \lambda_1 < 0$, so that both eigenvalues are real and negative. In this case, we conclude from Eqs. (9.32) that all trajectories tend to the origin as $t \to \infty$, so that the *node is stable.* In view of the definition of Sec. 9.2, the motion is clearly *asymptotically stable.* With the exception of the case in which $u_{10} = 0$, all trajectories approach the origin with zero slope. When $\lambda_2 > \lambda_1 > 0$ the arrowheads change direction and the *node is unstable.*

If the roots $\lambda_1$ and $\lambda_2$ are real but of opposite signs, one solution tends to zero while the other tends to infinity. In this case the equilibrium point is a *saddle point,* and the equilibrium is *unstable.* Figure 9.3b shows the phase portrait for $\lambda_2 < 0 < \lambda_1$.

2. *The eigenvalues $\lambda_1$ and $\lambda_2$ are real and equal,* in which case there are two Jordan forms possible,

$$[c] = \begin{bmatrix} \lambda_1 & 0 \\ 0 & \lambda_1 \end{bmatrix} \qquad (9.33)$$

and

$$[c] = \begin{bmatrix} \lambda_1 & 1 \\ 0 & \lambda_1 \end{bmatrix} \qquad (9.34)$$

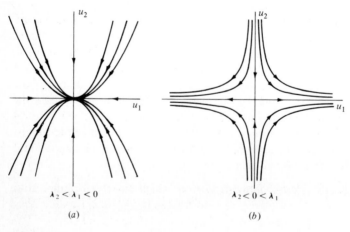

$\lambda_2 < \lambda_1 < 0$

(a)

$\lambda_2 < 0 < \lambda_1$

(b)

**Figure 9.3**

The case defined by Eq. (9.33) leads to

$$\dot{u}_1 = \lambda_1 u_1 \qquad \dot{u}_2 = \lambda_1 u_2 \tag{9.35}$$

having the solutions

$$u_1 = u_{10} e^{\lambda_1 t} \qquad u_2 = u_{20} e^{\lambda_1 t} \tag{9.36}$$

The trajectories are straight lines through the origin, and the equilibrium point is a *stable node* if $\lambda_1 < 0$ and an *unstable node* if $\lambda_1 > 0$. The case of Eq. (9.34) yields what is referred to as a *degenerate node*. We shall not pursue this subject any farther, as the case of equal eigenvalues is not very common.

3. *The eigenvalues $\lambda_1$ and $\lambda_2$ are complex conjugates*, in which case the Jordan form is simply

$$[c] = \begin{bmatrix} \lambda_1 & 0 \\ 0 & \lambda_1^* \end{bmatrix} \tag{9.37}$$

where $\lambda_2 = \lambda_1^*$ is the complex conjugate of $\lambda_1$. Letting $\lambda_1 = \alpha + i\beta$, $\lambda_1^* = \alpha - i\beta$, where $\alpha$ and $\beta$ are real, Eqs. (9.27) become

$$\dot{u}_1 = (\alpha + i\beta)u_1 \qquad \dot{u}_2 = (\alpha - i\beta)u_2 \tag{9.38}$$

from which we conclude that solutions $u_1$ and $u_2$ must also be complex conjugates, $u_2 = u_1^*$. Introducing the notation

$$u_1 = v_1 + iv_2 \qquad u_2 = v_1 - iv_2 \tag{9.39}$$

where $v_1$ and $v_2$ are real, we can write the solution for $u_1$ in the form

$$u_1 = (u_{10} e^{\alpha t}) e^{i\beta t} \tag{9.40}$$

which represents a logarithmic spiral. In this case the equilibrium point is known as a *spiral point*, or *focus*. Because the factor $e^{i\beta t}$ represents a vector of unit magnitude rotating with angular velocity $\beta$ in the complex plane, the magnitude of the complex vector $u_1$, and hence the stability of motion, is controlled by $e^{\alpha t}$. Indeed, for $\alpha < 0$ the *focal point is stable*, with the motion being *asymptotically stable*, and for $\alpha > 0$ it is *unstable*. The sign of $\beta$ merely gives the sense of rotation of the complex vector, counterclockwise for $\beta > 0$ and clockwise for $\beta < 0$. Figure 9.4a shows a typical trajectory for $\alpha < 0$ and $\beta > 0$.

When $\alpha = 0$ the magnitude of the radius vector is constant and the trajectories reduce to circles with the center at the origin (see Fig. 9.4b). In this case the equilibrium point is known as a *center*, or *vortex point*. The motion is periodic, and hence *stable*. This time, however, it is merely stable and not asymptotically stable.

The type of equilibrium points obtained for a given system can be determined, perhaps more directly, by examining the coefficients $a_{ij}$ $(i, j = 1, 2)$. To show this, we return to the characteristic equation (9.25) and write it in the form

$$\det ([a] - \lambda[1]) = \lambda^2 - (a_{11} + a_{22})\lambda + a_{11}a_{22} - a_{12}a_{21} = 0 \tag{9.41}$$

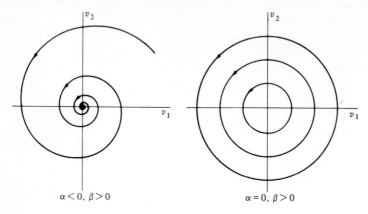

$\alpha < 0,\ \beta > 0$          $\alpha = 0,\ \beta > 0$

**Figure 9.4**

It will prove convenient to introduce the parameters

$$a_{11} + a_{22} = \text{tr}\,[a] = p$$

(9.42)

$$a_{11}a_{22} - a_{12}a_{21} = \det\,[a] = q$$

where $p$ and $q$ can be identified as the trace and determinant of the matrix $[a]$, respectively. With this notation, the characteristic equation becomes

$$\lambda^2 - p\lambda + q = 0$$

(9.43)

which has the roots

$$\begin{matrix}\lambda_1 \\ \lambda_2\end{matrix} = \tfrac{1}{2}(p \pm \sqrt{p^2 - 4q})$$

(9.44)

We again identify the cases discussed previously:

1. $p^2 > 4q$. In this case the eigenvalues are real and distinct. If $q$ is positive, both roots are of the same sign, and the equilibrium point is a stable node ($SN$) if $p$ is negative and an unstable node ($UN$) if $p$ is positive. If $q$ is negative, the roots are opposite in sign, and the equilibrium point is a saddle point ($SP$), irrespective of the sign of $p$.
2. $p^2 = 4q$. The roots are real and equal, in which case we obtain borderline nodes. From expressions (9.42) we conclude that this case is possible only if $a_{12}$ and $a_{21}$ are opposite in sign.
3. $p^2 < 4q$. For $q > 0$ the equilibrium point is a stable focus ($SF$) if $p < 0$ and an unstable focus ($UF$) if $p > 0$. When $p = 0$ the eigenvalues are pure imaginary complex conjugates and the equilibrium point is a center ($C$), which can be regarded as a borderline case separating stable and unstable foci.

The parameter plot $p$ versus $q$ shown in Fig. 9.5 gives a complete picture of the various possibilities. From this figure, it is obvious that the centers are indeed limiting cases obtained as the weakly stable and weakly unstable foci draw together. Hence, centers must be regarded as representing a mathematical concept more than a physical reality. It should be pointed out that centers are a characteristic of conservative systems. In a similar fashion, the case $p^2 = 4q$ appears in Fig. 9.5 as a parabola separating nodes and foci. Physically, the parabola $p^2 = 4q$ represents the curve separating aperiodic motion from oscillatory motion. The region designated by $SN$ is characterized by damped aperiodic motion, whereas that designated by $SF$ is characterized by damped oscillation. On the other hand, in the region denoted by $UN$ the motion is divergently aperiodic, whereas in the region marked by $UF$ the motion is divergently oscillatory. In the region denoted by $C$, consisting of the positive $q$ axis alone, the motion is harmonic. From Fig. 9.5, we conclude that the equilibrium point is stable if $p \leqslant 0$ and $q > 0$, and unstable for any other combination of $p$ and $q$.

From the above discussion, it appears that nodes and spiral points are either asymptotically stable or unstable, whereas saddle points are always unstable. On the other hand, centers are merely stable. We recall that *for asymptotic stability either the eigenvalues are real and negative, or they are complex conjugates with negative real parts. For instability at least one of the roots is real and positive, or complex with positive real part.* The cases of asymptotic stability and instability define what is known as *significant behavior*, whereas the case of mere stability constitutes what is referred to as *critical behavior*. These definitions enable us to discuss the circumstances under which the complete nonlinear system can be approximated by the linearized one. Indeed, *for significant behavior the nature of the equilibrium of the complete nonlinear system (9.21) is the same as that of the linearized system (9.22). The case of critical behavior is inconclusive,* and the complete nonlinear equations can yield either a center or a focal point (stable or unstable), as opposed to the center predicted by an infinitesimal analysis. In the case of critical behavior the linearized system cannot be used to draw conclusions about the behavior of the complete nonlinear system in the neighborhood of the equilibrium point, and higher-order terms contained in $\epsilon_1$ and $\epsilon_2$ must be examined. Although the case of critical behavior is obtained only for points on the positive $q$

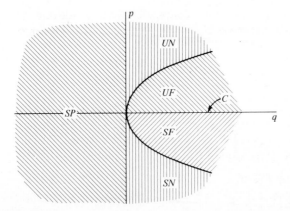

**Figure 9.5**

axis, which represents a relatively small region of the $q$, $p$ plane, this should not be construed as an indication that systems exhibiting critical behavior are very rare. On the contrary, mathematical models characterized by critical behavior are used quite extensively. This is precisely the case of the linear conservative mass-spring system discussed in Chap. 1, or the simple nonlinear conservative pendulum of Example 9.1.

**Example 9.2** Let us consider once again the pendulum of Example 9.1, governed by the differential equations

$$\dot{x}_1 = x_2 \qquad \dot{x}_2 = -\omega^2 \sin x_1 \qquad (a)$$

The equilibrium points were shown in that example to be defined by

$$x_1 = \pm j\pi \qquad j = 0, 1, 2, \ldots \qquad x_2 = 0 \qquad (b)$$

Because the equilibrium points $x_1 = 0$, $\pm 2\pi$, $\pm 4\pi$, ..., $x_2 = 0$ correspond to the same physical position, and a similar statement can be made concerning the equilibrium points $x_1 = \pm \pi$, $\pm 3\pi$, $\pm 5\pi$, ..., $x_2 = 0$, we shall consider only the equilibrium points $x_1 = x_2 = 0$ and $x_1 = \pi$, $x_2 = 0$, and then extend the conclusions to the other points.

In the neighborhood of $x_1 = x_2 = 0$, Eqs. ($a$) reduce to

$$\dot{x}_1 = x_2 \qquad \dot{x}_2 = -\omega^2 x_1 \qquad (c)$$

so that the matrix of the coefficients becomes

$$[a] = \begin{bmatrix} 0 & 1 \\ -\omega^2 & 0 \end{bmatrix} \qquad (d)$$

The corresponding characteristic equation is simply

$$\det([a] - \lambda[1]) = \lambda^2 + \omega^2 = 0 \qquad (e)$$

which has the roots

$$\begin{matrix} \lambda_1 \\ \lambda_2 \end{matrix} = \pm i\omega \qquad (f)$$

Because the roots are pure imaginary complex conjugates, we conclude that *the equilibrium point is a center*, so that *the motion in the neighborhood of the origin is stable*.

In the neighborhood of $x_1 = \pi$, $x_2 = 0$, Eqs. ($a$) become

$$\dot{x}_1 = x_2 \qquad \dot{x}_2 = \omega^2 x_1 \qquad (g)$$

and the matrix of the coefficients is

$$[a] = \begin{bmatrix} 0 & 1 \\ \omega^2 & 0 \end{bmatrix} \qquad (h)$$

so that the characteristic equation is

$$\det([a] - \lambda[1]) = \lambda^2 - \omega^2 = 0 \qquad (i)$$

The roots are

$$\begin{matrix} \lambda_1 \\ \lambda_2 \end{matrix} = \pm \omega \qquad (j)$$

Because the roots are real but opposite in sign, *the equilibrium point is a saddle point*. Clearly, *the motion in the neighborhood of* $x_1 = \pi$, which represents the upright position of the pendulum, *is unstable*.

The same problem can be discussed in terms of the parameters $p$ and $q$. Using Eqs. (9.42), we obtain for the equilibrium point $x_1 = x_2 = 0$

$$p = \text{tr } [a] \doteq 0 \qquad q = \det [a] = \omega^2 > 0 \qquad (k)$$

which coincides with the positive $q$ axis. Hence, as expected, the equilibrium point is a center. On the other hand, for $x_1 = \pi$, $x_2 = 0$, Eqs. (9.42) yield

$$p = \text{tr } [a] = 0 \qquad q = \det [a] = -\omega^2 < 0 \qquad (l)$$

which coincides with the negative $q$ axis. Again as expected, the equilibrium point is a saddle point.

We note that the two equilibrium points in question can be identified in Fig. 9.2 as the origin of the phase plane and the point $x_1 = \pi$, $x_2 = 0$, respectively. The motion considered here is in a small neighborhood of these points. In this particular case, however, the center predicted by means of the infinitesimal analysis remains a center for the complete nonlinear system. This conclusion is reached solely on physical grounds. Indeed, because there is no energy dissipated or added to the system, which would lead to either a weakly stable or a weakly unstable focus, respectively, the origin must remain a center.

It is clear that for $x_1 = 0$, $\pm 2\pi$, $\pm 4\pi$, ..., $x_2 = 0$ we obtain centers, and for $x_1 = \pm \pi$, $\pm 3\pi$, $\pm 5\pi$, ..., $x_2 = 0$ we obtain saddle points. The fact that the system possesses only centers and saddle points is no coincidence. Indeed, in Sec. 9.5 we shall see that this is a characteristic shared by all conservative systems.

## 9.4 ROUTH-HURWITZ CRITERION

From the Sec. 9.3, we conclude that the behavior of a nonlinear system in the neighborhood of an equilibrium point can be predicted on the basis of the linearized system, provided the system possesses significant behavior, i.e., if the roots of the characteristic equation

$$a_0 \lambda^m + a_1 \lambda^{m-1} + a_2 \lambda^{m-2} + \cdots + a_{m-1}\lambda + a_m = 0 \qquad (9.45)$$

are such that either all the real parts are negative or at least one of the real parts is positive; if some of the roots or all the roots are real, then the preceding statement applies to the roots themselves. Hence, significant behavior implies either asymptotic stability or instability, but not mere stability. In the above, $m = 2n$ for an $n$-degree-of-freedom system. Moreover, the coefficients $a_i$ $(i = 0, 1, 2, \ldots, m)$ are all real.

Significant behavior can be established, of course, by solving the characteristic equation for the system eigenvalues. For a second-order system, this presents no particular difficulty, as it amounts to finding the roots of a quadratic equation. For larger-order systems, however, this becomes a problem of increasing complexity. Hence, it appears desirable to be able to make a statement concerning the system stability without actually solving the characteristic equation. Because the imaginary parts of the eigenvalues do not affect the system stability, only the information concerning the real parts is necessary, and in particular the sign of the real parts.

There are two conditions necessary for none of the roots $\lambda_1, \lambda_2, \ldots, \lambda_m$ of Eq. (9.45) to have positive real parts. The conditions are:

1. All the coefficients $a_0, a_1, \ldots, a_m$ of the characteristic polynomial must have the same sign.
2. All the coefficients must be different from zero.

Assuming that $a_0 > 0$, the conditions imply that all the coefficients must be positive.

The above conditions are only necessary but not sufficient, so that their satisfaction does not guarantee stability. The conditions can be used, however, to identify unstable systems by inspection. Necessary and sufficient conditions for asymptotic stability were derived by both Routh and Hurwitz and they have come to be known as the *Routh-Hurwitz criterion.*

The coefficients $a_i$ $(i = 0, 1, 2, \ldots, m)$ of the characteristic polynomial can be used to construct the determinants

$$\Delta_1 = a_1 \qquad \Delta_2 = \begin{vmatrix} a_1 & a_0 \\ a_3 & a_2 \end{vmatrix} \qquad \Delta_3 = \begin{vmatrix} a_1 & a_0 & 0 \\ a_3 & a_2 & a_1 \\ a_5 & a_4 & a_3 \end{vmatrix} \cdots$$

$$\Delta_m = \begin{vmatrix} a_1 & a_0 & 0 & \cdots & 0 \\ a_3 & a_2 & a_1 & \cdots & 0 \\ a_5 & a_4 & a_3 & \cdots & 0 \\ \cdots\cdots\cdots\cdots\cdots\cdots\cdots\cdots \\ a_{2m-1} & a_{2m-2} & a_{2m-3} & \cdots & a_m \end{vmatrix} \tag{9.46}$$

where all the entries in the determinants corresponding to subscripts $r$ such that $r > m$ or $r < 0$ are to be replaced by zero. Then, assuming that $a_0 > 0$, the Routh-Hurwitz criterion states that *the necessary and sufficient conditions for all the roots $\lambda_j$ $(j = 1, 2, \ldots, m)$ of the characteristic equation to possess negative real parts is that all the determinants $\Delta_1, \Delta_2, \ldots, \Delta_m$ be positive.*† We note that the last two determinants are related by $\Delta_m = a_m \Delta_{m-1}$, so that it is only necessary to check the sign of the first $m - 1$ determinants.

As the number of degrees of freedom of the system increases, application of the Routh-Hurwitz criterion becomes increasingly laborious, as the computation of the

† For a proof of the crtierion, see N. G. Chetayev, *The Stability of Motion*, p. 75, Pergamon Press, New York, 1961.

large-order determinants in Eqs. (9.46) involves a large number of multiplications. The computation of large-order determinants can be avoided by considering the *Routh array*

| | | | | | |
|---|---|---|---|---|---|
| $\lambda^m$ | $a_0$ | $a_2$ | $a_4$ | $a_6$ | $\cdots$ |
| $\lambda^{m-1}$ | $a_1$ | $a_3$ | $a_5$ | $a_7$ | $\cdots$ |
| $\lambda^{m-2}$ | $c_1$ | $c_2$ | $c_3$ | $c_4$ | $\cdots$ |
| $\lambda^{m-3}$ | $d_1$ | $d_2$ | $d_3$ | $d_4$ | $\cdots$ |
| $\lambda^1$ | $m_1$ | $0$ | $0$ | $0$ | $\cdots$ |
| $\lambda^0$ | $n_1$ | $0$ | $0$ | $0$ | $\cdots$ |

where $a_0, a_1, \ldots, a_m$ are the coefficients of the characteristic polynomial and

$$c_1 = -\frac{1}{a_1}\begin{vmatrix} a_0 & a_2 \\ a_1 & a_3 \end{vmatrix} \qquad c_2 = -\frac{1}{a_1}\begin{vmatrix} a_0 & a_4 \\ a_1 & a_5 \end{vmatrix} \qquad c_3 = -\frac{1}{a_1}\begin{vmatrix} a_0 & a_6 \\ a_1 & a_7 \end{vmatrix} \cdots$$

$$(9.47)$$

are the entries in the row corresponding to $\lambda^{m-2}$,

$$d_1 = -\frac{1}{c_1}\begin{vmatrix} a_1 & a_3 \\ c_1 & c_2 \end{vmatrix} \qquad d_2 = -\frac{1}{c_1}\begin{vmatrix} a_1 & a_5 \\ c_1 & c_3 \end{vmatrix} \qquad d_3 = -\frac{1}{c_1}\begin{vmatrix} a_1 & a_7 \\ c_1 & c_4 \end{vmatrix} \cdots$$

$$(9.48)$$

are the entries in the row corresponding to $\lambda^{m-3}$, etc. Then, the Routh-Hurwitz criterion can be stated in terms of the Routh array as follows: *All the roots $\lambda_j$ ($j = 1, 2, \ldots, m$) of the characteristic equation possess negative real parts if all the entries in the first column of the Routh array have the same sign.*

Application of the Routh-Hurwitz criterion requires the coefficients $a_0, a_1, \ldots, a_m$, which in turn requires the derivation of the characteristic polynomial. This task also becomes increasingly difficult as the degree of the polynomial increases, so that the criterion can be used only for systems of moderate order.

**Example 9.3** Derive the Lagrange equations of motion for the two-degree-of-freedom system of Fig. 9.6, identify the equilibrium positions, derive the

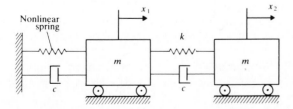

**Figure 9.6**

characteristic polynomial for each equilibrium position, and test the stability of the equilibrium positions by means of the Routh-Hurwitz criterion. The force in the nonlinear spring has the expression

$$f(x_1) = -kx_1 \left[ 1 - \left( \frac{x_1}{a} \right)^2 \right] \tag{a}$$

The Lagrange equations of motion for the system have the general form

$$\frac{d}{dt} \left( \frac{\partial T}{\partial \dot{x}_i} \right) + \frac{\partial \mathscr{F}}{\partial \dot{x}_i} + \frac{\partial V}{\partial x_i} = 0 \qquad i = 1, 2 \tag{b}$$

where

$$T = \tfrac{1}{2}m(\dot{x}_1^2 + \dot{x}_2^2) \tag{c}$$

is the kinetic energy,

$$\mathscr{F} = \tfrac{1}{2}c[\dot{x}_1^2 + (\dot{x}_2 - \dot{x}_1)^2] \tag{d}$$

is Rayleigh's dissipation function and

$$V = \int_{x_1}^{0} f(x_1)\, dx_1 + \tfrac{1}{2}k(x_2 - x_1)^2 = -k \int_{x_1}^{0} x_1 \left[ 1 - \left( \frac{x_1}{a} \right)^2 \right] dx_1$$

$$+ \tfrac{1}{2}k(x_2 - x_1)^2 = \tfrac{1}{2}k \left[ x_1^2 - \frac{a^2}{2} \left( \frac{x_1}{a} \right)^4 + (x_2 - x_1)^2 \right] \tag{e}$$

is the potential energy. Introducing Eqs. (c)–(e) into Eqs. (b), we obtain Lagrange's equations of motion

$$m\ddot{x}_1 + 2c\dot{x}_1 - c\dot{x}_2 + kx_1 \left[ 2 - \left( \frac{x_1}{a} \right)^2 \right] - kx_2 = 0$$

$$m\ddot{x}_2 - c\dot{x}_1 + c\dot{x}_2 - kx_1 + kx_2 = 0 \tag{f}$$

The equilibrium positions were defined in Sec. 9.2 as constant solutions of the equations of motion. Hence, they must satisfy the algebraic equations

$$kx_1 \left[ 2 - \left( \frac{x_1}{a} \right)^2 \right] - kx_2 = 0$$

$$-kx_1 + kx_2 = 0 \tag{g}$$

Equations (g) have three solutions, namely,

$$E_1: x_1 = x_2 = 0 \qquad E_2: x_1 = x_2 = a \qquad E_3: x_1 = x_2 = -a \tag{h}$$

To test the stability of the equilibrium point $E_1$, we linearize Eqs. (f) about the trivial solution $x_1 = x_2 = 0$, with the result

$$m\ddot{x}_1 + 2c\dot{x}_1 - c\dot{x}_2 + 2kx_1 - kx_2 = 0$$
$$m\ddot{x}_2 - c\dot{x}_1 + c\dot{x}_2 - kx_1 + kx_2 = 0$$

(i)

leading to the characteristic equation

$$\begin{vmatrix} m\lambda^2 + 2c\lambda + 2k & -c\lambda - k \\ -c\lambda - k & m\lambda^2 + c\lambda + k \end{vmatrix}$$

$$= (m\lambda^2 + 2c\lambda + 2k)(m\lambda^2 + c\lambda + k) - (c\lambda + k)^2$$
$$= m^2\lambda^4 + 3mc\lambda^3 + (3mk + c^2)\lambda^2 + 2ck\lambda + k^2 = 0 \quad (j)$$

Hence, the coefficients of the characteristic polynomial corresponding to the equilibrium position $E_1$ are

$$a_0 = m^2 \quad a_1 = 3mc \quad a_2 = 3mk + c^2 \quad a_3 = 2ck \quad a_4 = k^2 \quad (k)$$

To derive the characteristic equation corresponding to $E_2$, we introduce the transformation of coordinates

$$x_1 = a + y_1 \quad x_2 = a + y_2 \tag{l}$$

where $y_1$ and $y_2$ are small quantities. Inserting Eqs. (l) into Eqs. (f) and ignoring nonlinear terms in $y_1$, we obtain the linearized equations of motion about $E_2$

$$m\ddot{y}_1 + 2c\dot{y}_1 - c\dot{y}_2 - ky_1 - ky_2 = 0$$
$$m\ddot{y}_2 - c\dot{y}_1 + c\dot{y}_2 - ky_1 + ky_2 = 0$$

(m)

yielding the characteristic equation

$$\begin{vmatrix} m\lambda^2 + 2c\lambda - k & -c\lambda - k \\ -c\lambda - k & m\lambda^2 + c\lambda + k \end{vmatrix}$$

$$= (m\lambda^2 + 2c\lambda - k)(m\lambda^2 + c\lambda + k) - (c\lambda + k)^2$$
$$= m^2\lambda^4 + 3mc\lambda^3 + c^2\lambda^2 - ck\lambda - 2k^2 = 0 \quad (n)$$

so that the coefficients of the characteristic polynomial corresponding to the equilibrium position $E_2$ are

$$a_0 = m^2 \quad a_1 = 3mc \quad a_2 = c^2 \quad a_3 = -ck \quad a_4 = -2k^2 \quad (o)$$

It is not difficult to show that the characteristic polynomial corresponding to the equilibrium position $E_3$ is the same as for $E_2$.

We shall test the stability of the equilibrium positions both by means of the determinants $\Delta_1, \Delta_2, \Delta_3$, and $\Delta_4$ and by means of the Routh array. Hence, for

$E_1$, we have

$$\Delta_1 = a_1 = 3mc$$

$$\Delta_2 = \begin{vmatrix} a_1 & a_0 \\ a_3 & a_2 \end{vmatrix} = \begin{vmatrix} 3mc & m^2 \\ 2ck & 3mk + c^2 \end{vmatrix}$$

$$= 3mc(3mk + c^2) - 2m^2ck = 7m^2ck + 3mc^3$$

$$\Delta_3 = \begin{vmatrix} a_1 & a_0 & 0 \\ a_3 & a_2 & a_1 \\ 0 & a_4 & a_3 \end{vmatrix} = \begin{vmatrix} 3mc & m^2 & 0 \\ 2ck & 3mk + c^2 & 3mc \\ 0 & k^2 & 2ck \end{vmatrix} \qquad (p)$$

$$= 2ck \begin{vmatrix} 3mc & m^2 \\ 2ck & 3mk + c^2 \end{vmatrix} - k^2 \begin{vmatrix} 3mc & 0 \\ 2ck & 3mc \end{vmatrix}$$

$$= 2ck(7m^2ck + 3mc^2) - 9m^2c^2k^2 = 5m^2c^2k^2 + 6mc^4k$$

$$\Delta_4 = a_4\Delta_3 = k^2(5m^3c^2k^2 + 6mc^4k)$$

It is clear from Eqs. ($p$) that all the determinants are positive, so that all the eigenvalues have negative real parts, from which it follows that *the equilibrium position $E_1$ is asymptotically stable.*

Next, let us form the Routh array

$$
\begin{array}{c|ccc}
\lambda^4 & a_0 & a_2 & a_4 \\
\lambda^3 & a_1 & a_3 & 0 \\
\lambda^2 & c_1 & c_2 & 0 \\
\lambda & d_1 & 0 & 0 \\
1 & e_1 & 0 & 0
\end{array} \qquad (q)
$$

In the case of $E_1$, the coefficients of the characteristic polynomial are given by Eqs. ($k$). Moreover, using Eqs. (9.47), (9.48), etc., we compute the entries

$$c_1 = -\frac{1}{a_1}\begin{vmatrix} a_0 & a_2 \\ a_1 & a_3 \end{vmatrix} = -\frac{1}{3mc}\begin{vmatrix} m^2 & 3mk + c^2 \\ 3mc & 2ck \end{vmatrix} = \tfrac{1}{3}(7mk + 3c^2)$$

$$c_2 = -\frac{1}{a_1}\begin{vmatrix} a_0 & a_4 \\ a_1 & 0 \end{vmatrix} = a_4 = k^2$$

$$d_1 = -\frac{1}{c_1}\begin{vmatrix} a_1 & a_3 \\ c_1 & c_2 \end{vmatrix} = -\frac{3}{7mk + 3c^2}\begin{vmatrix} 3mc & 2ck \\ 7mk + 3c^2 & k^2 \\ 3 & \end{vmatrix} \qquad (r)$$

$$= \frac{5mck^2 + 6c^3k}{7mk + 3c^2}$$

$$e_1 = -\frac{1}{d_1}\begin{vmatrix} c_1 & c_2 \\ d_1 & 0 \end{vmatrix} = c_2 = k^2$$

Clearly, $a_0$, $a_1$, $c_1$, $d_1$, and $e_1$ are all positive, so that all the roots of the characteristic polynomial have negative real parts and $E_1$ is asymptotically stable, which we established already.

The same analysis can be used for $E_2$. This is not necessary, however. Indeed, the first of the necessary conditions for asymptotic stability requires that $a_0$, $a_1$, ..., $a_4$ have the same sign, which is clearly not the case here, as can be verified by examining Eqs. (o). Hence, *we cannot conclude that $E_2$ is asymptotically stable*. This should surprise no one, as the equilibrium position $E_2$ is unstable. Clearly, the same can be said for $E_3$.

## 9.5 CONSERVATIVE SYSTEMS. MOTION IN THE LARGE

In Secs. 9.3 and 9.4 we concerned ourselves with the motion in the neighborhood of equilibrium points. Such motion is sometimes referred to as *motion in the small*, as opposed to motion at some distance away from equilibrium points, called *motion in the large*. Although never stated specifically, Fig. 9.2 of Example 9.1 depicts motion in the large. In this section we propose to generalize and expand on the problem of Example 9.1. To this end, we confine ourselves to the simple second-order autonomous conservative system

$$\ddot{x} = f(x) \tag{9.49}$$

where $f(x)$ represents the conservative force per unit mass. From differential calculus, however, we obtain

$$\ddot{x} = \frac{d\dot{x}}{dt} = \frac{d\dot{x}}{dx}\frac{dx}{dt} = \dot{x}\frac{d\dot{x}}{dx} \tag{9.50}$$

so that Eq. (9.49) can be rewritten as

$$\dot{x}\,d\dot{x} = f(x)\,dx \tag{9.51}$$

yielding the integral

$$\tfrac{1}{2}\dot{x}^2 + V(x) = E = \text{const} \tag{9.52}$$

where $\tfrac{1}{2}\dot{x}^2$ is the kinetic energy, $V(x) = \int_x^0 f(x)\,dx$ the potential energy and $E$ the total energy, all per unit mass.

Introducing the notation $x = x_1$, $\dot{x} = x_2$, Eq. (9.52) becomes

$$\tfrac{1}{2}x_2^2 + V(x_1) = E = \text{const} \tag{9.53}$$

which represents a family of integral curves in the phase plane, where $E$ is the parameter of the family. The integral curves are symmetric with respect to axis $x_1$. If a third axis corresponding to $E$ is added, where the axis is normal to the phase plane defined by $x_1$ and $x_2$, then the integral curves (9.53) can be envisioned geometrically as the curves obtained as the intersections of the surfaces $E(x_1, x_2) = \tfrac{1}{2}x_2^2 + V(x_1)$ and the planes $E = \text{const}$. These intersections represent *level curves*, as any point on such a curve must belong to the plane $E = \text{const}$. Regarding the integral curves

(9.53) as level curves helps us rule out nodes and focal points as equilibrium points of system (9.49). This is so because the integral curves have points in common, namely, the equilibrium points, only when these points are nodes and foci. If the level curves given by Eq. (9.53) were to represent level curves with nodes and foci as equilibrium points, then $E(x_1, x_2)$ would have the same value at every point surrounding the equilibrium point, a fact that contradicts the concept of level curves, for which the value of $E(x_1, x_2)$ is different for different level curves. Hence, the only equilibrium points possible are centers and saddle points, so that *conservative systems cannot be asymptotically stable.*

As a simple example, let us consider a ball rolling on a frictionless track under gravity. Assuming that at any point $x_1$ the track is at height $h(x_1)$ above a given reference level (see Fig. 9.7a), then

$$V(x_1) = mgh(x_1) \tag{9.54}$$

Solving Eqs. (9.53) and (9.54) for $x_2$, we obtain

$$x_2 = \pm\sqrt{2[E - V(x_1)]} = \pm\sqrt{2[E - mgh(x_1)]} \tag{9.55}$$

which enables us to plot level curves corresponding to various values for $E$, as shown in Fig. 9.7b. Points corresponding to $dV/dx = 0$ are equilibrium points because at these points the force $f$ is zero. Hence, points 1, 2, and 3 are equilibrium points. Because the points are not in the same neighborhood, they are isolated equilibrium points. For $E < V_1$ no motion is possible. At $E = V_1$ we obtain a center, and for $V_1 < E < V_3$ there is periodic motion about point 1. Likewise for $E < V_2$ no motion in the neigborhood of point 2 is possible. Point 2, corresponding to $E = V_2$, is another center, and for $V_2 < E < V_3$, there can be periodic motion

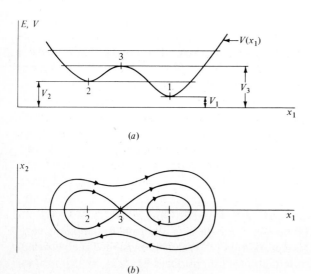

(a)

(b)

**Figure 9.7**

about that point. When the energy level reaches $E = V_3$, a saddle point is obtained at point 3. Whereas for $V_1 < E < V_3$ the motion is periodic about one of the two centers, for $E > V_3$ the motion is again periodic, but this time the trajectories enclose both centers and the saddle point. Hence, the motion corresponding to $E > V_3$ differs in nature from that corresponding to $V_1 < E < V_3$. The level curve corresponding to $E = V_3$ is a *separatrix*, which by definition separates regions characterized by different types of motion. In this particular case, the different types of motion are periodic motion about one center alone on the one hand, and periodic motion about two centers and one saddle point on the other. It is typical of conservative systems that a closed trajectory encloses an odd number of equilibrium points, with the number of centers exceeding the number of saddle points by one.

From Fig. 9.7 we can verify heuristically a theorem due to Lagrange that can be enunciated as follows: *An isolated equilibrium point corresponding to a minimum value of the potential energy is stable.* We can also verify another theorem due to Liapunov stating: *If the potential energy has no minimum at an equilibrium, then the equilibrium point is unstable.* These theorems can be proved rigorously by the Liapunov direct method discussed in Sec. 9.7.

## 9.6 LIMIT CYCLES

A question of particular interest in nonlinear systems is the existence of closed trajectories, as such trajectories imply periodic motion. From our past experience, we conclude that closed trajectories occur in conservative systems, with the closed trajectories enclosing an odd number of equilibrium points. The equilibrium points are centers and saddle points, and the number of centers exceeds the number of saddle points by one. It turns out that closed trajectories can occur also in nonlinear nonconservative systems, but the systems must be such that at the completion of one cycle the net energy change is zero. This implies that over parts of the cycle energy is dissipated, and over the balance of the cycle energy is imparted to the system. Such closed trajectories are referred to as *limit cycles of Poincaré*, or simply *limit cycles*. Limit cycles can be regarded as equilibrium motions in which the system performs periodic motion, as opposed to equilibrium points in which the system is at rest. Moreover, the amplitude of a given limit cycle depends on the system parameters alone, whereas the amplitude of a closed trajectory for a conservative system depends on the energy imparted to the system initially. In the case of limit cycles we must speak of orbital stability rather than stability in the sense of Liapunov.

It is very difficult to establish the existence of a limit cycle for a given system. There is the Poincaré-Bendixson classical theorem for the existence of limit cycles, and Bendixson's criterion for proving the lack of existence of a limit cycle, but their usefulness is limited.

A classical example of a system known to possess a limit cycle is *van der Pol's oscillator*. We shall use this example to examine some of the properties of limit cycles. The van der Pol oscillator is described by the differential equation

$$\ddot{x} + \mu(x^2 - 1)\dot{x} + x = 0 \qquad \mu > 0 \tag{9.56}$$

which can be regarded as an oscillator with variable damping. Indeed the term $\mu(x^2 - 1)$ can be regarded as an amplitude-dependent damping coefficient. For $|x| < 1$ the coefficient is negative and for $|x| > 1$ it is positive. Hence, for motions in the range $|x| < 1$ the negative damping tends to increase the amplitude, whereas for $|x| > 1$ the positive damping tends to reduce the amplitude, so that a limit cycle can be expected and is indeed obtained.

Letting $x = x_1, \dot{x} = x_2$, Eq. (9.56) can be replaced by the two first-orde differential equations

$$\dot{x}_1 = x_2 \qquad \dot{x}_2 = -x_1 + \mu(1 - x_1^2)x_2 \tag{9.57}$$

Clearly, the origin is an equilibrium point. To determine the nature of the equilibrium point, we form the matrix of the coefficients of the linearized system

$$[a] = \begin{bmatrix} 0 & 1 \\ -1 & \mu \end{bmatrix} \tag{9.58}$$

leading to the characteristic equation

$$\lambda^2 - \mu\lambda + 1 = 0 \tag{9.59}$$

which has the roots

$$\frac{\lambda_1}{\lambda_2} = \frac{\mu}{2} \pm \sqrt{\left(\frac{\mu}{2}\right)^2 - 1} \tag{9.60}$$

When $\mu > 2$ the roots $\lambda_1$ and $\lambda_2$ are both real and positive, so that the origin is an unstable node. On the other hand, when $\mu < 2$ the roots $\lambda_1$ and $\lambda_2$ are complex conjugates with positive real part, so that the origin is an unstable focus. In any event, the origin is an unstable equilibrium point, and any motion initiated in its neighborhood will tend to leave that neighborhood and reach the limit cycle.

To obtain the equation of the trajectories, we divide the second of Eqs. (9.57) by the first, with the result

$$\frac{dx_2}{dx_1} = \mu(1 - x_1^2) - \frac{x_1}{x_2} \tag{9.61}$$

A closed-form solution of this equation is not possible. The trajectories can be obtained by some graphical procedure, such as the method of isoclines,† or by numerical integration. The plots of Fig. 9.8 were obtained by numerical integration for the values $\mu = 0.2$ and $\mu = 1.0$. It is clear from Fig. 9.8 that the shape of the limit cycle depends on the parameter $\mu$. In fact, for $\mu \to 0$ the limit cycle tends to a circle. Because all trajectories approach the limit cycle, either from the inside or from the outside, the *limit cycle is stable*. Note that *for $\mu < 0$ an unstable limit cycle is*

---

† See, for example, C. Hayashi, *Nonlinear Oscillations in Physical Systems*, sec. 2–4, McGraw-Hill Book Co., New York, 1964.

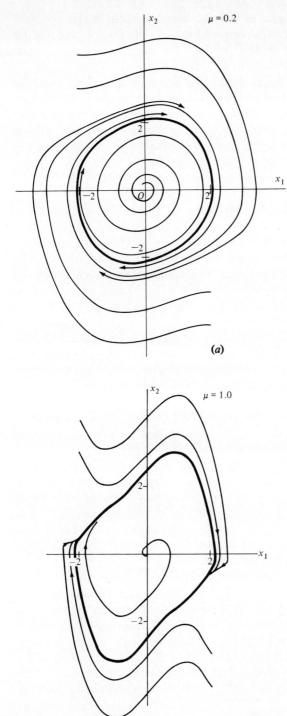

(a)

(b)    Figure 9.8

*obtained*. In view of the definitions of Sec. 9.2, *the limit cycle is asymptotically orbitally stable for $\mu > 0$, and orbitally unstable for $\mu < 0$*. We observe that a stable limit cycle encloses an unstable equilibrium point, and an unstable limit cycle encloses a stable equilibrium point.

Finally, we must point out the insufficiency of a linearized analysis about the origin for systems exhibiting limit cycles. A linearized analysis would have predicted instability for $\mu > 0$, with the motion increasing indefinitely. The term controlling the magnitude of the amplitude is the nonlinear one, namely $\mu x^2 \dot{x}$. A proper linearization in this case would have to be about the limit cycle, which would result in a linear system with periodic coefficients.

## 9.7 LIAPUNOV'S DIRECT METHOD

The *Liapunov direct method*, also called *Liapunov's second method*, can be regarded as an extension and generalization of the energy method of mechanics. It was inspired by a proof of Lagrange's theorem on the stability of dynamical systems in the neighborhood of an equilibrium point (see Sec. 9.5). The method proposes to *determine the system stability characteristics without actually carrying out the solution of the differential equations*. To this end, it is necessary to devise for the system a scalar function defined in the phase space and whose total time derivative is evaluated along a trajectory of the system. Not just any arbitrary function is suitable for a given system, but only a function possessing certain sign properties, as prescribed by one of a number of stability and instability theorems. If a testing function satisfying any one of these theorems can be found, then it represents a Liapunov function for the system. The fact that a Liapunov function cannot be found does not imply that the system is not stable. Indeed, the main drawback of the method is that there is no established procedure for producing a Liapunov function for any given dynamical system. For this reason, the Liapunov direct method must be regarded as more a philosophy of approach than a method. Liapunov functions can be constructed in a systematic manner for linear autonomous systems, reducing the stability problem to the solution of $n(2n + 1)$ algebraic equations for an $n$-degree-of-freedom system.[†] Moreover, there are classes of problems for which clues for devising Liapunov functions exist. Fortunately, this is the case with many problems that interest us.

In Sec. 4.5 we introduced the concepts of positive definite and positive semidefinite functions. In this section we wish to present these definitions in the context of the phase space, and use the opportunity to provide a geometric interpretation of a positive definite function. We shall be concerned with a system of order $m = 2n$, where $n$ is the number of degrees of freedom, and associate with it an $m$-dimensional phase space with coordinates $x_i$ $(i = 1, 2, ..., m)$. According to the definition introduced in Sec. 9.2, a spherical region of radius $h$ with the center at the origin is denoted symbolically by $\|\mathbf{x}\| = (\sum_{i=1}^{m} x_i^2)^{1/2} < h$ if the region does not

---

† See, for example, Meirovitch, op. cit., sec. 6.11.

include the boundary $\|\mathbf{x}\| = h$, and by $\|\mathbf{x}\| \leqslant h$ if the region does include the boundary. Next, we consider a real scalar function $U = U(x_1, x_2, \ldots, x_m)$ possessing continuous partial derivatives in $\|\mathbf{x}\| \leqslant h$ with respect to the variables $x_i$, where the function vanishes at the origin, $U(0, 0, \ldots, 0) = 0$. For such a function $U$, we introduce the following definitions:

1. The function $U(x_1, x_2, \ldots, x_m)$ is said to be *positive definite* in the spherical region $\|\mathbf{x}\| \leqslant h$ if $U(x_1, x_2, \ldots, x_m) > 0$ for any point such that $\mathbf{x} \neq \mathbf{0}$ and vanishes only at the origin.
2. The function $U(x_1, x_2, \ldots, x_m)$ is said to be *positive semidefinite* in the spherical region $\|\mathbf{x}\| \leqslant h$ if $U(x_1, x_2, \ldots, x_m) \geqslant 0$ and it can vanish also for some points in $\|\mathbf{x}\| \leqslant h$ other than the origin.
3. The function $U(x_1, x_2, \ldots, x_m)$ is said to be *indefinite* if it can take both positive and negative values in the spherical region $\|\mathbf{x}\| \leqslant h$, regardless of how small the radius $h$ is.

To obtain definitions for *negative definite* and *negative semidefinite* functions, we simply reverse the sense of the inequality signs in the first two definitions. The nature of a positive definite function can be interpreted geometrically by considering the phase plane shown in Fig. 9.9. If $c$ is any positive constant, then the equation

$$U(x_1, x_2) = c \qquad (9.62)$$

describes a curve in the phase plane. Considering a function $U$ such that $U(0, 0) = 0$, the curve $U = c$ reduces to a point coinciding with the origin as $c \to 0$. If $U(x_1, x_2)$ is positive definite, then for a small value of $c$, say $c = c_1$, the equation $U(x_1, x_2) = c_1$ represents a closed curve enclosing the origin. For another value of the constant say $c_2 > c_1$, the equation $U(x_1, x_2) = c_2$ represents a closed curve enclosing the curve $U(x_1, x_2) = c_1$ without intersecting it. Hence, the curves $U(x_1, x_2) = c$ represent a family of nonintersecting closed curves in the neighborhood of the origin that increase in size with $c$ and shrink to the origin for $c \to 0$. Considering the function $U(x_1, x_2) = k$, the circle $\|\mathbf{x}\| = \epsilon$ represents the smallest circle enclosing $U = k$, and the circle $\|\mathbf{x}\| = \delta$ the largest circle enclosed by $U = k$.

There remains the question as to how to test analytically whether a function is positive definite or not. If $U = U(x_1, x_2, \ldots, x_m)$ is a homogeneous function of

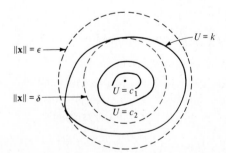

**Figure 9.9**

order $p$ in the variables $x_i$ $(i = 1, 2, ..., m)$ and $\beta$ is an arbitrary constant, then

$$U(\beta x_1, \beta x_2, ..., \beta x_m) = \beta^p U(x_1, x_2, ..., x_m) \tag{9.63}$$

Hence, if $p$ is an odd integer the function $U$ is indefinite. No conclusion can be drawn, however, if $p$ is an even integer.

A case of particular interest is that in which $U$ is a quadratic function, in which case it can be written in the matrix form

$$U = \sum_{i=1}^{m} \sum_{j=1}^{m} \alpha_{ij} x_i x_j = \{x\}^T [\alpha] \{x\} \tag{9.64}$$

where $[\alpha]$ is the symmetric matrix of the coefficients. Using a linear transformation of the type (9.26), we may be able to reduce the matrix $[\alpha]$ to a diagonal form, thus rendering $U$ free of cross products. In this case the requirement that $U$ be positive definite reduces to the requirement that all the coefficients of the resulting expression be positive, which is equivalent to requiring that all the eigenvalues of $[\alpha]$ be positive. The sign properties of $U$ can be checked more readily by means of the so-called Sylvester's theorem,[†] which states: *The necessary and sufficient conditions for the quadratic form (9.64) to be positive definite are that all the principal minor determinants associated with the matrix $[\alpha]$ be positive.* These conditions can be expressed in the mathematical form

$$\det [\alpha_{qr}] > 0 \qquad \begin{matrix} q, r = 1, 2, ..., s \\ s = 1, 2, ..., m \end{matrix} \tag{9.65}$$

As pointed out in Sec. 4.5, if $U$ is positive definite, then $[\alpha]$ is said to be a positive definite matrix.

Now we are in a position to introduce Liapunov's direct method. Under consideration is an $n$-degree-of-freedom autonomous system described by the $m = 2n$ first-order differential equations

$$\dot{x}_i = X_i(x_1, x_2, ..., x_m) \qquad i = 1, 2, ..., m \tag{9.66}$$

where the functions $X_i$ are continuous in the spherical region $\|x\| \leqslant h$. In addition, we assume that the origin of the phase space is an equilibrium point, so that $X_i(0, 0, ..., 0) = 0$ $(i = 1, 2, ..., m)$. Hence, we concern ourselves with the stability of the trivial solution. Next, we assume that we have a prospective Liapunov function $U(x_1, x_2, ..., x_m)$. By writing the total time derivative of $U$ in the form

$$\dot{U} = \frac{dU}{dt} = \sum_{i=1}^{m} \frac{\partial U}{\partial x_i} \dot{x}_i = \sum_{i=1}^{m} \frac{\partial U}{\partial x_i} X_i \tag{9.67}$$

we ensure that $\dot{U}$ is evaluated along a trajectory of system (9.66). With this in mind, we can state the following:

**Liapunov's stability theorem 1** *If there exists for system (9.66) a positive definite function $U(x_1, x_2, ..., x_m)$ whose total time derivative $\dot{U}(x_1, x_2, ..., x_m)$ is negative semidefinite along every trajectory of (9.66), then the trivial solution is stable.*

† See Chetayev, op. cit., sec. 20.

**Liapunov's stability theorem 2** *If there exists for system* (9.66) *a positive definite function* $U(x_1, x_2, ..., x_m)$ *whose total time derivative* $\dot{U}(x_1, x_2, ..., x_m)$ *is negative definite along every trajectory of* (9.66), *then the trivial solution is asymptotically stable.*

Proofs of the above theorems can be found in the text by Meirovitch.† Similarly, there are two instability theorems. We state only the first one.

**Liapunov's instability theorem 1.** *If there exists for the system* (9.66) *a function* $U(x_1, x_2, ..., x_m)$ *whose total time derivative* $\dot{U}(x_1, x_2, ..., x_m)$ *is positive definite along every trajectory of* (9.66) *and U itself can take positive values for arbitrarily small* $x_1, x_2, ..., x_m$, *then the trivial solution is unstable.*

In both stability theorems and the instability theorem it is possible to replace everywhere the words positive and negative by the words negative and positive, respectively, without altering the substance of the theorems. This is true because instead of considering the testing function $U$ it is possible to consider the function $-U$.

There are various generalizations of the above theorems. Some of the most important ones are due to Chetayev and Krasovskii. Chetayev's generalization of Liapunov's instability theorem 1 essentially states that $\dot{U}$ need not be positive definite in the entire neighborhood of the origin, but only in the subregion in which $U$ takes positive values. Krasovskii's generalization of Liapunov's stability theorem 2 states that $\dot{U}$ need be only negative semidefinite for the system to be asymptotically stable, provided $\dot{U}$ reduces to zero and stays zero for all subsequent times only at the origin. A similar generalization by Krasovskii exists for Liapunov's instability theorem 1.

Perhaps the connection between the above theorems and the definitions of stability in the sense of Liapunov, given in Sec. 9.2, can be revealed by a geometric interpretation of the theorems. To this end, we confine ourselves once again to a second-order system for which the phase space reduces to the phase plane. Introducing an axis $z$ normal to the phase plane defined by $x_1$ and $x_2$, the function $z = U(x_1, x_2)$ represents a three-dimensional surface. In the case of a positive definite function, the surface $z = U(x_1, x_2)$ resembles a cup tangent to the phase plane at the origin (see Fig. 9.10a). The intersections of the surface with the planes $z = c = $ const consist of level curves that, when projected on the phase plane, appear as nonintersecting closed curves surrounding the origin. Moreover, any path on the surface $z = U$ projects as a trajectory on the phase plane. Three distinct trajectories I, II, and III, representing integral curves of the system, are shown in Fig. 9.10, where the trajectories illustrate the two stability theorems and the first instability theorem of Liapunov, respectively. Curve I corresponds to a negative semidefinite $\dot{U}$. The trend of curve $I$ is downward, although it can also become stalled on a level curve and remain there for any subsequent time, which implies

---

† Meirovitch, op. cit., sec. 6.7.

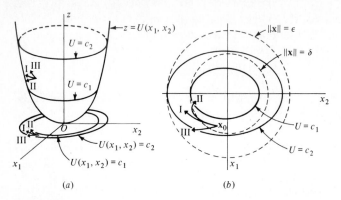

**Figure 9.10**

mere stability. Curve II, on the other hand, corresponds to negative definite $\dot{U}$ and cannot remain on a level curve, so that it approaches the origin; the curve corresponds to asymptotic stability. Curve III represents the opposite situation, namely, it corresponds to positive definite $\dot{U}$ and it moves away from the origin, which implies instability. Next let us assume that the trajectories are initiated at $\mathbf{x} = \mathbf{x}_0$, inside the circular region $\|\mathbf{x}\| < \delta$ but outside the curve $U = c_1$ enclosed by the circle $\|\mathbf{x}\| = \delta$ (see Fig. 9.10$b$). It is clear that curve II will cross the curve $U = c_1$, moving from the outside to the inside of $U = c_1$, in its way to the origin. On the other hand, curve III will cross the curve $U = c_2$ from the inside to the outside in its way to crossing the circle $\|\mathbf{x}\| = \epsilon$ enclosing $U = c_2$. There remains the question of curve I. Although curve I can increase its distance from the origin at various times, it will remain between the curves $U = c_1$ and $U = c_2$. Hence, it will never cross the circle $\|\mathbf{x}\| = \epsilon$, nor will it reach the origin.

In the case of conservative systems the total energy $E$ is constant, from which we conclude that its total time derivative is zero. Hence, if $E$ is positive definite (or negative definite) in the neighborhood of an equilibrium point, then we can choose $U = E$ (or $U = -E$) as a Liapunov function and conclude that the equilibrium is stable. But the total energy consists of the sum of the kinetic energy and potential energy, $U = E = T + V$, where by definition the kinetic energy $T$ is a positive function of the generalized velocities. It follows that if the potential energy $V$ is a positive definite function of the generalized coordinates in the neighborhood of a given equilibrium point, then $E$ is a positive definite function of the generalized coordinates and velocities and the equilibrium is stable. But for the potential energy to be a positive definite function of the generalized coordinates in the neighborhood of the equilibrium, it must have a minimum at this point, which proves Lagrange's theorem, introduced in Sec. 9.5.

The total energy $E$ can prove to be a suitable Liapunov function for nonconservative systems also, as shown in Example 9.4.

**Example 9.4** Consider the two-degree-of-freedom system of Example 9.3 and determine the nature of motion in the neighborhood of the equilibrium points by means of the Liapunov direct method.

The total energy of the system is

$$E = T + V \tag{a}$$

where, from Example 9.3, the kinetic energy is

$$T = \tfrac{1}{2}m(\dot{x}_1^2 + \dot{x}_2^2) \tag{b}$$

and the potential energy has the expression

$$V = \tfrac{1}{2}k\left[ x_1^2 - \frac{a^2}{2}\left(\frac{x_1}{a}\right)^4 + (x_2 - x_1)^2 \right] \tag{c}$$

Letting $U = E$ be our Liapunov function, we can write the time derivative of $U$ in the form

$$\dot{U} = m(\dot{x}_1\ddot{x}_1 + \dot{x}_2\ddot{x}_2) + k\left[ x_1\dot{x}_1 - a\left(\frac{x_1}{a}\right)^3\dot{x}_1 + (x_2 - x_1)(\dot{x}_2 - \dot{x}_1) \right]$$

$$= \left\{ m\ddot{x}_1 + kx_1\left[ 2 - \left(\frac{x_1}{a}\right)^2 \right] - kx_2 \right\}\dot{x}_1 + [m\ddot{x}_2 + k(x_2 - x_1)]\dot{x}_2 \tag{d}$$

so that, using the equations of motion, Eqs. ($f$) of Example 9.3, we obtain

$$\dot{U} = -c(2\dot{x}_1 - \dot{x}_2)\dot{x}_1 - c(\dot{x}_2 - \dot{x}_1)\dot{x}_2$$

$$= -c[\dot{x}_1^2 + (\dot{x}_2 - \dot{x}_1)^2] = -2\mathscr{F} < 0 \tag{e}$$

where $\mathscr{F}$ is Rayleigh's dissipation function. From Eq. ($e$), we conclude that $\dot{U}$ is negative semidefinite and, moreover, that it becomes identically zero only at equilibrium points. Hence, if $U$ is positive definite in the neighborhood of an equilibrium point, by Krasovskii's extension of Liapunov's stability theorem 2, the equilibrium is asymptotically stable. On the other hand, if $U$ is indefinite in the neighborhood of an equilibrium point, by Krasovskii's extension of Liapunov's instability theorem 1, the equilibrium is unstable. But $T$ is by definition a positive definite function of $\dot{x}_1$ and $\dot{x}_2$, so that if $V$ is a positive definite function of $x_1$ and $x_2$, then $U = E$ is a positive definite function of $x_1$, $x_2$, $\dot{x}_1$, and $\dot{x}_2$ and the equilibrium is asymptotically stable. On the other hand, if $V$ can take negative values in the neighborhood of an equilibrium point, then $U = E$ is indefinite and the equilibrium is unstable.

In the neighborhood of the equilibrium position $E_1$, the potential energy has the form

$$V = \tfrac{1}{2}k(x_1^2 + x_2^2) \tag{f}$$

which is a positive definite function of $x_1$ and $x_2$, so that the equilibrium is asymptotically stable.

To examine the equilibrium position $E_2$, we refer to Example 9.3 and introduce the coordinate transformation

$$x_1 = a + y_1 \qquad x_2 = a + y_2 \tag{g}$$

It is easy to verify that $\dot{U}$ is once again negative semidefinite and that $T$ is a positive definite function of $\dot{y}_1$ and $\dot{y}_2$. It remains to check the sign of $V$. Introducing Eqs. ($g$) into Eq. ($c$) and ignoring terms in $y_1$ of degree higher than 2, as well as a constant term, we obtain

$$V = \tfrac{1}{2}k[-2y_1^2 + (y_2 - y_1)^2] \tag{h}$$

which is indefinite. Hence, $U = E$ is indefinite, so that the equilibrium point $E_2$ is unstable. A similar analysis shows that the equilibrium point $E_3$ is equally unstable.

## PROBLEMS

**9.1** The differential equation of motion of a viscously damped pendulum can be written in the form

$$\ddot{\theta} + 2\zeta\omega\dot{\theta} + \omega^2 \sin\theta = 0$$

Transform the equation into two first-order differential equations and determine the equilibrium points. Compare the equilibrium points with those of the undamped pendulum and explain the results.

**9.2** Consider the mass-spring system described by the differential equation of motion

$$\ddot{x} + x - \frac{\pi}{2}\sin x = 0$$

transform the equation into two first-order differential equations and determine the equilibrium points.

**9.3** A bead of mass $m$ is free to slide along a circular hoop of radius $R$. Derive the differential equations of motion and find the positions of equilibrium of $m$ if the hoop is rotating about a vertical diametrical axis with the constant angular velocity $\omega$.

**9.4** Consider the damped pendulum of Prob. 9.1, choose a value for $\omega$, and plot phase portraits for the two cases $\zeta = 0.1$ and $\zeta = 2$. Examine the motion in the neighborhood of the equilibrium points and make a statement concerning their equilibrium.

**9.5** Integrate the equation of motion of Prob. 9.2 and obtain the equation of the trajectories. Note that a closed-form solution is possible, where the solution represents the system total energy $E$. Introduce a constant of integration that renders $E$ equal to zero at the origin, plot the trajectories corresponding to $E = -0.25$, 0, and 0.5 and discuss the various types of motion.

**9.6** Find the equation of the trajectories for the case in which the Jordan form is given by Eq. (9.34), and plot the corresponding phase portrait.

**9.7** Consider the system of Prob. 9.1, use the theory of Sec. 9.3, and derive matrices $[a]$ corresponding to the equilibrium points of the system. For equilibrium points not coinciding with the origin, use a coordinate transformation to translate the origin to the equilibrium under consideration. Use Eqs. (9.41) and (9.44) to determine the roots $\lambda_1$ and $\lambda_2$ for typical equilibrium points, establish the nature of the equilibrium points and make a statement as to whether the system exhibits significant or critical behavior in the neighborhood of the equilibrium points.

**9.8** Repeat Prob. 9.7 for the system of Prob. 9.2.

**9.9** Repeat Prob. 9.7 for the system of Prob. 9.3.

**9.10** Derive the differential equations of motion for the system shown in Fig. 9.11. The force in the spring is nonlinear and has the form $k[x - (\pi/2)\sin x]$, whereas damping is viscous and linear. Identify the system equilibrium positions, derive the characteristic equation associated with each equilibrium point and test the sign of the real part of the eigenvalues by means of the Routh-Hurwitz criterion.

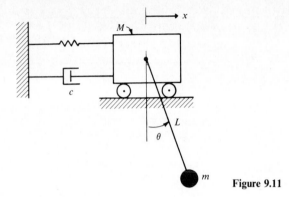

**Figure 9.11**

**9.11** Plot $f(x)$ versus $x$ for the system of Prob. 9.2, where $f(x)$ is the force in the spring. (Note that the force opposes the motion.) Derive the potential energy $V(x)$ and plot $V(x)$ versus $x$. The difference between the horizontal lines $E = \text{const}$ and $V(x)$ is the kinetic energy $T = \frac{1}{2}\dot{x}^2$. Use the plot $V(x)$ versus $x$ to sketch various level curves as demonstrated in Sec. 9.5. Calculate the energy level corresponding to the separatrix. What is the shape of the level curves for very large values of $E$?

**9.12** Consider the system of equations

$$\dot{x}_1 = x_2 + x_1(1 - x_1^2 - x_2^2) \qquad \dot{x}_2 = -x_1 + x_2(1 - x_1^2 - x_2^2)$$

use the coordinate transformation $x_1 = r \cos \theta$, $x_2 = r \sin \theta$ and derive the equation of the trajectories in terms of the polar coordinates $r$ and $\theta$. Integrate the equation, express $r$ as a function of $\theta$, and verify that $r = 1$ is a limit cycle of the system. Sketch several trajectories in the neighborhood of $r = 1$ for initial values corresponding to $r < 1$ and $r > 1$ and establish in this manner that $r = 1$ represents a stable limit cycle. What type of equilibrium point is the origin?

**9.13** Consider the system of Prob. 9.2, let the Liapunov function be the system total energy and test the stability of the equilibrium points by the Liapunov direct method.

**9.14** Consider the system of Prob. 9.3, let the Liapunov function be an integral of the motion (obtained by multiplying the differential equation of motion by $\dot{\theta}$ and integrating with respect to time) and derive stability criteria in terms of $\omega$, $R$, and $g$ for each equilibrium point by means of the Liapunov direct method.

**9.15** Test the stability of the equilibrium points of the system of Prob. 9.10 by means of the Liapunov direct method.

# TEN

# NONLINEAR SYSTEMS.
# PERTURBATION METHODS

## 10.1 GENERAL CONSIDERATIONS

Unlike linear systems, nonlinear systems do not lend themselves to general solutions. As a result, special methods of approach must be adopted in order to gain as much insight into the system behavior as possible. In Chap. 9 we used geometric theory to study the stability of nonlinear systems. The conclusions reached there were of a qualitative nature, in the sense that no time-dependent solutions were obtained, or even sought. Such qualitative results are far from being guaranteed, and the systems lending themselves to stability analysis are limited in number. Moreover, the systems are almost exclusively autonomous.

An entirely different approach to nonlinear systems can be taken if the terms rendering the system nonlinear are small. Under these circumstances, the small nonlinear terms are referred to as *perturbations*, and the system is said to be nearly linear. A similar statement can be made concerning nonautonomous or nonlinear nonautonomous systems. The perturbation terms are generally identified by means of a small parameter $\epsilon$. If the system is nearly linear, or nearly autonomous, then a solution is commonly sought in the form of a power series in the small parameter $\epsilon$. This is the so-called analytical approach, and the techniques used to obtain time-dependent solutions are known as *perturbation methods*.

Systems of special interest in vibrations are those possessing periodic solutions, as such solutions imply bounded motion. Of particular importance are systems that reduce to harmonic oscillators in the absence of perturbations. Such systems are referred to as quasi-harmonic.

In this chapter, we present a number of perturbation methods designed to produce periodic solutions to quasi-harmonic systems. In an attempt to learn as much as possible about the behavior of nonlinear systems, the solutions are used to explain certain phenomena associated with such systems. In particular, we shall show that: (1) for nonlinear systems, the period of oscillation depends on the amplitude, (2) for a given amplitude for the forcing function, nonlinear systems can experience three distinct response amplitudes and the associated "jump" phenomenon, and (3) for a harmonic excitation with a given frequency, nonlinear systems are characterized by a response consisting of harmonic components with a variety of frequencies.

## 10.2 THE FUNDAMENTAL PERTURBATION TECHNIQUE

Many physical systems are described by differential equations that can be separated into one part containing linear terms with constant coefficients and a second part, relatively small compared with the first, containing nonlinear terms or nonautonomous terms. Accordingly, the system is said to be *weakly nonlinear* or *weakly nonautonomous*. The small terms rendering the system nonlinear or nonautonomous, or both, are referred to as *perturbations*. A weakly nonlinear system is called *quasi-linear*. We are interested in systems that reduce to the harmonic oscillator in the absence of perturbations and refer to such systems as *quasi-harmonic*.

Let us consider the quasi-harmonic system described by the differential equation

$$\ddot{x} + \omega_0^2 x = f(x, \dot{x}) \tag{10.1}$$

where $f(x, \dot{x})$ is a nonlinear analytic function of $x$ and $\dot{x}$ which is sufficiently small that it can be regarded as a perturbation. To emphasize that $f(x, \dot{x})$ is small, it is convenient to introduce the small parameter $\epsilon$ and write Eq. (10.1) in the form

$$\ddot{x} + \omega_0^2 x = \epsilon f(x, \dot{x}) \tag{10.2}$$

For $\epsilon = 0$, Eq. (10.2) reduces to the equation of a harmonic oscillator, the solution of which we know, and for $\epsilon = 1$, Eq. (10.2) reduces to Eq. (10.1), the solution of which we seek. The presence of the parameter $\epsilon$ enables us to effect the transition between the known solution and the desired solution.

In general, Eq. (10.1), and hence Eq. (10.2), does not possess a closed-form solution. It is clear, however, that the solution of Eq. (10.2) must depend on $\epsilon$ in addition to the time $t$. Moreover, it must reduce to the solution of the differential equation for the harmonic oscillator as $\epsilon$ reduces to zero. Because $\epsilon$ is a small quantity, we seek a solution of Eq. (10.2) in the form of the power series in $\epsilon$

$$x(t, \epsilon) = x_0(t) + \epsilon x_1(t) + \epsilon^2 x_2(t) + \cdots \tag{10.3}$$

where the functions $x_i(t)$ ($i = 0, 1, 2, \ldots$) are independent of $\epsilon$. Moreover, $x_0(t)$ is the solution of the equation describing the motion of the harmonic oscillator, obtained by letting $\epsilon = 0$ in Eq. (10.2). Solution $x_0(t)$ is referred to as the *zero-order approximation*, or the *generating solution*, of Eq. (10.2). Because the left side of Eq.

(10.2) is linear, we can use Eq. (10.3) and write

$$\ddot{x} + \omega_0^2 x = (\ddot{x}_0 + \epsilon \ddot{x}_1 + \epsilon^2 \ddot{x}_2 + \cdots) + \omega_0^2(x_0 + \epsilon x_1 + \epsilon^2 x_2 + \cdots)$$

$$= \ddot{x}_0 + \omega_0^2 x_0 + \epsilon(\ddot{x}_1 + \omega_0^2 x_1) + \epsilon^2(\ddot{x}_2 + \omega_0^2 x_2) + \cdots \qquad (10.4)$$

Moreover, because $f(x, \dot{x})$ is an analytic function, we assume that it can be expanded into a power series in $\epsilon$ about the generating solution $(x_0, \dot{x}_0)$, so that, inserting Eq. (10.3) into $f(x, \dot{x})$ and collecting terms of like powers of $\epsilon$, we have

$$f(x, \dot{x}) = f(x_0, \dot{x}_0) + \epsilon\left[ x_1 \frac{\partial f(x_0, \dot{x}_0)}{\partial x} + \dot{x}_1 \frac{\partial f(x_0, \dot{x}_0)}{\partial \dot{x}} \right]$$

$$+ \epsilon^2 \left[ x_2 \frac{\partial f(x_0, \dot{x}_0)}{\partial x} + \dot{x}_2 \frac{\partial f(x_0, \dot{x}_0)}{\partial \dot{x}} + \frac{1}{2!} x_1^2 \frac{\partial^2 f(x_0, \dot{x}_0)}{\partial x^2} \right.$$

$$\left. + \frac{2}{2!} x_1 \dot{x}_1 \frac{\partial^2 f(x_0, \dot{x}_0)}{\partial x \partial \dot{x}} + \frac{1}{2!} \dot{x}_1^2 \frac{\partial^2 f(x_0, \dot{x}_0)}{\partial \dot{x}^2} \right] + \cdots \qquad (10.5)$$

where $\partial f(x_0, \dot{x}_0)/\partial x$ denotes $\partial f(x, \dot{x})/\partial x$ evaluated at $x = x_0$, $\dot{x} = \dot{x}_0$, etc. Inserting Eqs. (10.4) and (10.5) into Eq. (10.2), we obtain

$$\ddot{x}_0 + \omega_0^2 x_0 + \epsilon(\ddot{x}_1 + \omega_0^2 x_1) + \epsilon^2(\ddot{x}_2 + \omega_0^2 x_2) + \cdots$$

$$= \epsilon\left\{ f(x_0, \dot{x}_0) + \epsilon\left[ x_1 \frac{\partial f(x_0, \dot{x}_0)}{\partial x} + \dot{x}_1 \frac{\partial f(x_0, \dot{x}_0)}{\partial \dot{x}} \right] \right.$$

$$+ \epsilon^2 \left[ x_2 \frac{\partial f(x_0, \dot{x}_0)}{\partial x} + \dot{x}_2 \frac{\partial f(x_0, \dot{x}_0)}{\partial \dot{x}} + \frac{1}{2!} x_1^2 \frac{\partial^2 f(x_0, \dot{x}_0)}{\partial x^2} \right.$$

$$\left.\left. + \frac{2}{2!} x_1 \dot{x}_1 \frac{\partial^2 f(x_0, \dot{x}_0)}{\partial x \, \partial \dot{x}} + \frac{1}{2!} \dot{x}_1^2 \frac{\partial^2 f(x_0, \dot{x}_0)}{\partial \dot{x}^2} \right] + \cdots \right\} \qquad (10.6)$$

Because Eq. (10.6) must be satisfied for all values of $\epsilon$ and because the functions $x_i$ $(i = 0, 1, 2, \ldots)$ are independent of $\epsilon$, it follows that the coefficients of like powers of $\epsilon$ on both sides of Eq. (10.6) must be equal to one another. This leads to the system of equations

$$\ddot{x}_0 + \omega_0^2 x_0 = 0$$

$$\ddot{x}_1 + \omega_0^2 x_1 = f(x_0, \dot{x}_0)$$

$$\ddot{x}_2 + \omega_0^2 x_2 = x_1 \frac{\partial f(x_0, \dot{x}_0)}{\partial x} + \dot{x}_1 \frac{\partial f(x_0, \dot{x}_0)}{\partial \dot{x}} \qquad (10.7)$$

$$\cdots \cdots \cdots \cdots \cdots \cdots$$

which are all linear and can be solved recursively, because the right side of the equation for $x_n$ contains only variables and their derivatives through $x_{n-1}$ and $\dot{x}_{n-1}$ $(n = 1, 2, 3, \ldots)$.

Expression (10.3), representing the solution of Eq. (10.2) as a power series in the small parameter $\epsilon$, is referred to as a *formal solution*. The sequential solution of

Eqs. (10.7) gives rise to increasingly higher-order approximations for the solution of Eq. (10.2). In our case, the formal solution of Eq. (10.1) is obtained by setting $\epsilon = 1$ which, of course, stipulates that the function $f(x, \dot{x})$ on the right side of Eq. (10.1) is itself small.

The formal solutions need not converge. In fact, there is a real possibility that they may diverge. Nevertheless, such solutions are often more useful for numerical calculations than uniformly and absolutely convergent series, because such power series in $\epsilon$ may give a good approximation by using only a limited number of terms. For this reason they have been used widely in many problems of engineering and applied mathematics. Such series are referred to as asymptotic series and were first introduced by Poincaré (see the text by Ali H. Nayfeh†). The series, however, must reduce to the solution of the associated linear system as $\epsilon \to 0$.

**Example 10.1** Consider the van der Pol's oscillator of Sec. 9.6, assume that the parameter $\mu = \epsilon$ is small, and derive the first four differential equations corresponding to the set (10.7).

The van der Pol's oscillator is described by the differential equation (9.56). Consistent with the formulation of this section, we rewrite Eq. (9.56) in the form

$$\ddot{x} + x = \epsilon(1 - x^2)\dot{x} = \epsilon f(x, \dot{x}) \qquad (a)$$

and use Eq. (10.5) to obtain the expansion

$$f(x, \dot{x}) = (1 - x^2)\dot{x}$$
$$= (1 - x_0^2)\dot{x}_0 + \epsilon[-2x_0\dot{x}_0x_1 + (1 - x_0^2)\dot{x}_1]$$
$$+ \epsilon^2[-2x_0\dot{x}_0x_2 + (1 - x_0^2)\dot{x}_2 - \dot{x}_0x_1^2 - 2x_0x_1\dot{x}_1] + \cdots \qquad (b)$$

Recognizing that in our case $\omega^2 = 1$, Eq. (b) leads to the desired equations

$$\ddot{x}_0 + x_0 = 0$$
$$\ddot{x}_1 + x_1 = (1 - x_0^2)\dot{x}_0$$
$$\ddot{x}_2 + x_2 = -2x_0\dot{x}_0x_1 + (1 - x_0^2)\dot{x}_1 \qquad (c)$$
$$\ddot{x}_3 + x_3 = -2x_0\dot{x}_0x_2 + (1 - x_0^2)\dot{x}_2 - \dot{x}_0x_1^2 - 2x_0x_1\dot{x}_1$$

. . . . . . . . . . . . . . . . . . . . . . . . . . . . . . . . . . . . . . . . . . .

## 10.3 SECULAR TERMS

In seeking a solution in the form of the series (10.3), practical considerations dictate that the series be limited to the first several terms. This can produce an unbounded solution owing to the appearance in the solution of terms that grow

† A. H. Nayfeh, *Introduction to Perturbation Techniques*, sec. 1.5, John Wiley & Sons, Inc., New York, 1981.

indefinitely with time, where these terms are frequently referred to as *secular terms*. Such unbounded solutions can be obtained even for systems that are known to possess bounded solutions, such as conservative systems. Hence, a modification of the formal solution to prevent the formation of secular terms appears desirable. Before discussing such modifications, a closer look into the nature of secular terms is in order.

Let us consider a mass-spring system such as the one in Fig. 1.9a, but with $c = 0$ and with the spring exhibiting nonlinear behavior. In particular, we consider the case in which the restoring force in the spring can be regarded as the sum of two terms, one that varies linearly with the elongation plus another one that varies with the third power of the elongation. As mentioned in Sec. 9.1, such a spring is referred to as a "hardening spring." We shall be concerned with the case in which the cubic term is appreciably smaller than the linear one, so that the spring is nearly linear. Under these circumstances, the system is quasi-harmonic, and its differential equation can be written in the form

$$\ddot{x} + \omega_0^2(x + \epsilon x^3) = 0 \qquad \epsilon \ll 1 \tag{10.8}$$

where $\omega_0 = \sqrt{k/m}$ is the natural frequency of the associated harmonic oscillator, corresponding to $\epsilon = 0$. The symbol $m$ denotes the mass and $k$ can be identified as the slope of the spring force-displacement curve at $x = 0$, which is equal to the spring constant of the linearized system. Equation (10.8) is known as *Duffing's equation*.

If the solution of Eq. (10.8) is assumed in the form (10.3), then we can use Eqs. (10.7) and obtain the differential equations

$$\ddot{x}_0 + \omega_0^2 x_0 = 0$$
$$\ddot{x}_1 + \omega_0^2 x_1 = -\omega_0^2 x_0^3$$
$$\ddot{x}_2 + \omega_0^2 x_2 = -3\omega_0^2 x_0^2 x_1 \tag{10.9}$$

$$\cdots\cdots\cdots\cdots\cdots\cdots$$

which permit a sequential solution. Indeed, the solution of the first of Eqs. (10.9) is simply (see Sec. 1.6)

$$x_0 = A \cos(\omega_0 t + \phi) \tag{10.10}$$

where $A$ and $\phi$ are the constant amplitude and phase angle, respectively. Introducing solution (10.10) into the second of Eqs. (10.9) and recognizing that $\cos^3 a = \frac{1}{4}(3 \cos a + \cos 3a)$, we obtain

$$\ddot{x}_1 + \omega_0^2 x_1 = -\omega_0^2 A^3 \cos^3(\omega_0 t + \phi)$$
$$= -\tfrac{3}{4}\omega_0^2 A^3 \cos(\omega_0 t + \phi) - \tfrac{1}{4}\omega_0^2 A^3 \cos 3(\omega_0 t + \phi) \tag{10.11}$$

which can be verified to have the solution

$$x_1 = -\tfrac{3}{8}\omega_0 t A^3 \sin(\omega_0 t + \phi) + \tfrac{1}{32}A^3 \cos 3(\omega_0 t + \phi) \tag{10.12}$$

Examining solution (10.12), we observe that the first term becomes infinitely large as $t \to \infty$, so that the term is secular.

System (10.8) is conservative, however, and cannot admit an unbounded solution. In fact, the system is of the type studied qualitatively in Sec. 9.5. From Sec. 9.5, we conclude that the potential energy per unit mass has the expression

$$V(x) = -\omega_0^2 \int_x^0 (x + \epsilon x^3)\, dx = \tfrac{1}{2}\omega_0^2 \left( x^2 + \frac{\epsilon}{2} x^4 \right) \tag{10.13}$$

so that, introducing Eq. (10.13) into Eq. (9.52), we obtain

$$\tfrac{1}{2}\dot{x}^2 + \tfrac{1}{2}\omega_0^2 \left( x^2 + \frac{\epsilon}{2} x^4 \right) = E = \text{const} \tag{10.14}$$

or the total energy $E$ per unit mass is conserved. It is not difficult to show that the only equilibrium point of the system is at the origin of the phase space, $x = \dot{x} = 0$, and it is a center. For any given value of $E$, a value that depends on the initial conditions, the motion takes place along the level curve $E = \text{const}$, where the level curve represents a closed trajectory enclosing the center. Because the trajectories are closed, the motion must be periodic and bounded, so that the presence of secular terms in the solution deserves an explanation.

As it turns out, although secular terms increase indefinitely with time, they do not necessarily imply unbounded behavior. To explain this seeming paradox, let us consider the expansion

$$\sin (\omega_0 + \epsilon)t = \sin \omega_0 t \cos \epsilon t + \cos \omega_0 t \sin \epsilon t$$

$$= \left( 1 - \frac{1}{2!} \epsilon^2 t^2 + \frac{1}{4!} \epsilon^4 t^4 - \cdots \right) \sin \omega_0 t$$

$$+ \left( \epsilon t - \frac{1}{3!} \epsilon^3 t^3 + \frac{1}{5!} \epsilon^5 t^5 - \cdots \right) \cos \omega_0 t \tag{10.15}$$

If we assume that $\epsilon$ is small and retain only the first few terms in the series for $\sin \epsilon t$ and $\cos \epsilon t$, then the series will increase indefinitely with time, making it difficult to conclude that it represents the expansion of a bounded function. The function given by Eq. (10.15) is harmonic, but the same argument can be used for periodic functions, provided they are bounded and can be represented by Fourier series.

Periodic solutions are very important in the study of dynamical systems, and secular terms in a solution that is known to be periodic are undesirable. The question remains, however, as to how to produce a periodic solution by retaining only the first few terms by means of a perturbation method. From the above discussion, we conclude that a simple application of the perturbation technique, whereby *only the amplitude is altered*, may not always be satisfactory.

From Eq. (10.14), we conclude that for $\epsilon = 0$ the level curves are ellipses enclosing the origin. The period of motion is constant and equal to $T = 2\pi/\omega_0$. For $\epsilon \neq 0$ the level curves are ellipses of higher order with periods different from $2\pi/\omega_0$. In fact, for a given level curve, the period depends on the small parameter $\epsilon$ as well as on the total energy $E$. Hence, a perturbation method designed to seek periodic solutions must *alter both the amplitude and the period of oscillation*. In the following few sections we shall present several perturbation techniques designed to produce periodic solutions, irrespective of how few terms are used in the expansion.

## 10.4 LINDSTEDT'S METHOD

Let us concern ourselves with the quasi-harmonic system

$$\ddot{x} + \omega_0^2 x = \epsilon f(x, \dot{x}) \tag{10.16}$$

where $\epsilon$ is a small parameter, and $f(x, \dot{x})$ a nonlinear analytic function of $x$ and $\dot{x}$. The linear system obtained by setting $\epsilon = 0$ in Eq. (10.16) has the period $2\pi/\omega_0$. As pointed out in Sec. 10.3, however, the nonlinear term $\epsilon f(x, \dot{x})$ affects not only the amplitude but also the period of the system. Hence, in the presence of the nonlinear term, it is reasonable to expect that the system will no longer have the period $2\pi/\omega_0$ but will have the period $2\pi/\omega$, where $\omega$ is an unknown fundamental frequency depending on $\epsilon$, $\omega = \omega(\epsilon)$.

The essence of Lindstedt's method is to produce periodic solutions of Eq. (10.16) of every order of approximation by taking into account the fact that the period of oscillation is affected by the nonlinear term. According to this method, the solution of Eq. (10.16) is assumed in the form

$$x(t) = x_0(t) + \epsilon x_1(t) + \epsilon^2 x_2(t) + \cdots \tag{10.17}$$

with the stipulation that the solution $x(t)$ be periodic and of period $2\pi/\omega$, where the fundamental frequency $\omega$ is given by

$$\omega = \omega_0 + \epsilon \omega_1 + \epsilon^2 \omega_2 + \cdots \tag{10.18}$$

in which the parameters $\omega_i$ $(i = 1, 2, \ldots)$ are undetermined. They are determined by insisting that all $x_i(t)$ $(i = 0, 1, 2, \ldots)$ be periodic, as explained later. Instead of working with the fundamental frequency $\omega$ as an unknown quantity, it is more convenient to *alter the time scale by changing the independent variable from $t$ to $\tau$*, where *the period of oscillation in terms of the new variable $\tau$ is equal to $2\pi$*. Hence, introducing the substitution $\tau = \omega t$, $d/dt = \omega d/d\tau$ into Eq. (10.16), we obtain

$$\omega^2 x'' + \omega_0^2 x = \epsilon f(x, \omega x') \tag{10.19}$$

where primes designate differentiations with respect to $\tau$. Note that $\tau$ can be regarded as representing a dimensionless time. Next, we must expand $f(x, \omega x')$ in a power series in $\epsilon$. In view of Eqs. (10.17) and (10.18), this expansion has the form

$$f(x, \omega x') = f(x_0, \omega_0 x_0')$$

$$+ \epsilon \left[ x_1 \frac{\partial f(x_0, \omega_0 x_0')}{\partial x} + x_1' \frac{\partial f(x_0, \omega_0 x_0')}{\partial x'} + \omega_1 \frac{\partial f(x_0, \omega_0 x_0')}{\partial \omega} \right]$$

$$+ \epsilon^2 [\cdots] + \cdots \tag{10.20}$$

where $\partial f(x_0, \omega_0 x_0')/\partial x$ denotes $\partial f(x, \omega x')/\partial x$ evaluated at $x = x_0$, $x' = x_0'$, and $\omega = \omega_0$, etc. Introducing Eqs. (10.17), (10.18), and (10.20) into Eq. (10.19), we obtain the system of equations

$$\omega_0^2 x_0'' + \omega_0^2 x_0 = 0$$

$$\omega_0^2 x_1'' + \omega_0^2 x_1 = f(x_0, \omega_0 x_0') - 2\omega_0\omega_1 x_0''$$

$$\omega_0^2 x_2'' + \omega_0^2 x_2 = x_1 \frac{\partial f(x_0, \omega_0 x_0')}{\partial x} + x_1' \frac{\partial f(x_0, \omega_0 x_0')}{\partial x'} + \omega_1 \frac{\partial f(x_0, \omega_0 x_0')}{\partial \omega} \qquad (10.21)$$

$$- (2\omega_0\omega_2 + \omega_1^2)x_0'' - 2\omega_0\omega_1 x_1''$$

................................................

Equations (10.21) are solved recursively, as in Sec. 10.3. In contrast, however, here we have the additional task of determining the quantities $\omega_i$ $(i = 1, 2, \ldots)$, which is accomplished by requiring that each $x_i(t)$ $(i = 0, 1, 2, \ldots)$ be periodic and of period $2\pi$. The periodicity conditions have the mathematical form

$$x_i(\tau + 2\pi) = x_i(\tau) \qquad i = 0, 1, 2, \ldots \qquad (10.22)$$

The functions $x_i$ can be periodic only in the absence of secular terms. But, to ensure that $x_i$ are free of secular terms, we must prevent resonance, which requires that the right sides of Eqs. (10.21) do not contain harmonic terms in $\tau$ of unit frequency. This is guaranteed if the quantities $\omega_i$ $(i = 1, 2, \ldots)$ are so chosen as to render the coefficients of the harmonic terms of unit frequency equal to zero in $x_i$ $(i = 1, 2, \ldots)$. We note from the first of Eqs. (10.21) that no danger of secular terms exists in the case of $x_0$, as the equation for $x_0$ is homogeneous.

The procedure can be demonstrated by means of Duffing's equation discussed in Sec. 10.3, an equation known to possess a periodic solution. In terms of the present notation, we conclude from Eq. (10.8) that $f(x, \omega x') = f(x) = -\omega_0^2 x^3$, so that dividing through by $\omega_0^2$, Eqs. (10.21) reduce to

$$x_0'' + x_0 = 0$$

$$x_1'' + x_1 = -x_0^3 - 2\frac{\omega_1}{\omega_0} x_0''$$

$$x_2'' + x_2 = -3x_0^2 x_1 - \frac{1}{\omega_0^2}(2\omega_0\omega_2 + \omega_1^2)x_0'' - 2\frac{\omega_1}{\omega_0} x_1'' \qquad (10.23)$$

................................................

where the solutions $x_i$ $(i = 1, 2, \ldots)$ are subject to the periodicity conditions (10.22). The generating solution $x_0$ satisfies the periodicity condition automatically. Without loss of generality, we can assume that

$$x_i'(0) = 0 \qquad i = 0, 1, 2, \ldots \qquad (10.24)$$

which is equivalent to assuming that the initial velocity is zero. This can be done by including a phase angle in $\tau$, a procedure permissible by virtue of the fact that the system is autonomous.

Considering the initial condition corresponding to $i = 0$ in (10.24), the solution of the first of Eqs. (10.23) is simply

$$x_0 = A \cos \tau \tag{10.25}$$

Inserting solution (10.25) into the second of Eqs. (10.23) and using the trigonometric relation $\cos^3 \tau = \frac{1}{4}(3 \cos \tau + \cos 3\tau)$, we obtain

$$x_1'' + x_1 = \frac{1}{4} \frac{A}{\omega_0} (8\omega_1 - 3\omega_0 A^2) \cos \tau - \frac{1}{4}A^3 \cos 3\tau \tag{10.26}$$

It is easy to see that the first term on the right side of Eq. (10.26) can lead to resonance, and hence to secular terms. To suppress such terms, we invoke the periodicity condition corresponding to $i = 1$ in Eqs. (10.22), which amounts to simply setting the coefficient of $\cos \tau$ on the right side of Eq. (10.26) equal to zero. This establishes $\omega_1$ as having the value

$$\omega_1 = \frac{3}{8}\omega_0 A^2 \tag{10.27}$$

In addition, if we consider the initial condition corresponding to $i = 1$ in Eqs. (10.24), the particular solution of Eq. (10.26) can be shown to be

$$x_1 = \frac{1}{32}A^3 \cos 3\tau \tag{10.28}$$

where, for uniqueness, the homogeneous solution can be regarded as being accounted for in Eq. (10.25). Inserting Eqs. (10.25), (10.27), and (10.28) into the third of Eqs. (10.23) and using the trigonometric relation $\cos^2 \tau \cos 3\tau = \frac{1}{4}(\cos \tau + 2 \cos 3\tau + \cos 5\tau)$, we obtain

$$x_2'' + x_2 = \frac{1}{128} \frac{A}{\omega_0} (256\omega_2 + 15\omega_0 A^4) \cos \tau + \frac{21}{128}A^5 \cos 3\tau - \frac{3}{128}A^5 \cos 5\tau \tag{10.29}$$

Again, to prevent the formation of secular terms, we must have

$$\omega_2 = -\frac{15}{256}\omega_0 A^4 \tag{10.30}$$

so that, considering Eq. (10.30), the solution of Eq. (10.29) is simply

$$x_2 = -\frac{21}{1024}A^5 \cos 3\tau + \frac{1}{1024}A^5 \cos 5\tau \tag{10.31}$$

The procedure for obtaining higher-order approximations follows the same pattern, and at this point we conclude the discussion of the example by summarizing the results. Introducing Eqs. (10.25), (10.28), and (10.31) into Eq. (10.17), recalling that $\tau = \omega t$ and denoting the phase angle mentioned above by $\phi$, we can write the second-order approximation solution in the form

$$x(t) = A \cos (\omega t + \phi) + \epsilon \frac{1}{32}A^3 (1 - \epsilon \frac{21}{32}A^2) \cos 3(\omega t + \phi)$$

$$+ \epsilon^2 \frac{1}{1024}A^5 \cos 5(\omega t + \phi) \tag{10.32}$$

Moreover, inserting Eqs. (10.27) and (10.30) into Eq. (10.18), we conclude that the associated fundamental frequency is

$$\omega = \omega_0(1 + \epsilon\tfrac{3}{8}A^2 - \epsilon^2\tfrac{15}{256}A^4) \qquad (10.33)$$

so that the effect of the spring nonlinearity is reflected both in the amplitude and in the period of motion. For a given set of initial conditions $x(0)$, $\dot{x}(0)$, the values of $A$ and $\phi$ can be obtained from Eq. (10.32). Having $A$, the second-order approximation to the fundamental frequency $\omega$ is obtained from Eq. (10.33).

In conclusion, whereas the approach of Chap. 9, as illustrated by Example 9.1, helps us realize in a qualitative way that *the period of motion depends on the amplitude*, Eq. (10.33) indicates that this is the case in fact. It should be pointed out that the simple pendulum can be regarded under certain circumstances as a quasi-harmonic system of the type discussed here. Indeed, if the angle $\theta$ is such that the approximation $\sin\theta \approx \theta - \tfrac{1}{6}\theta^3$ is valid, then by setting $-\tfrac{1}{6} = \epsilon$ we recognize that the equation of the pendulum reduces to that of a quasi-harmonic mass-spring system with a softening spring. In fact, although solution (10.32) and (10.33) was obtained with the tacit assumption that $\epsilon$ was a positive quantity, the solution remains valid for negative values of $\epsilon$, provided these values are small.

It remains to show how the amplitude $A$ and phase angle $\phi$ are related to the initial conditions. To this end, let us define an initial time $t_0$ corresponding to $\tau = 0$, so that the initial conditions have the convenient form

$$x(t_0) = A_0 \qquad \dot{x}(t_0) = 0 \qquad (10.34)$$

where $A_0$ can be regarded as an initial displacement. We observe that solution (10.32) satisfies the second of conditions (10.34) automatically because it satisfies initial conditions (10.24). On the other hand, the first of Eqs. (10.34) yields

$$x(t_0) = A + \epsilon\tfrac{1}{32}A^3(1 - \epsilon\tfrac{21}{32}A^2) + \epsilon^2\tfrac{1}{1024}A^5 = A_0 \qquad (10.35)$$

Next, let us expand $A$ in a power series in $\epsilon$ of the form

$$A = A_0 + \epsilon A_1 + \epsilon^2 A_2 + \cdots \qquad (10.36)$$

so that Eq. (10.35) becomes

$$A_0 + \epsilon A_1 + \epsilon^2 A_2 + \cdots + \epsilon\tfrac{1}{32}(A_0 + \epsilon A_1 + \epsilon^2 A_2 + \cdots)^3$$
$$\times [1 - \epsilon\tfrac{21}{32}(A_0 + \epsilon A_1 + \epsilon^2 A_2 + \cdots)^2]$$
$$+ \epsilon^2\tfrac{1}{1024}(A_0 + \epsilon A_1 + \epsilon^2 A_2 + \cdots)^5 = A_0 \qquad (10.37)$$

Ignoring terms of order higher than two in $\epsilon$, and equating coefficients of like powers of $\epsilon$ on both sides, Eq. (10.37) yields the two algebraic equations

$$A_1 + \tfrac{1}{32}A_0^3 = 0 \qquad A_2 + \tfrac{3}{32}A_0^2A_1 - \tfrac{20}{1024}A_0^5 = 0 \qquad (10.38)$$

having the solutions

$$A_1 = -\tfrac{1}{32}A_0^3 \qquad A_2 = \tfrac{23}{1024}A_0^5 \qquad (10.39)$$

so that, to second-order approximation, we have

$$A = A_0 - \epsilon \tfrac{1}{32} A_0^3 + \epsilon^2 \tfrac{23}{1024} A_0^5 \tag{10.40}$$

Inserting Eq. (10.40) into Eq. (10.33), and again retaining only terms through second order in $\epsilon$, Eq. (10.32) reduces to

$$x(t) = A_0 \cos (\omega t + \phi) - \epsilon \tfrac{1}{32} A_0^3 [\cos (\omega t + \phi) - \cos 3(\omega t + \phi)]$$
$$+ \epsilon^2 \tfrac{1}{1024} A_0^5 [23 \cos (\omega t + \phi) - 24 \cos 3(\omega t + \phi) + \cos 5(\omega t + \phi)] \tag{10.41}$$

Moreover, inserting Eq. (10.40) into Eq. (10.33), and again retaining only terms through second order in $\epsilon$, the fundamental frequency becomes

$$\omega = \omega_0 (1 + \epsilon \tfrac{3}{8} A_0^2 - \epsilon^2 \tfrac{21}{256} A_0^4) \tag{10.42}$$

Recognizing that the phase angle $\phi$ is related to the initial time $t_0$ and the fundamental frequency $\omega$ by $\omega t_0 + \phi = 0$, the phase angle becomes

$$\phi = -\omega_0 t_0 (1 + \epsilon \tfrac{3}{8} A_0^2 - \epsilon^2 \tfrac{21}{256} A_0^4) \tag{10.43}$$

In the special case in which $\dot{x}(0) = 0$, the initial time $t_0$ is equal to zero and so is the phase angle $\phi$.

## 10.5 FORCED OSCILLATION OF QUASI-HARMONIC SYSTEMS. JUMP PHENOMENON

Let us consider a quasi-harmonic system consisting of a mass and a nonlinear spring subjected to a harmonic external force, where the system differential equation has the form

$$\ddot{x} + \omega^2 x = \epsilon[-\omega^2(\alpha x + \beta x^3) + F \cos \Omega t] \qquad \epsilon \ll 1 \tag{10.44}$$

in which $\omega$ is the natural frequency of the linearized system, $\alpha$ and $\beta$ are given parameters, $\epsilon F$ is the amplitude of the harmonic external force (per unit mass) and $\Omega$ is the driving frequency. Equation (10.44) is known as *Duffing's equation for an undamped system*, and is recognized as describing a *nonautonomous system*. The object is to explore the existence of periodic solutions of the equation.

Let us explore the possibility that Eq. (10.44) has a periodic solution of period $T = 2\pi/\Omega$. As in Sec. 10.4, we shall find it convenient to change the time scale so that the period of oscillation becomes $2\pi$. To this end, we introduce the substitution $\Omega t = \tau + \phi$, $d/dt = \Omega \, d/d\tau$, where $\tau$ is the new time variable and $\phi$ is an unknown phase angle. Because the system is nonautonomous, the time scale can no longer be shifted, with the implication that the phase angle cannot be chosen arbitrarily but must be determined as part of the solution. In terms of the new time, Eq. (10.44) becomes

$$\Omega^2 x'' + \omega^2 x = \epsilon[-\omega^2(\alpha x + \beta x^3) + F \cos (\tau + \phi)] \tag{10.45}$$

where primes denote differentiations with respect to $\tau$. To prevent secular terms, the solution of Eq. (10.45) must satisfy the periodicity condition

$$x(\tau + 2\pi) = x(\tau) \tag{10.46}$$

whereas the unknown phase angle permits the choice of the initial condition in the convenient form

$$x'(0) = 0 \tag{10.47}$$

We seek a solution of Eq. (10.45) in the form of a power series in $\epsilon$ not only for $x(\tau)$, but also for $\phi$. Hence, we let

$$x(\tau) = x_0(\tau) + \epsilon x_1(\tau) + \epsilon^2 x_2(\tau) + \cdots \tag{10.48}$$

and

$$\phi = \phi_0 + \epsilon \phi_1 + \epsilon^2 \phi_2 + \cdots \tag{10.49}$$

where $x_i(\tau)$ $(i = 1, 2, \ldots)$ are subject to the periodicity conditions

$$x_i(\tau + 2\pi) = x_i(\tau) \qquad i = 0, 1, 2, \ldots \tag{10.50}$$

and the initial conditions

$$x_i'(0) = 0 \qquad i = 0, 1, 2, \ldots \tag{10.51}$$

Introducing Eqs. (10.48) and (10.49) into Eq. (10.45) and equating coefficients of like powers of $\epsilon$ on both sides, we obtain the set of equations

$$\begin{aligned}
&\Omega^2 x_0'' + \omega^2 x_0 = 0 \\
&\Omega^2 x_1'' + \omega^2 x_1 = -\omega^2(\alpha x_0 + \beta x_0^3) + F \cos(\tau + \phi_0) \\
&\Omega^2 x_2'' + \omega^2 x_2 = -\omega^2(\alpha x_1 + 3\beta x_0^2 x_1) - F\phi_1 \sin(\tau + \phi_0)
\end{aligned} \tag{10.52}$$

$$\cdots\cdots\cdots\cdots\cdots\cdots\cdots\cdots\cdots\cdots\cdots\cdots\cdots\cdots\cdots\cdots\cdots$$

which are to be solved sequentially for $x_i(\tau)$ $(i = 0, 1, 2, \ldots)$, subject to the periodicity conditions (10.50) and initial conditions (10.51).

Considering the initial condition corresponding to $i = 0$, the solution of the first of Eqs. (10.52) is simply

$$x_0(\tau) = A_0 \cos \frac{\omega}{\Omega} \tau \tag{10.53}$$

where $A_0$ is constant. Solution (10.53) must satisfy the periodicity condition corresponding to $i = 0$, which is possible only if

$$\omega = \Omega \tag{10.54}$$

In further discussions this is assumed to be the case, and $\Omega$ will be replaced wherever appropriate by $\omega$. Introducing Eq. (10.53) into the second of Eqs. (10.52), dividing through by $\omega^2$ and recalling that $\cos^3 \tau = \frac{1}{4}(3 \cos \tau - \cos 3\tau)$, we obtain

$$x_1'' + x_1 = -\frac{F}{\omega^2} \sin \phi_0 \sin \tau$$

$$-\left(\alpha A_0 + \frac{3}{4} \beta A_0^3 - \frac{F}{\omega^2} \cos \phi_0\right) \cos \tau - \frac{1}{4}\beta A_0^3 \cos 3\tau \quad (10.55)$$

To satisfy the periodicity condition corresponding to $i = 1$, the coefficients of $\sin \tau$ and $\cos \tau$ on the right side of Eq. (10.55) must be zero. There are two ways in which these coefficients can be zero, namely,

$$\alpha A_0 + \tfrac{3}{4}\beta A_0^3 - \frac{F}{\omega^2} = 0 \qquad \phi_0 = 0 \qquad (10.56)$$

and

$$\alpha A_0 + \tfrac{3}{4}\beta A_0^3 + \frac{F}{\omega^2} = 0 \qquad \phi_0 = \pi \qquad (10.57)$$

From Eqs. (10.56) and (10.57), we conclude that for $\phi_0 = 0$ the zero-order response $x_0$ is in phase with the external force, whereas for $\phi_0 = \pi$ the zero-order response is 180° out of phase. But a 180°-out-of-phase response is equivalent to an in-phase response of negative amplitude. Hence, Eqs. (10.57) do not yield any information that cannot be obtained from Eqs. (10.56). Note that $A_0$ can be regarded as being fully determined by the first of Eqs. (10.56).

Considering Eqs. (10.56), as well as the initial condition corresponding to $i = 1$, the solution of Eq. (10.55) becomes

$$x_1(\tau) = A_1 \cos \tau + \tfrac{1}{32}\beta A_0^3 \cos 3\tau \qquad (10.58)$$

where $A_1$ remains to be determined. It is determined from the requirement that $x_2$ be periodic, in the same way as $A_0$ was determined from the periodicity condition imposed on $x_1$.

Introducing Eqs. (10.53) and (10.58) into the third of Eqs. (10.52), we obtain

$$x_2'' + x_2 = -\frac{F\phi_1}{\omega^2} \sin \tau - (\alpha A_1 + \tfrac{9}{4}\beta A_0^2 A_1 + \tfrac{3}{128}\beta^2 A_0^5) \cos \tau$$

$$- \tfrac{1}{4}\beta A_0^2(3A_1 + \tfrac{1}{8}\alpha A_0 + \tfrac{3}{16}\beta A_0^3) \cos 3\tau$$

$$- \tfrac{3}{128}\beta^2 A_0^5 \cos 5\tau \qquad (10.59)$$

For $x_2(\tau)$ to be periodic, the coefficients of $\sin \tau$ and $\cos \tau$ on the right side of Eq. (10.59) must be zero, from which we conclude that

$$A_1 = -\frac{3\beta^2 A_0^5}{32(4\alpha + 9\beta A_0^2)} \qquad \phi_1 = 0 \qquad (10.60)$$

In view of this, and considering the initial condition for $i = 2$, the solution of Eq.

(10.59) is simply

$$x_2(\tau) = A_2 \cos \tau + \tfrac{1}{32}\beta A_0^2(3A_1 + \tfrac{1}{8}\alpha A_0 + \tfrac{3}{16}\beta A_0^3) \cos 3\tau + \tfrac{1}{1024}\beta^2 A_0^5 \cos 5\tau$$

$$(10.61)$$

where $A_2$ is obtained from the next approximation.

The same procedure can be used to derive the higher-order approximations, although this is seldom necessary. Introducing Eqs. (10.53), (10.58), and (10.61) into Eq. (10.48), and changing the independent variable back to $t$, we can write the second-order approximation to the solution of Eq. (10.44) in the form

$$x(t) = (A_0 + \epsilon A_1 + \epsilon^2 A_2) \cos \omega t$$

$$+ \frac{\epsilon}{32}\, \beta A_0^2 [A_0 + \epsilon(3A_1 + \tfrac{1}{8}\alpha A_0 + \tfrac{3}{16}\beta A_0^3)] \cos 3\omega t$$

$$+ \tfrac{1}{1024}\beta^2 A_0^5 \cos 5\omega t \qquad\qquad (10.62)$$

We also note that to first-order approximation the phase angle is zero, $\phi = \phi_0 + \epsilon\phi_1 = 0$. The phase angle turns out to be zero to every order of approximation, a result that can be attributed to the fact that the system is undamped. Indeed, when the system is damped, $\phi \neq 0$, with the implication that the response is out of phase with the excitation.

The first of Eqs. (10.56) gives a relation between the amplitudes of the excitation and response, with the driving frequency $\omega$ playing the role of a parameter. We recall from Secs. 2.2 and 2.3 that the frequency response $G(i\omega)$ gives such a relation for linear systems. Hence, one can expect the first of Eqs. (10.56) to represent an analogous relation for nonlinear systems. Indeed, this is the case, and such an interpretation helps reveal a phenomenon typical of oscillators exhibiting nonlinear behavior, such as that described by Eq. (10.44). To show this, let us introduce the notation

$$\omega_0^2 = (1 + \epsilon\alpha)\omega^2 \qquad\qquad (10.63)$$

where $\omega_0$ can be identified as the natural frequency of the associated linear system, corresponding to the case $\beta = 0$ in Eq. (10.44). Using Eq. (10.63) to eliminate $\epsilon$ from the first of Eqs. (10.56), and recalling that $\epsilon$ is small, we obtain

$$\omega^2 = \omega_0^2(1 + \tfrac{3}{4}\epsilon\beta A_0^2) - \frac{\epsilon F}{A_0} \qquad\qquad (10.64)$$

If $\epsilon\beta$ is regarded as known, then Eq. (10.64) can be used to plot $A_0$ versus $\omega$ with $\epsilon F$ as a parameter and with $\omega$ expressed in terms of $\omega_0$. We note that for $\beta = 0$ the plot $A_0$ versus $\omega$ has two branches, one above and one below the $\omega$ axis, where both branches approach the vertical line $\omega = \omega_0$ asymptotically (see Fig. 10.1).

The vertical line through $\omega = \omega_0$ corresponds to the free-vibration case, $F = 0$. When $\epsilon\beta \neq 0$ but still small, the case $F = 0$ no longer yields the vertical line through $\omega = \omega_0$ but a parabola intersecting the $\omega$ axis at $\omega = \omega_0$. The plots $A_0$ versus $\omega$ corresponding to $\epsilon F \neq 0$ consist of two branches, one above the parabola and one between the $\omega$ axis and the lower half of the parabola, where both

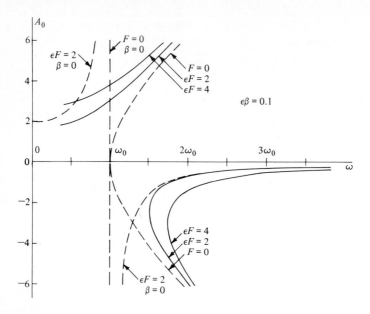

**Figure 10.1**

branches approach the parabola asymptotically, as shown in Fig. 10.1. Hence, the nonlinear effect consists of bending the asymptote $\omega = \omega_0$, corresponding to the linear oscillator, into a parabola. Moreover, the plots $A_0$ versus $\omega$ are also bent so as to cause them to approach the parabola asymptotically. Figure 10.1, corresponding to a hardening spring, shows that the parabola bends to the right. It is easy to prove that for negative values of $\epsilon\beta$, corresponding to a softening spring, the parabola bends to the left. The analogy with the linear oscillator becomes more evident if we plot $|A_0|$ versus $\omega$ instead of $A_0$ versus $\omega$. The first is obtained from the second by folding the lower half of the plane $(A_0, \omega)$ about the $\omega$ axis. Plots of $|A_0|$ versus $\omega$ corresponding to positive and negative values of $\epsilon\beta$ are shown in Fig. 10.2a and b, respectively. From the plots $|A_0|$ versus $\omega$, we observe that all curves are bent to the right for a hardening spring and to the left for a softening spring, compared with the linear oscillator.

In contrast with linear systems, the mass-nonlinear spring system exhibits no resonance. Considering again Fig. 10.2a, let us denote by $T$ the point at which a vertical axis is tangent to a given $|A_0|$ versus $\omega$ curve, and by $\omega_T$ the corresponding frequency. For a hardening spring, such a tangency point can lie only on the right branch of the plot. A vertical through any frequency $\omega$ such that $\omega < \omega_T$ intersects only the left branch of the plot, and only at one point. Hence for $\omega < \omega_T$, Eq. (10.64) has only one real root and two complex roots. On the other hand, for $\omega > \omega_T$, Eq. (10.64) has three distinct real roots, one on the left branch and two on the right branch. Hence, in a certain frequency range, the nonlinear theory predicts the existence of three distinct response amplitudes for a given amplitude of the

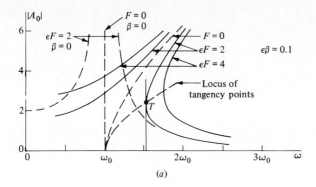

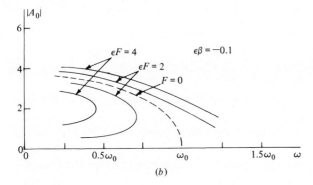

**Figure 10.2**

excitation force. The two roots on the right branch coalesce for $\omega = \omega_T$. As $\omega$ increases from a relatively small value, the amplitude $|A_0|$ increases, but there is no finite value of $\omega$ that renders $|A_0|$ infinitely large. The same conclusion can be reached for a system with a softening spring. Hence, resonance is not possible for mass-nonlinear spring systems, by contrast with mass-linear spring systems, which exhibit resonance at $\omega = \omega_0$.

For a damped system, Duffing's equation has the form

$$\ddot{x} + \omega^2 x = \epsilon[-2\zeta\omega\dot{x} - \omega^2(\alpha x + \beta x^3) + F \cos \Omega t] \qquad (10.65)$$

Following the same procedure as that for the undamped system, we conclude that $x_1$ is periodic if the following relations are satisfied:

$$2\zeta A_0 - \frac{F}{\omega^2} \sin \phi_0 = 0 \qquad (\alpha + \tfrac{3}{4}\beta A_0^2)A_0 - \frac{F}{\omega^2} \cos \phi_0 = 0 \qquad (10.66)$$

From Eqs. (10.66), we obtain the phase angle to the zero-order approximation

$$\phi_0 = \tan^{-1} \frac{2\zeta}{\alpha + \tfrac{3}{4}\beta A_0^2} \qquad (10.67)$$

so that the response is no longer in phase with the excitation. Moreover, using Eq.

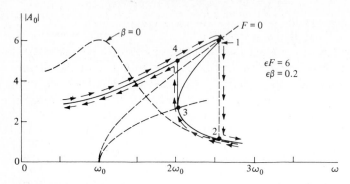

**Figure 10.3**

(10.63), and recalling once again that $\epsilon$ is small, we obtain from Eqs. (10.66)

$$[\omega_0^2(1 + \tfrac{3}{4}\epsilon\beta A_0^2) - \omega^2]^2 + (2\epsilon\zeta\omega_0^2)^2 = \left(\frac{\epsilon F}{A_0}\right)^2 \tag{10.68}$$

which can be used to plot $|A_0|$ versus $\omega$. Figure 10.3 shows such a plot for a damped system with a hardening spring. It is easy to see from Fig. 10.3 that in the presence of viscous damping the amplitude does not increase indefinitely with the driving frequency. Although the plot $|A_0|$ versus $\omega$ is now continuous, in the sense that it no longer consists of two branches, the possibility of discontinuities in the response exists. Indeed, as the driving frequency $\omega$ is increased from a relatively small value, the amplitude $|A_0|$ increases until it reaches point 1, at which point the tangent to the response curve $|A_0|$ versus $\omega$ is infinite and the amplitude experiences a sudden "jump" to point 2 on the lower limb of the response curve, from which point it decreases with an increase in the frequency. On the other hand, if the frequency is decreased from a relatively large value, the amplitude increases until point 3, where the tangent to the response curve becomes infinite again and the amplitude jumps to point 4 on the upper limb, from which point it decreases with a decrease in the frequency. The portion of the response curve between points 1 and 3 is never traversed and is to be regarded as unstable. Whether the system traverses the arc between 4 and 1 or that between 2 and 3 depends on the limb on which the system operates just prior to entering one of the two arcs, as the jump takes place after either one of the two arcs is traversed. Whereas the jump from 3 to 4 can take place also for undamped systems, the jump from 1 to 2 has no counterpart in undamped systems. The jump phenomenon can also occur for a damped system with a softening spring, but the jumps in amplitude take place in reverse directions.

## 10.6 SUBHARMONICS AND COMBINATION HARMONICS

When a linear oscillator is excited by a harmonic forcing function, the response is harmonic and has the same frequency as the excitation. In Sec. 10.5, we demonstrated in the case of a mass-nonlinear spring system, such as that described by Duffing's equation, that if the response to a given harmonic excitation is periodic, then the fundamental frequency of the response is equal to the natural frequency of the linearized system, and it must also be equal to the driving frequency. It turns out that, under certain circumstances, Duffing's equation possesses another periodic solution which has the fundamental frequency equal to one-third of the driving frequency.

Let us consider the equation

$$\ddot{x} + \omega^2 x = -\epsilon \omega^2 (\alpha x + \beta x^3) + F \cos \Omega t \qquad \epsilon \ll 1 \qquad (10.69)$$

where $F$ is not necessarily small. Otherwise, the equation has the same form as Eq. (10.44). Because the nonlinearity is due to the cubic term in $x$, we propose to explore the possibility of a periodic solution of Eq. (10.69) with the fundamental frequency $\omega = \Omega/3$. Letting the solution have the form

$$x(t) = x_0(t) + \epsilon x_1(t) + \epsilon^2 x_2(t) + \cdots \qquad (10.70)$$

inserting solution (10.70) into Eq. (10.69) and equating coefficients of like powers of $\epsilon$ on both sides of the resulting equation, we obtain the set of equations

$$\ddot{x}_0 + \left(\frac{\Omega}{3}\right)^2 x_0 = F \cos \Omega t$$

$$\ddot{x}_1 + \left(\frac{\Omega}{3}\right)^2 x_1 = -\left(\frac{\Omega}{3}\right)^2 (\alpha x_0 + \beta x_0^3)$$

$$\ddot{x}_2 + \left(\frac{\Omega}{3}\right)^2 x_2 = -\left(\frac{\Omega}{3}\right)^2 (\alpha x_1 + 3\beta x_0^2 x_1) \qquad (10.71)$$

$$\cdots \cdots \cdots \cdots \cdots \cdots \cdots \cdots \cdots \cdots \cdots \cdots \cdots$$

which is to be solved sequentially, where $x_i(t)$ $(i = 0, 1, 2, \ldots)$ are subject to the periodicity conditions

$$x_i\left(\frac{\Omega}{3} t + 2\pi\right) = x_i\left(\frac{\Omega}{3} t\right) \qquad i = 0, 1, 2, \ldots \qquad (10.72)$$

and the initial conditions

$$\dot{x}_i(0) = 0 \qquad i = 0, 1, 2, \ldots \qquad (10.73)$$

Taking into account the appropriate initial condition, the solution of the first of Eqs. (10.71) is simply

$$x_0(t) = A_0 \cos \frac{\Omega}{3} t - \frac{9F}{8\Omega^2} \cos \Omega t \qquad (10.74)$$

Introducing solution (10.74) into the second of Eqs. (10.71), and using the trigonometric relation $\cos a \cos b = \frac{1}{2}[\cos(a+b) + \cos(a-b)]$, we obtain

$$\ddot{x}_1 + \left(\frac{\Omega}{3}\right)^2 x_1$$

$$= -\left(\frac{\Omega}{3}\right)^2 \left\{ A_0 \left[ \alpha + \tfrac{3}{4}\beta A_0^2 - \tfrac{3}{4}\beta A_0 \frac{9F}{8\Omega^2} + \tfrac{3}{2}\beta \left(\frac{9F}{8\Omega^2}\right)^2 \right] \cos \frac{\Omega}{3} t \right.$$

$$- \left[ \alpha \frac{9F}{8\Omega^2} - \tfrac{1}{4}\beta A_0^3 + \beta A_0^2 \frac{9F}{8\Omega^2} + \tfrac{3}{4}\beta \left(\frac{9F}{8\Omega^2}\right)^3 \right] \cos \Omega t$$

$$- \tfrac{3}{4}\beta A_0 \frac{9F}{8\Omega^2} \left( A_0 - \frac{9F}{8\Omega^2} \right) \cos \frac{5\Omega}{3} t + \tfrac{3}{4}\beta A_0 \left(\frac{9F}{8\Omega^2}\right)^2 \cos \frac{7\Omega}{3} t$$

$$\left. - \tfrac{1}{4}\beta \left(\frac{9F}{8\Omega^2}\right)^3 \cos 3\Omega t \right\} \qquad (10.75)$$

To prevent the formation of secular terms, the coefficient of $\cos \Omega t/3$ on the right side of Eq. (10.75) must vanish, which yields the quadratic equation in $A_0$

$$A_0^2 - \frac{9F}{8\Omega^2} A_0 + 2\left(\frac{9F}{8\Omega^2}\right)^2 + \frac{4}{3}\frac{\alpha}{\beta} = 0 \qquad (10.76)$$

having the roots

$$A_0 = \frac{1}{2}\frac{9F}{8\Omega^2} \pm \frac{1}{2}\left[ \left(\frac{9F}{8\Omega^2}\right)^2 - 8\left(\frac{9F}{8\Omega^2}\right)^2 - \frac{16}{3}\frac{\alpha}{\beta} \right]^{1/2} \qquad (10.77)$$

Because $A_0$ is by definition a real quantity, a periodic solution of Eq. (10.69) with the fundamental frequency $\Omega/3$ is possible only if

$$-7\left(\frac{9F}{8\Omega^2}\right)^2 - \frac{16}{3}\frac{\alpha}{\beta} \geq 0 \qquad (10.78)$$

If we let $\omega = \Omega/3$ in Eq. (10.63), we obtain the relation

$$\Omega^2 = \frac{9}{\epsilon\alpha}\left( \omega_0^2 - \frac{\Omega^2}{9} \right) \qquad (10.79)$$

so that inequality (10.78) reduces to

$$\Omega^2 \geq 9\left[ \omega_0^2 + \tfrac{21}{16}\epsilon\beta \left(\frac{3F}{8\Omega}\right)^2 \right] \qquad (10.80)$$

Hence, if $\Omega$ satisfies inequality (10.80), then Eq. (10.69) admits a periodic solution with the fundamental frequency equal to $\Omega/3$.

Oscillations with frequencies that are a fraction of the driving frequency are known as *subharmonic oscillations*. Hence, Duffing's equation with no damping, Eq. (10.69), admits a subharmonic solution with the frequency $\Omega/3$. The subharmonic is said to be of order 3, and it should be pointed out that *the order of the subharmonic coincides* with the power of the nonlinear term.

When a linear oscillator is excited by two harmonic forcing functions with distinct frequencies, say $\Omega_1$ and $\Omega_2$, the response is a superposition of two harmonic components with frequencies equal to the excitation frequencies $\Omega_1$ and $\Omega_2$. By contrast, if a mass-nonlinear spring system is excited by two harmonic forcing functions with distinct frequencies $\Omega_1$ and $\Omega_2$, then the response consists of harmonic components with frequencies in the form of integer multiples of $\Omega_1$ and $\Omega_2$ as well as linear combinations of $\Omega_1$ and $\Omega_2$, where the type of harmonics obtained depends on the nature of the nonlinear term. To substantiate this statement, let us consider Duffing's equation in the form

$$\ddot{x} + \omega^2 x = -\epsilon\beta_0 x^3 + F_1 \cos \Omega_1 t + F_2 \cos \Omega_2 t \qquad \epsilon \ll 1 \qquad (10.81)$$

which differs from Eq. (10.69) only to the extent that $\alpha = 0$, and $\beta_0 = \beta\omega^2 = \beta\omega_0^2$. Of course, the excitation now consists of two harmonic forces with distinct frequencies, $\Omega_1 \neq \Omega_2$. Assuming a solution in the form (10.70), we obtain the set of equations

$$\ddot{x}_0 + \omega_0^2 x_0 = F_1 \cos \Omega_1 t + F_2 \cos \Omega_2 t$$

$$\ddot{x}_1 + \omega_0^2 x_1 = -\beta_0 x_0^3$$

$$\ddot{x}_2 + \omega_0^2 x_2 = -3\beta_0 x_0^2 x_1 \qquad (10.82)$$

$$\cdots\cdots\cdots\cdots\cdots\cdots\cdots\cdots\cdots$$

which is to be solved in sequence. For convenience, we require that $x_i(t)$ ($i = 0, 1, 2, \ldots$) satisfy the initial conditions (10.73). To demonstrate the existence of harmonic solutions with frequencies that are integer multiples of $\Omega_1$ and $\Omega_2$ as well as linear combinations of $\Omega_1$ and $\Omega_2$, it is possible to ignore the homogeneous solutions. Hence, the solution of the first of Eqs. (10.82) can be written in the form

$$x_0(t) = G_1 \cos \Omega_1 t + G_2 \cos \Omega_2 t \qquad (10.83)$$

which represents the steady-state response of a harmonic oscillator to two harmonic forces, where

$$G_1 = \frac{F_1}{\omega_0^2 - \Omega_1^2} \qquad G_2 = \frac{F_2}{\omega_0^2 - \Omega_2^2} \qquad (10.84)$$

Introducing solution (10.83) into the second of Eqs. (10.82), and using again the formula $\cos a \cos b = \frac{1}{2}[\cos (a + b) + \cos (a - b)]$, we obtain

$$\ddot{x}_1 + \omega_0^2 x_1 = H_1 \cos \Omega_1 t + H_2 \cos \Omega_2 t$$

$$+ H_3[\cos (2\Omega_1 + \Omega_2)t + \cos (2\Omega_1 - \Omega_2)t]$$

$$+ H_4[\cos (\Omega_1 + 2\Omega_2)t + \cos (\Omega_1 - 2\Omega_2)t]$$

$$+ H_5 \cos 3\Omega_1 t + H_6 \cos 3\Omega_2 t \qquad (10.85)$$

in which

$$H_1 = -\tfrac{3}{4}\beta_0 G_1(G_1^2 + 2G_2^2) \qquad H_2 = -\tfrac{3}{4}\beta_0 G_2(2G_1^2 + G_2^2)$$
$$H_3 = -\tfrac{3}{4}\beta_0 G_1^2 G_2 \qquad\qquad H_4 = -\tfrac{3}{4}\beta_0 G_1 G_2^2 \qquad (10.86)$$
$$H_5 = -\tfrac{1}{4}\beta_0 G_1^3 \qquad\qquad\quad H_6 = -\tfrac{1}{4}\beta_0 G_2^3$$

It is evident from the nature of the right side of Eq. (10.85) that the solution $x_1(t)$ has harmonic components of frequencies $\Omega_1, \Omega_2, 2\Omega_1 \pm \Omega_2, \Omega_1 \pm 2\Omega_2, 3\Omega_1$, and $3\Omega_2$. Hence, in contrast with linear systems, the response of the mass-nonlinear spring system described by Eq. (10.81) consists not only of harmonic components of frequencies $\Omega_1$ and $\Omega_2$ but also of higher harmonics of frequencies $3\Omega_1$ and $3\Omega_2$, as well as harmonics of frequencies $2\Omega_1 \pm \Omega_2$ and $\Omega_1 \pm 2\Omega_2$, where the latter are known as *combination harmonics*. Because the terms with higher harmonics and combination harmonics appear only in the first-order component $x_1(t)$ and not in the zero-order solution $x_0(t)$, they are generally smaller in magnitude than the terms with frequencies equal to the driving frequencies $\Omega_1$, and $\Omega_2$. However, when the value of one of the frequencies $2\Omega_1 \pm \Omega_2, \Omega_1 \pm 2\Omega_2, 3\Omega_1$, and $3\Omega_2$ is in the neighborhood of $\omega_0$, higher amplitudes can be expected.

It should be pointed out that the frequencies $2\Omega_1 \pm \Omega_2, \Omega_1 \pm 2\Omega_2, 3\Omega_1$, and $3\Omega_2$ are peculiar to Eq. (10.81), because the nonlinear term is cubic in $x$. For systems with different types of nonlinearity different higher frequencies and combination frequencies will be obtained.

## 10.7 SYSTEMS WITH TIME-DEPENDENT COEFFICIENTS. MATHIEU'S EQUATION

Let us consider the pendulum illustrated in Fig. 10.4 and denote by $\theta$ the angular displacement of the pendulum and by $u$ the vertical motion of the support when acted upon by the force $F$. The interest lies in the stability characteristics of the

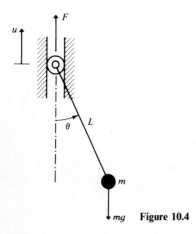

**Figure 10.4**

system as the support executes harmonic motion. For convenience, we describe the motion of the system by the coordinates $\theta$ and $u$, although later we shall treat $u$ as known.

First, let us derive Lagrange's equations of motion. To this end, we write the expression for the kinetic energy, potential energy and virtual work of the system in the form

$$T = \tfrac{1}{2}m[(L\dot\theta \cos \theta)^2 + (\dot u + L\dot\theta \sin \theta)^2]$$

$$= \tfrac{1}{2}m(L^2\dot\theta^2 + 2L\dot u\dot\theta \sin \theta + \dot u^2) \tag{10.87}$$

$$V = mg[L(1 - \cos \theta) + u] \tag{10.88}$$

and

$$\overline{\delta W} = F\delta u \tag{10.89}$$

respectively. Letting $q_1 = \theta$, $q_2 = u$, Lagrange's equations, Eqs. (6.43), become

$$\frac{d}{dt}\left(\frac{\partial T}{\partial \dot\theta}\right) - \frac{\partial T}{\partial \theta} + \frac{\partial V}{\partial \theta} = \Theta$$

$$\frac{d}{dt}\left(\frac{\partial T}{\partial \dot u}\right) - \frac{\partial T}{\partial u} + \frac{\partial V}{\partial u} = U \tag{10.90}$$

where $\Theta$ and $U$ are generalized forces. Writing the virtual work in the form

$$\overline{\delta W} = \Theta\delta\theta + U\delta u \tag{10.91}$$

and comparing Eq. (10.91) with Eq. (10.89), we conclude that

$$\Theta = 0 \qquad U = F \tag{10.92}$$

Finally, inserting Eqs. (10.87) and (10.88) into Eqs. (10.90), performing the indicated differentiations and considering Eqs. (10.92), we obtain the explicit form of Lagrange's equations

$$mL^2\ddot\theta + mL\ddot u \sin \theta + mgL \sin \theta = 0$$

$$mL\ddot\theta \sin \theta + mL^2\dot\theta^2 \cos \theta + m\ddot u + mg = F \tag{10.93}$$

In the neighborhood of $\theta = 0$, Eqs. (10.93) reduce to

$$\ddot\theta + \left(\frac{g}{L} + \frac{\ddot u}{L}\right)\theta = 0 \tag{10.94}$$

$$\ddot u + g = \frac{F}{m}$$

Next, let us assume that the motion of the support is harmonic and of the form

$$u = A \cos \omega t \tag{10.95}$$

Then, the second of Eqs. (10.94) yields

$$F = m(g - A\omega^2 \cos \omega t) \tag{10.96}$$

which represents the force necessary to produce the motion (10.95). On the other hand, the first of Eqs. (10.94) becomes

$$\ddot{\theta} + \left(\frac{g}{L} - \frac{A\omega^2}{L} \cos \omega t\right)\theta = 0 \tag{10.97}$$

which is a *nonautonomous equation*. Equation (10.97) is linear and the coefficient of $\theta$ is a harmonic function of time. Such an equation is known as *Mathieu's equation* and is encountered in various forms in many problems in mathematical physics.

We note that when $A = 0$ Eq. (10.97) reduces to the equation of a simple harmonic oscillator, so that when $A \neq 0$ but small the system is quasi-harmonic. We observe that Eq. (10.97) admits the equilibrium position $\theta = 0$, which is stable for the simple pendulum, $A = 0$. For certain values of the parameters $g/L$ and $A\omega^2/L$ the same equilibrium position can be rendered unstable by the moving support. We propose to study the stability properties of the system for small $A\omega^2/L$ compared with $g/L$. In this case the system behavior can be studied by a perturbation technique. To this end, we shall find it convenient to introduce the notation

$$\theta = x \qquad \frac{g}{L} = \delta \qquad -\frac{A\omega^2}{L} = 2\epsilon \tag{10.98}$$

Moreover, it is customary to let $\omega = 2$, so that Eq. (10.97) reduces to the standard form of Mathieu's equation,

$$\ddot{x} + (\delta + 2\epsilon \cos 2t)x = 0 \qquad \epsilon \ll 1 \tag{10.99}$$

The stability characteristics of Eq. (10.99) can be studied conveniently by means of the parameter plane $(\delta, \epsilon)$. The plane can be divided into regions of stability and instability by the so-called *boundary curves*, or *transition curves*, separating these regions. These transition curves are such that a point belonging to any one curve represents a periodic solution of Eq. (10.99). We shall determine a number of boundary curves by means of Lindstedt's method under the assumption that $\epsilon$ is a small parameter. Hence, let the solution of Eq. (10.99) have the form

$$x(t) = x_0(t) + \epsilon x_1(t) + \epsilon^2 x_2(t) + \cdots \tag{10.100}$$

and, moreover, assume that

$$\delta = n^2 + \epsilon\delta_1 + \epsilon^2\delta_2 + \cdots \qquad n = 0, 1, 2, \ldots \tag{10.101}$$

Inserting Eqs. (10.100) and (10.101) into Eq. (10.99) and equating coefficients of like powers of $\epsilon$ to zero, we obtain the sets of equations

$$\ddot{x}_0 + n^2 x_0 = 0$$
$$\ddot{x}_1 + n^2 x_1 = -(\delta_1 + 2 \cos 2t)x_0$$
$$\ddot{x}_2 + n^2 x_2 = -(\delta_1 + 2 \cos 2t)x_1 - \delta_2 x_0 \qquad n = 0, 1, 2, \ldots \tag{10.102}$$

$$\cdots\cdots\cdots\cdots\cdots\cdots\cdots\cdots\cdots$$

that must be solved sequentially for the various values of $n$ ($n = 0, 1, 2, \ldots$). The zero-order approximation is given by

$$x_0 = \begin{cases} \cos nt \\ \sin nt \end{cases} \qquad n = 0, 1, 2, \ldots \tag{10.103}$$

The boundary curves are obtained by introducing the solutions $x_0 = \cos nt$ and $x_0 = \sin nt$ ($n = 0, 1, 2, \ldots$) into Eqs. (10.102) and insisting that the solutions $x_i(t)$ ($i = 1, 2, \ldots$) be periodic.

Equations (10.102) yield an infinite number of solution pairs, one pair for every value of $n$, with the exception of the case $n = 0$ for which there is only one solution. Considering first the case $n = 0$, in which case $x_0 = 1$, the second of Eqs. (10.102) reduces to

$$\ddot{x}_1 = -\delta_1 - 2 \cos 2t \tag{10.104}$$

For $x_1$ to be periodic, we must have $\delta_1 = 0$. In view of this, the solution of Eq. (10.104) is simply

$$x_1 = \tfrac{1}{2} \cos 2t \tag{10.105}$$

so that the third of Eqs. (10.102) becomes

$$\ddot{x}_2 = (-2 \cos 2t)(\tfrac{1}{2} \cos 2t) - \delta_2 = -(\tfrac{1}{2} + \delta_2) - \tfrac{1}{2} \cos 4t \tag{10.106}$$

For $x_2$ to be periodic, the constant term on the right side of Eq. (10.106) must be set equal to zero, which yields $\delta_2 = -\tfrac{1}{2}$. Hence, corresponding to $n = 0$ there is only one transition curve, namely,

$$\delta = -\tfrac{1}{2}\epsilon^2 + \cdots \tag{10.107}$$

which is a parabola passing through the origin of the parameter plane ($\delta, \epsilon$).

Next, let us consider the case $n = 1$. Corresponding to $x_0 = \cos t$, the second of Eqs. (10.102) becomes

$$\ddot{x}_1 + x_1 = -(\delta_1 + 2 \cos 2t) \cos t = -(\delta_1 + 1) \cos t - \cos 3t \tag{10.108}$$

where we used the relation $\cos a \cos b = \tfrac{1}{2}[\cos (a + b) + \cos (a - b)]$. To prevent the formation of secular terms, we must have $\delta_1 = -1$, from which it follows that the solution of Eq. (10.108) is

$$x_1 = \tfrac{1}{8} \cos 3t \tag{10.109}$$

Inserting $x_0$, $x_1$, and $\delta_1$ into the third of Eqs. (10.102), corresponding to $n = 1$, we obtain

$$\ddot{x}_2 + x_2 = -\tfrac{1}{8}(-1 + 2 \cos 2t) \cos 3t - \delta_2 \cos t$$
$$= -(\tfrac{1}{8} + \delta_2) \cos t + \tfrac{1}{8} \cos 3t - \tfrac{1}{8} \cos 5t \tag{10.110}$$

For $x_2$ to be periodic, the coefficient of $\cos t$ must be zero, or $\delta_2 = -\tfrac{1}{8}$. Hence, the transition curve corresponding to $x_0 = \cos t$ is

$$\delta = 1 - \epsilon - \tfrac{1}{8}\epsilon^2 + \cdots \tag{10.111}$$

For $x_0 = \sin t$, the second of Eqs. (10.102) becomes

$$\ddot{x}_1 + x_1 = -(\delta_1 + 2 \cos 2t) \sin t = -(\delta_1 - 1) \sin t - \sin 3t \qquad (10.112)$$

where we used the relation $\sin a \cos b = \frac{1}{2}[\sin (a + b) + \sin (a - b)]$. The solution $x_1$ is periodic if $\delta_1 = 1$ and it has the form

$$x_1 = \frac{1}{8} \sin 3t \qquad (10.113)$$

In view of this, the third of Eqs. (10.102) becomes

$$\ddot{x}_2 + x_2 = -\frac{1}{8}(1 + 2 \cos 2t) \sin 3t - \delta_2 \sin t$$

$$= -(\tfrac{1}{8} + \delta_2) \sin t - \tfrac{1}{8} \sin 3t + \tfrac{1}{8} \sin 5t \qquad (10.114)$$

so that we must have $\delta_2 = -\frac{1}{8}$. It follows that the transition curve corresponding to $x_0 = \sin t$ is

$$\delta = 1 + \epsilon - \tfrac{1}{8}\epsilon^2 + \cdots \qquad (10.115)$$

In a similar manner, it can be shown that the transition curve corresponding to $x_0 = \cos 2t$ is

$$\delta = 4 + \tfrac{5}{12}\epsilon^2 + \cdots \qquad (10.116)$$

and that corresponding to $x_0 = \sin 2t$ is

$$\delta = 4 - \tfrac{1}{12}\epsilon^2 + \cdots \qquad (10.117)$$

Following the same pattern, transition curves can be obtained for $x_0 = \cos nt$ and $x_0 = \sin nt$ $(n = 3, 4, \ldots)$.

The boundary curves (10.107), (10.111), (10.115), (10.116), and (10.117) can be verified to be the same as those obtained by other methods.† The curves are plotted in Fig. 10.5. For pairs of parameters defining points inside shaded areas the motion is unstable. The region terminating at $\delta = 1$, $\epsilon = 0$ is known as the *principal instability region* and is appreciably wider than those terminating at $\delta = n^2$, $\epsilon = 0$ $(n = 2, 3, \ldots)$, as the latter become progressively narrower as $n$ increases. In discussing the solution of Eq. (10.99) we regarded $\epsilon$ as a positive quantity, although nowhere was such a restriction placed on $\epsilon$. Indeed, the results are valid for both positive and negative values of $\epsilon$, as reflected in Fig. 10.5. We may mention, in passing, that Fig. 10.5 represents what is commonly known as a *Strutt diagram*.

We observe from Fig. 10.5 that stability is possible also for negative values of $\delta$, which corresponds to the equilibrium position $\theta = 180°$. Hence, for the right choice of parameters, the pendulum can be stabilized in the upright position by moving the support harmonically.

---

† See, for example, L. Meirovitch, *Methods of Analytical Dynamics*, sec. 8.5, McGraw-Hill Book Co., New York, 1970.

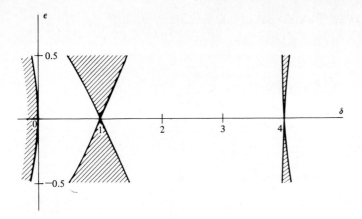

**Figure 10.5**

# PROBLEMS

**10.1** Consider the damped linear oscillator

$$\ddot{x} + 2\epsilon\omega_0\dot{x} + \omega_0^2 x = 0$$

and obtain a perturbation solution of the form (10.3). Include terms through second order in $\epsilon$ in the solution, compare the result with solution (1.54) and draw conclusions. Note that solution (1.54) must be expanded in a power series in $\epsilon$ before a comparison is possible.

**10.2** Consider the quasi-harmonic system described by the differential equation

$$\ddot{x} + x = \epsilon x^2 \qquad \epsilon \ll 1$$

and use Lindstedt's method to obtain a second-order approximation periodic solution. Let the initial conditions be $x(0) = A_0$, $\dot{x}(0) = 0$.

**10.3** Consider the van der Pol equation

$$\ddot{x} + x = \epsilon(1 - x^2)\dot{x} \qquad \epsilon \ll 1$$

and obtain a first-order approximation periodic solution by means of Lindstedt's method. Note that the amplitude is not arbitrary but determined by the periodicity condition. Let $\epsilon = 0.2$ and plot the periodic solution in the phase plane. Draw conclusions as to the meaning of the plot.

**10.4** Consider the differential equation

$$\ddot{x} + \omega^2 x = \epsilon[-\omega^2 \alpha x + \dot{x}(1 - x^2) + F \cos \Omega t] \qquad \epsilon \ll 1$$

describing the behavior of the van der Pol oscillator subjected to harmonic excitation. Use the method of Sec. 10.5 to obtain a first-order approximation periodic solution with period $2\pi/\Omega$.

**10.5** Equation (10.65) is known as Duffing's equation with small damping. Use the method of Sec. 10.5 to obtain a first-order approximation periodic solution with period $2\pi/\Omega$, and verify Eqs. (10.66), (10.67), and (10.68) in the process. Use Eq. (10.68) to plot the response curve for the parameters $\epsilon\zeta = 0.1$, $\epsilon\beta = -0.2$, $\epsilon F = 4$.

**10.6** Consider the differential equation

$$\ddot{x} + \omega^2 x = -\epsilon\omega^2(\alpha x - \beta x^2) + F \cos \Omega t \qquad \epsilon \ll 1$$

and obtain a subharmonic solution.

**10.7** Use the method of Sec. 10.7 to verify Eqs. (10.116) and (10.117).

# ELEVEN

## RANDOM VIBRATIONS

## 11.1 GENERAL CONSIDERATIONS

In our preceding study of vibrations, it was possible to distinguish between three types of excitation functions, namely, harmonic, periodic, and nonperiodic, where the latter is also known as transient. The common characteristic of these functions is that their values can be given in advance for any time $t$. Such functions are said to be *deterministic*, and typical examples are shown in Fig. 11.1$a$, $b$, and $c$. The response of systems to deterministic excitation is also deterministic. For linear systems, there is no difficulty in expressing the response to any arbitrary deterministic excitation in some closed form, such as the convolution integral. The theory of nonlinear systems is not nearly as well developed, and the response to arbitrary excitations cannot be obtained even in the form of a convolution integral. Nevertheless, even for nonlinear systems, the response can be obtained in terms of time by means of numerical integration.

There are many physical phenomena, however, that do not lend themselves to explicit time description. Examples of such phenomena are jet engine noise, the height of waves in a rough sea, the intensity of an earthquake, etc. The implication is that the value at some future time of the variables describing these phenomena cannot be predicted. If the intensity of earth tremors is measured as a function of time, then the record of one tremor will be different from that of another one. The reasons for the difference are many and varied, and they may have little or nothing to do with the measuring instrument. The main reason may be that there are simply too many factors affecting the outcome. Phenomena whose outcome at a future instant of time cannot be predicted are classified as *nondeterministic*, and referred to as *random*. A typical random function is shown in Fig. 11.1$d$.

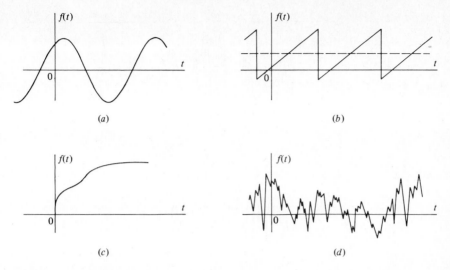

Figure 11.1

The response of a system to random excitation is also a random phenomenon. Because of the complexity involved, the description of random phenomena as functions of time does not appear as a particularly meaningful approach, and new methods of analysis must be adopted. Many random phenomena exhibit a certain pattern, in the sense that the data can be described in terms of certain averages. This characteristic of random phenomena is called *statistical regularity*. If the excitation exhibits statistical regularity, so does the response. In such cases it is more feasible to describe the excitation and response in terms of *probabilities* of occurrence than to seek a deterministic description. In this chapter we develop the tools for the statistical approach to vibration analysis, and then use these tools to derive the response of linear systems to random excitation.

## 11.2 ENSEMBLE AVERAGES. STATIONARY RANDOM PROCESSES

Let us consider an experiment consisting of measuring the displacement of the landing gear (regarded as rigid) of an aircraft taxiing on a given rough runway, and denote by $x_1(t)$ the time history corresponding to that displacement. If at some other time the same aircraft taxies on the same runway under similar conditions, then the associated time history $x_2(t)$ will in general be different from $x_1(t)$ because there may be a slight variation in the tire pressure, the wind conditions may be slightly different, etc. Next let us assume that the experiment is repeated a large number of times, and plot the corresponding time histories $x_k(t)$ ($k = 1, 2, \ldots$), as shown in Fig. 11.2. These time histories are generally different from one another, as can be concluded from Fig. 11.2. The reasons for the time histories being different

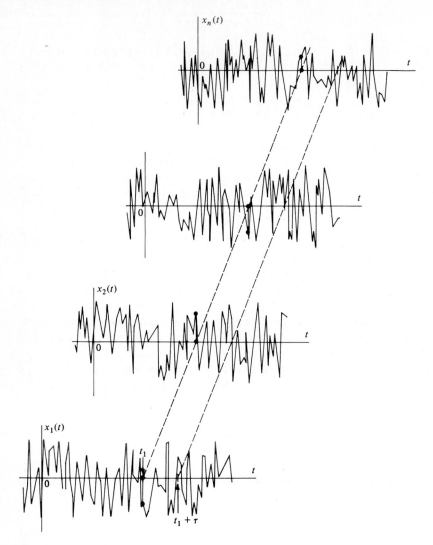

**Figure 11.2**

are very complex, perhaps because not all the factors affecting them were taken into account or are even completely understood. This implies that the time histories cannot be expressed explicitly in terms of known functions of time. Because of this, the displacement under consideration must be regarded as a *random phenomenon*. Random phenomena are quite common in the physical world, and their mathematical treatment can yield meaningful results when the data possess certain regularity, as discussed later.

An individual time history, say $x_k(t)$, describing a random phenomenon is called a *sample function*, and the variable $x_k(t)$ itself is referred to as a *random variable.* The entire collection or *ensemble* of all possible time histories that might result from the experiment is known as a *random process* or *stochastic process*, and denoted by $\{x_k(t)\}$. Note that, although the functions $x_k(t)$ can be regarded as the components of a vector, in this case braces denote a random process and not a vector.

The displacement of the landing gear discussed above plays the role of an excitation to which the aircraft is subjected. Because the excitation is not deterministic, but a random process, the question arises as to how to calculate the response of the system. The simplest approach might be to calculate the response to every sample function in the ensemble. Such an approach would not be very efficient, however, as there may be hundreds of sample functions in the excitation random process, and the amount of work involved in handling the data would, most likely, be prohibitive. Moreover, since the excitation is a random process, the response is also a random process, so that the same difficulty would arise in handling the response data. In addition, there remains the question as to how to interpret the results. Hence, a more efficient and more meaningful way of describing the excitation and response random processes appears highly desirable. To this end, it is necessary to abandon the description of the excitation and response in terms of time in favor of a description based on certain averages. These averages are sometimes referred to as *statistics.* When the averages tend to recognizable limits as the number of sample functions becomes large, the random process is said to exhibit *statistical regularity.*

Let us assume that the random process depicted in Fig. 11.2 consists of $n$ sample functions $x_k(t)$ ($k = 1, 2, \ldots, n$), and compute average values over the collection of sample functions, where such quantities are referred to as *ensemble averages.* The *mean value* of the random process at a given time $t = t_1$ is obtained by simply summing up the values corresponding to the time $t_1$ of all the individual sample functions in the ensemble and dividing the result by the number of sample functions. The implication is that every sample function is assigned equal weight. Moreover, it is assumed that the system possesses statistical regularity. Hence, the mean value at the arbitrary fixed time $t_1$ can be written mathematically as

$$\mu_x(t_1) = \lim_{n \to \infty} \frac{1}{n} \sum_{k=1}^{n} x_k(t_1) \tag{11.1}$$

A different type of ensemble average is obtained by summing up the products of the instantaneous values of the sample functions at two times $t = t_1$ and $t = t_1 + \tau$ (see Fig. 11.2), and dividing the result by the number of sample functions. Such an average is called *autocorrelation function,* and its mathematical expression is

$$R_x(t_1, t_1 + \tau) = \lim_{n \to \infty} \frac{1}{n} \sum_{k=1}^{n} x_k(t_1)x_k(t_1 + \tau) \tag{11.2}$$

By fixing three or more times, such as $t_1, t_1 + \tau, t_1 + \sigma$, etc., we can calculate high-order averages. Such averages are seldom needed, however.

In general, the mean value and the autocorrelation function depend on the time $t_1$. When $\mu_x(t_1)$ and $R_x(t_1, t_1 + \tau)$ do depend on $t_1$, the random process $\{x_k(t)\}$ is said to be *nonstationary*. In the special case in which $\mu_x(t_1)$ and $R_x(t_1, t_1 + \tau)$ do not depend on $t_1$, the random process is said to be *weakly stationary*. Hence, for a weakly stationary random process the mean value is constant, $\mu_x(t_1) = \mu_x = \text{const}$, and the autocorrelation function depends on the time shift $\tau$ alone, $R_x(t_1 t_1 + \tau) = R_x(\tau)$. When all possible averages over $\{x_k(t)\}$ are independent of $t_1$, the random process is said to be *strongly stationary*. In many practical applications, strong stationarity can be assumed if weak stationarity is established. This will be shown to be the case for Gaussian random processes (Sec. 11.5). In view of this, we shall not insist on distinguishing between the two, and refer to a process as simply *stationary*.

## 11.3 TIME AVERAGES. ERGODIC RANDOM PROCESSES

Ensemble averages, such as the mean value and autocorrelation function discussed in Sec. 11.2, generally require a large number of sample functions. Under certain circumstances, however, it is possible to obtain the same mean value and autocorrelation function for a random process $\{x_k(t)\}$ by using a single "representative" sample function and averaging over the time $t$. Such averages are called *time averages* or *temporal averages*, as opposed to ensemble averages. Considering the sample function $x_k(t)$, the *temporal mean value* is defined as

$$\mu_x(k) = \lim_{T \to \infty} \frac{1}{T} \int_{-T/2}^{T/2} x_k(t)\, dt \tag{11.3}$$

whereas the *temporal autocorrelation function* has the expression

$$R_x(k, \tau) = \lim_{T \to \infty} \frac{1}{T} \int_{-T/2}^{T/2} x_k(t) x_k(t + \tau)\, dt \tag{11.4}$$

If the random process $\{x_k(t)\}$ is stationary, and if the temporal mean value $\mu_x(k)$ and the temporal autocorrelation function $R_x(k, \tau)$ are the same, irrespective of the time history $x_k(t)$ over which these averages are calculated, then the process is said to be *ergodic*. Hence, for ergodic processes the temporal mean value and autocorrelation function calculated over a representative sample function must by necessity be equal to the ensemble mean value and autocorrelation function, respectively, so that $\mu_x(k) = \mu_x = \text{const}$ and $R_x(k, \tau) = R_x(\tau)$. As with the stationarity property, we can distinguish between *weakly ergodic* processes, for which the mean value and autocorrelation function are the same regardless of the sample function used, and *strongly ergodic* processes, for which all possible statistics possess this property. Again there is no such distinction for Gaussian random processes (Sec. 11.5), so that we shall assume that weak ergodicity implies also strong ergodicity. It should be pointed out that *any ergodic process is by necessity a stationary process, but a stationary process is not necessarily ergodic.*

The ergodicity assumption permits the use of a single sample function to

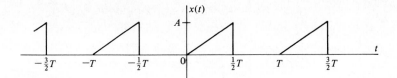

**Figure 11.3**

calculate averages describing a given random process instead of having to use the entire ensemble. The implication is that the chosen sample function is representative of the entire random process. In view of this, the subscript $k$ identifying the particular time history used will be dropped in the future. A great many stationary random processes associated with physical phenomena are ergodic, so that we shall be concerned for the most part with ergodic processes. If a given process is not ergodic but merely stationary, then we must simply work with ensemble averages instead of time averages.

Note that the above time averages are defined for all functions of time, including deterministic functions.

**Example 11.1** Calculate the temporal mean value and autocorrelation function of the function depicted in Fig. 11.3 and plot the autocorrelation function.

Because the function is periodic, averages calculated over a long time duration approach those calculated by considering one period alone. Concentrating on the period $-T/2 < t < T/2$, the function can be described analytically by

$$x(t) = \begin{cases} 0 & -\dfrac{T}{2} < t < 0 \\[2mm] \dfrac{2A}{T} t & 0 < t < \dfrac{T}{2} \end{cases} \tag{a}$$

Hence, using Eq. (11.3), the mean value is simply

$$\mu_x = \frac{1}{T} \int_{-T/2}^{T/2} x(t)\, dt = \frac{1}{T} \int_{0}^{T/2} \frac{2A}{T} t\, dt = \frac{A}{4} \tag{b}$$

To calculate the autocorrelation function, we distinguish between the time shifts $0 < \tau < T/2$ and $T/2 < \tau < T$, as shown in Figs. 11.4$a$ and $b$, respectively. Using Eq. (11.4), and considering Fig. 11.4$a$, we obtain for $0 < \tau < T/2$

$$R_x(\tau) = \frac{1}{T} \int_{-T/2}^{T/2} x(t)x(t+\tau)\, dt = \frac{1}{T} \int_{0}^{(T/2)-\tau} \frac{2A}{T} t\, \frac{2A}{T} (t+\tau)\, dt$$

$$= \frac{A^2}{6} \left[ 1 - 3\frac{\tau}{T} + 4\left(\frac{\tau}{T}\right)^3 \right] \qquad 0 < \tau < \frac{T}{2} \tag{c}$$

where the limits of integration are defined by the overlapping portions of $x(t)$

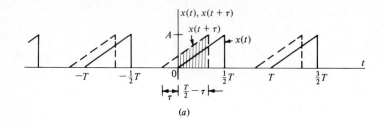

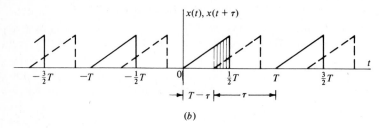

(b)

**Figure 11.4**

and $x(t + \tau)$ (note shaded area in Fig. 11.4a). On the other hand, from Fig. 11.4b, we obtain for $T/2 < \tau < T$

$$R_x(\tau) = \frac{1}{T} \int_{T-\tau}^{T/2} \frac{2A}{T} t \frac{2A}{T} [t - (T - \tau)] \, dt$$

$$= \frac{A^2}{6} \left[ 1 - \frac{3}{T} (T - \tau) + \frac{4}{T^3} (T - \tau)^3 \right] \qquad \frac{T}{2} < \tau < T \qquad (d)$$

The expressions for any other time shifts can be deduced from those above. From Fig. 11.4, it is not difficult to conclude that the autocorrelation function $R_x(\tau)$ must be periodic in $\tau$ with period $T$. Hence, from Eqs. (c) and (d), and the fact that $R_x(\tau)$ is periodic, we can plot the autocorrelation function as shown in Fig. 11.5.

## 11.4 MEAN SQUARE VALUES

The *mean square value* of a random variable provides a measure of the energy associated with the vibration described by that variable. The definition of the mean square value $x(t)$ is

$$\psi_x^2 = \lim_{T \to \infty} \frac{1}{T} \int_{-T/2}^{T/2} x^2(t) \, dt \qquad (11.5)$$

The positive square root of the mean square value is known as the *root mean square*, or the rms. Note that definition (11.5) applies to any arbitrary function $x(t)$, although we are interested in sample functions from an ergodic random process.

For an ergodic process, the mean value $\mu_x$ is constant. In this case, $\mu_x$ can be regarded as the *static component* of $x(t)$, and $x(t) - \mu_x$ as the *dynamic component*. In many applications, the interest lies in the mean square value of the dynamic component. This quantity is simply the mean square value about the mean, and is known as the *variance*. Its expression is

$$\sigma_x^2 = \lim_{T \to \infty} \frac{1}{T} \int_{-T/2}^{T/2} [x(t) - \mu_x]^2 \, dt \tag{11.6}$$

The positive square root of the variance is known as the *standard deviation*. Expanding Eq. (11.6), we obtain

$$\sigma_x^2 = \lim_{T \to \infty} \frac{1}{T} \int_{-T/2}^{T/2} x^2(t) \, dt - 2\mu_x \lim_{T \to \infty} \frac{1}{T} \int_{-T/2}^{T/2} x(t) \, dt + \mu_x^2 \tag{11.7}$$

and, recalling definitions (11.3) and (11.5), Eq. (11.7) reduces to

$$\sigma_x^2 = \psi_x^2 - \mu_x^2 \tag{11.8}$$

or *the variance is equal to the mean square value minus the square of the mean value.*

**Example 11.2** Calculate the mean square value, the variance and the standard deviation of the function of Example 11.1.

Comparing Eqs. (11.4) and (11.5), we conclude that $\psi_x^2 = R_x(0)$, or the mean square value is equal to the autocorrelation function evaluated at $\tau = 0$. Hence, from Eq. (c) of Example 11.1, we obtain simply the mean square value

$$\psi_x^2 = R_x(0) = \frac{A^2}{6} \tag{a}$$

Introducing the above and Eq. (b) of Example 11.1 into Eq. (11.8), we obtain the variance

$$\sigma_x^2 = \psi_x^2 - \mu_x^2 = \frac{A^2}{6} - \left(\frac{A}{4}\right)^2 = \tfrac{5}{48}A^2 \tag{b}$$

The standard deviation is simply

$$\sigma_x = \sqrt{\tfrac{5}{48}}A \tag{c}$$

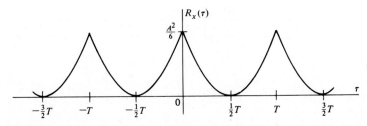

**Figure 11.5**

## 11.5 PROBABILITY DENSITY FUNCTIONS

We have shown that for an ergodic process, averages describing a given random process can be calculated by using a single representative sample function. Information concerning the properties of the random variable in the amplitude domain can be gained by means of *probability density functions*. To introduce the concept, let us consider the time history $x(t)$ depicted in Fig. 11.6a and denote by $\Delta t_1, \Delta t_2, \ldots$ the time intervals during which the amplitude $x(t)$ is smaller than a given value $x$. Denoting by Prob $[x(t) < x]$ the probability that $x(t)$ is smaller than $x$, we observe that Prob $[x(t) < x]$ is equal to the probability that $t$ lies in one of the time intervals $\Delta t_1, \Delta t_2, \ldots$. Considering a given large time interval $T$ such that $0 < t < T$ and assuming that $t$ has an equal chance of taking any value from 0 to $T$, we obtain an estimate of the desired probability in the form

$$\text{Prob}\,[x(t) < x] = \lim_{T \to \infty} \frac{1}{T} \sum_i \Delta t_i \tag{11.9}$$

Letting $x$ vary, we obtain the function

$$P(x) = \text{Prob}\,[x(t) < x] \tag{11.10}$$

which is known as the *probability distribution function* associated with the random variable $x(t)$. The function $P(x)$ is plotted in Fig. 11.6b as a function of $x$. The probability distribution function is a monotonically increasing function possessing the properties

$$P(-\infty) = 0 \qquad 0 \leqslant P(x) \leqslant 1 \qquad P(\infty) = 1 \tag{11.11}$$

Next, let us consider the probability that the amplitude of the random variable is smaller than the value $x + \Delta x$ and denote that probability by $P(x + \Delta x)$. Clearly, the probability that $x(t)$ takes values between $x$ and $x + \Delta x$ is $P(x + \Delta x) - P(x)$. This enables us to introduce the *probability density function*, defined as

$$p(x) = \lim_{\Delta x \to 0} \frac{P(x + \Delta x) - P(x)}{\Delta x} = \frac{dP(x)}{dx} \tag{11.12}$$

Geometrically, $p(x)$ represents the tangent to the probability distribution function

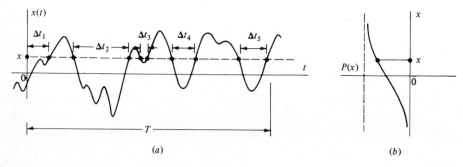

(a)                                                          (b)

**Figure 11.6**

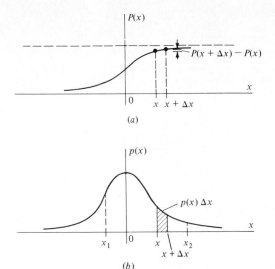

**Figure 11.7**

$P(x)$. Typical functions $P(x)$ and $p(x)$ are shown in Figs. 11.7a and b, respectively. From Eq. (11.12) and Figs. 11.7a and b, we conclude that the area under the curve $p(x)$ versus $x$ corresponding to the amplitude increment $\Delta x$ is equal to the change in $P(x)$ corresponding to the same increment. From Eq. (11.12), it is clear that the probability that $x(t)$ lies between the values $x_1$ and $x_2$ is

$$\text{Prob}\,(x_1 < x < x_2) = \int_{x_1}^{x_2} p(x)\,dx \tag{11.13}$$

which is equivalent to saying that the probability in question is equal to the area under the curve $p(x)$ versus $x$ bounded by the vertical lines through $x = x_1$ and $x = x_2$. The function $p(x)$ has the properties

$$p(x) \geqslant 0 \qquad p(-\infty) = 0 \qquad p(\infty) = 0$$

$$P(x) = \int_{-\infty}^{x} p(\xi)\,d\xi \qquad P(\infty) = \int_{-\infty}^{\infty} p(x)\,dx = 1 \tag{11.14}$$

where $\xi$ is a mere dummy variable of integration.

As an illustration, let us consider first the function $x(t)$ depicted in Fig. 11.8a. The fact that the function is deterministic does not detract from the usefulness of the example. From Fig. 11.8a, we conclude that the probability that $x(t)$ takes values smaller than $-A$ is zero. Similarly, the probability that $x(t)$ takes values smaller than $A$ is equal to unity, because the event is a certainty. Due to the nature of the function $x(t)$, the probability increases linearly from zero at $x = -A$ to unity at $x = A$. The plot $P(x)$ versus $x$ is shown in Fig. 11.8b. Using Eq. (11.12), it is possible to plot $p(x)$ versus $x$, as shown in Fig. 11.8c. The probability density function $p(x)$ is known as the *rectangular*, or *uniform*, *distribution*, for obvious reasons.

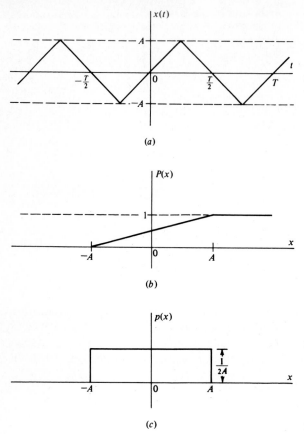

(a)

(b)

(c)

**Figure 11.8**

Of particular interest in our study is the probability distribution associated with a random variable, such as that shown in Fig. 11.9a. According to the *central limit theorem*,† if the random variable is the sum of a large number of independent random variables, none of which contributes significantly to the sum, then under very general conditions the distribution approaches the *normal*, or *Gaussian*, *distribution*. This is true even when the individual distributions of the independent random variables may not be specified, may all be different and may not be Gaussian. The normal distribution is described by the expressions

$$P(x) = \frac{1}{\sqrt{2\pi}} \int_{-\infty}^{x} e^{-\xi^2/2} \, d\xi$$

$$p(x) = \frac{1}{\sqrt{2\pi}} e^{-x^2/2}$$

(11.15)

† See, for example, W. Feller, *Probability Theory and Its Applications*, vol. 1, p. 202, John Wiley & Sons, Inc., New York, 1950.

The functions $P(x)$ versus $x$ and $p(x)$ versus $x$ are plotted in Figs. 11.9b and c, respectively. Figure 11.9c represents the so-called "standardized" normal distribution, in the sense that its mean value is zero and its standard deviation is unity. Normal distributions that are not standardized will be discussed later in this chapter. The probability distribution function $P(x)$ is also known as the *error function*, and appears in tabulated form in many mathematical handbooks, although the definition may vary slightly from table to table.

Another probability distribution of interest is the *Rayleigh distribution*, obtained when the random variable is restricted to positive values. The Rayleigh distribution is defined by

$$P(x) = \begin{cases} 1 - e^{-x^2/2} & x > 0 \\ 0 & x < 0 \end{cases}$$

$$p(x) = \begin{cases} xe^{-x^2/2} & x > 0 \\ 0 & x < 0 \end{cases}$$

(11.16)

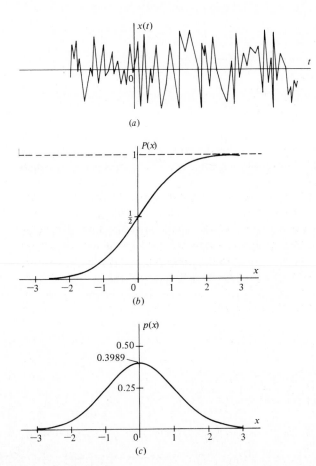

(a)

(b)

(c)

**Figure 11.9**

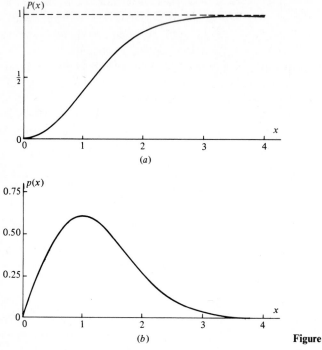

(a)

(b)

**Figure 11.10**

Plots $P(x)$ versus $x$ and $p(x)$ versus $x$ are shown in Figs. 11.10$a$ and $b$, respectively. The Rayleigh distribution discussed here can also be regarded as standardized.

A problem that occurs frequently is to determine the probability density function $p(x)$ associated with the random variable $x = x(y)$ for the case in which the probability density function $p(y)$ associated with the random variable $y$ is known. Let us consider the random variable $x(y)$ depicted in Fig. 11.11, and draw horizontal lines corresponding to $x = x_0$ and $x = x_0 + \Delta x_0$. The intersections of these lines with the curve $x(y)$ versus $y$ define the increments in $y$ bounded by $y_1$ and $y_1 + \Delta y_1$, $y_2$ and $y_2 + \Delta y_2$, etc. But the probability that $x(y)$ lies in the interval bounded by $x_0$ and $x_0 + \Delta x_0$ must be equal to the probability that $y$ lies in any one of the increments bounded by $y_i$ and $y_i + \Delta y_i$ $(i = 1, 2, \ldots)$ so that

$$\text{Prob}\,(x_0 < x < x_0 + \Delta x_0) = \sum_i \text{Prob}\,(y_i < y < y_i + \Delta y_i) \qquad (11.17)$$

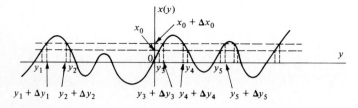

**Figure 11.11**

For a sufficiently small increment $\Delta x_0$, Eq. (11.17) implies that

$$p(x_0)\,\Delta x_0 = \sum_i p(y_i)|\Delta y_i| \tag{11.18}$$

where, because $p(x_0)$ and $p(y_i)$ are positive quantities, the absolute values of $\Delta y_i$ must be used to account for the fact that to a given increment $\Delta x_0$ there may correspond a negative increment $\Delta y_i$, as is the case with $\Delta y_2$, $\Delta y_4$, etc. Letting $x_0$ vary, dropping the identifying subscript 0 that is no longer needed and taking the limit as $\Delta x \to 0$, we obtain the probability density function $p(x)$ in the form

$$p(x) = \sum_i \frac{p(y_i)}{|dx/dy_i|} = \sum_i \left[\frac{p(y)}{|dx/dy|}\right]_{y=y_i} \tag{11.19}$$

where $y_i$ are all the values of $y$ corresponding to $x(y)$. It is clear from Fig. 11.11 that for a given value $x(y) = x$, there can be many values $y = y_i$.

As an illustration, let us consider a sine wave of given amplitude $A$ and frequency $\omega$ but random phase angle $\phi$. For a fixed value $t_0$ of the time $t$, the sine wave can be considered as a random function of $\phi$ and represented as follows:

$$x(\phi) = A \sin(\omega t_0 + \phi) \tag{11.20}$$

The function $x(\phi)$ is plotted in Fig. 11.12. Assuming that $\phi$ has a uniform probability density function, as defined earlier in this section, and considering only the interval $0 < \phi < 2\pi$, we can write

$$p(\phi) = \begin{cases} \dfrac{1}{2\pi} & 0 < \phi < 2\pi \\ 0 & \phi < 0 \qquad \phi > 2\pi \end{cases} \tag{11.21}$$

But from Fig. 11.12 we see that for each value of $x$ in the interval $0 < \phi < 2\pi$ there are two values of $\phi$. Moreover, because the magnitudes of the slopes at these two points are equal, we have

$$p(x) = 2\frac{1}{2\pi}\frac{1}{|dx/d\phi|} = \frac{1}{\pi}\frac{1}{A\cos(\omega t_0 + \phi)} = \frac{1}{\pi}\frac{1}{A[1 - \sin^2(\omega t_0 + \phi)]^{1/2}} \tag{11.22}$$

Inserting Eq. (11.20) into (11.22), and considering the fact that $x$ cannot exceed $A$ in

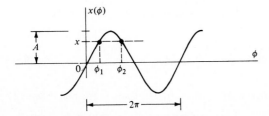

**Figure 11.12**

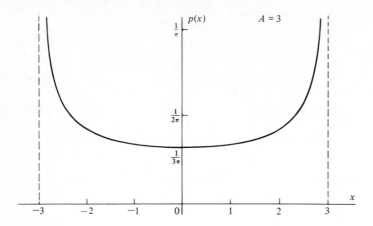

**Figure 11.13**

magnitude, we obtain

$$p(x) = \begin{cases} \dfrac{1}{\pi} \dfrac{1}{(A^2 - x^2)^{1/2}} & |x| < A \\ 0 & |x| > A \end{cases} \tag{11.23}$$

The probability density function $p(x)$ is plotted in Fig. 11.13 for $A = 3$.

## 11.6 DESCRIPTION OF RANDOM DATA IN TERMS OF PROBABILITY DENSITY FUNCTIONS

If a sample time history $x(t)$ from a stationary random process is given, it is often convenient to reduce it to a probability density function $p(x)$. This is done by converting the function $x(t)$ into a voltage signal and feeding it into an analog amplitude probability density analyzer.† If we have the probability density function $p(x)$, various averages can be calculated.

Next, let us consider a real single-valued continuous function $g(x)$ of the random variable $x(t)$. Then, by definition, the *mathematical expectation of $g(x)$*, or the *expected value of $g(x)$*, is given by

$$E[g(x)] = \overline{g(x)} = \int_{-\infty}^{\infty} g(x)p(x)\,dx \tag{11.24}$$

In the special case in which $g(x) = x$, we obtain the *mean value*, or *expected value*, of $x$ in the form

$$E[x] = \bar{x} = \int_{-\infty}^{\infty} xp(x)\,dx \tag{11.25}$$

---

† See J. S. Bendat and A. G. Piersol, *Random Data: Analysis and Measurement Procedures*, sec. 8.2, Interscience-Wiley, New York, 1971.

Note that this definition involves integration with respect to $x$, whereas definition (11.3) involves integration with respect to $t$. When $g(x) = x^2$, definition (11.24) yields

$$E[x^2] = \overline{x^2} = \int_{-\infty}^{\infty} x^2 p(x)\, dx \tag{11.26}$$

which is called the *mean square value* of $x$. As in Sec. 11.4, its square root is known as the *root mean square* value, or rms value.

Following the same pattern, the *variance* of $x$ is

$$\sigma_x^2 = E[(x - \bar{x})^2] = \int_{-\infty}^{\infty} (x - \bar{x})^2 p(x)\, dx$$

$$= \int_{-\infty}^{\infty} x^2 p(x)\, dx - 2\bar{x} \int_{-\infty}^{\infty} x p(x)\, dx + (\bar{x})^2 \int_{-\infty}^{\infty} p(x)\, dx \tag{11.27}$$

Recalling Eqs. (11.25) and (11.26), as well as the fact that $\int_{-\infty}^{\infty} p(x)\, dx = 1$, Eq. (11.27) yields

$$\sigma_x^2 = \overline{x^2} - (\bar{x})^2 \tag{11.28}$$

As in Sec. 11.4, the square root of the variance is known as the *standard deviation*.

The above results can be given a geometric interpretation. To this end, we consider Fig. 11.14, showing the plot $p(x)$ versus $x$, and recall that the area under the curve is equal to unity. Then, if $p(x)\, dx = dA$ is a differential element of area, as indicated in Fig. 11.14, $\bar{x}$ is simply the centroidal distance of the total area under the curve. It also follows that the variance $\sigma_x^2$ is equal to the centroidal moment of inertia of the area, and the standard deviation $\sigma_x$ plays the role of the radius of gyration. Moreover, Eq. (11.28) represents the "parallel axis theorem," according to which the centroidal moment of inertia is equal to the moment of inertia about the point 0 minus the total area times the centroidal distance squared.

The normal probability density function can be expressed in terms of the mean value $\bar{x}$ and standard deviation $\sigma_x$ in the form

$$p(x) = \frac{1}{\sigma_x \sqrt{2\pi}} \exp\left[-\frac{(x - \bar{x})^2}{2\sigma_x^2}\right] \tag{11.29}$$

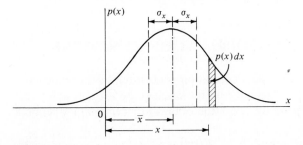

**Figure 11.14**

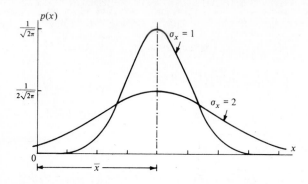

**Figure 11.15**

From Eq. (11.29), we conclude that for small $\sigma_x$ the curve $p(x)$ versus $x$ is sharply peaked at $x = \bar{x}$, whereas for large $\sigma_x$ the curve tends to be flatter and more spread out. Plots of $p(x)$ versus $x$ are shown in Fig. 11.15 for two different values of $\sigma_x$.

**Example 11.3** Calculate the mean value and mean square value of the function $x(t)$ of Example 11.1 by using the probability density function of $x(t)$.

Using the analogy with the function of Fig. 11.8a, it can be shown that the function of Example 11.1 has the probability density function (see Prob. 11.10)

$$p(x) = \begin{cases} \dfrac{1}{2A}[A\delta(x) + 1] & 0 \leqslant x < A \\ 0 & \text{everywhere else} \end{cases} \tag{a}$$

where $\delta(x)$ is the Dirac delta function.

Inserting Eq. (a) into (11.25), we obtain the mean value

$$E[x] = \int_{-\infty}^{\infty} xp(x)\,dx = \int_{0}^{A} x\,\frac{1}{2A}[A\delta(x) + 1]\,dx = \frac{A}{4} \tag{b}$$

Moreover, introducing Eq. (a) into (11.26), we arrive at the mean square value

$$E[x^2] = \int_{-\infty}^{\infty} x^2 p(x)\,dx = \int_{0}^{A} x^2\,\frac{1}{2A}[A\delta(x) + 1]\,dx = \frac{A^2}{6} \tag{c}$$

Note that the mean value and mean square value obtained here agree with those obtained in Example 11.2 by using time averages, which is to be expected.

## 11.7 PROPERTIES OF AUTOCORRELATION FUNCTIONS

The autocorrelation function provides information concerning the dependence of the value of a random variable at one time on the value of the variable at another time. We recall from Sec. 11.3 that the definition of the autocorrelation function is

$$R_x(\tau) = \lim_{T \to \infty} \frac{1}{T} \int_{-T/2}^{T/2} x(t)x(t + \tau)\,dt \tag{11.30}$$

Next, let us consider

$$R_x(-\tau) = \lim_{T \to \infty} \frac{1}{T} \int_{-T/2}^{T/2} x(t)x(t - \tau)\, dt$$

$$= \lim_{T \to \infty} \frac{1}{T} \int_{(-T/2)-\tau}^{(T/2)-\tau} x(\lambda)x(\tau + \lambda)\, d\lambda \tag{11.31}$$

where we made the substitution $t - \tau = \lambda$, $dt = d\lambda$. Because both limits of integration in the last integral are shifted in the same direction and by the same amount $\tau$, the interval of integration remains equal to $T$. It is easy to see that, as $T \to \infty$, the shift in the location of the interval of integration becomes inconsequential, so that

$$R_x(-\tau) = \lim_{T \to \infty} \frac{1}{T} \int_{-T/2}^{T/2} x(t)x(t + \tau)\, dt \tag{11.32}$$

Comparing Eqs. (11.30) and (11.32), we conclude that

$$R_x(\tau) = R_x(-\tau) \tag{11.33}$$

or *the autocorrelation is an even function of $\tau$.*

Another property of the autocorrelation function can be revealed by considering

$$\lim_{T \to \infty} \frac{1}{T} \int_{-T/2}^{T/2} [x(t) \pm x(t + \tau)]^2\, dt$$

$$= \lim_{T \to \infty} \frac{1}{T} \int_{-T/2}^{T/2} [x^2(t) \pm 2x(t)x(t + \tau) + x^2(t + \tau)]\, dt$$

$$= \lim_{T \to \infty} \frac{2}{T} \int_{-T/2}^{T/2} x^2(t)\, dt \pm \lim_{T \to \infty} \frac{2}{T} \int_{-T/2}^{T/2} x(t)x(t + \tau)\, dt$$

$$= 2R_x(0) \pm 2R_x(\tau) \geqslant 0 \tag{11.34}$$

The above inequality is true because the first integral cannot be negative. From inequality (11.34), it follows that

$$R_x(0) \geqslant |R_x(\tau)| \tag{11.35}$$

which implies that *the maximum value of the autocorrelation function is obtained at $\tau = 0$.* From definition (11.5) we conclude that $R_x(0)$ is equal to the mean square value of the random variable $x(t)$, namely,

$$R_x(0) = \psi_x^2 \tag{11.36}$$

Hence, the maximum value of the autocorrelation function is equal to the mean square value.

Note that if $x(t)$ is periodic, then $R_x(t)$ is also periodic, and the maximum value of $R_x(\tau)$ is obtained not only at $\tau = 0$ but also for values of $\tau$ that are integer multiples of the period. An illustration of this fact can be seen in Fig. 11.5.

## 11.8 RESPONSE TO RANDOM EXCITATION. FOURIER TRANSFORMS

Throughout this chapter, we computed various statistical averages by carrying out integrations in the time domain. In random vibrations, it is often convenient to describe the excitation and response in terms of functions in the frequency domain. This requires the introduction of new concepts, and in particular the *Fourier transform*.

In Sec. 2.12, we demonstrated that a periodic function of period $T$, such as that shown in Fig. 2.21, can be represented by a Fourier series, namely, an infinite series of harmonic functions of frequencies $p\omega_0$ ($p = 0, \pm 1, \pm 2, \ldots$), where $\omega_0 = 2\pi/T$ is the fundamental frequency. Letting the period $T$ approach infinity, the function becomes nonperiodic. In the process, the discrete frequencies $p\omega_0$ draw closer and closer together until they become continuous, at which time the Fourier series becomes a Fourier integral.

Let us return to the periodic function illustrated in Fig. 2.21 and represent it by the Fourier series in its complex form,

$$f(t) = \sum_{p=-\infty}^{\infty} C_p e^{ip\omega_0 t} \qquad \omega_0 = \frac{2\pi}{T} \tag{11.37}$$

where the coefficients $C_p$ are given by

$$C_p = \frac{1}{T} \int_{-T/2}^{T/2} f(t) e^{-ip\omega_0 t} \, dt \qquad p = 0, \pm 1, \pm 2, \ldots \tag{11.38}$$

provided the integrals exist. The Fourier expansion, Eqs. (11.37) and (11.38), provides the information concerning the frequency composition of the periodic function $f(t)$. Introducing the notation $p\omega_0 = \omega_p$, $(p + 1)\omega_0 - p\omega_0 = \omega_0 = 2\pi/T = \Delta\omega_p$, Eqs. (11.37) and (11.38) can be rewritten as

$$f(t) = \sum_{p=-\infty}^{\infty} \frac{1}{T}(TC_p) e^{i\omega_p t} = \frac{1}{2\pi} \sum_{p=-\infty}^{\infty} (TC_p) e^{i\omega_p t} \, \Delta\omega_p \tag{11.39}$$

$$TC_p = \int_{-T/2}^{T/2} f(t) e^{-i\omega_p t} \, dt \tag{11.40}$$

Letting the period increase indefinitely, $T \to \infty$, dropping the subscript $p$ so that the discrete variable $\omega_p$ simply becomes the continuous variable $\omega$ and taking the limit, we can replace the summation in Eq. (11.39) by integration and obtain

$$f(t) = \lim_{\substack{T \to \infty \\ \Delta\omega_p \to 0}} \frac{1}{2\pi} \sum_{p=-\infty}^{\infty} (TC_p) e^{i\omega_p t} \, \Delta\omega_p = \frac{1}{2\pi} \int_{-\infty}^{\infty} F(\omega) e^{i\omega t} \, d\omega \tag{11.41}$$

$$F(\omega) = \lim_{\substack{T \to \infty \\ \Delta\omega_p \to 0}} (TC_p) = \int_{-\infty}^{\infty} f(t) e^{-i\omega t} \, dt \tag{11.42}$$

Equation (11.41) implies that any arbitrary function $f(t)$ can be described by an

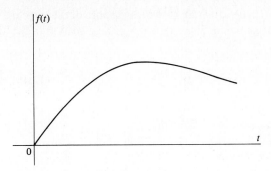

**Figure 11.16**

integral representing contributions of harmonic components having a *continuous frequency spectrum* ranging from $-\infty$ to $+\infty$. The quantity $F(\omega)\,d\omega$ can be regarded as the contribution to the function $f(t)$ of the harmonics in the frequency interval from $\omega$ to $\omega + d\omega$.

Equation (11.41) is the Fourier integral representation of an arbitrary function $f(t)$, such as that shown in Fig. 11.16, which is obtained from Fig. 2.21 by stretching the period $T$ indefinitely. Moreover, the function $F(\omega)$ in Eq. (11.42) is known as the *Fourier transform of $f(t)$*, so that the integrals

$$F(\omega) = \int_{-\infty}^{\infty} f(t)e^{-i\omega t}\,dt \qquad (11.43)$$

$$f(t) = \frac{1}{2\pi}\int_{-\infty}^{\infty} F(\omega)e^{i\omega t}\,d\omega \qquad (11.44)$$

represent simply a Fourier transform pair, where $f(t)$ is known as the *inverse Fourier transform of $F(\omega)$*. By analogy with the Fourier series expansion of a periodic function, Eqs. (11.37) and (11.38), the Fourier transform pair, Eqs. (11.43) and (11.44), also provides the information concerning the frequency composition of $f(t)$, where this time $f(t)$ is nonperiodic.

The representation of $f(t)$ by an integral is possible provided the integral (11.43) exists. The existence is ensured if $f(t)$ satisfies Dirichlet's conditions† in the domain $-\infty < t < \infty$ and if the integral $\int_{-\infty}^{\infty} |f(t)|\,dt$ is convergent. If the integral $\int_{-\infty}^{\infty} |f(t)|\,dt$ is not convergent, then the Fourier transform $F(\omega)$ need not exist. This is indeed the case for $f(t) = \sin \alpha t$, for which the integral $\int_{-\infty}^{\infty} |f(t)|\,dt$ is divergent.

From Sec. 2.12, we conclude that if Eq. (11.37) represents an excitation function, then the response of the system can be written in the form

$$x(t) = \sum_{p=-\infty}^{\infty} G_p C_p e^{ip\omega_0 t} \qquad (11.45)$$

where $G_p$ is the frequency response associated with the frequency $p\omega_0$. Following a

---

† The function $f(t)$ is said to satisfy Dirichlet's conditions in the interval $(a, b)$ if (1) $f(t)$ has only a finite number of maxima and minima in $(a, b)$ and (2) $f(t)$ has only a finite number of finite discontinuities in $(a, b)$, and no infinite discontinuities.

procedure similar to that used for $f(t)$, we conclude that the response of the system to an arbitrary excitation of the type shown in Fig. 11.16 can also be written in the form of a Fourier transform pair, as follows:

$$X(\omega) = \int_{-\infty}^{\infty} x(t)e^{-i\omega t}\, dt \tag{11.46}$$

$$x(t) = \frac{1}{2\pi} \int_{-\infty}^{\infty} X(\omega)e^{i\omega t}\, d\omega \tag{11.47}$$

where the Fourier transform of the response is

$$X(\omega) = G(\omega)F(\omega) \tag{11.48}$$

which is simply the product of the frequency response and the Fourier transform of the excitation. Note that, for consistency of notation, we dropped $i$ from the argument of $G$.

To obtain the system response as a function of time, it is necessary to evaluate the definite integral in Eq. (11.47), which can lead to contour integrations in the complex plane, a delicate task at best. However, when the frequency composition rather than the time dependence of the response is of interest, Fourier transforms are of great value. This is certainly true when the excitation is nondeterministic, as is the case in random vibration.

**Example 11.4** Calculate the response $x(t)$ of an undamped single-degree-of-freedom system to the excitation $f(t)$ in the form of a rectangular pulse, such as that shown in Fig. 2.27, by using an approach based on the Fourier transform. Plot the frequency spectra associated with $f(t)$ and $x(t)$.

Recalling that $f(t) = F(t)/k$, the function $f(t)$ can be defined by

$$f(t) = \begin{cases} \dfrac{F_0}{k} & \text{for } -T < t < T \\ 0 & \text{for } t < T, t > T \end{cases} \tag{a}$$

and we note that $f(t)$ has only two finite discontinuities and no infinite discontinuities, so that $f(t)$ satisfies Dirichlet's conditions. Hence, it is possible to write a Fourier transform for $f(t)$ as follows:

$$F(\omega) = \int_{-\infty}^{\infty} f(t)e^{-i\omega t}\, dt = \frac{F_0}{k}\int_{-T}^{T} e^{-i\omega t}\, dt = \frac{F_0}{k}\frac{1}{i\omega}\left(e^{i\omega T} - e^{-i\omega T}\right) \tag{b}$$

For $\zeta = 0$ the frequency response, Eq. (2.46), reduces to

$$G(\omega) = \frac{1}{1 - (\omega/\omega_n)^2} \tag{c}$$

so that, inserting Eqs. (b) and (c) into Eq. (11.48), we obtain

$$X(\omega) = G(\omega)F(\omega) = \frac{F_0}{k}\frac{e^{i\omega T} - e^{-i\omega T}}{i\omega[1 - (\omega/\omega_n)^2]} \tag{d}$$

Hence, the response $x(t)$ can be written in the form of the inverse Fourier transform

$$x(t) = \frac{1}{2\pi} \int_{-\infty}^{\infty} X(\omega)e^{i\omega t}\, d\omega = \frac{F_0}{k}\frac{1}{2\pi i} \int_{-\infty}^{\infty} \frac{e^{i\omega T} - e^{-i\omega T}}{\omega[1 - (\omega/\omega_n)^2]}\, e^{i\omega t}\, d\omega \qquad (e)$$

Before attempting the evaluation of the above integral, it will prove convenient to consider the following partial fractions expansion:

$$\frac{1}{\omega[1 - (\omega/\omega_n)^2]} = \frac{1}{\omega} - \frac{1}{2(\omega - \omega_n)} - \frac{1}{2(\omega + \omega_n)} \qquad (f)$$

so that Eq. $(e)$ becomes

$$x(t) = \frac{F_0}{k}\frac{1}{2\pi i} \int_{-\infty}^{\infty} \left[\frac{1}{\omega} - \frac{1}{2(\omega - \omega_n)} - \frac{1}{2(\omega + \omega_n)}\right][e^{i\omega(t + T)} - e^{i\omega(t - T)}]\, d\omega$$

$$\qquad (g)$$

To evaluate the integrals involved in $(g)$ it is necessary to perform contour integrations in the complex plane. As this exceeds the scope of this text, we present here pertinent results only, namely,

$$\int_{-\infty}^{\infty} \frac{e^{i\omega\lambda}}{\omega}\, d\omega = \begin{cases} 0 & \text{for } \lambda < 0 \\ 2\pi i & \text{for } \lambda > 0 \end{cases}$$

$$\int_{-\infty}^{\infty} \frac{e^{i\omega\lambda}}{\omega - \omega_n}\, d\omega = \begin{cases} 0 & \text{for } \lambda < 0 \\ 2\pi i e^{i\omega_n\lambda} & \text{for } \lambda > 0 \end{cases} \qquad (h)$$

$$\int_{-\infty}^{\infty} \frac{e^{i\omega\lambda}}{\omega + \omega_n}\, d\omega = \begin{cases} 0 & \text{for } \lambda < 0 \\ 2\pi i e^{-i\omega_n\lambda} & \text{for } \lambda > 0 \end{cases}$$

From Eq. $(g)$ we note that $\lambda$ takes the values $t + T$ and $t - T$. Hence, we must distinguish between the time domains defined by $t + T < 0$ and $t - T < 0$, $t + T > 0$ and $t - T < 0$, and $t + T > 0$ and $t - T > 0$, which are the same as the domains $t < -T$, $-T < t < T$, and $t > T$, respectively. Inserting the integrals $(h)$ with proper $\lambda$ into $(g)$, we obtain

$$x(t) = 0 \qquad \text{for } t < -T$$

$$x(t) = \frac{F_0}{k}\frac{1}{2\pi i}(2\pi i - \tfrac{1}{2}2\pi i e^{i\omega_n(t + T)} - \tfrac{1}{2}2\pi i e^{-i\omega_n(t + T)})$$

$$= \frac{F_0}{k}[1 - \cos \omega_n(t + T)] \qquad \text{for } -T < t < T$$

$$\qquad (i)$$

$$x(t) = \frac{F_0}{k}\frac{1}{2\pi i}[(2\pi i - \tfrac{1}{2}2\pi i e^{i\omega_n(t + T)} - \tfrac{1}{2}2\pi i e^{-i\omega_n(t + T)})$$

$$- (2\pi i - \tfrac{1}{2}2\pi i e^{i\omega_n(t - T)} - \tfrac{1}{2}2\pi i e^{-i\omega_n(t - T)})]$$

$$= \frac{F_0}{k}[\cos \omega_n(t - T) - \cos \omega_n(t + T)] \qquad \text{for } t > T$$

Note that $x(t)$ is the same as that given by Eq. $(c)$ of Example 2.5.

The frequency spectrum associated with $f(t)$ is given by Eq. (b). Recalling that $(e^{i\omega T} - e^{-i\omega T})/2i = \sin \omega T$, Eq. (b) becomes

$$F(\omega) = \frac{2F_0}{k} \frac{\sin \omega T}{\omega} \qquad (j)$$

Figure 11.17a shows the plot $F(\omega)$ versus $\omega$. Moreover, the frequency spectrum associated with $x(t)$ is given by Eq. (d). In a similar manner, the equation can be reduced to

$$X(\omega) = \frac{2F_0}{k} \frac{\sin \omega T}{\omega[1 - (\omega/\omega_n)^2]} \qquad (k)$$

Figure 11.17b shows the plot $X(\omega)$ versus $\omega$. Note that Figs. 11.17a and b represent continuous frequency spectra, as opposed to Figs. 2.23a and b, which represent discrete frequency spectra.

Comparing the method of solution of this example to that of Example 2.5, it is easy to see that the use of the convolution integral provides a simpler approach to the problem of obtaining the response $x(t)$ than the Fourier transform approach. This is particularly true in view of the fact that the question of contour integrations in the complex plane has really been avoided in this example. In random vibration, however, the time-domain response plays no particular role and the interest lies primarily in frequency-domain analyses for which Fourier transforms are indispensable. The preceding statement refers to spectral analysis, a basic tool in the treatment of random vibration.

## 11.9 POWER SPECTRAL DENSITY FUNCTIONS

The autocorrelation function provides information concerning properties of a random variable in the time domain. On the other hand, the *power spectral density function* provides similar information in the frequency domain. Although for ergodic random processes the power spectral density function furnishes essentially no information not furnished by the autocorrelation function, in certain applications the first form is more convenient than the second.

Let us consider the representative sample function $f(t)$ from the ergodic random process $\{f(t)\}$ and write the autocorrelation function of the process in the form

$$R_f(\tau) = \lim_{T \to \infty} \frac{1}{T} \int_{-T/2}^{T/2} f(t)f(t + \tau)\, dt \qquad (11.49)$$

Then, let us define the power spectral density function $S_f(\omega)$ as the Fourier transform of $R_f(\tau)$, namely,

$$S_f(\omega) = \int_{-\infty}^{\infty} R_f(\tau)e^{-i\omega\tau}\, d\tau \qquad (11.50)$$

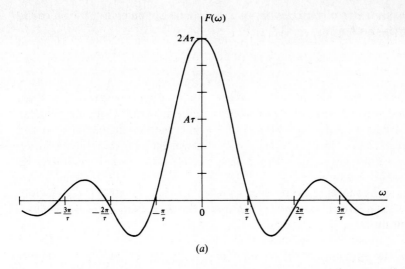

(a)

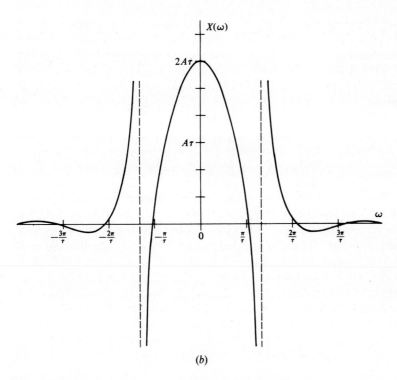

(b)

**Figure 11.17**

which implies that the autocorrelation function can be obtained in the form of the inverse Fourier transform

$$R_f(\tau) = \frac{1}{2\pi} \int_{-\infty}^{\infty} S_f(\omega) e^{i\omega\tau} \, d\omega \qquad (11.51)$$

The conditions for the existence of the power spectral density function $S_f(\omega)$ are that the function $R_f(\tau)$ satisfy Dirichlet's conditions and that the integral $\int_{-\infty}^{\infty} |R_f(\tau)| \, dt$ be convergent (see Sec. 11.8). Various authors define $S_f(\omega)$ as the quantity given by Eq. (11.50) divided by $2\pi$. As will be seen shortly, this latter definition has certain advantages. However, in this case $S_f(\omega)$ would no longer be the Fourier transform of $R_f(\tau)$.

Next, we wish to explore the physical significance of the function $S_f(\omega)$. To this end, we let $\tau = 0$ in Eqs. (11.49) and (11.51), and write the mean square value of $f(t)$ in the two forms

$$R_f(0) = \lim_{T \to \infty} \frac{1}{T} \int_{-T/2}^{T/2} f^2(t) \, dt = \frac{1}{2\pi} \int_{-\infty}^{\infty} S_f(\omega) \, d\omega \qquad (11.52)$$

Assuming that $f(t)$ describes a voltage, the mean square value of $f(t)$ represents the mean power dissipated in a 1-ohm resistor. In view of this, Eq. (11.52) can be regarded as the statement that the integral of $S_f(\omega)/2\pi$ with respect to $\omega$ over the entire range of frequencies, $-\infty < \omega < \infty$, gives the total mean power of $f(t)$. Hence, it follows that $S_f(\omega)$ (divided by $2\pi$) is the *power spectral density function*, or the *power density spectrum of $f(t)$*. The function $S_f(\omega)$ is also known as the *mean square spectral density*. As can be inferred from the name, the power spectral density function represents a continuous spectrum, so that in terms of electrical terminology the average power dissipated in a 1-ohm resistor by the frequency components of a voltage lying in an infinitesimal band between $\omega$ and $\omega + d\omega$ is proportional to $S_f(\omega) \, d\omega$ (again divided by the factor $2\pi$). If for a given random process the mean square spectral density $S_f(\omega)$ is known, perhaps obtained through measurement, then Eq. (11.52) can be used to evaluate the mean square value of an ergodic random process. The function $S_f(\omega)$ has certain properties that can be used to render the evaluation of averages easier. These properties will now be discussed.

In view of its physical interpretation, we must conclude that $S_f(\omega)$ *is always nonnegative*, i.e., it is either positive or zero, $S_f(\omega) \geq 0$. We have shown in Sec. 11.7 that $R_f(\tau)$ is an even function of $\tau$, $R_f(\tau) = R_f(-\tau)$. From Eq. (11.50), it follows that

$$S_f(\omega) = \int_{-\infty}^{\infty} R_f(\tau) e^{-i\omega\tau} \, d\tau = \int_{-\infty}^{\infty} R_f(-\tau) e^{-i\omega\tau} \, d\tau$$

$$= -\int_{\infty}^{-\infty} R_f(\sigma) e^{i\omega\sigma} \, d\sigma = S_f(-\omega) \qquad (11.53)$$

where $\sigma$ is a dummy variable of integration. Equation (11.53) states that the power spectral density $S_f(\omega)$ *is an even function of $\omega$*. Because $R_f(\tau)$ is an even function of

$\tau$, Eq. (11.50) leads to

$$S_f(\omega) = \int_{-\infty}^{\infty} R_f(\tau)e^{i\omega\tau}\,d\tau = \int_{-\infty}^{\infty} R_f(\tau)(\cos\omega\tau - i\sin\omega\tau)\,d\tau$$

$$= \int_{-\infty}^{\infty} R_f(\tau)\cos\omega\tau\,d\tau = 2\int_0^{\infty} R_f(\tau)\cos\omega\tau\,d\tau \qquad (11.54)$$

But the autocorrelation $R_f(\tau)$ is a real function, so that from the last integral in (11.54) it follows that $S_f(\omega)$ *is a real function*. As a result of $S_f(\omega)$ being an even, real function of $\omega$, Eq. (11.51) can be reduced to

$$R_f(\tau) = \frac{1}{\pi}\int_0^{\infty} S_f(\omega)\cos\omega\tau\,d\omega \qquad (11.55)$$

Equations (11.54) and (11.55) are called the *Wiener-Khintchine equations,* and except for a factor of 2 they represent what is known as a Fourier cosine transform pair. It follows from Eq. (11.55) that

$$R_f(0) = \frac{1}{\pi}\int_0^{\infty} S_f(\omega)\,d\omega \qquad (11.56)$$

which provides a convenient formula for the calculation of the mean square value of a stationary random process if the power spectral density is given. The advantage of Eqs. (11.55) and (11.56) over Eqs. (11.51) and (11.52), respectively, is that Eqs. (11.55) and (11.56) contain no negative frequencies.

## 11.10 NARROWBAND AND WIDEBAND RANDOM PROCESSES

The mean square spectral density provides a measure of the representation of given frequencies in a random process. For convenience, we present our discussion in terms of ergodic random processes. Random processes are often identified by the shape of the power density spectra. In particular, we distinguish between narrowband and wideband random processes. The terminology used is not precise, and it provides only a qualitative description of a given process. A *narrowband process* is characterized by a sharply peaked power density spectrum $S_f(\omega)$, in the sense that $S_f(\omega)$ has significant values only in a short band of frequencies centered around the frequency corresponding to the peak. A sample time history representative of a narrowband process contains only a narrow range of frequencies. In the case of a *wideband process,* on the other hand, the power density spectrum $S_f(\omega)$ has significant values over a wide band of frequencies whose width is of the same order of magnitude as the center frequency of the band. A sample time history representative of a wideband process contains a wide range of frequencies. At the two extremes we find a power density spectrum consisting of two symmetrically placed delta functions, corresponding to a sinusoidal sample function, and a uniform power density spectrum, corresponding to a sample function in which all the frequencies are equally represented. The first, of course, is

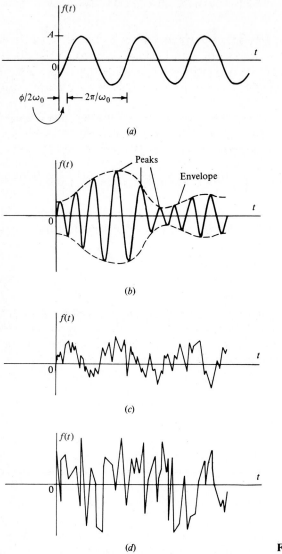

(a)

(b)

(c)

(d)                                        **Figure 11.18**

a deterministic function, but it can be regarded as random if the phase angle is randomly distributed (see Sec. 11.5). The second is known as *white noise* by analogy with white light, which has a flat spectrum over the visible range. If the frequency band is infinite, then we speak of *ideal white noise*. This concept represents a physical impossibility because it implies an infinite mean square value, and hence infinite power. A judicious use of the concept, however, can lead to meaningful results. For comparison purposes, it may prove of interest to plot some sample functions and the autocorrelation functions, probability density functions, and power density spectra corresponding to these sample functions.

Figure 11.18 shows plots of possible time histories. Figure 11.18a shows the simple sinusoidal function $f(t) = A \sin(\omega_0 t + \phi)$, whereas Figs. 11.18b, c, and d show time histories corresponding to a narrowband random process, a wideband random process and an ideal white noise, respectively. Note that Fig. 11.18b has the appearance of a sinusoidal function with randomly varying amplitude. Figures 11.18c and d look somewhat similar because both time histories contain a wide range of frequencies.

Figure 11.19 shows plots of possible probability density functions. Figure 11.19a depicts the probability density function for a sinusoidal wave. This function was obtained in Sec. 11.5 by regarding the phase angle as random, and was plotted

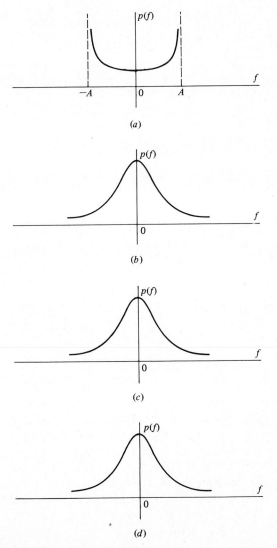

(a)

(b)

(c)

(d)

**Figure 11.19**

in Fig. 11.13. It is not possible to give analytical expressions for the probability density functions associated with a narrowband process, a wideband process, and an ideal white noise. However, they all approach the Gaussian distribution, as shown in Figs. 11.19b, c, and d, respectively.

Plots of the autocorrelation function corresponding to a sinusoidal wave, a narrowband process, a wideband process and ideal white noise are shown in Figs. 11.20a, b, c, and d, respectively. The autocorrelation function for the sinusoidal wave $f(t) = A \sin(\omega_0 t + \phi)$ can be calculated as follows:

$$R_f(\tau) = \lim_{T \to \infty} \frac{A^2}{T} \int_{-T/2}^{T/2} \sin(\omega_0 t + \phi) \sin[\omega_0(t + \tau) + \phi] \, dt$$

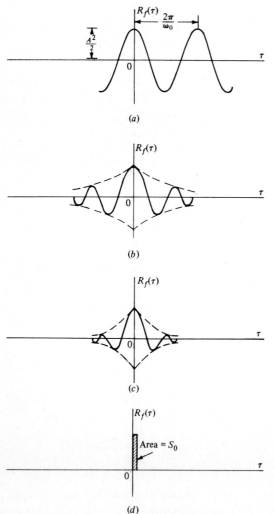

(a)

(b)

(c)

(d)

**Figure 11.20**

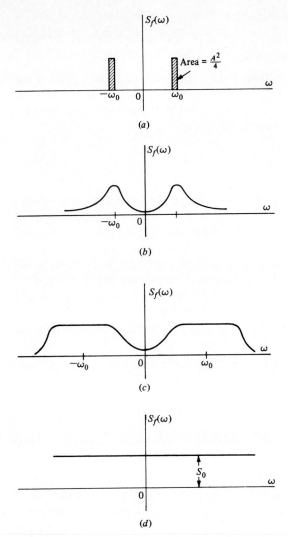

$$= \frac{A^2}{2\pi} \int_{-\pi}^{\pi} (\cos \omega_0 \tau \sin^2 \alpha + \sin \omega_0 \tau \sin \alpha \cos \alpha) \, d\alpha$$

$$= \frac{A^2}{2} \cos \omega_0 \tau \tag{11.57}$$

which is a cosine function with the same frequency as the sine wave but with zero phase angle. The autocorrelation function for the narrowband process appears as a cosine function of decaying amplitude, and that of a wideband process appears sharply peaked and decaying rapidly to zero. In the limit, as the width of the frequency band increases indefinitely, the autocorrelation function reduces to that for the ideal white noise, having the form

$$R_f(\tau) = S_0\, \delta(\tau) \tag{11.58}$$

where $\delta(\tau)$ is the Dirac delta function. This can be verified by substituting Eq. (11.58) into Eq. (11.50).

Figure 11.21$a$ shows a plot of the power density spectrum for the sine wave. It can be verified that its mathematical expression is

$$S_f(\omega) = \frac{\pi A^2}{2}\left[\delta(\omega + \omega_0) + \delta(\omega - \omega_0)\right] \tag{11.59}$$

The power spectral densities for the narrowband and wideband process are shown in Figs. 11.21$b$ and $c$, respectively, which justifies the terminology used to describe these processes. Figure 11.21$d$ depicts the power density spectrum for the ideal white noise, indicating that all frequencies are equally represented.

A more realistic random process than the ideal white noise is the *band-limited white noise*. The corresponding power density spectrum, shown in Fig. 11.22, is flat over the band of frequencies $\omega_1 < \omega < \omega_2$ (and $-\omega_2 < \omega < -\omega_1$), where $\omega_1$ and $\omega_2$ are known as the *lower cutoff* and *upper cutoff* frequencies, respectively. The band-limited white noise can serve at times as a reasonable approximation for the power density spectrum of a wideband process. The associated autocorrelation function can be obtained from Eq. (11.55) in the form

$$R_f(\tau) = \frac{1}{\pi}\int_0^{\infty} S_f(\omega)\cos\omega\tau\, d\omega = \frac{S_0}{\pi}\int_{\omega_1}^{\omega_2}\cos\omega\tau\, d\omega$$

$$= \frac{S_0}{\pi}\frac{\sin\omega_2\tau - \sin\omega_1\tau}{\tau} \tag{11.60}$$

The autocorrelation function is shown in Fig. 11.23$a$. As a matter of interest, let $\omega_1 = 0$ and $\omega_2 \to \infty$, so that the band-limited white noise approaches the ideal white noise. In this case, Fig. 11.23$b$ approaches a Dirac delta function in the form of a triangle with the base equal to $2\pi/\omega_2$ and the height equal to $S_0\omega_2/\pi$. The area of the triangle is equal to $S_0$, thus verifying Eq. (11.58).

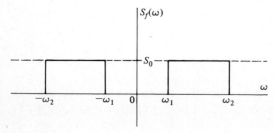

**Figure 11.22**

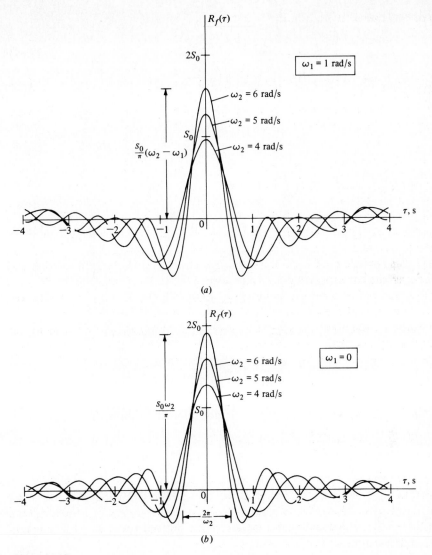

**Figure 11.23**

Narrowband processes that are stationary and Gaussian lend themselves to further characterization. Before we can show this, it is necessary to develop an expression for the power spectral density of a derived process. In particular, the interest lies in an expression for the power spectral density $S_{\dot{f}}(\omega)$ of a stationary process $\dot{f}(t)$ under the assumption that the power spectral density $S_f(\omega)$ of the stationary process $f(t)$ is known. To this end, we recall Eq. (11.2) and recognize that for a stationary process the autocorrelation function does not depend on the time $t_1$, so that replacing $t_1$ by the arbitrary time $t$ the autocorrelation function

$R_f(\tau)$ of $f(t)$ can be written in the form

$$R_f(\tau) = \lim_{n \to \infty} \frac{1}{n} \sum_{k=1}^{n} f_k(t) f_k(t + \tau) \tag{11.61}$$

Differentiating Eq. (11.61) with respect to $\tau$, we obtain

$$\frac{dR_f(\tau)}{d\tau} = \lim_{n \to \infty} \frac{1}{n} \sum_{k=1}^{n} \frac{d}{d\tau} [f_k(t) f_k(t + \tau)] \tag{11.62}$$

But,

$$\frac{d}{d\tau} [f_k(t) f_k(t + \tau)] = f_k(t) \frac{d}{d\tau} [f_k(t + \tau)]$$

$$= f_k(t) \frac{d}{d(t + \tau)} [f_k(t + \tau)] \frac{d(t + \tau)}{d\tau} = f_k(t) \dot{f}_k(t + \tau) \tag{11.63}$$

so that

$$\frac{dR_f(\tau)}{d\tau} = \lim_{n \to \infty} \frac{1}{n} \sum_{k=1}^{n} f_k(t) \dot{f}_k(t + \tau) \tag{11.64}$$

For stationary processes, however, the value of the summation is independent of time, so that we can also write

$$\frac{dR_f(\tau)}{d\tau} = \lim_{n \to \infty} \frac{1}{n} \sum_{k=1}^{n} f_k(t - \tau) \dot{f}_k(t) \tag{11.65}$$

Using the above approach once more, it is not difficult to show that

$$\frac{d^2 R_f(\tau)}{d\tau^2} = -\lim_{n \to \infty} \frac{1}{n} \sum_{k=1}^{n} \dot{f}_k(t - \tau) \dot{f}_k(t)$$

$$= -\lim_{n \to \infty} \frac{1}{n} \sum_{k=1}^{n} \dot{f}_k(t) \dot{f}_k(t + \tau) = -R_{\dot{f}}(\tau) \tag{11.66}$$

where $R_{\dot{f}}(\tau)$ is the autocorrelation function of the derived process $\dot{f}(t)$. From Eq. (11.51), however, we can write

$$\frac{d^2 R_f(\tau)}{d\tau^2} = -\frac{1}{2\pi} \int_{-\infty}^{\infty} \omega^2 S_f(\omega) e^{i\omega\tau} d\omega \tag{11.67}$$

Moreover,

$$R_{\dot{f}}(\tau) = \frac{1}{2\pi} \int_{-\infty}^{\infty} S_{\dot{f}}(\omega) e^{i\omega\tau} d\omega \tag{11.68}$$

where $S_{\dot{f}}(\omega)$ is the power spectral density of $\dot{f}$. Hence, inserting Eqs. (11.67) and (11.68) into Eq. (11.66), we conclude that

$$S_{\dot{f}}(\omega) = \omega^2 S_f(\omega) \tag{11.69}$$

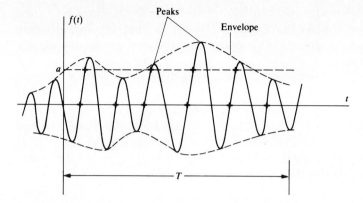

**Figure 11.24**

or, the power spectral density of the derived process $\dot{f}$ can be obtained by merely multiplying the known power spectral density of $f$ by $\omega^2$.

For a stationary process with zero mean value, if we let $\tau = 0$ and use Eqs. (11.26), (11.28), (11.52), and (11.61), we obtain

$$\sigma_f^2 = R(0) = E[f^2] = \frac{1}{2\pi} \int_{-\infty}^{\infty} S_f(\omega) \, d\omega \tag{11.70}$$

where $\sigma_f$ is the standard deviation. Similarly, letting $\tau = 0$ in Eq. (11.68) and using Eq. (11.69), we can write

$$\sigma_{\dot{f}}^2 = R_{\dot{f}}(0) = E[\dot{f}^2] = \frac{1}{2\pi} \int_{-\infty}^{\infty} \omega^2 S_f(\omega) \, d\omega \tag{11.71}$$

Now we return to the characterization of a narrowband process. To this end, we consider a typical sample function $f(t)$ of an ensemble $\{f\}$, as shown in Fig. 11.24. The function appears as a sinusoid with slowly varying random amplitude and random phase. The interest lies in characterizing the expected frequency and amplitude. To characterize the expected frequency, we define the expected number of crossings at the level $f = a$ with positive slope per unit time as follows

$$v_a^+ = \lim_{n \to \infty} \frac{1}{n} \sum_{k=1}^{n} \frac{1}{T} N_{ka}^+(T) \tag{11.72}$$

where $N_{ka}^+(T)$ represents the number of crossing with positive slope in the time interval $T$. Note that crossings with positive slope are marked by crosses in Fig. 11.24. It can be shown† that for a stationary process

$$v_a^+ = \int_0^{\infty} \dot{f} p(a, \dot{f}) \, d\dot{f} \tag{11.73}$$

† See S. H. Crandall and W. D. Mark, *Random Vibration in Mechanical Systems*, p. 47, Academic Press, Inc., New York, 1963.

where $p(a, \dot{f})$ is the intersection of the joint probability density function $p(f, \dot{f})$ and the plane $f = a$ (see Sec. 11.13). Equation (11.73) is valid for any arbitrary stationary process. If the process is Gaussian, then the joint probability density function has the form

$$p(f, \dot{f}) = \frac{1}{2\pi\sigma_f\sigma_{\dot{f}}} \exp\left[-\frac{1}{2}\left(\frac{f^2}{\sigma_f^2} + \frac{\dot{f}^2}{\sigma_{\dot{f}}^2}\right)\right] \tag{11.74}$$

where the standard deviations $\sigma_f$ and $\sigma_{\dot{f}}$ can be obtained from the power spectral density $S_f(\omega)$ by means of Eqs. (11.70) and (11.71), and we note that Eq. (11.74) reflects the fact that $f$ and $\dot{f}$ are uncorrelated. Inserting Eq. (11.74) with $f = a$ into Eq. (11.73) and carrying out the integration, we obtain

$$v_a^+ = \frac{1}{2\pi}\frac{\sigma_{\dot{f}}}{\sigma_f} e^{-a^2/2\sigma_f^2} \tag{11.75}$$

Then, the *average frequency*, or *expected frequency*, $\omega_0$ is defined as the expected number of zero crossings with positive slope per unit time multiplied by $2\pi$, so that letting $a = 0$ in Eq. (11.75) and considering Eqs. (11.70) and (11.71), we can write

$$\omega_0 = 2\pi v_0^+ = \frac{\sigma_{\dot{f}}}{\sigma_f} = \left[\frac{\displaystyle\int_{-\infty}^{\infty} \omega^2 S_f(\omega)\, d\omega}{\displaystyle\int_{-\infty}^{\infty} S_f(\omega)\, d\omega}\right]^{1/2} \tag{11.76}$$

It can also be shown† that for a narrowband stationary Gaussian random process the *probability density function of the envelope* is

$$p(a) = \frac{a}{\sigma_f^2} e^{-a^2/2\sigma_f^2} \tag{11.77}$$

which can be identified as the Rayleigh distribution. The probability density function of the peaks is also given by the Rayleigh distribution of Eq. (11.77).

## 11.11 RESPONSE OF LINEAR SYSTEMS TO STATIONARY RANDOM EXCITATION

In Chap. 2, we showed that the response $x(t)$ of a linear system to the arbitrary excitation $f(t)$ can be written in the form of the convolution integral

$$x(t) = \int_0^t f(\lambda)g(t - \lambda)\, d\lambda \tag{11.78}$$

where $g(t)$ is the impulse response, and $\lambda$ merely a dummy variable. The function $f(t)$ is defined only for $t > 0$ and is zero for $t < 0$. Likewise, Eq. (11.78) defines the response $x(t)$ only for $t > 0$. Random variables, however, are not restricted to

† See S. H. Crandall and W. D. Mark, op. cit. pp. 48–53.

positive times, so that we wish to modify Eq. (11.78) to accommodate functions $f(t)$ of negative argument. To this end, it can be shown that the lower limit of the convolution integral can be merely extended to $-\infty$, so that

$$x(t) = \int_{-\infty}^{t} f(\lambda)g(t - \lambda)\, d\lambda \tag{11.79}$$

However, from the definition of the impulse response (see Sec. 2.13), $g(t - \lambda)$ is zero for $t < \lambda$. Because the variable of integration in (11.79) is $\lambda$ and not $t$, a slight change in emphasis permits us to restate the above by saying that $g(t - \lambda)$ is zero for $\lambda > t$. It follows that the upper limit of the integral in (11.79) can be changed to any value larger than $t$ without affecting the value of the integral. Choosing the upper limit as infinity, to preserve the symmetry of the integral, we can write the convolution integral in the form

$$x(t) = \int_{-\infty}^{\infty} f(\lambda)g(t - \lambda)\, d\lambda \tag{11.80}$$

Using the change of variable $t - \lambda = \tau$, $d\lambda = -d\tau$, with an appropriate change in the integration limits, it is easy to demonstrate that the convolution integral remains symmetric in $f(t)$ and $g(t)$, or

$$x(t) = \int_{-\infty}^{\infty} f(\lambda)g(t - \lambda)\, d\lambda = \int_{-\infty}^{\infty} g(\lambda)f(t - \lambda)\, d\lambda \tag{11.81}$$

Next let us denote by $X(\omega)$ the Fourier transform of $x(t)$, so that using Eq. (11.80) we can write

$$X(\omega) = \int_{-\infty}^{\infty} x(t)e^{-i\omega t}\, dt = \int_{-\infty}^{\infty} f(\lambda)\left[\int_{-\infty}^{\infty} g(t - \lambda)e^{-i\omega t}\, dt\right] d\lambda$$

$$= \int_{-\infty}^{\infty} f(\lambda)e^{-i\omega\lambda}\, d\lambda \int_{-\infty}^{\infty} g(\sigma)e^{-i\omega\sigma}\, d\sigma \tag{11.82}$$

where the substitution $t - \lambda = \sigma$, $dt = d\sigma$ has been used. But,

$$\int_{-\infty}^{\infty} f(\lambda)e^{-i\omega\lambda}\, d\lambda = F(\omega) \tag{11.83}$$

is the Fourier transform of the excitation and

$$\int_{-\infty}^{\infty} g(\sigma)e^{-i\omega\sigma}\, d\sigma = G(\omega) \tag{11.84}$$

is the Fourier transform of the impulse response, so that Eq. (11.83) yields

$$X(\omega) = G(\omega)F(\omega) \tag{11.85}$$

Comparing Eq. (11.85) with Eq. (11.48), we conclude that *the frequency response $G(\omega)$ can be identified as the Fourier transform of the impulse response*. Equations (11.81) and (11.85) state that *the convolution of $f(t)$ and $g(t)$ and the product*

$G(\omega)F(\omega)$ represent a *Fourier transform pair*. This statement is known as the *time-domain convolution theorem*.

The above relations are valid for any arbitrary excitation $f(t)$. Our interest lies in the case in which the excitation is in the form of the stationary random process $\{f(t)\}$. Then the response random process $\{x(t)\}$ will also be stationary. We shall be interested in calculating first- and second-order statistics for the response random process, given the corresponding statistics for the excitation random process.

Let us consider the stationary excitation and response random process $\{f(t)\}$ and $\{x(t)\}$, respectively. Averaging Eq. (11.81) over the ensemble, we can write the mean value of the response random process as

$$E[x(t)] = E\left[\int_{-\infty}^{\infty} g(\lambda)f(t - \lambda)\, d\lambda\right] \tag{11.86}$$

Assuming that the order of the ensemble averaging and integration operations are interchangeable, Eq. (11.86) can be written as

$$E[x(t)] = \int_{-\infty}^{\infty} g(\lambda)E[f(t - \lambda)]\, d\lambda \tag{11.87}$$

But for stationary random processes, the mean value of the process is constant, $E[f(t - \lambda)] = E[f(t)] = \text{const}$, so that

$$E[x(t)] = E[f(t)] \int_{-\infty}^{\infty} g(\lambda)\, d\lambda \tag{11.88}$$

Letting $\omega = 0$ in Eq. (11.84), and changing the dummy variable from $\sigma$ to $\lambda$, we obtain

$$\int_{-\infty}^{\infty} g(\lambda)\, d\lambda = G(0) \tag{11.89}$$

so that Eq. (11.88) reduces to

$$E[x(t)] = G(0)E[f(t)] = \text{const} \tag{11.90}$$

which implies that the mean value of the response to an excitation in the form of a stationary random process is constant and proportional to the mean value of the excitation process. It follows that *if the excitation mean value is zero, then the response mean value is also zero*.

Next, let us evaluate the autocorrelation function of the response random process. To this end, it will prove convenient to introduce two new dummy variables $\lambda_1$ and $\lambda_2$, and write the convolution integrals

$$x(t) = \int_{-\infty}^{\infty} g(\lambda_1)f(t - \lambda_1)\, d\lambda_1$$

$$x(t + \tau) = \int_{-\infty}^{\infty} g(\lambda_2)f(t + \tau - \lambda_2)\, d\lambda_2 \tag{11.91}$$

Using Eqs. (11.91) to form the response autocorrelation function $R_x(\tau)$ and assuming once again that the order of ensemble averaging and integration is interchangeable, we can write

$$R_x(\tau) = E[x(t)x(t + \tau)]$$

$$= E\left[\int_{-\infty}^{\infty} g(\lambda_1)f(t - \lambda_1)\,d\lambda_1 \int_{-\infty}^{\infty} g(\lambda_2)f(t + \tau - \lambda_2)\,d\lambda_2\right]$$

$$= E\left[\int_{-\infty}^{\infty}\int_{-\infty}^{\infty} g(\lambda_1)g(\lambda_2)f(t - \lambda_1)f(t + \tau - \lambda_2)\,d\lambda_1\,d\lambda_2\right]$$

$$= \int_{-\infty}^{\infty}\int_{-\infty}^{\infty} g(\lambda_1)g(\lambda_2)E[f(t - \lambda_1)f(t + \tau - \lambda_2)]\,d\lambda_1\,d\lambda_2 \tag{11.92}$$

Because the excitation random process is stationary, we have

$$E[f(t - \lambda_1)f(t + \tau - \lambda_2)] = E[f(t)f(t + \tau + \lambda_1 - \lambda_2)]$$

$$= R_f(\tau + \lambda_1 - \lambda_2) \tag{11.93}$$

where $R_f(\tau + \lambda_1 - \lambda_2)$ is the autocorrelation function of the excitation process. Hence, the response autocorrelation function, Eq. (11.92), reduces to

$$R_x(\tau) = \int_{-\infty}^{\infty}\int_{-\infty}^{\infty} g(\lambda_1)g(\lambda_2)R_f(\tau + \lambda_1 - \lambda_2)\,d\lambda_1\,d\lambda_2 \tag{11.94}$$

We note that Eq. (11.94) does not depend on $t$, which implies that the value of the response autocorrelation function is also insensitive to a translation in time, thus corroborating the statement made earlier that *if for a linear system the excitation is a stationary random process, then the response is also a stationary random process.*

Quite often information concerning the response random process can be obtained more readily by calculating first the response power spectral density instead of the response autocorrelation function, particularly if the excitation random process is given in terms of the power spectral density. To demonstrate this, let us use Eq. (11.94) and express the response mean square spectral density as the Fourier transform of the response autocorrelation in the form

$$S_x(\omega) = \int_{-\infty}^{\infty} R_x(\tau)e^{-i\omega\tau}\,d\tau$$

$$= \int_{-\infty}^{\infty} e^{-i\omega\tau}\left[\int_{-\infty}^{\infty}\int_{-\infty}^{\infty} g(\lambda_1)g(\lambda_2)R_f(\tau + \lambda_1 - \lambda_2)\,d\lambda_1\,d\lambda_2\right]d\tau \tag{11.95}$$

But $R_f(\tau + \lambda_1 - \lambda_2)$ can be expressed as the inverse Fourier transform

$$R_f(\tau + \lambda_1 - \lambda_2) = \frac{1}{2\pi}\int_{-\infty}^{\infty} S_f(\omega)e^{i\omega(\tau + \lambda_1 - \lambda_2)}\,d\omega \tag{11.96}$$

so that, inserting Eq. (11.96) into (11.95), considering Eq. (11.84), interchanging the order of integration, and rearranging, we obtain

$$S_x(\omega) = \int_{-\infty}^{\infty} e^{-i\omega\tau} \left\{ \int_{-\infty}^{\infty} \int_{-\infty}^{\infty} g(\lambda_1)g(\lambda_2) \right.$$

$$\left. \times \left[ \frac{1}{2\pi} \int_{-\infty}^{\infty} S_f(\omega)e^{i\omega(\tau + \lambda_1 - \lambda_2)} d\omega \right] d\lambda_1 \, d\lambda_2 \right\} d\tau$$

$$= \int_{-\infty}^{\infty} e^{-i\omega\tau} \left\{ \frac{1}{2\pi} \int_{-\infty}^{\infty} S_f(\omega) \left[ \int_{-\infty}^{\infty} g(\lambda_1)e^{i\omega\lambda_1} \, d\lambda_1 \right. \right.$$

$$\left. \left. \times \int_{-\infty}^{\infty} g(\lambda_2)e^{-i\omega\lambda_2} \, d\lambda_2 \right] e^{i\omega\tau} \, d\omega \right\} d\tau$$

$$= \int_{-\infty}^{\infty} e^{-i\omega\tau} \left[ \frac{1}{2\pi} \int_{-\infty}^{\infty} S_f(\omega)G(-\omega)G(\omega)e^{i\omega\tau} \, d\omega \right] d\tau$$

$$= \int_{-\infty}^{\infty} e^{-i\omega\tau} \left[ \frac{1}{2\pi} \int_{-\infty}^{\infty} S_f(\omega)|G(\omega)|^2 e^{i\omega\tau} \, d\omega \right] d\tau \qquad (11.97)$$

where use has been made of the fact that $G(-\omega)$ is the complex conjugate of the frequency response $G(\omega)$. Comparing the first integral in Eq. (11.95) with the last in Eq. (11.97), and recognizing that the response autocorrelation function $R_x(\tau)$ must be equal to the inverse Fourier transform of the response mean square spectral density $S_x(\omega)$, we conclude that

$$S_x(\omega) = |G(\omega)|^2 S_f(\omega) \qquad (11.98)$$

and

$$R_x(\tau) = \frac{1}{2\pi} \int_{-\infty}^{\infty} S_x(\omega)e^{i\omega\tau} \, d\omega = \frac{1}{2\pi} \int_{-\infty}^{\infty} |G(\omega)|^2 S_f(\omega)e^{i\omega\tau} \, d\omega \qquad (11.99)$$

constitute a Fourier transform pair. Equation (11.98) represents a simple algebraic expression relating the power spectral densities of the excitation and response random processes, whereas Eq. (11.99) gives the response autocorrelation in the form of an inverse Fourier transform involving the excitation power spectral density. From Eq. (11.98), we conclude that in the case of a lightly *damped single-degree of freedom system*, for which the frequency response has a sharp peak at $\omega = \omega_n(1 - 2\zeta^2)^{1/2}$, *if the excitation power spectral density function represents a wideband random process, then the response power spectral density function is a narrowband random process*, where $\zeta$ is the damping factor and $\omega_n$ the frequency of undamped oscillation.

The mean square value of the response random process can be obtained by letting $\tau = 0$ in Eq. (11.99). The result is simply

$$R_x(0) = E[x^2(t)] = \frac{1}{2\pi} \int_{-\infty}^{\infty} |G(\omega)|^2 S_f(\omega) \, d\omega \qquad (11.100)$$

Examining Eqs. (11.98), (11.99), and (11.100), it appears that if the system is linear and the excitation random process is stationary, then the response mean square spectral density, autocorrelation function, and mean square value, can all be calculated from the knowledge of the mean square spectral density $S_f(\omega)$ of the excitation random process and the magnitude $|G(\omega)|$ of the frequency response.

It should be pointed out that *if the excitation random process is Gaussian and the system is linear, then the response random process is also Gaussian. This implies that for stationary processes the response probability distribution is completely defined by the response mean value and mean square value.*

It is not difficult to show that the above relations and conclusions concerning response random processes remain valid if the excitation random process is not merely stationary but ergodic. The only difference is that for ergodic random processes the averages are time averages, calculated by using a single representative sample function from the entire process, instead of ensemble averages over the collection of sample functions.

## 11.12 RESPONSE OF SINGLE-DEGREE-OF-FREEDOM SYSTEMS TO RANDOM EXCITATION

Let us consider a mass-damper-spring system traveling with the uniform velocity $v$ on a rough road, so that its support is imparted a vertical motion, as shown in Fig. 11.25. If the road roughness is described by the random variable $y(s)$, then the vertical motion of the support is $y(t)$, where $t = s/v$. From Eq. (2.86), we conclude that the differential equation of motion for the mass $m$ is

$$\ddot{x}(t) + 2\zeta\omega_n\dot{x}(t) + \omega_n^2 x(t) = \omega_n^2 f(t) \tag{11.101}$$

where
$$f(t) = \frac{2\zeta}{\omega_n}\dot{y}(t) + y(t) \tag{11.102}$$

is an equivalent displacement excitation, in which $\zeta$ is the damping factor and $\omega_n$ the undamped frequency of oscillation. We assume that the random process associated with $f(t)$ is ergodic and Gaussian, so that the response $x(t)$ is also an ergodic and Gaussian process. Hence, both the excitation and response random processes are fully described by the mean value and mean square value.

For a stationary process the mean value is constant. Because a constant

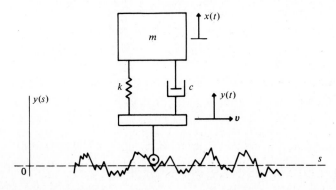

**Figure 11.25**

component of the excitation merely leads to a constant component of the response, a problem that can be treated separately, we can assume without loss of generality that this constant is zero,

$$E[f(t)] = 0 \tag{11.103}$$

It follows immediately that the response mean value is also zero,

$$E[x(t)] = 0 \tag{11.104}$$

Next, we wish to calculate various basic statistics describing the response random process, such as the autocorrelation function, the power spectral density function and the mean square value. This requires knowledge of certain statistics describing the excitation random process. We consider two related cases, namely, that of the ideal white noise and that of band-limited white noise.

In Sec. 11.10 it was indicated that the autocorrelation function corresponding to the ideal white noise power spectral density $S_f(\omega) = S_0$ is

$$R_f(\tau) = S_0 \, \delta(\tau) \tag{11.105}$$

where $\delta(\tau)$ is the Dirac delta function. Moreover, from Example 2.3, we conclude that the impulse response of the single-degree-of-freedom system described by Eq. (11.101) has the form

$$g(t) = \frac{\omega_n^2}{\omega_d} e^{-\zeta\omega_n t} \sin \omega_d t \, \mathcal{u}(t) \tag{11.106}$$

where the unit step function $\mathcal{u}(t)$ ensures that $g(t) = 0$ for $t < 0$. Hence, introducing Eqs. (11.105) and (11.106) into Eq. (11.94), we obtain the response autocorrelation function

$$R_x(\tau) = \frac{S_0\omega_n^4}{\omega_d^2} \int_{-\infty}^{\infty} \int_{-\infty}^{\infty} \delta(\tau + \lambda_1 - \lambda_2) e^{-\zeta\omega_n(\lambda_1 + \lambda_2)}$$

$$\times \sin \omega_d\lambda_1 \sin \omega_d\lambda_2 \, \mathcal{u}(\lambda_1)\mathcal{u}(\lambda_2) \, d\lambda_1 \, d\lambda_2$$

$$= \frac{S_0\omega_n^4}{\omega_d^2} \int_{0}^{\infty} \int_{0}^{\infty} \delta(\tau + \lambda_1 - \lambda_2) e^{-\zeta\omega_n(\lambda_1 + \lambda_2)} \sin \omega_d\lambda_1 \sin \omega_d\lambda_2 \, d\lambda_1 \, d\lambda_2 \tag{11.107}$$

In our evaluation of $R_x(\tau)$, we assume that $\tau > 0$. The value of $R_x(-\tau)$ can be obtained by using the fact that the autocorrelation is an even function of $\tau$. Due to the nature of the delta function, if we integrate with respect to $\lambda_2$, we obtain

$$R_x(\tau) = \frac{S_0\omega_n^4}{\omega_d^2} \int_{0}^{\infty} e^{-\zeta\omega_n(\tau + 2\lambda_1)} \sin \omega_d\lambda_1 \sin \omega_d(\tau + \lambda_1) \, d\lambda_1$$

$$= \frac{S_0\omega_n^4}{\omega_d^2} e^{-\zeta\omega_n\tau} \left( \sin \omega_d\tau \int_{0}^{\infty} e^{-2\zeta\omega_n\lambda_1} \sin \omega_d\lambda_1 \cos \omega_d\lambda_1 \, d\lambda_1 \right.$$

$$\left. + \cos \omega_d\tau \int_{0}^{\infty} e^{-2\zeta\omega_n\lambda_1} \sin^2 \omega_d\lambda_1 \, d\lambda_1 \right) \qquad \tau > 0 \tag{11.108}$$

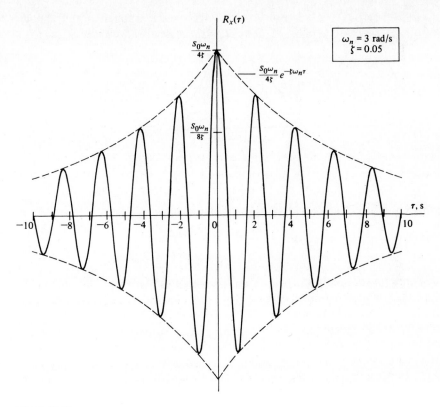

**Figure 11.26**

But the value of the integrals in Eq. (11.108) can be found in standard integral tables,† so that Eq. (11.108) reduces to

$$R_x(\tau) = \frac{S_0 \omega_n}{4\zeta} e^{-\zeta \omega_n \tau} \left[ \cos \omega_d \tau + \frac{\zeta}{(1 - \zeta^2)^{1/2}} \sin \omega_d \tau \right] \qquad \tau > 0 \quad (11.109)$$

Using the fact that $R_x(-\tau) = R_x(\tau)$, we can write directly

$$R_x(\tau) = \frac{S_0 \omega_n}{4\zeta} e^{\zeta \omega_n \tau} \left[ \cos \omega_d \tau - \frac{\zeta}{(1 - \zeta^2)^{1/2}} \sin \omega_d \tau \right] \qquad \tau < 0 \quad (11.110)$$

The autocorrelation function is plotted in Fig. 11.26 for the case of light damping. It is easy to see that the response autocorrelation function is that of a narrowband process (see Sec. 11.10).

The response power density spectrum is quite simple to obtain. We recall that the frequency response for the system in question was obtained in Sec. 2.3. Hence,

† See, for example, B. O. Peirce and R. M. Foster, *A Short Table of Integrals*, nos. 430 and 435, Ginn and Company, Boston, 1956.

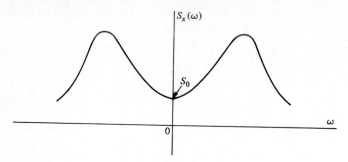

**Figure 11.27**

inserting $S_f(\omega) = S_0$ and Eq. (2.48) into (11.98), we obtain simply

$$S_x(\omega) = |G(\omega)|^2 S_f(\omega) = \frac{S_0}{[1 - (\omega/\omega_n)^2]^2 + (2\zeta\omega/\omega_n)^2} \qquad (11.111)$$

The response power spectral density $S_x(\omega)$ is plotted in Fig. 11.27. Once again we conclude that the plot $S_x(\omega)$ versus $\omega$ is typical of a narrowband process. Because, according to Eqs. (11.98) and (11.99), $R_x(\tau)$ and $S_x(\omega)$ represent a Fourier transform pair, no essentially new information can be derived from $S_x(\omega)$ that cannot be derived from $R_x(\tau)$, or from $R_x(\tau)$ that cannot be derived from $S_x(\omega)$.

The mean square value can be obtained in two ways, namely, by letting $\tau = 0$ in Eq. (11.109), or by integrating Eq. (11.111) with respect to $\omega$. From Eq. (11.109), we can write simply

$$R_x(0) = E[x^2(t)] = \frac{S_0\omega_n}{4\zeta} \qquad (11.112)$$

On the other hand, inserting Eq. (11.111) into (11.100), we can write

$$R_x(0) = E[x^2(t)] = \frac{S_0}{2\pi} \int_{-\infty}^{\infty} \frac{d\omega}{[1 - (\omega/\omega_n)^2]^2 + (2\zeta\omega/\omega_n)^2} \qquad (11.113)$$

The integration of Eq. (11.113) can be performed by converting the real variable $\omega$ into a complex variable, and the real line integral into a contour integral in the complex plane, where the latter can be evaluated by the residue theorem. Following this procedure, it can be shown[†] that Eq. (11.113) leads to the same mean square value as that given by Eq. (11.112).

Because the random process is Gaussian with zero mean value, the mean square value, Eq. (11.112), is sufficient to determine the shape of the response probability density function, thus making it possible to evaluate the probability that the response $x(t)$ might exceed a given displacement. The mean square value also determines the probability density function of the Rayleigh distribution for the envelope and peaks of the response (see Fig. 11.24).

† See L. Meirovitch, *Analytical Methods in Vibrations*, pp. 503–505, The Macmillan Co., New York, 1973.

When the excitation power spectral density is in the form of band-limited white noise, with lower and upper cutoff frequencies $\omega_1$ and $\omega_2$, respectively, the response power spectral density has the form depicted in Fig. 11.28. Then, if the system is lightly damped and the excitation frequency band $\omega_1 < \omega < \omega_2$ includes the system natural frequency $\omega_n$ as well as its bandwidth $\Delta\omega = 2\zeta\omega_n$ (see Sec. 2.3 for definition), and if the excitation bandwidth is large compared to the system bandwidth, the response mean square value, which is equal to the area under the curve $S_x(\omega)$ versus $\omega$ divided by $2\pi$, can be approximated by $S_0\omega_n/4\zeta$. Hence, under these circumstances, the ideal white noise assumption leads to meaningful results.

Returning to Eq. (11.111), we observe that, whereas the excitation power spectral density $S_f(\omega)$ is flat, the response power spectral density $S_x(\omega)$ is not, and in fact is sharply peaked in the vicinity of $\omega = \omega_n$ for light damping. Moreover, the response spectrum has the value $S_0$ for relatively small frequencies, and it vanishes for very large frequencies, as can be seen from Fig. 11.28. This behavior can be attributed entirely to $|G(\omega)|$, which prescribes the amount of energy transmitted by the system at various frequencies. Hence, the linear system considered acts like a *linear filter*. For very light damping the system can be regarded as a *narrowband filter*.

## 11.13 JOINT PROBABILITY DISTRIBUTION OF TWO RANDOM VARIABLES

The preceding discussion was confined to properties of a single random process. Yet in many instances it is necessary to describe certain joint properties of two or more random processes. For example, these random processes may consist of the vibration of two or more distinct points in a structure. The statistics discussed in Secs. 11.2 through 11.10 can be calculated independently for the various random processes involved, but in addition there may be important information contained in certain joint statistics. In this section we confine ourselves to two random variables, and in Sec. 11.14 we discuss random processes.

There are three basic types of statistical functions describing joint properties of sample time histories representative of two random processes, namely, joint probability density functions, cross-correlation functions and cross-spectral density functions. These functions provide information concerning joint properties of two processes in the amplitude domain, time domain and frequency domain respectively.

Let us consider the two random variables $x(t)$ and $y(t)$, and define the *joint*, or

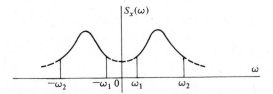

**Figure 11.28**

*second-order, probability distribution function* $P(x, y)$ associated with the probability that $x(t) \leqslant x$ and $y(t) \leqslant y$ as follows:

$$P(x, y) = \text{Prob}\,[x(t) \leqslant x;\; y(t) \leqslant y] \tag{11.114}$$

The above joint probability distribution function can be described in terms of a *joint probability density function* $p(x, y)$ according to

$$P(x, y) = \int_{-\infty}^{x} \int_{-\infty}^{y} p(\xi, \eta)\, d\xi\, d\eta \tag{11.115}$$

where the function $p(x, y)$ is given by the surface shown in Fig. 11.29. Note that $\xi$ and $\eta$ in Eq. (11.115) are mere dummy variables. The probability that $x_1 < x \leqslant x_2$ and $y_1 < y \leqslant y_2$ is given by

$$\text{Prob}\,(x_1 < x \leqslant x_2;\; y_1 < y \leqslant y_2) = \int_{x_1}^{x_2} \int_{y_1}^{y_2} p(x, y)\, dx\, dy \tag{11.116}$$

and represented by the shaded volume in Fig. 11.29.

The joint probability density function $p(x, y)$ possesses the property

$$p(x, y) \geqslant 0 \tag{11.117}$$

which implies that the joint probability is a *nonnegative* number. Moreover, the probability that $x$ is any real number and that $y$ is any real number is unity because the event is a certainty. This is expressed by

$$\int_{-\infty}^{\infty} \int_{-\infty}^{\infty} p(x, y)\, dx\, dy = 1 \tag{11.118}$$

First-order probabilities can be obtained from second-order joint probabilities. Indeed, the probability that $x$ lies within the open interval $x_1 < x < x_2$ regardless of the value of $y$ is

$$\text{Prob}\,(x_1 < x < x_2;\; -\infty < y < \infty) = \int_{x_1}^{x_2} \left[ \int_{-\infty}^{\infty} p(x, y)\, dy \right] = \int_{x_1}^{x_2} p(x)\, dx$$

where $\hspace{10cm}$ (11.119)

$$p(x) = \int_{-\infty}^{\infty} p(x, y)\, dy \tag{11.120}$$

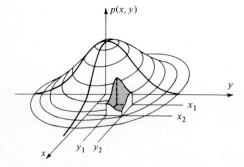

**Figure 11.29**

is the first-order probability density of $x$ alone. Similarly,

$$p(y) = \int_{-\infty}^{\infty} p(x, y)\, dy \tag{11.121}$$

is the first-order probability density of $y$ alone. The two random variables $x$ and $y$ are said to be *statistically independent* if

$$p(x, y) = p(x)p(y) \tag{11.122}$$

Next let us consider the *mathematical expectation* of a real continuous function $g(x, y)$ of the random variables $x(t)$ and $y(t)$ in the form

$$E[g(x, y)] = \int_{-\infty}^{\infty} \int_{-\infty}^{\infty} g(x, y)p(x, y)\, dx\, dy \tag{11.123}$$

The *mean values* of $x(t)$ and $y(t)$ alone are simply

$$\bar{x} = E[x] = \int_{-\infty}^{\infty} \int_{-\infty}^{\infty} xp(x, y)\, dx\, dy = \int_{-\infty}^{\infty} xp(x)\, dx$$

$$\bar{y} = E[y] = \int_{-\infty}^{\infty} \int_{-\infty}^{\infty} yp(x, y)\, dx\, dy = \int_{-\infty}^{\infty} yp(y)\, dy \tag{11.124}$$

In the case in which $g(x, y) = (x - \bar{x})(y - \bar{y})$, Eq. (11.123) defines the *covariance* between $x$ and $y$ in the form

$$C_{xy} = E[(x - \bar{x})(y - \bar{y})] = \int_{-\infty}^{\infty} \int_{-\infty}^{\infty} (x - \bar{x})(y - \bar{y})p(x, y)\, dx\, dy$$

$$= E[xy] - E[x]E[y] \tag{11.125}$$

Recalling Eq. (11.27), we conclude that $C_x = E[(x - \bar{x})^2] = \sigma_x^2$ represents the variance of $x$, whereas $C_y = E[(y - \bar{y})^2] = \sigma_y^2$ is the variance of $y$. The square roots of the variances, namely, $\sigma_x$ and $\sigma_y$, are the standard deviations of $x$ and $y$, respectively.

A relation between the covariance $C_{xy}$ and the standard deviations $\sigma_x$ and $\sigma_y$ can be revealed by considering the integral

$$\int_{-\infty}^{\infty} \int_{-\infty}^{\infty} \left( \frac{x - \bar{x}}{\sigma_x} \pm \frac{y - \bar{y}}{\sigma_y} \right)^2 p(x, y)\, dx\, dy$$

$$= \frac{1}{\sigma_x^2} \int_{-\infty}^{\infty} \int_{-\infty}^{\infty} (x - \bar{x})^2 p(x, y)\, dx\, dy$$

$$\pm \frac{2}{\sigma_x \sigma_y} \int_{-\infty}^{\infty} \int_{-\infty}^{\infty} (x - \bar{x})(y - \bar{y})p(x, y)\, dx\, dy$$

$$+ \frac{1}{\sigma_y^2} \int_{-\infty}^{\infty} \int_{-\infty}^{\infty} (y - \bar{y})^2 p(x, y)\, dx\, dy = 2 \pm 2\frac{C_{xy}}{\sigma_x \sigma_y} \geq 0 \quad (11.126)$$

where the inequality is valid because the first integral cannot be negative. It follows that

$$\sigma_x \sigma_y \geqslant |C_{xy}| \tag{11.127}$$

or the product of the standard deviations of $x$ and $y$ is larger than or equal to the magnitude of the covariance between $x$ and $y$. The normalized quantity

$$\rho_{xy} = \frac{C_{xy}}{\sigma_x \sigma_y} \tag{11.128}$$

is known as the *correlation coefficient*. Its value lies between $-1$ and $+1$, as can be concluded from inequality (11.126).

When the covariance $C_{xy}$ is equal to zero the random variables $x$ and $y$ are said to be *uncorrelated. Statistically independent random variables are also uncorrelated, but uncorrelated random variables are not necessarily statistically independent,* although they can be. To show this, let us introduce $p(x, y) = p(x)p(y)$ into Eq. (11.125) and obtain

$$C_{xy} = \int_{-\infty}^{\infty} \int_{-\infty}^{\infty} (x - \bar{x})(y - \bar{y})p(x, y)\, dx\, dy$$

$$= \int_{-\infty}^{\infty} xp(x)\, dx \int_{-\infty}^{\infty} yp(y)\, dy - E[x]E[y] = 0 \tag{11.129}$$

On the other hand, in the general case in which $p(x, y) \neq p(x)p(y)$ the fact that the covariance is zero merely implies that

$$E[xy] = E[x]E[y] \tag{11.130}$$

However, *in the very important case in which $p(x, y)$ represents the joint normal probability density function, uncorrelated random variables are also statistically independent.* Indeed, the joint normal probability density function has the expression

$$p(x, y) = \frac{1}{2\pi \sigma_x \sigma_y \sqrt{1 - \rho_{xy}^2}} \exp \left\{ -\frac{1}{2\sqrt{1 - \rho_{xy}^2}} \left[ \left( \frac{x - \bar{x}}{\sigma_x} \right)^2 \right.\right.$$

$$\left.\left. - 2\rho_{xy} \frac{x - \bar{x}}{\sigma_x} \frac{y - \bar{y}}{\sigma_y} + \left( \frac{y - \bar{y}}{\sigma_y} \right)^2 \right] \right\} \tag{11.131}$$

so that when the correlation coefficient is zero, Eq. (11.131) reduces to the product of the individual normal probability density functions

$$p(x) = \frac{1}{\sqrt{2\pi}\, \sigma_x} \exp \left[ -\frac{(x - \bar{x})^2}{2\sigma_x^2} \right]$$

$$p(y) = \frac{1}{\sqrt{2\pi}\, \sigma_y} \exp \left[ -\frac{(y - \bar{y})^2}{2\sigma_y^2} \right] \tag{11.132}$$

thus satisfying Eq. (11.122), with the implication that the random variables $x$ and $y$

are statistically independent. Note that this result is not valid for arbitrary joint probability distributions.

## 11.14 JOINT PROPERTIES OF STATIONARY RANDOM PROCESSES

Let us consider two arbitrary random processes $\{x_k(t)\}$ and $\{y_k(t)\}$ of the type discussed in Sec. 11.2. The time histories $x_k(t)$ and $y_k(t)$ $(k = 1, 2, \ldots)$ resemble those depicted in Fig. 11.2. The object is to calculate certain *ensemble averages*. In particular, let us calculate the *mean values* at the arbitrary fixed time $t_1$ as follows:

$$\mu_x(t_1) = \lim_{n \to \infty} \frac{1}{n} \sum_{k=1}^{n} x_k(t_1) \qquad \mu_y(t_1) = \lim_{n \to \infty} \frac{1}{n} \sum_{k=1}^{n} y_k(t_1) \qquad (11.133)$$

For arbitrary random processes, the mean values at different times, say $t_1 \neq t_2$, are different, so that

$$\mu_x(t_1) \neq \mu_x(t_2) \qquad \mu_y(t_1) \neq \mu_y(t_2) \qquad (11.134)$$

Next let us calculate the *covariance functions* at the arbitrary fixed times $t_1$ and $t_1 + \tau$ as follows:

$$C_x(t_1, t_1 + \tau) = \lim_{n \to \infty} \frac{1}{n} \sum_{k=1}^{n} [x_k(t_1) - \mu_x(t_1)][x_k(t_1 + \tau) - \mu_x(t_1 + \tau)]$$

$$C_y(t_1, t_1 + \tau) = \lim_{n \to \infty} \frac{1}{n} \sum_{k=1}^{n} [y_k(t_1) - \mu_y(t_1)][y_k(t_1 + \tau) - \mu_y(t_1 + \tau)] \qquad (11.135)$$

$$C_{xy}(t_1, t_1 + \tau) = \lim_{n \to \infty} \frac{1}{n} \sum_{k=1}^{n} [x_k(t_1) - \mu_x(t_1)][y_k(t_1 + \tau) - \mu_y(t_1 + \tau)]$$

The values of the covariance functions depend in general on the times $t_1$ and $t_1 + \tau$.

To provide a more detailed description of the random processes, higher-order statistics should be calculated, which involves the values of the time histories evaluated at three or more times, such as $t_1, t_1 + \tau, t_1 + \sigma$, etc. For reasons to be explained shortly, this is actually not necessary.

In the special case in which the mean values $\mu_x(t_1)$ and $\mu_y(t_1)$ and the covariance functions $C_x(t_1, t_1 + \tau)$, $C_y(t_1, t_1 + \tau)$ and $C_{xy}(t_1, t_1 + \tau)$ do not depend on $t_1$, the random processes $\{x_k(t)\}$ and $\{y_k(t)\}$ are said to be *weakly stationary*. Otherwise they are nonstationary. Hence, for weakly stationary random processes the mean values are constant, $\mu_x(t_1) = \mu_x = \text{const}$ and $\mu_y(t_1) = \mu_y = \text{const}$, and the covariance functions depend on the time shift $\tau$ alone, $C_x(t_1, t_1 + \tau) = C_x(\tau)$, $C_y(t_1, t_1 + \tau) = C_y(\tau)$ and $C_{xy}(t_1, t_1 + \tau) = C_{xy}(\tau)$. If all possible statistics are independent of $t_1$, then the random processes $\{x_k(t)\}$ and $\{y_k(t)\}$ are said to be *strongly stationary*. For normal, or Gaussian, random processes, however, higher-order averages can be derived from the mean values and covariance functions. It follows that *for Gaussian random processes, weak stationarity implies also strong stationarity.* Because our interest lies primarily in normal random processes, there is no need to calculate higher-order statistics, and random processes will be referred to as merely *stationary* if the mean values and covariance functions are insensitive to a translation in the time $t_1$. The remainder of this section is devoted exclusively to stationary random processes.

Ensemble averages can be calculated conveniently in terms of probability density functions. To this end, let us introduce the notation $x_1 = x_k(t)$, $x_2 = x_k(t + \tau), y_1 = y_k(t), y_2 = y_k(t + \tau)$, where $x_1$ and $x_2$ represent random variables from the stationary random process $\{x_k(t)\}$ and $y_1$ and $y_2$ represent random variables from the stationary random process $\{y_k(t)\}$. Then, the joint probability density functions $p(x_1, x_2), p(y_1, y_2)$ and $p(x_1, y_2)$ are independent of $t$. In view of this, the *mean values* can be written as

$$\mu_x = E[x] = \int_{-\infty}^{\infty} \int_{-\infty}^{\infty} x_1 p(x_1, x_2) \, dx_1 \, dx_2 = \int_{-\infty}^{\infty} x_1 p(x_1) \, dx_1 = \text{const}$$

$$\mu_y = E[y] = \int_{-\infty}^{\infty} \int_{-\infty}^{\infty} y_1 p(y_1, y_2) \, dy_1 \, dy_2 = \int_{-\infty}^{\infty} y_1 p(y_1) \, dy_1 = \text{const}$$

$$(11.136)$$

and the *correlation functions* have the expressions

$$R_x(\tau) = E[x_1 x_2] = \int_{-\infty}^{\infty} \int_{-\infty}^{\infty} x_1 x_2 p(x_1, x_2) \, dx_1 \, dx_2$$

$$R_y(\tau) = E[y_1 y_2] = \int_{-\infty}^{\infty} \int_{-\infty}^{\infty} y_1 y_2 p(y_1, y_2) \, dy_1 \, dy_2 \qquad (11.137)$$

$$R_{xy}(\tau) = E[x_1 y_2] = \int_{-\infty}^{\infty} \int_{-\infty}^{\infty} x_1 y_2 p(x_1, y_2) \, dx_1 \, dy_2$$

where $R_x(\tau)$ and $R_y(\tau)$ represent *autocorrelation functions*, and $R_{xy}(\tau)$ is a *cross-correlation function*. Moreover, the *covariance functions* can be written as

$$C_x(\tau) = E[(x_1 - \mu_x)(x_2 - \mu_x)]$$
$$= \int_{-\infty}^{\infty} \int_{-\infty}^{\infty} (x_1 - \mu_x)(x_2 - \mu_x) p(x_1, x_2) \, dx_1 \, dx_2$$
$$= R_x(\tau) - \mu_x^2$$

$$C_y(\tau) = E[(y_1 - \mu_y)(y_2 - \mu_y)]$$
$$= \int_{-\infty}^{\infty} \int_{-\infty}^{\infty} (y_1 - \mu_y)(y_2 - \mu_y) p(y_1, y_2) \, dy_1 \, dy_2 \qquad (11.138)$$
$$= R_y(\tau) - \mu_y^2$$

$$C_{xy}(\tau) = E[(x_1 - \mu_x)(y_2 - \mu_y)]$$
$$= \int_{-\infty}^{\infty} \int_{-\infty}^{\infty} (x_1 - \mu_x)(y_2 - \mu_y) p(x_1, y_2) \, dx_1 \, dy_2$$
$$= R_{xy}(\tau) - \mu_x \mu_y$$

From Eqs. (11.138), we conclude that the covariance functions are identical to the correlation functions only when the mean values are zero. When the covariance function $C_{xy}(\tau)$ is equal to zero for all $\tau$, the stationary random processes $\{x_k(t)\}$ and $\{y_k(t)\}$ are said to be uncorrelated. From the last of Eqs. (11.138), we conclude that this can happen only if the cross-correlation function $R_{xy}(\tau)$ is equal to zero for all $\tau$ and, in addition, either $\mu_x$ or $\mu_y$ is equal to zero.

Next let us denote $x_1 = x_k(t - \tau)$, $x_2 = x_k(t)$, $y_1 = y_k(t - \tau)$ and $y_2 = y_k(t)$. Then, because for stationary random processes $p(x_1, x_2)$, $p(y_1, y_2)$ and $p(x_1, y_2)$ are independent of a translation in the time $t$, it follows that the autocorrelation functions are even functions of $\tau$, that is,

$$R_x(-\tau) = R_x(\tau) \qquad R_y(-\tau) = R_y(\tau) \tag{11.139}$$

whereas the cross-correlation function merely satisfies

$$R_{xy}(-\tau) = R_{yx}(\tau) \tag{11.140}$$

By using the same approach as that used in Sec. 11.8, it can be shown that

$$R_x(0) \geqslant |R_x(\tau)| \qquad R_y(0) \geqslant |R_y(\tau)| \tag{11.141}$$

In contrast, however, $R_{xy}(\tau)$ does not necessarily have a maximum at $\tau = 0$. Bounds on the cross-correlation function can be established by considering

$$\int_{-\infty}^{\infty} \int_{-\infty}^{\infty} (x_1 \pm y_2)^2 p(x_1, y_2)\, dx_1\, dy_2$$

$$= \int_{-\infty}^{\infty} \int_{-\infty}^{\infty} x_1^2 p(x_1, y_2)\, dx_1\, dy_2 \pm 2 \int_{-\infty}^{\infty} \int_{-\infty}^{\infty} x_1 y_2 p(x_1, y_2)\, dx_1\, dy_2$$

$$+ \int_{-\infty}^{\infty} \int_{-\infty}^{\infty} y_2^2 p(x_1, y_2)\, dx_1\, dy_2$$

$$= R_x(0) \pm 2R_{xy}(\tau) + R_y(0) \geqslant 0 \tag{11.142}$$

where the inequality is valid because the first integral in Eq. (11.142) cannot be negative. Note that the dependence on the time shift $\tau$ appears only when the variables with different subscripts are involved. It follows from Eq. (11.142) that

$$\tfrac{1}{2}[R_x(0) + R_y(0)] \geqslant |R_{xy}(\tau)| \tag{11.143}$$

Moreover, considering the integral

$$\int_{-\infty}^{\infty} \int_{-\infty}^{\infty} \left[ \frac{x_1}{\sqrt{R_x(0)}} \pm \frac{y_2}{\sqrt{R_y(0)}} \right]^2 p(x_1, y_2)\, dx_1\, dy_2 \tag{11.144}$$

which is also nonnegative, it can be shown that

$$R_x(0)R_y(0) \geqslant |R_{xy}(\tau)|^2 \tag{11.145}$$

From the above, we conclude that the correlation properties of the two stationary random processes $\{x_k(t)\}$ and $\{y_k(t)\}$ can be described by the correlation functions $R_x(\tau)$, $R_y(\tau)$, $R_{xy}(\tau)$ and $R_{yx}(\tau)$. Moreover, in view of relations (11.139) and (11.140), these functions need be calculated only for values of $\tau$ larger than or equal to zero.

At this point, it is possible to introduce power spectral densities and cross-spectral densities associated with the two random processes $\{x_k(t)\}$ and $\{y_k(t)\}$. We defer, however, the discussion to the next section, when these concepts are discussed in the context of ergodic random processes.

## 11.15 JOINT PROPERTIES OF ERGODIC RANDOM PROCESSES

Let us consider the two stationary random processes $\{x_k(t)\}$ and $\{y_k(t)\}$ of Sec. 11.14, but instead of calculating ensemble averages, we select two arbitrary time histories $x_k(t)$ and $y_k(t)$ from these processes and calculate time averages. In general, the averages calculated by using these sample functions will be different for different $x_k(t)$ and $y_k(t)$, so that we shall identify these averages by the index $k$.

The *temporal mean values* can be written in the form

$$\mu_x(k) = \lim_{T \to \infty} \frac{1}{T} \int_{-T/2}^{T/2} x_k(t)\, dt \qquad \mu_y(k) = \lim_{T \to \infty} \frac{1}{T} \int_{-T/2}^{T/2} y_k(t)\, dt \quad (11.146)$$

whereas the *temporal covariance functions* have the expressions

$$C_x(\tau, k) = \lim_{T \to \infty} \frac{1}{T} \int_{-T/2}^{T/2} [x_k(t) - \mu_x(k)][x_k(t + \tau) - \mu_x(k)]\, dt$$

$$C_y(\tau, k) = \lim_{T \to \infty} \frac{1}{T} \int_{-T/2}^{T/2} [y_k(t) - \mu_y(k)][y_k(t + \tau) - \mu_y(k)]\, dt \quad (11.147)$$

$$C_{xy}(\tau, k) = \lim_{T \to \infty} \frac{1}{T} \int_{-T/2}^{T/2} [x_k(t) - \mu_x(k)][y_k(t + \tau) - \mu_y(k)]\, dt$$

If the temporal mean values and covariance functions calculated by using the sample functions $x_k(t)$ and $y_k(t)$ are equal to the ensemble mean values and covariance functions, regardless of the pair of sample functions used, then the stationary random processes $\{x_k(t)\}$ and $\{y_k(t)\}$ are said to be *weakly ergodic*. If all ensemble averages can be deduced from temporal averages, then the stationary random processes are said to be *strongly ergodic*. Because Gaussian processes are fully described by first- and second-order statistics alone, no distinction need be made for such processes, and we shall refer to them as merely *ergodic*. Again, *ergodicity implies stationarity, but stationarity does not imply ergodicity.* Hence, the processes $\{x_k(t)\}$ and $\{y_k(t)\}$ are ergodic if

$$\mu_x(k) = \mu_x = \text{const} \qquad \mu_y(k) = \mu_y = \text{const} \quad (11.148)$$

and

$$C_x(\tau, k) = C_x(\tau) \qquad C_y(\tau, k) = C_y(\tau) \qquad C_{xy}(\tau, k) = C_{xy}(\tau) \quad (11.149)$$

The covariance functions are related to the correlation functions $R_x(\tau)$, $R_y(\tau)$, and $R_{xy}(\tau)$ by

$$C_x(\tau) = R_x(\tau) - \mu_x^2 \qquad C_y(\tau) = R_y(\tau) - \mu_y^2 \qquad C_{xy}(\tau) = R_{xy}(\tau) - \mu_x\mu_y \quad (11.150)$$

in which the correlation functions have the expressions

$$R_x(\tau) = \lim_{T \to \infty} \frac{1}{T} \int_{-T/2}^{T/2} x(t)x(t + \tau)\, dt$$

$$R_y(\tau) = \lim_{T \to \infty} \frac{1}{T} \int_{-T/2}^{T/2} y(t)y(t + \tau)\, dt \qquad (11.151)$$

$$R_{xy}(\tau) = \lim_{T \to \infty} \frac{1}{T} \int_{-T/2}^{T/2} x(t)y(t + \tau)\, dt$$

where the index identifying the sample functions $x_k(t)$ and $y_k(t)$ has been omitted because the correlation functions are the same for any pair of sample functions. In view of the fact that ergodicity implies stationarity, properties (11.139) and (11.140) and inequalities (11.141), (11.143), and (11.145) continue to be valid.

Next, let us assume that the autocorrelation functions $R_x(\tau)$ and $R_y(\tau)$ and the cross-correlation function $R_{xy}(\tau)$ exist, and define the power spectral density functions as the Fourier transforms

$$S_x(\omega) = \int_{-\infty}^{\infty} R_x(\tau)e^{-i\omega\tau}\, d\tau \qquad S_y(\omega) = \int_{-\infty}^{\infty} R_y(\tau)e^{-i\omega\tau}\, d\tau \qquad (11.152)$$

and the cross-spectral density function as the Fourier transform

$$S_{xy}(\omega) = \int_{-\infty}^{\infty} R_{xy}(\tau)e^{-i\omega\tau}\, d\tau \qquad (11.153)$$

Then, if the power spectral and cross-spectral density functions are given for the two processes, the autocorrelation and cross-correlation functions can be obtained from the inverse Fourier transforms

$$R_x(\tau) = \frac{1}{2\pi} \int_{-\infty}^{\infty} S_x(\omega)e^{i\omega\tau}\, d\omega \qquad R_y(\tau) = \frac{1}{2\pi} \int_{-\infty}^{\infty} S_y(\omega)e^{i\omega\tau}\, d\omega$$

$$(11.154)$$

$$R_{xy}(\tau) = \frac{1}{2\pi} \int_{-\infty}^{\infty} S_{xy}(\omega)e^{i\omega\tau}\, d\omega$$

Using properties (11.139), it can be shown that the power spectral density functions are even functions of $\omega$,

$$S_x(-\omega) = S_x(\omega) \qquad S_y(-\omega) = S_y(\omega) \qquad (11.155)$$

whereas using property (11.140) it follows that

$$S_{xy}(-\omega) = S_{yx}(\omega) \qquad (11.156)$$

from which we conclude that if $S_{xy}(\omega)$ and $S_{yx}(\omega)$ are given for $\omega > 0$, then Eq.

(11.156) can be used to obtain $S_{yx}(\omega)$ and $S_{xy}(\omega)$ for $\omega < 0$, respectively. In view of properties (11.155), Eqs. (11.152) reduce to

$$S_x(\omega) = 2 \int_0^\infty R_x(\tau) \cos \omega\tau \, d\tau \qquad S_y(\omega) = 2 \int_0^\infty R_y(\tau) \cos \omega\tau \, d\tau \quad (11.157)$$

and the first two of Eqs. (11.154) become

$$R_x(\tau) = \frac{1}{\pi} \int_0^\infty S_x(\omega) \cos \omega\tau \, d\omega \qquad R_y(\tau) = \frac{1}{\pi} \int_0^\infty S_y(\omega) \cos \omega\tau \, d\omega$$

$$(11.158)$$

Equations (11.157) and (11.158) are known as the *Wiener-Khintchine equations*. Note that $S_x(\omega)$ and $S_y(\omega)$ are nonnegative on physical grounds, and they are real because $R_x(\tau)$ and $R_y(\tau)$ are real.

## 11.16 RESPONSE CROSS-CORRELATION FUNCTIONS FOR LINEAR SYSTEMS

Let us consider two linear systems defined in the time domain by the impulse response $g_r(t)$ and $g_s(t)$ and in the frequency domain by the frequency responses $G_r(\omega)$ and $G_s(\omega)$, where the latter are the Fourier transforms of the former, namely,

$$G_r(\omega) = \int_{-\infty}^\infty g_r(t)e^{-i\omega t} \, dt \qquad G_s(\omega) = \int_{-\infty}^\infty g_s(t)e^{-i\omega t} \, dt \quad (11.159)$$

The relations between the excitations $f_r(t)$ and $f_s(t)$ and the corresponding responses $q_r(t)$ and $q_s(t)$ can be given in the form of the block diagrams of Fig. 11.30a, whereas those between the transformed excitations $F_r(\omega)$ and $F_x(\omega)$ and the corresponding transformed responses $Q_r(\omega)$ and $Q_s(\omega)$ can be given in the form of the block diagrams of Fig. 11.30b, where $F_r(\omega)$ is the Fourier transform of $f_r(t)$, etc.

Assuming that the excitation and response processes are ergodic, the cross-correlation function between the response processes $q_r(t)$ and $q_s(t)$ can be written in the form

$$R_{q_r q_s}(\tau) = \lim_{T \to \infty} \int_{-T/2}^{T/2} q_r(t)q_s(t + \tau) \, dt \quad (11.160)$$

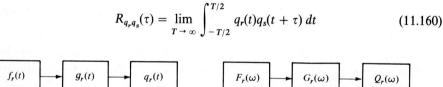

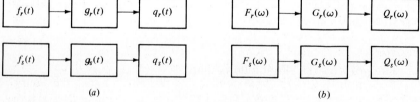

(a)            (b)

**Figure 11.30**

But, for linear systems the relation between the excitation and response can be expressed in terms of the convolution integral, Eq. (11.81). Hence, we can write

$$
q_r(t) = \int_{-\infty}^{\infty} g_r(\lambda_r) f_r(t - \lambda_r) \, d\lambda_r
$$

$$
q_s(t) = \int_{-\infty}^{\infty} g_s(\lambda_s) f_s(t - \lambda_s) \, d\lambda_s
$$

(11.161)

where $\lambda_r$ and $\lambda_s$ are corresponding dummy variables. Inserting Eqs. (11.161) into (11.160) and changing the order of integration, we obtain

$$
R_{q_r q_s}(\tau) = \lim_{T \to \infty} \frac{1}{T} \int_{-T/2}^{T/2} \left[ \int_{-\infty}^{\infty} g_r(\lambda_r) f_r(t - \lambda_r) \, d\lambda_r \right]
$$

$$
\times \left[ \int_{-\infty}^{\infty} g_s(\lambda_s) f_s(t + \tau - \lambda_s) \, d\lambda_s \right] dt
$$

$$
= \int_{-\infty}^{\infty} \int_{-\infty}^{\infty} g_r(\lambda_r) g_s(\lambda_s)
$$

$$
\times \left[ \lim_{T \to \infty} \frac{1}{T} \int_{-T/2}^{T/2} f_r(t - \lambda_r) f_s(t + \tau - \lambda_s) \, dt \right] d\lambda_r \, d\lambda_s
$$

(11.162)

Because the excitation processes are ergodic, and hence stationary, we recognize that

$$
\lim_{T \to \infty} \frac{1}{T} \int_{-T/2}^{T/2} f_r(t - \lambda_r) f_s(t + \tau - \lambda_s) \, dt
$$

$$
= \lim_{T \to \infty} \frac{1}{T} \int_{-T/2}^{T/2} f_r(t) f_s(t + \tau + \lambda_r - \lambda_s) \, dt = R_{f_r f_s}(\tau + \lambda_r - \lambda_s) \quad (11.163)
$$

is the cross-correlation function between the excitation processes. Hence, Eq. (11.162) can be written in the form

$$
R_{q_r q_s}(\tau) = \int_{-\infty}^{\infty} \int_{-\infty}^{\infty} g_r(\lambda_r) g_s(\lambda_s) R_{f_r f_s}(\tau + \lambda_r - \lambda_s) \, d\lambda_r \, d\lambda_s \quad (11.164)
$$

which represents an expression relating the cross-correlation function between the response processes to the cross-correlation function between the excitation processes in the time domain. Note the analogy between Eq. (11.164) and Eq. (11.94), where the latter is an expression relating the autocorrelation function of a single response to the autocorrelation function of a single excitation.

The interest lies in an expression analogous to Eq. (11.164) but in the frequency domain instead of the time domain. To this end, we take the Fourier transform of both sides of Eq. (11.164). But, the Fourier transform of $R_{q_r q_s}(\tau)$ is the cross-spectral density function associated with the response processes $q_r(t)$ and

$q_s(t)$, or

$$S_{q_r q_s}(\omega) = \int_{-\infty}^{\infty} R_{q_r q_s}(\tau) e^{-i\omega\tau} \, d\tau$$

$$= \int_{-\infty}^{\infty} e^{-i\omega\tau} \left[ \int_{-\infty}^{\infty} \int_{-\infty}^{\infty} g_r(\lambda_r) g_s(\lambda_s) R_{f_r f_s}(\tau + \lambda_r - \lambda_s) \, d\lambda_r \, d\lambda_s \right] d\tau$$

$$(11.165)$$

Moreover, $R_{f_r f_s}(\tau + \lambda_r - \lambda_s)$ can be expressed as the inverse Fourier transform

$$R_{f_r f_s}(\tau + \lambda_r - \lambda_s) = \frac{1}{2\pi} \int_{-\infty}^{\infty} S_{f_r f_s}(\omega) e^{i\omega(\tau + \lambda_r - \lambda_s)} \, d\omega \qquad (11.166)$$

where $S_{f_r f_s}(\omega)$ is the cross-spectral density function associated with the excitation processes $f_r(t)$ and $f_s(t)$. Inserting Eq. (11.166) into Eq. (11.165), considering Eqs. (11.159), changing the integration order and rearranging, we obtain

$$S_{q_r q_s}(\omega) = \int_{-\infty}^{\infty} e^{-i\omega\tau} \left\{ \int_{-\infty}^{\infty} \int_{-\infty}^{\infty} g_r(\lambda_r) g_s(\lambda_s) \right.$$

$$\left. \times \left[ \frac{1}{2\pi} \int_{-\infty}^{\infty} S_{f_r f_s}(\omega) e^{i\omega(\tau + \lambda_r - \lambda_s)} \, d\omega \right] d\lambda_r \, d\lambda_s \right\} d\tau$$

$$= \int_{-\infty}^{\infty} e^{-i\omega\tau} \left\{ \frac{1}{2\pi} \int_{-\infty}^{\infty} S_{f_r f_s}(\omega) \left[ \int_{-\infty}^{\infty} g_r(\lambda_r) e^{i\omega\lambda_r} \, d\lambda_r \right. \right.$$

$$\left. \left. \times \int_{-\infty}^{\infty} g_s(\lambda_s) e^{-i\omega\lambda_s} \, d\lambda_s \right] e^{i\omega\tau} \, d\omega \right\} d\tau$$

$$= \int_{-\infty}^{\infty} e^{-i\omega\tau} \left[ \frac{1}{2\pi} \int_{-\infty}^{\infty} S_{f_r f_s}(\omega) G_r^*(\omega) G_s(\omega) e^{i\omega\tau} \, d\omega \right] d\tau \qquad (11.167)$$

where $G_r^*(\omega) = G_r(-\omega)$ is the complex conjugate of $G_r(\omega)$. Comparing the first integral in Eq. (11.165) with the last one in Eq. (11.167), and recognizing that the cross-correlation function $R_{q_r q_s}(\tau)$ between the response process $q_r(t)$ and $q_s(t)$ must be equal to the inverse Fourier transform of the cross-spectral density function $S_{q_r q_s}(\omega)$ associated with these response processes, we must conclude that

$$S_{q_r q_s}(\omega) = G_r^*(\omega) G_s(\omega) S_{f_r f_s}(\omega) \qquad (11.168)$$

$$R_{q_r q_s}(\tau) = \frac{1}{2\pi} \int_{-\infty}^{\infty} S_{q_r q_s}(\omega) e^{i\omega\tau} \, d\omega = \frac{1}{2\pi} \int_{-\infty}^{\infty} G_r^*(\omega) G_s(\omega) S_{f_r f_s}(\omega) e^{i\omega\tau} \, d\omega$$

$$(11.169)$$

represent a Fourier transform pair. The algebraic expression (11.168) relates the cross-spectral density functions associated with the excitation and response processes in the frequency domain. Note the analogy between Eq. (11.168) and Eq. (11.98).

For any two time histories $f_r(t)$ and $f_s(t)$ corresponding to two stationary random signals, the cross-spectral density function $S_{f_r f_s}(\omega)$ can be obtained by means of an analog cross-spectral density analyzer.†

## 11.17 RESPONSE OF MULTI-DEGREE-OF-FREEDOM SYSTEMS TO RANDOM EXCITATION

We showed in Sec. 4.3 that the equations of motion of a damped $n$-degree-of-freedom system can be written in the matrix form

$$[m]\{\ddot{x}(t)\} + [c]\{\dot{x}(t)\} + [k]\{x(t)\} = \{F(t)\} \tag{11.170}$$

where $[m]$, $[c]$, and $[k]$ are $n \times n$ symmetric matrices called the inertia, damping, and stiffness matrices, respectively. The $n$-dimensional vector $\{x(t)\}$ contains the generalized coordinates $x_i(t)$, whereas the $n$-dimensional vector $\{F(t)\}$ contains the associated generalized forces $F_i(t)$ ($i = 1, 2, ..., n$). The interest lies in the case in which the excitations $F_i(t)$ represent ergodic random processes, from which it follows that the responses $x_i(t)$ are also ergodic random processes.

The general response of a damped multi-degree-of-freedom system to external excitation cannot be obtained readily, even when the excitation is deterministic. The difficulty lies in the fact that classical modal analysis cannot generally be used to uncouple the system of equations (11.170). However, as shown in Sec. 4.14, in the special case in which the damping matrix is a linear combination of the inertia and stiffness matrices, the modal matrix associated with the undamped linear system can be used as a linear transformation uncoupling the system of equations. Similarly, when damping is light, a reasonable approximation can be obtained by simply ignoring the coupling terms in the transformed equations. For simplicity, we shall confine ourselves to the case in which the classical modal matrix $[u] = [\{u\}_1 \{u\}_2 \cdots \{u\}_n]$ associated with the undamped system can be used as a transformation matrix uncoupling the set (11.170), either exactly or approximately. Following the procedure of Sec. 4.14, let us write the solution of Eq. (11.170) in the form

$$\{x(t)\} = [u]\{q(t)\} \tag{11.171}$$

where the components $q_r(t)$ ($r = 1, 2, ..., n$) of the vector $\{q(t)\}$ are generalized coordinates consisting of linear combinations of the random process $x_i(t)$ ($i = 1, 2, ..., n$). Inserting Eq. (11.171) into (11.170), premultiplying the result by $[u]^T$, using the orthonormality relations

$$[u]^T[m][u] = [1] \qquad [u]^T[k][u] = [\omega^2] \tag{11.172}$$

as well as assuming that

$$[u]^T[c][u] = [2\zeta\omega] \tag{11.173}$$

---

† See Bendat and Piersol, op. cit., sec. 8.5.

where $[2\zeta\omega]$ is a diagonal matrix, we obtain the set of independent equations for the natural coordinates

$$\ddot{q}_r(t) + 2\zeta_r\omega_r\dot{q}_r(t) + \omega_r^2 q_r(t) = \omega_r^2 f_r(t) \qquad r = 1, 2, \dots, n \qquad (11.174)$$

where $\zeta_r$ is a damping factor associated with the $r$th mode, $\omega_r$ is the $r$th frequency of the undamped system and

$$f_r(t) = \sum_{i=1}^{n} \frac{1}{\omega_r^2} u_{ri}F_i(t) = \frac{1}{\omega_r^2} \{u\}_r^T\{F(t)\} \qquad r = 1, 2, \dots, n \qquad (11.175)$$

is a generalized random force, in which $\{u\}_r$ represents the $r$th modal vector of the undamped system. Note that $f_r(t)$ actually has units $LM^{1/2}$, where $L$ denotes length and $M$ denotes mass.

Our first objective is to calculate the cross-correlation function between two response processes. To this end, we introduce the Fourier transforms of $q_r(t)$ and $f_r(t)$, respectively, in the form

$$Q_r(\omega) = \int_{-\infty}^{\infty} q_r(t)e^{-i\omega t}\, dt$$

$$F_r(\omega) = \int_{-\infty}^{\infty} f_r(t)e^{-i\omega t}\, dt = \sum_{i=1}^{n} \frac{1}{\omega_r^2} u_{ri} \int_{-\infty}^{\infty} F_i(t)e^{-i\omega t}\, dt \qquad (11.176)$$

Then, transforming both sides of Eqs. (11.174), we obtain

$$Q_r(\omega)(-\omega^2 + i2\zeta_r\omega\omega_r + \omega_r^2) = \omega_r^2 F_r(\omega) \qquad r = 1, 2, \dots, n \qquad (11.177)$$

Equations (11.177) can be solved for $Q_r(\omega)$ with the result

$$Q_r(\omega) = G_r(\omega)F_r(\omega) \qquad r = 1, 2, \dots, n \qquad (11.178)$$

where

$$G_r(\omega) = \frac{1}{1 - (\omega/\omega_r)^2 + i2\zeta_r\omega/\omega_r} \qquad r = 1, 2, \dots, n \qquad (11.179)$$

is the frequency response associated with the $r$th natural mode. Note the analogy between Eqs. (11.178) and Eq. (11.85).

Next, we wish to calculate the cross-correlation function between the response processes $x_i(t)$ and $x_j(t)$. But, any two elements $x_i(t)$ and $x_j(t)$ of the response vector can be obtained from Eq. (11.171) in the form

$$x_i(t) = \sum_{r=1}^{n} u_{ir}q_r(t) \qquad i = 1, 2, \dots, n$$

$$(11.180)$$

$$x_j(t) = \sum_{s=1}^{n} u_{js}q_s(t) \qquad j = 1, 2, \dots, n$$

Note that because the cross-correlation function $R_{x_i x_j}(\tau)$ between the response processes $x_i(t)$ and $x_j(t + \tau)$ involves the product of these processes, different dummy indices $r$ and $s$ were used in (11.180). Hence, let us write

$$R_{x_i x_j}(\tau) = \lim_{T \to \infty} \frac{1}{T} \int_{-T/2}^{T/2} x_i(t) x_j(t + \tau) \, dt$$

$$= \lim_{T \to \infty} \frac{1}{T} \int_{-T/2}^{T/2} \sum_{r=1}^{n} \sum_{s=1}^{n} u_{ir} u_{js} q_r(t) q_s(t + \tau) \, dt$$

$$= \sum_{r=1}^{n} \sum_{s=1}^{n} u_{ir} u_{js} R_{q_r q_s}(\tau) \tag{11.181}$$

where

$$R_{q_r q_s}(\tau) = \lim_{T \to \infty} \frac{1}{T} \int_{-T/2}^{T/2} q_r(t) q_s(t + \tau) \, dt \tag{11.182}$$

is the cross-correlation function between the generalized responses $q_r(t)$ and $q_s(t)$. But the cross-correlation function $R_{q_r q_s}(\tau)$ is related to the cross-spectral density function $S_{f_r f_s}(\omega)$ by Eq. (11.169). Hence, introducing Eq. (11.169) into (11.181), we obtain

$$R_{x_i x_j}(\tau) = \frac{1}{2\pi} \sum_{r=1}^{n} \sum_{s=1}^{n} u_{ir} u_{js} \int_{-\infty}^{\infty} G_r^*(\omega) G_s(\omega) S_{f_r f_s}(\omega) e^{i\omega\tau} \, d\tau \tag{11.183}$$

In general, however, we are given not the cross-spectral density function $S_{f_r f_s}(\omega)$ between the generalized excitations $f_r(t)$ and $f_s(t)$ but the cross-spectral density function $S_{F_i F_j}(\omega)$ between the excitations $F_i(t)$ and $F_j(t)$. This presents no particular difficulty, because the two of them are related, as can easily be shown. To this end, let us express $S_{f_r f_s}(\omega)$ as the Fourier transform

$$S_{f_r f_s}(\omega) = \int_{-\infty}^{\infty} R_{f_r f_s}(\tau) e^{-i\omega\tau} \, d\tau \tag{11.184}$$

where $R_{f_r f_s}(\tau)$ is the cross-correlation function between the generalized excitations $f_r(t)$ and $f_s(t)$. Recalling Eq. (11.175) for $f_r(t)$, and considering a companion equation for $f_s(t + \tau)$, we can write

$$R_{f_r f_s}(\tau) = \lim_{T \to \infty} \frac{1}{T} \int_{-T/2}^{T/2} f_r(t) f_s(t + \tau) \, dt$$

$$= \lim_{T \to \infty} \frac{1}{T} \int_{-T/2}^{T/2} \sum_{i=1}^{n} \sum_{j=1}^{n} \frac{1}{\omega_r^2} \frac{1}{\omega_s^2} u_{ir} u_{js} F_i(t) F_j(t + \tau) \, dt$$

$$= \sum_{i=1}^{n} \sum_{j=1}^{n} \frac{1}{\omega_r^2} \frac{1}{\omega_s^2} u_{ir} u_{js} \lim_{T \to \infty} \frac{1}{T} \int_{-T/2}^{T/2} F_i(t) F_j(t + \tau) \, dt$$

$$= \sum_{i=1}^{n} \sum_{j=1}^{n} \frac{1}{\omega_r^2} \frac{1}{\omega_s^2} u_{ir} u_{js} R_{F_i F_j}(\tau) \tag{11.185}$$

where

$$R_{F_i F_j}(\tau) = \lim_{T \to \infty} \frac{1}{T} \int_{-T/2}^{T/2} F_i(t) F_j(t + \tau) \, dt \tag{11.186}$$

is the cross-correlation function between the forces $F_i(t)$ and $F_j(t)$. Introducing Eq. (11.185) into (11.184), we obtain

$$
S_{f_r f_s}(\omega) = \int_{-\infty}^{\infty} \sum_{i=1}^{n} \sum_{j=1}^{n} \frac{1}{\omega_r^2} \frac{1}{\omega_s^2} u_{ir} u_{js} R_{F_i F_j}(\tau) e^{-i\omega\tau} \, d\tau
$$

$$
= \sum_{i=1}^{n} \sum_{j=1}^{n} \frac{1}{\omega_r^2} \frac{1}{\omega_s^2} u_{ir} u_{sj} \int_{-\infty}^{\infty} R_{F_i F_j}(\tau) e^{-i\omega\tau} \, d\tau
$$

$$
= \sum_{i=1}^{n} \sum_{j=1}^{n} \frac{1}{\omega_r^2} \frac{1}{\omega_s^2} u_{ir} u_{sj} S_{F_i F_j}(\omega) \tag{11.187}
$$

where

$$
S_{F_i F_j}(\omega) = \int_{-\infty}^{\infty} R_{F_i F_j}(\tau) e^{-i\omega\tau} \, d\tau \tag{11.188}
$$

is the cross-spectral density function between the excitation processes $F_i(t)$ and $F_j(t)$. For any two time histories $F_i(t)$ and $F_j(t)$ describing stationary random variables, the function $S_{F_i F_j}(\omega)$ can be obtained by means of an analog cross-spectral density analyzer.† The cross-correlation function between the response random processes $x_i(t)$ and $x_j(t)$ is obtained by introducing Eq. (11.187) into (11.183).

For $j = i$, the response cross-correlation function reduces to the autocorrelation function

$$
R_{x_i}(\tau) = \frac{1}{2\pi} \sum_{r=1}^{n} \sum_{s=1}^{n} u_{ir} u_{is} \int_{-\infty}^{\infty} G_r^*(\omega) G_s(\omega) S_{f_r f_s}(\omega) e^{i\omega\tau} \, d\omega \tag{11.189}
$$

In addition, letting $\tau = 0$ in Eq. (11.189), we obtain the mean square value

$$
R_{x_i}(0) = \frac{1}{2\pi} \sum_{r=1}^{n} \sum_{s=1}^{n} u_{ir} u_{is} \int_{-\infty}^{\infty} G_r^*(\omega) G_s(\omega) S_{f_r f_s}(\omega) \, d\omega \tag{11.190}
$$

associated with the response random process $x_i(t)$.

We assume for simplicity that the mean values of the response processes $x_i(t)$ ($i = 1, 2, \ldots, n$) are all equal to zero. Moreover, the positive square roots of the mean square values $R_{x_i}(0)$ represent the standard deviations $\sigma_{x_i}$ associated with the probability density functions of $x_i(t)$ ($i = 1, 2, \ldots, n$). Hence, if the excitation processes are known to be Gaussian, then the response processes are also Gaussian, with the probability density functions $p(x_i)$ fully defined by $\sigma_{x_i}$.

Before concluding this section, it will prove of interest to reformulate the problem in matrix notation. Indeed, recognizing that there are $n \times n$ cross-correlation functions $R_{x_i x_j}(\tau)$ corresponding to every pair of indices $i$ and $j$, we can introduce the *response correlation matrix*

$$
[R_x(\tau)] = \lim_{T \to \infty} \frac{1}{T} \int_{-T/2}^{T/2} \{x(t)\} \{x(t + \tau)\}^T \, dt \tag{11.191}
$$

† See Bendat and Piersol, op. cit., sec. 8.5.

But the vectors $\{x(t)\}$ and $\{q(t)\}$ are related by Eq. (11.171). Moreover, we can write

$$\{x(t + \tau)\}^T = \{q(t + \tau)\}^T[u]^T \qquad (11.192)$$

so that, inserting Eqs. (11.171) and (11.192) into Eq. (11.191), we obtain

$$[R_x(\tau)] = \lim_{T \to \infty} \frac{1}{T} \int_{-T/2}^{T/2} [u]\{q(t)\}\{q(t + \tau)\}^T[u]^T \, dt$$

$$= [u] \left[ \lim_{T \to \infty} \frac{1}{T} \int_{-T/2}^{T/2} \{q(t)\}\{q(t + \tau)\}^T \, dt \right] [u]^T$$

$$= [u][R_q(\tau)][u]^T \qquad (11.193)$$

where

$$[R_q(\tau)] = \lim_{T \to \infty} \frac{1}{T} \int_{-T/2}^{T/2} \{q(t)\}\{q(t + \tau)\}^T \, dt \qquad (11.194)$$

is the response correlation matrix associated with the coordinates $q_r(t)$ $(r = 1, 2, \ldots, n)$. Denoting by $[G(\omega)]$ the diagonal matrix of the frequency response functions, Eqs. (11.179), and considering Eq. (11.169), we can write the correlation matrix in the form

$$[R_q(\tau)] = \frac{1}{2\pi} \int_{-\infty}^{\infty} [G^*(\omega)][S_f(\omega)][G(\omega)]e^{i\omega\tau} \, d\omega \qquad (11.195)$$

where $[S_f(\omega)]$ is the $n \times n$ *excitation spectral matrix* associated with the generalized forces $f_r(t)$. The matrix $[S_f(\omega)]$ can be written as the Fourier transform of the excitation correlation matrix $[R_f(\tau)]$ associated with $f_r(t)$ as follows:

$$[S_f(\omega)] = \int_{-\infty}^{\infty} [R_f(\tau)]e^{-i\omega\tau} \, d\tau \qquad (11.196)$$

But $[R_f(\tau)]$ has the form

$$[R_f(\tau)] = \lim_{T \to \infty} \frac{1}{T} \int_{-T/2}^{T/2} \{f(t)\}\{f(t + \tau)\}^T \, dt \qquad (11.197)$$

where $\{f(t)\}$ is the vector of the generalized forces $f_r(t)$. Considering Eqs. (11.175), we can write

$$\{f(t)\} = [\omega^2]^{-1}[u]\{F(t)\} \qquad \{f(t + \tau)\}^T = \{F(t + \tau)\}^T[u]^T[\omega^2]^{-1} \qquad (11.198)$$

so that, introducing Eqs. (11.198) into (11.197), we obtain

$$[R_f(\tau)] = \lim_{T \to \infty} \frac{1}{T} \int_{-T/2}^{T/2} [\omega^2]^{-1}[u]\{F(t)\}\{F(t + \tau)\}^T[u]^T[\omega^2]^{-1} \, dt$$

$$= [\omega^2]^{-1}[u] \left[ \lim_{T \to \infty} \frac{1}{T} \int_{-T/2}^{T/2} \{F(t)\}\{F(t + \tau)\}^T \, dt \right] [u]^T[\omega^2]^{-1}$$

$$= [\omega^2]^{-1}[u][R_F(\tau)][u]^T[\omega^2]^{-1} \qquad (11.199)$$

in which

$$[R_F(\tau)] = \lim_{T \to \infty} \frac{1}{T} \int_{-T/2}^{T/2} \{F(t)\}\{F(t+\tau)\}^T \, dt \qquad (11.200)$$

is the correlation matrix associated with the forces $F_i(t)$ $(i = 1, 2, \ldots, n)$. Introduction of Eq. (11.199) into (11.196) yields

$$[S_f(\omega)] = \int_{-\infty}^{\infty} [\omega^2]^{-1}[u][R_F(\tau)][u]^T[\omega^2]^{-1}e^{-i\omega\tau} \, d\tau$$

$$= [\omega^2]^{-1}[u] \int_{-\infty}^{\infty} [R_F(\tau)]e^{-i\omega\tau} \, d\tau [u]^T[\omega^2]^{-1}$$

$$= [\omega^2]^{-1}[u][S_F(\omega)][u]^T[\omega^2]^{-1} \qquad (11.201)$$

where

$$[S_F(\omega)] = \int_{-\infty}^{\infty} [R_F(\omega)]e^{-i\omega\tau} \, d\tau \qquad (11.202)$$

is the excitation spectral matrix associated with the forces $F_i(t)$ $(i = 1, 2, \ldots, n)$. The matrix $[S_F(\omega)]$ lends itself to evaluation by means of an analog cross-spectral density analyzer. The response correlation matrix is obtained by simply introducing Eq. (11.195) into (11.193). The result is

$$[R_x(\tau)] = \frac{1}{2\pi} [u] \int_{-\infty}^{\infty} [G^*(\omega)][S_f(\omega)][G(\omega)]e^{i\omega\tau} \, d\omega[u]^T \qquad (11.203)$$

where $[S_f(\omega)]$ is given by Eq. (11.201).

Denoting by $[u]_i$ the $i$th row matrix of the modal matrix $[u]$, namely,

$$[u]_i = [u_{i1} \quad u_{i2} \quad \cdots \quad u_{in}] \qquad (11.204)$$

the autocorrelation function associated with the response random process $x_i(t)$ can be written from Eq. (11.203) in the form

$$R_{x_i}(\tau) = \frac{1}{2\pi} [u]_i \int_{-\infty}^{\infty} [G^*(\omega)][S_f(\omega)][G(\omega)]e^{i\omega\tau} \, d\omega[u]_i^T \qquad (11.205)$$

which for $\tau = 0$ reduces to the mean square value

$$R_{x_i}(0) = \frac{1}{2\pi} [u]_i \int_{-\infty}^{\infty} [G^*(\omega)][S_f(\omega)][G(\omega)] \, d\omega[u_i]^T \qquad (11.206)$$

**Example 11.5** Consider the system shown in Fig. 11.31, where the force $F_1(t)$ can be regarded as an ergodic random process with zero mean and with ideal white noise power spectral density, $S_{F_1}(\omega) = S_0$, and obtain the mean square values associated with the responses $x_1(t)$ and $x_2(t)$.

The mean square values associated with $x_1(t)$ and $x_2(t)$ will be obtained by the modal analysis outlined in this section. The differential equations of motion

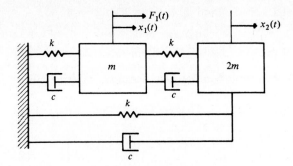

**Figure 11.31**

associated with the system can be shown to be

$$m\ddot{x}_1 + 2c\dot{x}_1 - c\dot{x}_2 + 2kx_1 - kx_2 = F_1(t)$$

$$2m\ddot{x}_2 - c\dot{x}_1 + 2c\dot{x}_2 - kx_1 + 2kx_2 = 0$$

$$(a)$$

so that the eigenvalue problem associated with the undamped free vibration of the system has the form

$$\omega^2 m \begin{bmatrix} 1 & 0 \\ 0 & 2 \end{bmatrix} \begin{Bmatrix} u_1 \\ u_2 \end{Bmatrix} = k \begin{bmatrix} 2 & -1 \\ -1 & 2 \end{bmatrix} \begin{Bmatrix} u_1 \\ u_2 \end{Bmatrix} \qquad (b)$$

The solution of the eigenvalue problem $(b)$ was obtained in Example 4.5, and the modes were normalized according to Eq. (11.172) in Example 4.6. The modal matrix is

$$[u] = \frac{1}{\sqrt{m}} \begin{bmatrix} 0.4597 & 0.8881 \\ 0.6280 & -0.3251 \end{bmatrix} \qquad (c)$$

and the matrix of the natural frequencies squared is

$$[\omega^2] = \frac{k}{m} \begin{bmatrix} 0.6340 & 0 \\ 0 & 2.3660 \end{bmatrix} \qquad (d)$$

The excitation spectral matrix associated with the actual coordinates $x_1(t)$ and $x_2(t)$ is

$$[S_F(\omega)] = \begin{bmatrix} S_0 & 0 \\ 0 & 0 \end{bmatrix} \qquad (e)$$

so that, using Eq. (11.201), we obtain the excitation spectral matrix associated with the normal coordinates $q_1(t)$ and $q_2(t)$ in the form

$$[S_f(\omega)] = [\omega^2]^{-1}[u][S_F(\omega)][u]^T[\omega^2]^{-1} = \frac{mS_0}{k^2} \begin{bmatrix} 0.5258 & 0.1925 \\ 0.1925 & 0.0704 \end{bmatrix} \qquad (f)$$

Moreover, the frequency response functions associated with the normal coordinates $q_1(t)$ and $q_2(t)$ have the form

$$G_r(\omega) = \frac{1}{1 - (\omega/\omega_r)^2 + i2\zeta_r\omega/\omega_r} \qquad r = 1, 2 \qquad (g)$$

where $\omega_1^2$ and $\omega_2^2$ are obtained from $(d)$ and $2\zeta_1\omega_1$ and $2\zeta_2\omega_2$ from

$$[2\zeta\omega] = [u]^T[c][u] = \frac{c}{m}\begin{bmatrix} 0.6340 & 0 \\ 0 & 2.3660 \end{bmatrix} \qquad (h)$$

where the matrix $[2\zeta\omega]$ is diagonal because the damping matrix $[c]$ is proportional to the stiffness matrix $[k]$.

The response mean square values are given by Eq. (11.206). First, let us form

$$[G^*(\omega)][S_f(\omega)][G(\omega)] = \frac{mS_0}{k^2}\begin{bmatrix} 0.5258|G_1|^2 & 0.1925G_1^*G_2 \\ 0.1925G_1G_2^* & 0.0704|G|_2^2 \end{bmatrix} \qquad (i)$$

Using the notation (11.204), we can write

$$[u]_1 = \frac{1}{\sqrt{m}}[0.4597 \quad 0.8881] \qquad [u]_2 = \frac{1}{\sqrt{m}}[0.6280 \quad -0.3251] \qquad (j)$$

so that we can form

$$[u]_1[G^*(\omega)][S_f(\omega)][G(\omega)][u]_1^T$$

$$= \frac{S_0}{k^2}[0.1111|G_1|^2 + 0.0786(G_1^*G_2 + G_1G_2^*) + 0.0556|G_2|^2]$$

$$[u]_2[G^*(\omega)][S_f(\omega)][G(\omega)][u]_2^T \qquad (k)$$

$$= \frac{S_0}{k^2}[0.2074|G_1|^2 - 0.0393(G_1^*G_2 + G_1G_2^*) + 0.0074|G_2|^2]$$

Hence, using formula (11.206), we can write the mean square values

$$R_{x_1}(0) = \frac{S_0}{2\pi k^2}\left[ 0.1111 \int_{-\infty}^{\infty}|G_1|^2\,d\omega + 0.0786 \int_{-\infty}^{\infty}(G_1^*G_2 + G_1G_2^*)\,d\omega \right.$$

$$\left. + 0.0556 \int_{-\infty}^{\infty}|G_2|^2\,d\omega \right]$$

$$(l)$$

$$R_{x_2}(0) = \frac{S_0}{2\pi k^2}\left[ 0.2074 \int_{-\infty}^{\infty}|G_1|^2\,d\omega - 0.0393 \int_{-\infty}^{\infty}(G_1^*G_2 + G_1G_2^*)\,d\omega \right.$$

$$\left. + 0.0074 \int_{-\infty}^{\infty}|G_2|^2\,d\omega \right]$$

Note that the brackets on the right side of Eqs. $(k)$ and $(l)$ do not denote matrices.

Equations $(l)$ give the mean square values $R_{x_i}(0)$ $(i = 1, 2)$ in terms of integrals involving the frequency response functions $G_1(\omega)$ and $G_2(\omega)$ and their complex conjugates. From Sec. 11.12, we obtain

$$\int_{-\infty}^{\infty}|G_r|^2\,d\omega = \int_{-\infty}^{\infty}\frac{d\omega}{[1 - [\omega/\omega_r)^2]^2 + [2\zeta_r\omega/\omega_r]^2} = \frac{\pi\omega_r}{2\zeta_r} \qquad r = 1, 2 \qquad (m)$$

On the other hand, the integral

$$\int_{-\infty}^{\infty} (G_1^* G_2 + G_1 G_2^*)\, d\omega$$

$$= \int_{-\infty}^{\infty} \frac{\{[1 - (\omega/\omega_1)^2][1 - (\omega/\omega_2)^2] + (2\zeta_1 \omega/\omega_1)(2\zeta_2 \omega/\omega_2)\}\, d\omega}{\{[1 - (\omega/\omega_1)^2]^2 + (2\zeta_1 \omega/\omega_1)^2\}\{[1 - (\omega/\omega_2)^2]^2 + (2\zeta_2 \omega/\omega_2)^2\}} \quad (n)$$

requires once again the use of the residue theorem. Because no new knowledge is gained from the evaluation of the integral, we shall not pursue the subject any further.

## 11.18 RESPONSE OF CONTINUOUS SYSTEMS TO RANDOM EXCITATION

The response of continuous systems to random excitation can also be conveniently obtained by means of modal analysis. In fact, the procedure is entirely analogous to that for discrete systems. The procedure can be best illustrated by considering a specific system. For convenience, let us choose the uniform bar in bending discussed in Sec. 5.9. The boundary-value problem is described by the differential equation

$$m \frac{\partial^2 y(x, t)}{\partial t^2} + c \frac{\partial y(x, t)}{\partial t} + EI \frac{\partial^4 y(x, t)}{\partial x^4} = f(x, t) \qquad 0 < x < L \quad (11.207)$$

where $f(x, t)$ is an ergodic distributed random excitation and $y(x, t)$ is the ergodic random response. Note that the second term on the left side of Eq. (11.207) represents a distributed damping force. Moreover, the vibration $y(x, t)$ is subject to four boundary conditions, two at each end. Let us assume that the solution of the eigenvalue problem associated with the undamped system consists of the natural frequencies $\omega_r$ and natural modes $Y_r(x)$ $(r = 1, 2, \ldots)$, and that the solution is known; the modes are orthogonal. Moreover, let us assume that the natural modes are normalized so as to satisfy the orthonormality relations

$$\int_0^L m Y_r(x) Y_s(x)\, dx = \delta_{rs}$$
$$\qquad\qquad\qquad\qquad\qquad\qquad r, s = 1, 2, \ldots \qquad (11.208)$$
$$\int_0^L Y_r(x) EI \frac{d^4 Y_s(x)}{dx^4}\, dx = \omega_r^2 \delta_{rs}$$

where $\delta_{rs}$ is the Kronecker delta. In addition, the damping is such that

$$\int_0^L c Y_r(x) Y_s(x)\, dx = 2\zeta_r \omega_r \delta_{rs} \qquad r, s = 1, 2, \ldots \qquad (11.209)$$

Then, using the transformation

$$y(x, t) = \sum_{r=1}^{\infty} Y_r(x) q_r(t) \qquad (11.210)$$

in conjunction with the standard modal analysis, we obtain the independent set of ordinary differential equations

$$\ddot{q}_r(t) + 2\zeta_r \omega_r \dot{q}_r(t) + \omega_r^2 q_r(t) = \omega_r^2 f_r(t) \qquad r = 1, 2, \ldots \qquad (11.211)$$

where

$$f_r(t) = \frac{1}{\omega_r^2} \int_0^L Y_r(x) f(x, t) \, dx \qquad r = 1, 2, \ldots \qquad (11.212)$$

are generalized random forces. As for the discrete systems of Sec. 11.17, the forces $f_r(t)$ actually have units $LM^{1/2}$.

Equations (11.211) for the continuous system possess precisely the same structure as Eqs. (11.174) for the discrete system. Hence, the remaining part of the analysis resembles entirely that of Sec. 11.17. Indeed, using Eq. (11.212) and a similar equation for $f_s(t + \tau)$, we can write

$$
\begin{aligned}
R_{f_r f_s}(\tau) &= \lim_{T \to \infty} \frac{1}{T} \int_{-T/2}^{T/2} f_r(t) f_s(t + \tau) \, dt \\
&= \lim_{T \to \infty} \frac{1}{T} \int_{-T/2}^{T/2} \left[ \frac{1}{\omega_r^2} \int_0^L Y_r(x) f(x, t) \, dx \right] \\
&\qquad \times \left[ \frac{1}{\omega_s^2} \int_0^L Y_s(x') f(x', t + \tau) \, dx' \right] dt \\
&= \frac{1}{\omega_r^2} \frac{1}{\omega_s^2} \int_0^L \int_0^L Y_r(x) Y_s(x') \\
&\qquad \times \left[ \lim_{T \to \infty} \frac{1}{T} \int_{-T/2}^{T/2} f(x, t) f(x', t + \tau) \, dt \right] dx \, dx' \\
&= \frac{1}{\omega_r^2} \frac{1}{\omega_s^2} \int_0^L \int_0^L Y_r(x) Y_s(x') R_{f_x f_{x'}}(x, x', \tau) \, dx \, dx' \qquad (11.213)
\end{aligned}
$$

where $x$ and $x'$ are dummy variables denoting two different points of the domain $0 < x < L$, and

$$R_{f_x f_{x'}}(x, x', \tau) = \lim_{T \to \infty} \frac{1}{T} \int_{-T/2}^{T/2} f(x, t) f(x', t + \tau) \, dt \qquad (11.214)$$

is the *distributed cross-correlation function* between the distributed forces $f(x, t)$ and $f(x', t)$. Note that $R_{f_x f_{x'}}(x, x', \tau)$ has units of distributed force squared. Introducing Eq. (11.213) into (11.184), we obtain the cross-spectral density function

$$
\begin{aligned}
S_{f_r f_s}(\omega) &= \int_{-\infty}^{\infty} \left[ \frac{1}{\omega_r^2} \frac{1}{\omega_s^2} \int_0^L \int_0^L Y_r(x) Y_s(x') R_{f_x f_{x'}}(x, x', \tau) \, dx \, dx' \right] e^{-i\omega\tau} \, d\tau \\
&= \frac{1}{\omega_r^2} \frac{1}{\omega_s^2} \int_0^L \int_0^L Y_r(x) Y_s(x') \left[ \int_{-\infty}^{\infty} R_{f_x f_{x'}}(x, x', \tau) e^{-i\omega\tau} \, d\tau \right] dx \, dx' \\
&= \frac{1}{\omega_r^2} \frac{1}{\omega_s^2} \int_0^L \int_0^L Y_r(x) Y_s(x') S_{f_x f_{x'}}(x, x', \omega) \, dx \, dx' \qquad (11.215)
\end{aligned}
$$

where

$$S_{f_x f_{x'}}(x, x', \omega) = \int_{-\infty}^{\infty} R_{f_x f_{x'}}(x, x', \tau)e^{-i\omega\tau} d\tau \qquad (11.216)$$

is the *distributed cross-spectral density function* between the excitation processes $f(x, t)$ and $f(x', t)$.

The cross-correlation function between the response at $x$ and $x'$ can be written in the form

$$R_{y_x y_{x'}}(x, x', \tau) = \lim_{T \to \infty} \frac{1}{T} \int_{-T/2}^{T/2} y(x, t)y(x', t + \tau) \, dt$$

$$= \lim_{T \to \infty} \frac{1}{T} \int_{-T/2}^{T/2} \left[ \sum_{r=1}^{\infty} Y_r(x)q_r(t) \right] \left[ \sum_{s=1}^{\infty} Y_s(x')q_s(t + \tau) \right] dt$$

$$= \sum_{r=1}^{\infty} \sum_{s=1}^{\infty} Y_r(x) Y_s(x') R_{q_r q_s}(\tau) \qquad (11.217)$$

where $R_{q_r q_s}(\tau)$ is the cross-correlation function between the generalized responses $q_r(t)$ and $q_s(t)$, and has the form indicated by Eq. (11.182). However, $R_{q_r q_s}(\tau)$ is related to the cross-spectral density function $S_{f_r f_s}(\omega)$ between the generalized excitations $f_r(t)$ and $f_s(t)$ by Eq. (11.169), so that, inserting that equation into (11.217), we obtain

$$R_{y_x y_{x'}}(x, x', \tau) = \frac{1}{2\pi} \sum_{r=1}^{\infty} \sum_{s=1}^{\infty} Y_r(x) Y_s(x') \int_{-\infty}^{\infty} G_r^*(\omega)G_s(\omega)S_{f_r f_s}(\omega)e^{i\omega\tau} d\omega \qquad (11.218)$$

where $S_{f_r f_s}(\omega)$ is given by Eq. (11.215). Note that in Eq. (11.215) $x$ and $x'$ play the role of dummy variables of integration, whereas in Eq. (11.218) $x$ and $x'$ identify the points between which the cross-correlation function is evaluated, in the same way as the indices $i$ and $j$ in Eq. (11.183) do for discrete systems.

For $x = x'$, the response cross-correlation function reduces to the autocorrelation function

$$R_y(x, \tau) = \frac{1}{2\pi} \sum_{r=1}^{\infty} \sum_{s=1}^{\infty} Y_r(x) Y_s(x) \int_{-\infty}^{\infty} G_r^*(\omega)G_s(\omega)S_{f_r f_s}(\omega)e^{i\omega\tau} d\omega \qquad (11.219)$$

and letting $\tau = 0$ in Eq. (11.219), we obtain the mean square value of the response at point $x$ in the form

$$R_y(x, 0) = \frac{1}{2\pi} \sum_{r=1}^{\infty} \sum_{s=1}^{\infty} Y_r(x) Y_s(x) \int_{-\infty}^{\infty} G_r^*(\omega)G_s(\omega)S_{f_r f_s}(\omega) \, d\omega \qquad (11.220)$$

The square root of $R_y(x, 0)$ is the standard deviation associated with the probability density function of $y(x, t)$. Hence, assuming that $S_{f_x f_{x'}}(x, x', \omega)$ is given, Eq. (11.220) can be used in conjunction with Eq. (11.215) to calculate the standard deviation.

If the excitation process is Gaussian with zero mean, then so is the response process. In this case the standard deviation $\sqrt{R_y(x, 0)}$ determines fully the probability density function associated with the vibration $y(x, t)$.

The above formulation calls for an infinite number of natural modes $Y_r(x)$ $(r = 1, 2, \ldots)$. Of course, in practice only a finite number of modes need and should be taken, as Eq. (11.207) ceases to be valid for higher modes (see Sec. 5.6). It was implicit in the above discussion that a closed-form solution of the eigenvalue problem of the system is possible. A similar approach can be used also when only an approximate solution of the eigenvalue problem can be obtained. In such a case, a Rayleigh-Ritz procedure leads to a formulation resembling in structure that of a multi-degree-of-freedom system (see Prob. 11.25).

## PROBLEMS

**11.1** Calculate and plot the temporal autocorrelation function for the sinusoid $x(t) = A \sin (2\pi/T)t$.

**11.2** Calculate the temporal mean value and autocorrelation function for the periodic function shown in Fig. 11.32. Plot the autocorrelation function.

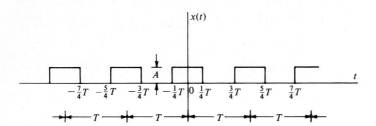

**Figure 11.32**

**11.3** Calculate and plot the temporal autocorrelation function for the periodic function shown in Fig. 11.8a.

**11.4** The function $x(t) = A|\sin (2\pi/T)t|$ is known as a *rectified sinusoid*. Its period is $T/2$, as opposed to $T$ for the ordinary sinusoid, as can be seen from Fig. 11.33. Calculate the mean value and the autocorrelation function for the rectified sinusoid.

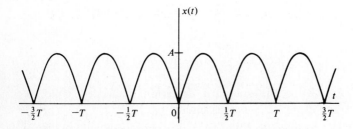

**Figure 11.33**

**11.5** Calculate and plot the autocorrelation function for the pulse-width-modulated wave shown in Fig. 11.34.

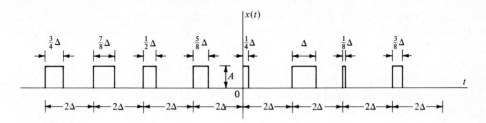

**Figure 11.34**

**11.6** Calculate the mean square value for the function of Prob. 11.1.

**11.7** Calculate the mean square value, the variance and the standard deviation for the function of Prob. 11.2.

**11.8** Calculate the mean square value for the function of Prob. 11.3.

**11.9** Calculate the mean value, the mean square value, the variance, and the standard deviation for the rectified sinusoid of Prob. 11.4.

**11.10** Use definition (11.9) and obtain the probability distribution $P(x)$ for the function of Example 11.1. Then use Eq. (11.12) and derive the probability density function $p(x)$. Plot $P(x)$ versus $x$ and $p(x)$ versus $x$.

**11.11** Assume that the time $t$ is uniformly distributed, and use Eq. (11.19) to verify the probability density function shown in Fig. 11.8c.

**11.12** Consider a rectified sinusoid with the constant amplitude $A$ and constant frequency $\omega$ but random phase angle $\phi$. For a fixed value $t_0$ of time, the rectified sinusoid can be regarded as a function of the random variable $\phi$ given by

$$x(\phi) = A|\sin(\omega t_0 + \phi)|$$

Let $\phi$ have a uniform probability density function $p(\phi)$, and calculate the probability density function $p(x)$ by the method of Sec. 11.5.

**11.13** Calculate the mean square value for the function shown in Fig. 11.8a by using Eq. (11.26).

**11.14** Calculate the mean value and the mean square value for the rectified sinusoid by using the probability density function $p(x)$ derived in Prob. 11.12.

**11.15** Calculate the power spectral density for the function of Example 11.2.

**11.16** Consider an ergodic random process with zero power spectral density at $\omega = 0$, and show that the autocorrelation function $R_f(\tau)$ must satisfy $\int_{-\infty}^{\infty} R_f(\tau) \, d\tau = 0$.

**11.17** Verify that the mathematical expression for the power density spectrum of the sine wave $f(t) = A \sin(2\pi/T)t$ is

$$S_f(\omega) = \frac{\pi A^2}{2}\left[\delta\left(\omega + \frac{2\pi}{T}\right) + \delta\left(\omega - \frac{2\pi}{T}\right)\right]$$

where $\delta(\omega + 2\pi/T)$ and $\delta(\omega - 2\pi/T)$ are Dirac delta functions acting at $\omega = -2\pi/T$ and $\omega = 2\pi/T$, respectively.

**11.18** A damped single-degree-of-freedom system is excited by a random process whose power density spectrum is as shown in Fig. 11.35. Let $\zeta = 0.05$ and $\omega_n = \omega_0/2$, and plot the response power density spectrum.

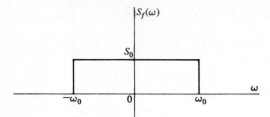

Figure 11.35

**11.19** Prove Eq. (11.112) by performing the integration indicated in Eq. (11.113).

**11.20** Prove inequality (11.145).

**11.21** Calculate the cross-correlation function between the functions of Prob. 11.1 and 11.2.

**11.22** Let $x(t)$ be the response of a linear system to the excitation $f(t)$, and show that

$$S_{fx}(\omega) = G(\omega)S_f(\omega)$$

where $S_{fx}(\omega)$ is the cross-spectral density function between the excitation and response, $G(\omega)$ is the frequency response and $S_f(\omega)$ is the excitation power spectral density function. [*Hint:* Begin by writing the cross-correlation function between the excitation and response in the form

$$R_{fx}(\tau) = \lim_{T \to \infty} \frac{1}{T} \int_{-T/2}^{T/2} f(t)x(t + \tau)\, dt$$

and recall that the response is related to the excitation by the convolution integral, Eq. (11.81)].

**11.23** Consider the system shown in Fig. 11.36, and derive the equations of motion. Let $c = 0.02\sqrt{km}$, and derive general equations for the cross-correlation function between $x_1(t)$ and $x_2(t)$ by observing that the modal matrix uncouples the equations of motion. Obtain the response mean square values for $x_1(t)$ and $x_2(t)$. The excitation $F_1(t)$ can be assumed to be an ergodic random process possessing an ideal white noise power density spectrum, whereas $F_2(t) = 0$.

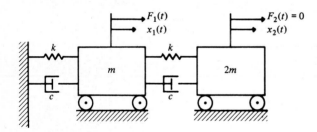

**Figure 11.36**

**11.24** Consider a uniform bar in bending simply supported at both ends and subjected to the excitation

$$f(x, t) = F(t)\delta\left(x - \frac{L}{2}\right)$$

where $F(t)$ is an ergodic random process with ideal white noise power spectral density, and $\delta(x - L/2)$ a spatial Dirac delta function. Use the method of Sec. 11.18, and derive expressions for the cross-correlation function between the responses at the points $x = L/4$ and $x' = 3L/4$ and for the mean square value of the response at $x = L/4$.

**11.25** Formulate the response problem of a continuous system to random excitation by means of an approximate method whereby the formulation is reduced to that of a multi-degree-of-freedom discrete system.

# TWELVE

## COMPUTATIONAL TECHNIQUES

## 12.1 INTRODUCTION

In earlier chapters, we showed how to derive the response of dynamical systems to a variety of excitations. In many cases, the nature of the system or the nature of the excitation makes it difficult to produce quantitative results. In such cases, numerical results can be obtained by means of a high-speed computer. In this chapter, we propose to present techniques that are especially suited for digital computers. To this end, we introduce certain approaches not discussed earlier.

In our treatment of vibration problems, we regarded the time as a continuous independent variable. Digital computers, however, do not accept continuous quantities, so that we must alter our formulation of the problems so as to make them suitable for digital computation. In particular, both the dependent and independent variables must be regarded as discrete, and in fact as digital, quantities. Systems in which the formulation is entirely in terms of discrete quantities are known as discrete-time systems. One approach permitting an efficient formulation of response problems in discrete time is that based on the transition matrix. This approach can be used for general linear dynamical systems, such as systems with arbitrary viscous damping. In the case of nonlinear systems with relatively strong nonlinearities, the response can only be obtained by numerical integration. One family of methods used widely for the integration of nonlinear ordinary differential equations consists of the Runge-Kutta methods. The Runge-Kutta methods can be cast conveniently in discrete-time form.

In random vibrations, frequency-domain techniques play an important role, as can be concluded from Chap. 11. In particular, the Fourier transform is an essential tool in spectral analysis. Quite often, however, Fourier transforms are difficult to evaluate in analytical form, so that they must be evaluated numerically. In recent years, numerical techniques were developed for the evaluation of Fourier transforms on digital computers. Such techniques are known under the general term of fast Fourier transforms and involve discretization in the frequency domain.

We begin this chapter with a discussion of the derivation of the system response by means of the transition matrix, an approach widely used in the linear system theory. It is shown that the response of general linear damped systems can be derived by this approach. Then, a discussion of discrete-time systems follows, in which the derivation of the system response by means of both the convolution sum and the discrete-time transition matrix is presented. The discussion of discrete-time systems is completed with a presentation of the discrete-time version of the fourth-order Runge-Kutta method for nonlinear systems. From discretization in the time domain, attention turns to discretization in the frequency domain. To this end, some supporting material, such as the convolution theorem, is presented. This permits the discretization of Fourier transforms in the frequency domain, leading ultimately to efficient computational techniques such as the fast Fourier transforms.

## 12.2 RESPONSE OF LINEAR SYSTEMS BY THE TRANSITION MATRIX

In Sec. 4.14, we derived the response of multi-degree-of-freedom linear systems by means of convolution integrals, where the response is to be interpreted as the displacement vector. A different approach permits the calculation of both the displacement and the velocity vectors at the same time. This latter approach proves very convenient in numerical computation of the response.

As indicated in Sec. 9.2, the displacement vector $\{x(t)\}$ and velocity vector $\{\dot{x}(t)\}$ of an $n$-degree-of-freedom system define the so-called *state* of the system. They can be arranged in a $2n$-dimensional vector of the form

$$\{y(t)\} = \left\{ \begin{matrix} \{x(t)\} \\ \{\dot{x}(t)\} \end{matrix} \right\} \tag{12.1}$$

where $\{y(t)\}$ is known as the *state vector*. Similarly, one can introduce the $2n$-dimensional excitation vector

$$\{Y(t)\} = \left\{ \begin{matrix} \{0\} \\ \{F(t)\} \end{matrix} \right\} \tag{12.2}$$

where $\{F(t)\}$ is the force vector. Then, the equation of motion of an $n$-degree-of-freedom linear system can be written in the general matrix form

$$\{\dot{y}(t)\} = [A]\{y(t)\} + [B]\{Y(t)\} \tag{12.3}$$

where $[A]$ and $[B]$ are $2n \times 2n$ matrices of coefficients depending on the nature of the system. In most cases considered in this text, $[A]$ and $[B]$ are constant matrices. The implication of Eq. (12.3) is that $n$ second-order ordinary differential equations for the displacements of an $n$-degree-of-freedom system can be replaced by $2n$

simultaneous first-order ordinary differential equations for the $2n$ components of the state vector. As can be concluded from Sec. 9.2, this requires the adjoining of certain identities to the original equations of motion.

To obtain the solution of Eq. (12.3), we consider first the homogeneous equation

$$\{\dot{y}(t)\} = [A]\{y(t)\} \tag{12.4}$$

The matrix equation (12.4) is similar in structure to the scalar first-order differential equation discussed in Sec. 1.5, so that the solution also must be similar in structure. Indeed, letting $\{y(0)\}$ be the initial state vector, the solution of the homogeneous equation (12.4) can be verified to be

$$\{y(t)\} = e^{[A]t}\{y(0)\} \tag{12.5}$$

where $e^{[A]t}$ is a matrix having the form of the series

$$e^{[A]t} = [1] + t[A] + \frac{t^2}{2!}[A]^2 + \frac{t^3}{3!}[A]^3 + \cdots \tag{12.6}$$

Turning our attention to the nonhomogeneous equation (12.3), let us introduce a $2n \times 2n$ matrix $[K(t)]$, premultiply Eq. (12.3) by $[K(t)]$, and obtain

$$[K(t)]\{\dot{y}(t)\} = [K(t)][A]\{y(t)\} + [K(t)][B]\{Y(t)\} \tag{12.7}$$

Then, let us consider

$$\frac{d}{dt}\{[K(t)]\{y(t)\}\} = [\dot{K}(t)]\{y(t)\} + [K(t)]\{\dot{y}(t)\} \tag{12.8}$$

so that Eq. (12.7) can be rewritten as

$$\frac{d}{dt}\{[K(t)]\{y(t)\}\} - [\dot{K}(t)]\{y(t)\} = [K(t)][A]\{y(t)\} + [K(t)][B]\{Y(t)\} \tag{12.9}$$

Next, we choose $[K(t)]$ so as to satisfy

$$[\dot{K}(t)] = -[A][K(t)] \tag{12.10}$$

which has the solution

$$[K(t)] = e^{-[A]t}[K(0)] \tag{12.11}$$

where

$$e^{-[A]t} = [1] - t[A] + \frac{t^2}{2!}[A]^2 - \frac{t^3}{3!}[A]^3 + \cdots \tag{12.12}$$

For convenience, we choose $[K(0)]$ as the identity matrix, or

$$[K(0)] = [1] \tag{12.13}$$

so that Eq. (12.11) reduces to

$$[K(t)] = e^{-[A]t} \tag{12.14}$$

From Eq. (12.12), however, we observe that the matrices $[K(t)]$ and $[A]$ commute, or

$$[A][K(t)] = [K(t)][A] \tag{12.15}$$

Inserting Eq. (12.15) into Eq. (12.10), we conclude that the matrix $[K(t)]$ also satisfies

$$[\dot{K}(t)] = -[K(t)][A] \tag{12.16}$$

Hence, in view of Eq. (12.16), Eq. (12.9) can be reduced to

$$\frac{d}{dt}\{[K(t)]\{y(t)\}\} = [K(t)][B]\{Y(t)\} \tag{12.17}$$

To complete the solution of Eq. (12.3), it remains to solve Eq. (12.17), which amounts to a simple integration yielding

$$[K(t)]\{y(t)\} = [K(0)]\{y(0)\} + \int_0^t [K(\tau)][B]\{Y(\tau)\}\, d\tau$$

$$= \{y(0)\} + \int_0^t [K(\tau)][B]\{Y(\tau)\}\, d\tau \tag{12.18}$$

Premultiplying Eq. (12.18) by $[K(t)]^{-1}$, we obtain finally the solution of the nonhomogeneous equation (12.3) in the form

$$\{y(t)\} = [K(t)]^{-1}\{y(0)\} + \int_0^t [K(t)]^{-1}[K(\tau)][B]\{Y(\tau)\}\, d\tau$$

$$= e^{[A]t}\{y(0)\} + \int_0^t e^{[A](t-\tau)}[B]\{Y(\tau)\}\, d\tau \tag{12.19}$$

which contains both the homogeneous and the particular solution. Clearly, the homogeneous solution is the same as the solution (12.5) obtained earlier and it involves the initial state $\{y(0)\}$. On the other hand, the particular solution involves the excitation $\{Y(t)\}$ and has the form of a convolution integral.

The matrix

$$[\Phi(t, \tau)] = e^{[A](t-\tau)} \tag{12.20}$$

is often referred to as the *transition matrix*. It can be obtained from Eq. (12.6) by simply replacing $t$ by $t - \tau$. The transition matrix possesses a very important property known as the *group property*, defined mathematically by

$$[\Phi(t_3, t_1)] = [\Phi(t_3, t_2)][\Phi(t_2, t_1)] \tag{12.21}$$

The group property can be used to advantage in computing the transition matrix by breaking a time interval into smaller subintervals, thus permitting the convergence of the series (12.6) with fewer terms. The group property also implies that

$$[\Phi(t_2, t_1)] = [\Phi(t_1, t_2)]^{-1} \tag{12.22}$$

Equations (12.21) and (12.22) can be derived from Eq. (12.20).

**Example 12.1** Derive a general expression for the response of a mass-spring system by means of the transition matrix. Then, use this expression to calculate the response of the system to the excitation

$$F(t) = f_0 t \alpha(t) \tag{a}$$

where $\alpha(t)$ is the unit step function.

Letting $c = 0$ in Eq. (1.14), the differential equation of motion of an undamped single-degree-of-freedom system reduces to

$$m\ddot{x}(t) + kx(t) = F(t) \tag{b}$$

Dividing Eq. (b) through by $m$ and rearranging, we obtain

$$\ddot{x}(t) = -\omega_n^2 x(t) + \frac{1}{m} F(t) \tag{c}$$

Moreover, introducing the identity

$$\dot{x}(t) = \dot{x}(t) \tag{d}$$

the second-order differential equation of motion (a) can be reduced to the state form (12.3), in which

$$\{y(t)\} = \begin{Bmatrix} x(t) \\ \dot{x}(t) \end{Bmatrix} \qquad \{Y(t)\} = \begin{Bmatrix} 0 \\ F(t) \end{Bmatrix}$$

$$[A] = \begin{bmatrix} 0 & 1 \\ -\omega_n^2 & 0 \end{bmatrix} \qquad [B] = \begin{bmatrix} 1 & 0 \\ 0 & 1/m \end{bmatrix} \tag{e}$$

To derive the solution by the transition matrix, we must expand the series for $e^{[A]t}$. Introducing the third of Eqs. (e) into Eq. (12.6), we can write

$$e^{[A]t} = \begin{bmatrix} 1 & 0 \\ 0 & 1 \end{bmatrix} + t \begin{bmatrix} 0 & 1 \\ -\omega_n^2 & 0 \end{bmatrix} - \frac{(\omega_n t)^2}{2!} \begin{bmatrix} 1 & 0 \\ 0 & 1 \end{bmatrix}$$

$$- \frac{\omega_n^2 t^3}{3!} \begin{bmatrix} 0 & 1 \\ -\omega_n^2 & 0 \end{bmatrix} + \frac{(\omega_n t)^4}{4!} \begin{bmatrix} 1 & 0 \\ 0 & 1 \end{bmatrix} + \cdots \tag{f}$$

Then, recalling the series expansions

$$\sin \omega_n t = \omega_n t - \frac{1}{3!} (\omega_n t)^3 + \frac{1}{5!} (\omega_n t)^5 - \cdots$$

$$\cos \omega_n t = 1 - \frac{1}{2!} (\omega_n t)^2 + \frac{1}{4!} (\omega_n t)^4 - \cdots \tag{g}$$

Eq. (f) can be rewritten as

$$e^{[A]t} = \begin{bmatrix} \cos \omega_n t & \omega_n^{-1} \sin \omega_n t \\ -\omega_n \sin \omega_n t & \cos \omega_n t \end{bmatrix} \tag{h}$$

Hence, inserting Eqs. (e) in conjunction with Eq. (h) into Eq. (12.19), the general solution of Eq. (*b*) can be written in the state form

$$
\begin{Bmatrix} x(t) \\ \dot{x}(t) \end{Bmatrix} = \begin{bmatrix} \cos \omega_n t & \omega_n^{-1} \sin \omega_n t \\ -\omega_n \sin \omega_n t & \cos \omega_n t \end{bmatrix} \begin{Bmatrix} x(0) \\ \dot{x}(0) \end{Bmatrix}
$$

$$
+ \int_0^t \begin{bmatrix} \cos \omega_n(t-\tau) & \omega_n^{-1} \sin \omega_n(t-\tau) \\ -\omega_n \sin \omega_n(t-\tau) & \cos \omega_n(t-\tau) \end{bmatrix} \begin{bmatrix} 1 & 0 \\ 0 & 1/m \end{bmatrix} \begin{Bmatrix} 0 \\ F(\tau) \end{Bmatrix} d\tau
$$

$$
= \begin{bmatrix} \cos \omega_n t & \omega_n^{-1} \sin \omega_n t \\ -\omega_n \sin \omega_n t & \cos \omega_n t \end{bmatrix} \begin{Bmatrix} x(0) \\ \dot{x}(0) \end{Bmatrix}
$$

$$
+ \frac{1}{m\omega_n} \int_0^t \begin{Bmatrix} \sin \omega_n(t-\tau) \\ \omega_n \cos \omega_n(t-\tau) \end{Bmatrix} F(\tau)\, d\tau \tag{i}
$$

which contains both the solution for the displacement $x(t)$ and the velocity $\dot{x}(t)$ in terms of the initial displacement $x(0)$, the initial velocity $\dot{x}(0)$ and the external excitation $F(t)$.

Inserting Eq. (*a*) into Eq. (*i*), letting the initial state be zero, using the change of variables $\omega_n(t-\tau) = \lambda$ and integrating, we obtain the desired response

$$
\begin{Bmatrix} x(t) \\ \dot{x}(t) \end{Bmatrix} = \frac{f_0}{m\omega_n} \int_0^t \begin{Bmatrix} \sin \omega_n(t-\tau) \\ \omega_n \cos \omega_n(t-\tau) \end{Bmatrix} \tau\, d\tau = \frac{f_0}{k} \int_0^{\omega_n t} \begin{Bmatrix} (t - \omega_n^{-1}\lambda)\sin\lambda \\ (\omega_n t - \lambda)\cos\lambda \end{Bmatrix} d\lambda
$$

$$
= \frac{f_0}{k} \begin{Bmatrix} -t\cos\lambda - \omega_n^{-1}(\sin\lambda - \lambda\cos\lambda) \\ \omega_n t \cos\lambda - (\cos\lambda - \lambda\sin\lambda) \end{Bmatrix} \Bigg|_0^{\omega_n t}
$$

$$
= \frac{f_0}{k\omega_n} \begin{Bmatrix} \omega_n t - \sin \omega_n t \\ \omega_n(1 - \cos \omega_n t) \end{Bmatrix} \tag{j}
$$

## 12.3 COMPUTATION OF THE TRANSITION MATRIX

In Sec. 12.2, we derived the general response of a system by the transition matrix, and in Example 12.1 we obtained the response of an undamped single-degree-of-freedom system to some given excitation. In that example, the transition matrix was a relatively simple $2 \times 2$ matrix with components in the form of trigonometric functions, leading to a closed-form solution for the response. Being able to produce a transition matrix in terms of simple known function is more the exception than the rule, and in general it is necessary to evaluate the transition matrix numerically.

Using Eqs. (12.6) and (12.20), we can write the transition matrix in the form of the infinite series

$$
[\Phi(t, 0)] = e^{[A]t} = [1] + t[A] + \frac{t^2}{2!}[A]^2 + \frac{t^3}{3!}[A]^3 + \cdots \tag{12.23}
$$

The inclusion of an infinite number of terms in numerical computation is not practical, so that the series must be truncated, which implies that the transition matrix can only be computed approximately. An approximation including terms through $n$th power in $[A]$ only has the form

$$[\Phi]_n = [1] + t[A] + \frac{t^2}{2!} [A]^2 + \cdots + \frac{t^n}{n!} [A]^n \qquad (12.24)$$

The computation can be performed efficiently by rewriting Eq. (12.24) as

$$[\Phi]_n = [1] + t[A]\left([1] + \frac{t}{2}[A]\left([1]\right.\right.$$

$$\left.\left. + \frac{t}{3}[A]\left([1] + \cdots + \frac{t}{n-t}[A]\left([1] + \frac{t}{n}[A]\right)\cdots\right)\right)\right) \qquad (12.25)$$

and by carrying out the recursive computations

$$[\psi]_1 = [1] + \frac{t}{n}[A]$$

$$[\psi]_2 = [1] + \frac{t}{n-1}[A][\psi]_1$$

$$[\psi]_3 = [1] + \frac{t}{n-2}[A][\psi]_2 \qquad (12.26)$$

$$\cdots\cdots\cdots\cdots\cdots\cdots\cdots\cdots$$

$$[\Phi]_n = [1] + t[A][\psi]_{n-1}$$

The computation of $[\Phi]_n$ by means of Eqs. (12.26) requires $n-1$ matrix multiplications.

Before the computation of $[\Phi]_n$ can be carried out, it is necessary to specify a time interval $t$. If $t$ is too large, however, then the number $n$ of terms must be relatively large for convergence. In fact, the number of terms required depends not only on $t$ but also on the matrix $[A]$. The number can be reduced by breaking the time interval $t$ into the smaller intervals $\Delta t_i = t_i - t_{i-1}$ $(i = 1, 2, \ldots, k)$ and using Eq. (12.21) to write

$$[\Phi]_n = [\Phi(t, 0)]_n$$

$$= [\Phi(t_k, t_{k-1})]_n[\Phi(t_{k-1}, t_{k-2})]_n \cdots [\Phi(t_2, t_1)]_n[\Phi_n(t_1, t_0)]_n \qquad (12.27)$$

where

$$[\Phi(t_i, t_{i-1})]_n = [\Phi(\Delta t_i, 0)]_n \qquad i = 1, 2, \ldots, k \qquad (12.28)$$

in which $t_0 = 0$ and $t_k = t$.

## 12.4 ALTERNATIVE COMPUTATION OF THE TRANSITION MATRIX

In Sec. 12.3, we pointed out that the number of terms required for the computation of the transition matrix depends on the time interval $t$ and on the matrix $[A]$. In this section, we consider another procedure for the computation of the transition matrix, one that permits us to make the preceding statement more explicit by connecting the convergence of the computation of the transition matrix to the product of the time $t$ and the eigenvalue of $[A]$ of largest modulus.

Let us assume that $[A]$ is an $m \times m$ matrix and consider the eigenvalue problem

$$[A]\{u\}_i = \lambda_i\{u\}_i \qquad i = 1, 2, \ldots, m \tag{12.29}$$

where $\lambda_i$ and $\{u\}_i$ $(i = 1, 2, \ldots, m)$ are the eigenvalues and eigenvectors of $[A]$, both complex in general. Note that in the case of vibrating systems $m = 2n$, in which $n$ is the number of degrees of freedom of the system. Because in general the matrix $[A]$ is not symmetric, the eigenvectors $\{u\}_i$ $(i = 1, 2, \ldots, m)$ are not mutually orthogonal. Nevertheless, they do satisfy some type of orthogonality relations. To demonstrate this, let us consider the eigenvalue problem associated with $[A]^T$ and write it in the form

$$[A]^T\{v\}_j = \lambda_j\{v\}_j \qquad j = 1, 2, \ldots, m \tag{12.30}$$

where $\lambda_j$ and $\{v\}_j$ $(j = 1, 2, \ldots, m)$ are the eigenvalues and eigenvectors of $[A]^T$. Equation (12.30) represents the so-called *adjoint* eigenvalue problem. But, because

$$\det [A]^T = \det [A] \tag{12.31}$$

$[A]$ and $[A]^T$ have the same characteristic polynomial, so that *the eigenvalues of $[A]^T$ are the same as the eigenvalues of $[A]$.* On the other hand, the eigenvectors of $[A]^T$ are different from the eigenvectors of $[A]$. The set of eigenvectors $\{v\}_j$ $(j = 1, 2, \ldots, m)$ is known as the *adjoint* of the set of eigenvectors $\{u\}_i$ $(i = 1, 2, \ldots, m)$. The eigenvalue problem (12.30) can also be written in the form

$$\{v\}_j^T[A] = \lambda_j\{v\}_j^T \qquad j = 1, 2, \ldots, m \tag{12.32}$$

Because of their position relative to the matrix $[A]$ in Eqs. (12.29) and (12.32), $\{u\}_i$ are called *right eigenvectors* of $[A]$ and $\{v\}_j$ are referred to as *left eigenvectors* of $[A]$. Multiplying Eq. (12.29) on the left by $\{v\}_j^T$ and Eq. (12.32) on the right by $\{u\}_i$, we obtain

$$\{v\}_j^T[A]\{u\}_i = \lambda_i\{v\}_j^T\{u\}_i \tag{12.33a}$$

$$\{v\}_j^T[A]\{u\}_i = \lambda_j\{v\}_j^T\{u\}_i \tag{12.33b}$$

so that, subtracting Eq. (12.33b) from Eq. (12.33a), we can write

$$(\lambda_i - \lambda_j)\{v\}_j^T\{u\}_i = 0 \tag{12.34}$$

But, if $\lambda_i \neq \lambda_j$, we must have

$$\{v\}_j^T\{u\}_i = 0 \qquad \lambda_i \neq \lambda_j \qquad i, j = 1, 2, \ldots, m \qquad (12.35)$$

or, *the left eigenvectors and right eigenvectors of* $[A]$ *corresponding to distinct eigenvalues are orthogonal.* It should be emphasized that the eigenvectors are *not mutually orthogonal* in the same ordinary sense as for symmetric matrices. The type of orthogonality described by Eq. (12.35) is referred to as *biorthogonality* and the two sets of eigenvectors are said to be *biorthogonal.* Inserting Eq. (12.35) into Eqs. (12.33), we conclude that

$$\{v\}_j^T[A]\{u\}_i = 0 \qquad \lambda_i \neq \lambda_j \qquad i, j = 1, 2, \ldots, m \qquad (12.36)$$

so that *the eigenvectors* $\{u\}_i$ *and* $\{v\}_j$ *are biorthogonal with respect to the matrix* $[A]$ *as well.* When $i = j$, the products $\{v\}_i^T\{u\}_i$ and $\{v\}_i^T[A]\{u\}_i$ are not zero. It is convenient to normalize the two sets of eigenvectors by letting

$$\{v\}_j^T\{u\}_i = \delta_{ij} \qquad i, j = 1, 2, \ldots, m \qquad (12.37)$$

where $\delta_{ij}$ is the Kronecker delta. Then, from Eqs. (12.33), we conclude that

$$\{v\}_j^T[A]\{u\}_i = \lambda_i\delta_{ij} \qquad i, j = 1, 2, \ldots, m \qquad (12.38)$$

Hence, in this case the eigenvectors $\{u\}_i$ and $\{v\}_j$ are *biorthonormal*, both in an ordinary sense and with respect to the matrix $[A]$.

Next, let us introduce the $m \times m$ matrices of right and left eigenvectors

$$[U] = [\{u\}_1 \quad \{u\}_2 \quad \cdots \quad \{u\}_m] \qquad [V] = [\{v\}_1 \quad \{v\}_2 \quad \cdots \quad \{v\}_m] \qquad (12.39)$$

as well as the $m \times m$ matrix of eigenvalues

$$\Lambda = \text{diag}\,[\lambda_1 \quad \lambda_2 \quad \cdots \quad \lambda_m] \qquad (12.40)$$

where it was assumed that all the eigenvalues of $[A]$ are distinct. Then, Eqs. (12.37) and (12.38) can be written in the compact form

$$[V]^T[U] = [1] \qquad (12.41a)$$

$$[V]^T[A][U] = [\Lambda] \qquad (12.41b)$$

Equations (12.41) can be used to express the transition matrix in a computationally attractive form. To this end, we use Eq. (12.41a) and write

$$[V]^T = [U]^{-1} \qquad [U] = ([V]^T)^{-1} \qquad (12.42)$$

so that, multiplying Eq. (12.41a) on the left by $[U]$ and on the right by $[V]^T$ and considering Eq. (12.43), we have

$$[U][V]^T = [1] \qquad (12.43)$$

Similarly, multiplying Eq. (12.41b) on the left by $[U]$ and on the right by $[V]^T$ and considering Eq. (12.43), we have

$$[U][\Lambda][V]^T = [A] \qquad (12.44)$$

Equations (12.43) and (12.44) can be used to express the transition matrix in the desired form. To this end, we recall Eq. (12.6) and write

$$e^{[A]t} = [1] + t[A] + \frac{t^2}{2!}[A]^2 + \frac{t^3}{3!}[A]^3 + \cdots$$

$$= [U][V]^T + t[U][\Lambda][V]^T + \frac{t^2}{2!}[U][\Lambda][V]^T[U][\Lambda][V]^T$$

$$+ \frac{t^3}{3!}[U][\Lambda][V]^T[U][\Lambda][V]^T[U][\Lambda][V]^T + \cdots$$

$$= [U][V]^T + t\,[U][\Lambda][V]^T + \frac{t^2}{2!}[U][\Lambda]^2[V]^T + \frac{t^3}{3!}[U][\Lambda]^3[V]^T + \cdots$$

$$= [U]\left([1] + t[\Lambda] + \frac{t^2}{2!}[\Lambda]^2 + \frac{t^3}{3!}[\Lambda]^3 + \cdots\right)[V]^T = [U]e^{[\Lambda]t}[V]^T$$

$$\tag{12.45}$$

where use was made of Eq. (12.41a) and where it was recognized that

$$[1] + t[\Lambda] + \frac{t^2}{2!}[\Lambda]^2 + \frac{t^3}{3!}[\Lambda]^3 + \cdots = e^{[\Lambda]t} \tag{12.46}$$

Hence, it follows from Eq. (12.20) that the transition matrix has the form

$$[\Phi(t, \tau)] = [U]e^{[\Lambda](t-\tau)}[V]^T \tag{12.47}$$

The advantage of the form (12.47) for the transition matrix compared to the form (12.20) lies in that in Eq. (12.47) the series involves raising $[\Lambda]$ instead of $[A]$ to the indicated powers. Indeed, because $[\Lambda]$ is diagonal, raising it to a given power amounts to raising the diagonal elements to the same power. On the other hand, before one can use Eq. (12.47), it is necessary to solve the eigenvalue problems associated with $[A]$ and $[A]^T$, as well as to normalize the associated eigenvectors so as to satisfy Eq. (12.41a).

The form (12.47) of the transition matrix is useful in a different respect also. Due to the fact that $[\Lambda]$ is diagonal, we can write

$$e^{[\Lambda](t-\tau)} = \text{diag}\,[e^{\lambda_i(t-\tau)}] \tag{12.48}$$

so that each diagonal term is in the form of an exponential. Hence, the convergence of $e^{[\Lambda](t-\tau)}$ depends on the convergence of all $e^{\lambda_i(t-\tau)}$ $(i = 1, 2, ..., m)$. Clearly, convergence depends on max $|\lambda_i|(t-\tau)$, where max $|\lambda_i|$ denotes the magnitude of the eigenvalue of $[A]$ of largest modulus. Of course, for smaller max $|\lambda_i|(t-\tau)$, fewer terms are required in the series for $e^{[\Lambda](t-\tau)}$. In this regard, we recall that the time interval $(t, \tau) = t - \tau$ can be divided into a number of smaller subintervals by using the group property, as reflected in Eq. (12.27).

The transition matrix in the form (12.47) can also be used to derive the system response. Indeed, inserting Eq. (12.45) into Eq. (12.19), we obtain

$$\{y(t)\} = [U]e^{[\Lambda]t}[V]^T\{y(0)\} + \int_0^t [U]e^{[\Lambda](t-\tau)}[V]^T[B]\{Y(\tau)\}\,d\tau \tag{12.49}$$

The procedure described by Eq. (12.49) can be regarded as representing a modal analysis for the response of general linear dynamic systems.

## 12.5 RESPONSE OF GENERAL DAMPED SYSTEMS BY THE TRANSITION MATRIX

The approach to the system response based on the transition matrix is valid for any general linear dynamic system, provided the equations of motion can be reduced to the form (12.3). The response can be obtained by means of Eq. (12.19) or by means of Eq. (12.49). In this section, we show how the approach can be used in the case of a general viscously damped system, for which the classical modal analysis of Sec. 4.14 fails.

From Sec. 4.3, the equations of motion of an $n$-degree-of-freedom damped system can be written in the matrix form (see Eq. (4.16))

$$[m]\{\ddot{q}(t)\} + [c]\{\dot{q}(t)\} + [k]\{q(t)\} = \{Q(t)\} \tag{12.50}$$

Adjoining the identities

$$\{\dot{q}(t)\} = \{\dot{q}(t)\} \tag{12.51}$$

and introducing the $2n$-dimensional state and excitation vectors

$$\{y(t)\} = \left\{\begin{matrix}\{q(t)\}\\\{\dot{q}(t)\}\end{matrix}\right\} \qquad \{Y(t)\} = \left\{\begin{matrix}\{0\}\\\{Q(t)\}\end{matrix}\right\} \tag{12.52}$$

as well as the $2n \times 2n$ coefficient matrices

$$[A] = \left[\begin{matrix}[0] & [1]\\-[m]^{-1}[k] & -[m]^{-1}[c]\end{matrix}\right] \qquad [B] = \left[\begin{matrix}[0] & [0]\\[0] & [m]^{-1}\end{matrix}\right] \tag{12.53}$$

Eqs. (12.50) and (12.51) can be cast in the state form (12.3).

The response of the system can be obtained directly by means of Eq. (12.19). Alternatively, one can solve the eigenvalue problems associated with $[A]$ and $[A]^T$, obtain the right and left eigenvector matrices $[U]$ and $[V]$, as well as the matrix $[\Lambda]$ of eigenvalues, and derive the response by means of Eq. (12.49).

**Example 12.2** Determine the response of the system shown in Fig. 12.1 to the excitation

$$Q_1(t) = 0 \qquad Q_2(t) = \hat{Q}\delta(t) \tag{a}$$

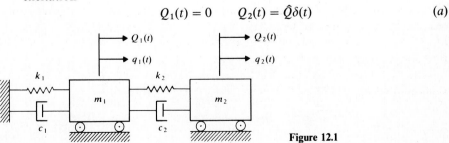

**Figure 12.1**

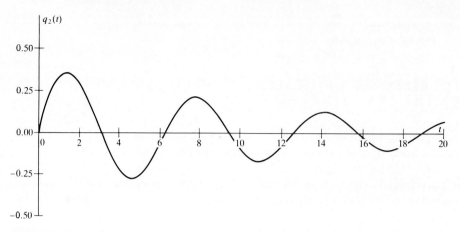

**Figure 12.2**

by the approach based on the transition matrix. The initial conditions are zero and the system parameters have the values

$$m_1 = m \qquad m_2 = 2m \qquad c_1 = c_2 = 0.8m\omega \qquad k_1 = m\omega^2 \qquad k_2 = 4m\omega^2 \quad (b)$$

The equations of motion have the matrix form (12.50), in which

$$[m] = \begin{bmatrix} m_1 & 0 \\ 0 & m_2 \end{bmatrix} = m \begin{bmatrix} 1 & 0 \\ 0 & 2 \end{bmatrix}$$

$$[c] = \begin{bmatrix} c_1 + c_2 & -c_2 \\ -c_2 & c_2 \end{bmatrix} = m\omega \begin{bmatrix} 1.6 & -0.8 \\ -0.8 & 0.8 \end{bmatrix} \qquad (c)$$

$$[k] = \begin{bmatrix} k_1 + k_2 & -k_2 \\ -k_2 & k_2 \end{bmatrix} = m\omega^2 \begin{bmatrix} 5 & -4 \\ -4 & 4 \end{bmatrix}$$

The equations can be written in the state form (12.3), in which the state and excitation vectors have the form

$$\{y(t)\} = \begin{Bmatrix} y_1(t) \\ y_2(t) \\ y_3(t) \\ y_4(t) \end{Bmatrix} = \begin{Bmatrix} q_1(t) \\ q_2(t) \\ \dot{q}_1(t) \\ \dot{q}_2(t) \end{Bmatrix} \qquad \{Y(t)\} = \begin{Bmatrix} Y_1(t) \\ Y_2(t) \\ Y_3(t) \\ Y_4(t) \end{Bmatrix} = \begin{Bmatrix} 0 \\ 0 \\ Q_1(t) \\ Q_2(t) \end{Bmatrix} = \hat{Q}\delta(t) \begin{Bmatrix} 0 \\ 0 \\ 0 \\ 1 \end{Bmatrix}$$

$$(d)$$

Moreover, inserting Eqs. (c) into Eqs. (12.53), we obtain the coefficient matrices

$$[A] = \begin{bmatrix} 0 & 0 & 1 & 0 \\ 0 & 0 & 0 & 1 \\ -5\omega^2 & 4\omega^2 & -1.6\omega & 0.8\omega \\ 2\omega^2 & -2\omega^2 & 0.4\omega & -0.4\omega \end{bmatrix} \qquad [B] = \frac{1}{m} \begin{bmatrix} 0 & 0 & 0 & 0 \\ 0 & 0 & 0 & 0 \\ 0 & 0 & 1 & 0 \\ 0 & 0 & 0 & 0.5 \end{bmatrix}$$

$$(e)$$

The response was obtained by using Eq. (12.19). The displacement of $m_2$ is plotted in Fig. 12.2.

## 12.6 DISCRETE-TIME SYSTEMS

In Sec. 2.15, we showed how the response of a system to arbitrary excitation can be evaluated by means of a convolution integral. But, except for excitations that can be described by relatively simple functions of time, the evaluation of convolution integrals can cause difficulties. Hence, in general one must rely on numerical evaluation of the convolution integrals, which can be carried out most conveniently on digital computers. However, all functions encountered in our study of vibrations until now were continuous in time and digital computers cannot handle such functions. This leads naturally to the concept of *discrete-time systems*, whereby the excitation and response are being treated as discrete functions of time, in contrast with *continuous-time systems*, in which they are continuous functions of time.

In system analysis terminology, the excitation is often referred to as the *input signal* and the response as the *output signal*. In continuous-time systems the signals are continuous functions of the time $t$, so that they are *continuous signals*. An example of a continuous signal is shown in Fig. 12.3a. On the other hand, in discrete-time systems the signals are defined only for discrete values of time $t_k$ ($k = 0, \pm 1, \pm 2, \ldots$). In such cases, $t$ is said to be a *discrete-time variable* and the signals are called *discrete signals*. Figure 12.3b shows an example of a discrete signal. Discrete signals do not arise naturally in vibrations, but are the result of discretization in time of continuous signals. As pointed out above, discretization in time is necessary because digital computers work with discrete signals.

Conversion of a continuous signal into a discrete one is carried out by means of a *sampler*, as shown in Fig. 12.4a. The input to the sampler is the continuous signal $f(t)$ and the output is a sequence of numbers $f(t_k)$ spaced in time, where $f(t_k)$ are the values of $f(t)$ at the sampling instances $t_k$. The sampler can be represented schematically by a switch, as shown in Fig. 12.4b, where the switch is open for all times except at the sampling instances $t_k$, when it closes instantaneously to permit the signal to pass through. The samplings are taken ordinarily at equal time interval, so that $t_k = kT$, where $T$ is the *sampling period*.

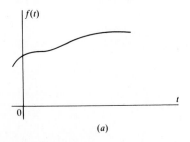

(a)

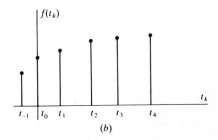

(b)

**Figure 12.3**

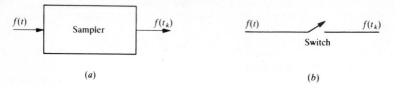

(a)                                          (b)

**Figure 12.4**

To convert discrete signals into continuous ones, the discrete signal must be passed through a *data hold circuit*. There are various kinds of holds, but the simplest one is the *zero-order hold*, defined mathematically by

$$f(\tau) = f(nT) \qquad nT \leqslant \tau \leqslant nT + T \tag{12.54}$$

The zero-order hold generates a continuous function having the form of a *staircase*, as shown in Fig. 12.5.

Our object is to show how the response of systems can be processed on digital computers. Although strictly speaking a digital computer accepts not merely discrete signals but discrete digital signals, the distinction is not mathematically significant and we shall refer to them simply as discrete-time signals. All signals involved in discrete-time systems can be regarded as *sequences of sample values resulting from sampling continuous-time signals*. As an illustration, assuming that the continuous-time signal $f(t)$ shown in Fig. 12.3a is sampled every $T$ seconds beginning at $t = 0$, the discrete-time signal $f(nT) = f(n)$ consists of the sequence $f(0), f(1), f(2), \ldots$, where for simplicity of notation we omitted the sampling period $T$ from the argument. To describe the sequence mathematically, it is convenient to introduce the *discrete-time unit impulse*, or *unit sample*, as the discrete-time Kronecker delta

$$\delta(n - k) = \begin{cases} 1 & n = k \\ 0 & n \neq k \end{cases} \tag{12.55}$$

The unit impulse is shown in Fig. 12.6. Then, the discrete-time signal $f(n)$ can be

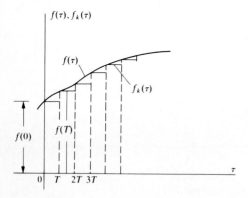

**Figure 12.5**

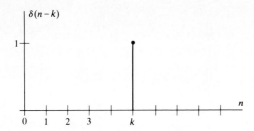

Figure 12.6

represented mathematically by the series

$$f(n) = \sum_{k=0}^{\infty} f(k)\delta(n - k) \tag{12.56}$$

The discrete-time signal $f(n)$ corresponding to the continuous-time signal of Fig. 12.3a is exhibited in Fig. 12.7.

Next, we propose to present the counterpart of the convolution integral, Eq. (2.170), in discrete time. By analogy with continuous-time systems, we can define the *discrete-time impulse response $g(n)$ as the response of a linear discrete-time system to a discrete-time impulse $\delta(n)$ applied at $k = 0$, with all the initial conditions being equal to zero.* The relation between $g(n)$ and $\delta(n)$ is shown in the block diagram of Fig. 12.8a. Note that the above definition implies that $g(n) = 0$ for $n < 0$, as there cannot be any response before the system is excited. Because the system is linear, if the excitation is delayed by $k$ periods, then the response is also delayed by $k$ periods. The relation between $g(n - k)$ and $\delta(n - k)$ is shown in Fig. 12.8b. Moreover, if the excitation has the form of an impulse of magnitude $F(k)$ applied at $n = k$, where the impulse is denoted by $F(k)\delta(n - k)$, then the response is simply $F(k)g(n - k)$, as shown in Fig. 12.8c. Now, let us assume that the excitation is in the form of the discrete-time signal

$$F(n) = \sum_{k=0}^{\infty} F(k)\delta(n - k) \tag{12.57}$$

Then, denoting the discrete-time response by $x(n)$, we can write simply

$$x(n) = \sum_{k=0}^{\infty} F(k)g(n - k) = \sum_{k=0}^{n} F(k)g(n - k) \tag{12.58}$$

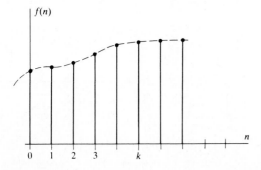

Figure 12.7

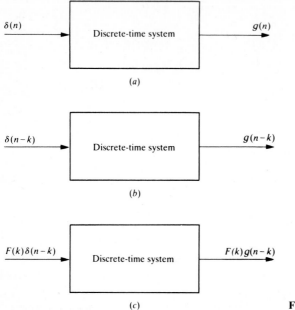

$\delta(n)$ → Discrete-time system → $g(n)$

(a)

$\delta(n-k)$ → Discrete-time system → $g(n-k)$

(b)

$F(k)\delta(n-k)$ → Discrete-time system → $F(k)g(n-k)$

(c)

**Figure 12.8**

where we replaced the upper limit in the series by $n$ in recognition of the fact that $g(n - k) = 0$ for $k > n$. Equation (12.58) expresses the response of a linear discrete-time system in the form of a *convolution sum* and it represents the discrete-time counterpart of the convolution integral given by Eq. (2.170).

The convolution sum has the drawback that the response $x(n)$ at any time $t_n = nT$ must be computed without being able to take advantage of the values of the response computed at preceding times $t_i = iT (i = 1, 2, \ldots, n - 1)$. Moreover, the sum becomes progressively longer with increasing $n$. A more efficient algorithm is one permitting recursive computation of the response, in the sense that the computation of $x(k + 1)$ is based only on $x(k)$ and $F(k)$, that is, on the response and excitation at the current time and not at earlier times. To derive the response recursively, we consider the approach based on the transition matrix discussed in Sec. 12.2. Letting $t = kT$ in Eq. (12.19), we obtain the state at that particular sampling instance in the form

$$\{y(kT)\} = e^{[A]kT}\{y(0)\} + \int_0^{kT} e^{[A](kT+T-\tau)}[B]\{Y(\tau)\}\, d\tau \qquad (12.59)$$

At the next sampling, we have

$$\{y(kT + T)\} = e^{[A](kT+T)}\{y(0)\} + \int_0^{kT+T} e^{[A](kT+T-\tau)}[B]\{Y(\tau)\}\, d\tau$$

$$= e^{[A]T}\left[ e^{[A]kT}\{y(0)\} + \int_0^{kT} e^{[A](kT-\tau)}[B]\{Y(\tau)\}\, d\tau \right]$$

$$+ \int_{kT}^{kT+T} e^{[A](kT+T-\tau)}[B]\{Y(\tau)\}\, d\tau \qquad (12.60)$$

Assuming that $\{Y(t)\}$ is piecewise constant, we can write

$$\int_{kT}^{kT+T} e^{[A](kT+T-\tau)}[B]\{Y(\tau)\}\, d\tau = \left[ \int_{kT}^{kt+T} e^{[A](kT+T-\tau)}\, d\tau \right] [B]\{Y(kT)\} \qquad (12.61)$$

Moreover, introducing the change of variables $kT + T - \tau = t$, the integral on the right side of Eq. (12.61) can be reduced to

$$\int_{kT}^{kT+T} e^{[A](kT+T-\tau)}\, d\tau = \int_{T}^{0} e^{[A]t}\, (-dt) = \int_{0}^{T} e^{[A]t}\, dt \qquad (12.62)$$

Finally, introducing the notation,

$$[\Phi] = e^{[A]T} \qquad [\Gamma] = \int_{0}^{T} e^{[A]t}\, dt\, [B] \qquad (12.63)$$

as well as

$$\{y(kT)\} = \{y(k)\} \qquad \{y(kT+T)\} = \{y(k+1)\} \qquad \{Y(kT)\} = \{Y(k)\} \qquad (12.64)$$

and using Eqs. (12.59)–(12.62), we obtain the sequence

$$\{y(k+1)\} = [\Phi]\{y(k)\} + [\Gamma]\{Y(k)\} \qquad k = 0, 1, 2, \ldots \qquad (12.65)$$

where we note that $[\Phi]$ represents the transition matrix for the discrete-time system, whose value can be computed by replacing $t$ by $T$ in series (12.6). Formula (12.65) is relatively easy to program on a digital computer. To compute $\{y(k+1)\}$ only the current state and force vectors $\{y(k)\}$ and $\{Y(k)\}$ are needed, and the earlier state and force vectors can be discarded. Moreover, the length of the computation is the same at each step.

**Example 12.3** Derive the response of the mass-spring system of Example 12.1 to the excitation

$$F(t) = f_0 t u(t) \qquad (a)$$

by means of the convolution sum.

The excitation can be written in the form (12.57), in which

$$F(k) = f_0 kT \qquad k = 0, 1, 2, \ldots \qquad (b)$$

where $T$ is the sampling period. The impulse response for a mass-spring system is

$$g(t) = \frac{1}{m\omega_n} \sin \omega_n t\, u(t) \qquad (c)$$

where $m$ is the mass and $\omega_n$ the natural frequency. The impulse response of the equivalent discrete-time system can be shown to be (see Prob. 12.7)

$$g(k) = g(kT) = \frac{T}{m\omega_n} \sin k\omega_n T \qquad k = 0, 1, 2, \ldots \qquad (d)$$

Inserting Eqs. (b) and (d) into Eq. (12.58), we obtain the response in the form of the sequence

$$x(1) = F(0)g(1) + F(1)g(0) = 0$$

$$x(2) = F(0)g(2) + F(1)g(1) + F(2)g(0) = \frac{f_0 T^2}{m\omega_n} \sin \omega_n T$$

$$x(3) = F(0)g(3) + F(1)g(2) + F(2)g(1) + F(3)g(0)$$

$$= \frac{f_0 T^2}{m\omega_n} (2 \sin \omega_n T + \sin 2\omega_n T) \qquad (e)$$

$$x(4) = F(0)g(4) + F(1)g(3) + F(2)g(2) + F(3)g(1) + F(4)g(0)$$

$$= \frac{f_0 T^2}{m\omega_n} (3 \sin \omega_n T + 2 \sin 2\omega_n T + \sin 3\omega_n T)$$

. . . . . . . . . . . . . . . . . . . . . . . . . . . . . . . . . . . . . . .

The response sequence for $\omega_n = 1$ rad/s and $T = 0.1$ s is plotted in Fig. 12.9 in the form of points marked by circles.

**Example 12.4** Solve the problem of Examples 12.1 and 12.3 by means of the approach based on the transition matrix for discrete-time systems.

The response sequence is given by Eq. (12.65), which requires the matrices $[\Phi]$ and $[\Gamma]$. To compute these matrices, we make use of some results obtained in Example 12.1. Letting $t = T$ in Eq. (h) of Example 12.1, we can write the transition matrix for the discrete-time system in the form

$$[\Phi] = e^{[A]T} = \begin{bmatrix} \cos \omega_n T & \omega_n^{-1} \sin \omega_n T \\ -\omega_n \sin \omega_n T & \cos \omega_n T \end{bmatrix} \qquad (a)$$

Moreover, inserting Eq. (h) and the fourth of Eqs. (e) of Example 12.1 into the second of Eqs. (12.63) and integrating, we obtain

$$[\Gamma] = \int_0^T e^{[A]t} \, dt [B] = \int_0^T \begin{bmatrix} \cos \omega_n t & \omega_n^{-1} \sin \omega_n t \\ -\omega_n \sin \omega_n t & \cos \omega_n t \end{bmatrix} dt \begin{bmatrix} 1 & 0 \\ 0 & 1/m \end{bmatrix}$$

$$= \frac{1}{\omega_n} \begin{bmatrix} \sin \omega_n t & -(m\omega_n)^{-1} \cos \omega_n t \\ \omega_n \cos \omega_n t & m^{-1} \sin \omega_n t \end{bmatrix} \Bigg|_0^T$$

$$= \frac{1}{\omega_n} \begin{bmatrix} \sin \omega_n T & (m\omega_n)^{-1}(1 - \cos \omega_n T) \\ -\omega_n(1 - \cos \omega_n T) & m^{-1} \sin \omega_n T \end{bmatrix} \qquad (b)$$

Moreover, the initial state vector and the excitation vector are given by

$$\{y(0)\} = \begin{Bmatrix} 0 \\ 0 \end{Bmatrix} \qquad \{Y(k)\} = \begin{Bmatrix} 0 \\ f_0 kT \end{Bmatrix} \qquad k = 0, 1, 2, \ldots \tag{c}$$

The response sequence is obtained from Eq. (12.65) in the form

$$\{y(k+1)\} = [\Phi]\{y(k)\} + [\Gamma]\{Y(k)\} = [\Phi]\{y(k)\} + f_0 kT\{\Gamma\}_2 \qquad k = 0, 1, 2, \ldots \tag{d}$$

where $\{\Gamma\}_2$ is the second column of the matrix $[\Gamma]$. Inserting Eqs. (a)–(c) into Eq. (d), the response sequence is computed as follows:

$$\{y(1)\} = [\Phi]\{y(0)\} + 0 \times \{\Gamma\}_2 = \begin{Bmatrix} 0 \\ 0 \end{Bmatrix}$$

$$\{y(2)\} = [\Phi]\{y(1)\} + f_0 T\{\Gamma\}_2 = \frac{f_0 T}{k} \begin{Bmatrix} 1 - \cos \omega_n T \\ \omega_n \sin \omega_n T \end{Bmatrix}$$

$$\{y(3)\} = [\Phi]\{y(2)\} + 2f_0 T\{\Gamma\}_2$$

$$= \frac{f_0 T}{k} \begin{bmatrix} \cos \omega_n T & \omega_n^{-1} \sin \omega_n T \\ -\omega_n \sin \omega_n T & \cos \omega_n T \end{bmatrix} \begin{Bmatrix} 1 - \cos \omega_n T \\ \omega_n \sin \omega_n T \end{Bmatrix} \tag{e}$$

$$+ \frac{2f_0 T}{k} \begin{Bmatrix} 1 - \cos \omega_n T \\ \omega_n \sin \omega_n T \end{Bmatrix}$$

$$= \frac{f_0 T}{k} \begin{Bmatrix} 2 - \cos \omega_n T - \cos 2\omega_n T \\ \omega_n(\sin \omega_n T + \sin 2\omega_n T) \end{Bmatrix}$$

. . . . . . . . . . . . . . . . . . . . . . . . . . . . . . . . . . . . . . . . . . . . . . . . .

where we note that the symbol $k$ appearing in the denominator is the spring constant of the system, and should not be confused with the integer $k$ used to denote the time. In actual numerical work, one does not really compute the sequence as indicated by Eqs. (e), but programs Eq. (d) as a "DO LOOP" on a digital computer. To this end, one must decide on a sampling period $T$ and evaluate $[\Phi]$ and $[\Gamma]_2$ numerically by means of Eqs. (a) and (b). Then, with Eqs. (c) as input, the response sequence is computed numerically. The output is a sequence of two-dimensional vectors representing the numerical values of the vectors $\{y(1)\}, \{y(2)\}, \{y(3)\}, \ldots$ given by Eqs. (e).

A comparison of the results obtained in Example 12.3 with those obtained here is in order. We first note that the values of Eqs. (e) of Example 12.3 correspond to the values of the top component of the vectors $\{y(1)\}, \{y(2)\}, \{y(3)\}, \ldots$ in Eqs. (e) of this example. The results do not appear to coincide and indeed they are somewhat different. This is to be expected, because both discrete-time approaches are only approximate. For a quantitative comparison, the response sequence for $\omega_n = 1$ rad/s and $T = 0.1$ s and corresponding to the top component of $\{y(k)\}$ ($k = 1, 2, \ldots$) is plotted in Fig. 12.9 in the form of points marked by black circles. Moreover, to develop a better feel for

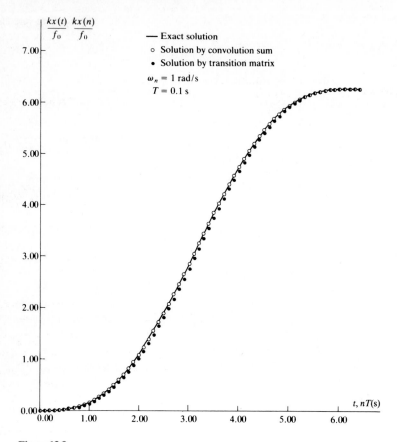

**Figure 12.9**

the approximate solutions, the exact solution obtained in Example 12.1 is plotted in Fig. 12.9 in solid line. We observe that both approximate solutions fall
below the exact solution, but the solution obtained by the convolution sum is
more accurate than that obtained by the transition matrix. In fact, the solution
by the convolution sum almost coincides with the exact solution. This can be
explained by the fact that in the solution by the convolution sum the response
is exact and the only approximation is in the excitation. On the other hand, in
the solution by the transition matrix, the excitation is approximated by means
of zero-order hold, and the matrix $[\Gamma]$ involves integration of $e^{[A]t}$ over the
sampling period $T$, so that the convolution integral in (12.19) is approximated
with a lower degree of accuracy. Of course, the accuracy of both approximate
solutions, particularly of that obtained by the transition matrix, can be
improved by reducing the sampling period $T$, but this is likely to increase the
computer costs. Hence, the question is how large the sampling period $T$ can be
made and still obtain accurate results. For vibrating systems, the size of the

sampling period $T$, so that the convolution integral in (12.19) is approximated period should be only a fraction of the period of the system, depending on the desired accuracy.

The conclusions concerning the accuracy of the solutions obtained by the convolution sum and by the transition matrix are for this particular example and should not be regarded as generally valid. Indeed, it can be verified that in the case of sectionally constant excitation, the solution by the transition matrix is more accurate; in fact, the solution is exact.

## 12.7 THE RUNGE-KUTTA METHODS

The various computational algorithms discussed in Secs. 12.2–12.6 were concerned with linear systems exclusively. Nonlinear systems were discussed in Chap. 9 qualitatively, by examining the stability characteristics in the neighborhood of equilibrium points. Then, in Chap. 10, we presented several perturbation techniques for the response of weakly nonlinear systems, i.e., systems for which the nonlinearity is sufficiently small that it can be regarded as a higher-order effect. In this section, we finally address the response of nonlinear systems for which the nonlinearity is not necessarily small. Evaluation of the response of nonlinear systems almost invariably involves some type of numerical integration, so that we consider a discrete-time solution directly. Numerical integration is carried out most conveniently in terms of first-order equations, so that the dynamical equations of motion must be recast in state form.

Numerical integration provides only an approximate solution, the accuracy of the solution depending on the order of the approximation. We consider here the *Runge-Kutta methods*, a family of methods characterized by different orders of approximation, where the order is related to the number of terms in a Taylor series expansion. The most widely used is the fourth-order Runge-Kutta method. Derivation of the algorithm is very complex, and lies beyond the scope of this text. To develop a feel for the approach, we propose to derive the second-order method and only give the basic equations for the fourth-order method. We introduce the ideas by considering a first-order nonlinear system described by the differential equation

$$\dot{x}(t) = f[x(t), t] \tag{12.66}$$

where $f$ is a nonlinear function of $x(t)$ and $t$. Actually, in most applications $f$ does not involve the time $t$ explicitly, but only implicitly through $x(t)$. Expanding the solution of Eq. (12.66) in a Taylor series, we can write

$$x(t + T) = x(t) + T\dot{x}(t) + \frac{T^2}{2!}\ddot{x}(t) + \frac{T^3}{3!}\dddot{x}(t) + \cdots \tag{12.67}$$

where $T$ is a small time increment. According to Eq. (12.66), however, $\dot{x} = f$. In addition

$$\ddot{x} = \frac{d\dot{x}}{dt} = \frac{df}{dt} = \frac{\partial f}{\partial x}\dot{x} + \frac{\partial f}{\partial t} = f\frac{\partial f}{\partial x} + \frac{\partial f}{\partial t}$$

$$\dddot{x} = \frac{d\ddot{x}}{dt} = \frac{\partial}{\partial x}\left(f\frac{\partial f}{\partial x} + \frac{\partial f}{\partial t}\right)\dot{x} + \frac{\partial}{\partial t}\left(f\frac{\partial f}{\partial x} + \frac{\partial f}{\partial t}\right)$$

$$= \left[\left(\frac{\partial f}{\partial x}\right)^2 + f\frac{\partial^2 f}{\partial x^2} + \frac{\partial^2 f}{\partial x\,\partial t}\right]f + \frac{\partial f}{\partial t}\frac{\partial f}{\partial x} + f\frac{\partial^2 f}{\partial t\,\partial x} + \frac{\partial^2 f}{\partial t^2} \qquad (12.68)$$

$$= f^2\frac{\partial^2 f}{\partial x^2} + f\left(\frac{\partial f}{\partial x}\right)^2 + 2f\frac{\partial^2 f}{\partial x\,\partial t} + \frac{\partial f}{\partial x}\frac{\partial f}{\partial t} + \frac{\partial^2 f}{\partial t^2}$$

. . . . . . . . . . . . . . . . . . . . . . . . . . . . . . . . . . . . . . . .

so that Eq. (12.67) becomes

$$x(t+T) = x(t) + Tf + \frac{T^2}{2!}\left(f\frac{\partial f}{\partial x} + \frac{\partial f}{\partial t}\right) + \frac{T^3}{3!}\left[f^2\frac{\partial^2 f}{\partial x^2} + f\left(\frac{\partial f}{\partial x}\right)^2\right.$$

$$\left. + 2f\frac{\partial^2 f}{\partial x\,\partial t} + \frac{\partial f}{\partial x}\frac{\partial f}{\partial t} + \frac{\partial^2 f}{\partial t^2}\right] + \cdots \qquad (12.69)$$

Equation (12.69) forms the basis for the Runge-Kutta methods.

For our numerical algorithm, we wish to derive the discrete-time version of Eq. (12.69). To this end, we consider the discrete times $t = t_k = kT$, $t + T = t_{k+1} = (k+1)T$ ($k = 0, 1, 2, \ldots$) and introduce the notation

$$x(t) = x(t_k) = x(k) \qquad x(t+T) = x(t_{k+1}) = x(k+1)$$

$$f[x(t), t] = f[x(t_k), t_k] = f(k) \tag{12.70}$$

Then, we can rewrite Eq. (12.69) in the discrete-time form

$$x(k+1) = x(k) + Tf(k) + \frac{T^2}{2!}\left[f(k)\frac{\partial f(k)}{\partial x} + \frac{\partial f(k)}{\partial t}\right]$$

$$+ \frac{T^3}{3!}\left\{f^2(k)\frac{\partial^2 f(k)}{\partial x^2} + f(k)\left[\frac{\partial f(k)}{\partial x}\right]^2 + 2f(k)\frac{\partial^2 f(k)}{\partial x\,\partial t}\right.$$

$$\left. + \frac{\partial f(k)}{\partial x}\frac{\partial f(k)}{\partial t} + \frac{\partial^2 f(k)}{\partial t^2}\right\} + \cdots \qquad k = 0, 1, 2, \ldots \quad (12.71)$$

More often than not $f$ does not depend explicitly on time, in which case Eqs. (12.71) reduce to

$$x(k+1) = x(k) + Tf(k) + \frac{T^2}{2!}f(k)\frac{\partial f(k)}{\partial x}$$

$$+ \frac{T^3}{3!}\left\{f^2(k)\frac{\partial^2 f(k)}{\partial x^2} + f(k)\left[\frac{\partial f(k)}{\partial x}\right]^2\right\} + \cdots \qquad k = 0, 1, 2, \ldots \quad (12.72)$$

Equations (12.71), or Eqs. (12.72), can be used to derive solutions to any desired order of approximation. In the remaining part of this section, we shall work with Eqs. (12.72).

The lowest order of approximation is obtained by retaining the first-order term in Eqs. (12.72) and it has the form

$$x(k+1) = x(k) + Tf(k) \qquad k = 0, 1, 2, \ldots \qquad (12.73)$$

The method of computing the first-order approximation by means of Eqs. (12.73) is known as *Euler's method*. It represents a linearization of the nonlinear system and tends to be very inaccurate when the nonlinearity is pronounced, so that the method is not recommended in general.

To develop the second-order Runge-Kutta method, we assume an approximation having the expression

$$x(k+1) = x(k) + c_1 g_1 + c_2 g_2 \qquad k = 0, 1, 2, \ldots \qquad (12.74)$$

where $c_1$ and $c_2$ are constants and

$$g_1 = Tf(k) \qquad g_2 = Tf[x(k) + \alpha_2 g_1] \qquad (12.75)$$

Note that $x(k) + \alpha_2 g_1$ merely represents the argument of the function $f$ in the expression for $g_2$, in which $\alpha_2$ is a constant. The constants $c_1, c_2,$ and $\alpha_2$ are determined by insisting that Eqs. (12.72) and (12.74) agree through terms of second order in $T$. From Eqs. (12.75) we can write the Taylor series expansion

$$g_2 = Tf[x(k) + \alpha_2 g_1] = Tf[x(k) + \alpha_2 Tf(k)]$$

$$= T\left[ f(k) + \alpha_2 Tf(k) \frac{\partial f(k)}{\partial x} + \cdots \right] \qquad (12.76)$$

so that, using the first of Eqs. (12.75) and Eq. (12.76), Eqs. (12.74) become

$$x(k+1) = x(k) + c_1 Tf(k) + c_2 T\left[ f(k) + \alpha_2 Tf(k) \frac{\partial f(k)}{\partial x} + \cdots \right]$$

$$= x(k) + (c_1 + c_2)Tf(k) + c_2\alpha_2 T^2 f(k)\frac{\partial f(k)}{\partial x} + \cdots \qquad k = 0, 1, 2, \ldots$$

$$(12.77)$$

Equating terms through second order in $T$ in Eqs. (12.72) and (12.77), we conclude that the constants $c_1, c_2,$ and $\alpha_2$ must satisfy the equations

$$c_1 + c_2 = 1 \qquad c_2\alpha_2 = \tfrac{1}{2} \qquad (12.78)$$

so that there are two equations and three unknowns. This implies that Eqs. (12.78) do not have a unique solution, so that one of the constants can be chosen arbitrarily, provided the choice corresponding to $c_2 = 0$ is excluded. One satisfactory choice is $c_2 = 1/2$, which yields

$$c_1 = c_2 = \tfrac{1}{2} \qquad \alpha_2 = 1 \qquad (12.79)$$

Inserting Eqs. (12.79) into Eqs. (12.74), in conjunction with Eqs. (12.75) and (12.76), we obtain a computational algorithm defining the *second-order Runge-Kutta method* in the form

$$x(k + 1) = x(k) + \tfrac{1}{2}(g_1 + g_2) \qquad k = 0, 1, 2, \ldots \qquad (12.80)$$

where

$$g_1 = Tf(k) \qquad g_2 = Tf[x(k) + g_1] \qquad k = 0, 1, 2, \ldots \qquad (12.81)$$

and we note that the choice (12.78) leads to a symmetric form for the algorithm.

Following the same procedure, we can derive higher-order Runge-Kutta approximations. The derivations become increasingly complex, however, so that they are omitted. Instead, we present simply the results. The most widely used is the *fourth-order Runge-Kutta method*, defined by the algorithm

$$x(k + 1) = x(k) + \tfrac{1}{6}(g_1 + 2g_2 + 2g_3 + g_4) \qquad k = 0, 1, 2, \ldots \qquad (12.82)$$

where

$$g_1 = Tf(k) \qquad g_2 = Tf[x(k) + 0.5g_1] \qquad g_3 = Tf[x(k) + 0.5g_2]$$

$$g_4 = Tf[x(k) + g_3] \qquad k = 0, 1, 2, \ldots \qquad (12.83)$$

In the vibration of single- and multi-degree-of-freedom systems, Eq. (12.66) is a vector equation instead of a scalar equation. Adopting the vector notion, instead of the matrix notation, the state equations for a system of order $m$ can be written as follows:

$$\dot{\mathbf{y}}(t) = \mathbf{f}[\mathbf{y}(t)] \qquad (12.84)$$

where $\mathbf{y}$ and $\mathbf{f}$ are $m$-dimensional vectors. Then, the fourth-order Runge-Kutta method can be defined by the algorithm

$$\mathbf{y}(k + 1) = \mathbf{y}(k) + \tfrac{1}{6}(\mathbf{g}_1 + 2\mathbf{g}_2 + 2\mathbf{g}_3 + \mathbf{g}_4) \qquad k = 0, 1, 2, \ldots \qquad (12.85)$$

where

$$\mathbf{g}_1 = T\mathbf{f}(k) \qquad \mathbf{g}_2 = T\mathbf{f}[\mathbf{y}(k) + 0.5\mathbf{g}_1] \qquad \mathbf{g}_3 = T\mathbf{f}[\mathbf{y}(k) + 0.5\mathbf{g}_2]$$

$$\mathbf{g}_4 = T\mathbf{f}[\mathbf{y}(k) + \mathbf{g}_3] \qquad k = 0, 1, 2, \ldots \qquad (12.86)$$

are $m$-dimensional vectors.

The fourth-order Runge-Kutta method involves four evaluations of the vector $\mathbf{f}$ for each integration step, so that it requires a large amount of computer time. The method is extremely accurate, however, so that it requires fewer steps for a desired level of accuracy than other methods. This makes it a favorite in numerical integration of nonlinear differential equations.

**Example 12.5** Consider a mass-nonlinear spring system governed by the differential equation of motion

$$\ddot{x} + 4(x + x^3) = 0 \qquad (a)$$

and subject to the initial conditions

$$x(0) = 0.8 \qquad \dot{x}(0) = 0 \qquad (b)$$

and obtain the response by the fourth-order Runge-Kutta method using the sampling period $T = 0.01$ s. Plot $x(t)$ versus $t$ for $0 < t < 5$ s.

Introducing the notation

$$x(t) = y_1(t) \qquad \dot{x}(t) = y_2(t) \tag{c}$$

Eq. ($a$) can be replaced by the state equations

$$\dot{y}_1 = y_2 \qquad \dot{y}_2 = -4(y_1 + y_1^3) \tag{d}$$

so that the components of the vector $\mathbf{f}$ are

$$f_1' = y_2 \quad f_2 = -4(y_1 + y_1^3) \tag{e}$$

Inserting Eqs. ($e$) into Eqs. (12.85) and (12.86) we obtain the equations defining the computational algorithm in the form

$$y_1(k + 1) = y_1(k) + \tfrac{1}{6}[g_{11}(k) + 2g_{21}(k) + 2g_{31}(k) + g_{41}(k)]$$

$$y_2(k + 1) = y_2(k) + \tfrac{1}{6}[g_{12}(k) + 2g_{22}(k) + 2g_{32}(k) + g_{42}(k)]$$

$$k = 0, 1, 2, \ldots \tag{f}$$

where $g_{1i}, g_{2i}, g_{3i}$, and $g_{4i}$ $(i = 1, 2)$ are the components of the vectors $\mathbf{g}_1, \mathbf{g}_2, \mathbf{g}_3$, and $\mathbf{g}_4$, respectively, and are given by the expressions

$$g_{11}(k) = Tf_1[y_1(k), y_2(k)] = Ty_2(k)$$

$$g_{12}(k) = Tf_2[y_1(k), y_2(k)] = -4T[y_1(k) + y_1^3(k)]$$

$$g_{21}(k) = Tf_1[y_1(k) + 0.5g_{11}(k), y_2(k) + 0.5g_{12}(k)]$$
$$= T[y_2(k) + 0.5g_{12}(k)]$$

$$g_{22}(k) = Tf_2[y_1(k) + 0.5g_{11}(k), y_2(k) + 0.5g_{12}(k)]$$
$$= -4T\{y_1(k) + 0.5g_{11}(k) + [y_1(k) + 0.5g_{11}(k)]^3\}$$

$$g_{31}(k) = Tf_1[y_1(k) + 0.5g_{21}(k), y_2(k) + 0.5g_{22}(k)]$$
$$= T[y_2(k) + 0.5g_{22}(k)]$$

$$g_{32}(k) = Tf_2[y_1(k) + 0.5g_{21}(k), y_2(k) + 0.5g_{22}(k)]$$
$$= -4T\{y_1(k) + 0.5g_{21}(k) + [y_1(k) + 0.5g_{21}(k)]^3\}$$

$$g_{41}(k) = Tf_1[y_1(k) + g_{31}(k), y_2(k) + g_{32}(k)]$$
$$= T[y_2(k) + g_{32}(k)]$$

$$g_{42}(k) = Tf_2[y_1(k) + g_{31}(k), y_2(k) + g_{32}(k)]$$
$$= -4T\{y_1(k) + g_{31}(k) + [y_1(k) + g_{31}(k)]^3\}$$

$$k = 0, 1, 2, \ldots \tag{g}$$

The response is shown in Fig. 12.10 and we note that the period of the system is approximately 2.6 s, which is different from the period of $\pi$ s of the linear system.

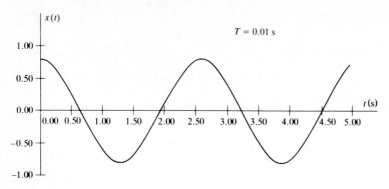

**Figure 12.10**

In this regard, we recall from Chap. 9 that the period of the nonlinear system depends on the initial conditions, whereas that of the linear system is constant and independent of the initial conditions. Because the nonlinear spring in this example is a hardening spring, we can expect a higher frequency than that of the linear system, which implies a smaller period.

## 12.8 THE FREQUENCY-DOMAIN CONVOLUTION THEOREM

In the convolution theorem presented in Sec. 11.11 the convolution takes place in the time domain. A similar convolution theorem exists for the frequency domain. Consider the Fourier transform of the product $f(t)g(t)$, or

$$X(\sigma) = \int_{-\infty}^{\infty} f(t)g(t)e^{-i\sigma t} \, dt \qquad (12.87)$$

But $g(t)$ can be written as the inverse Fourier transform

$$g(t) = \frac{1}{2\pi} \int_{-\infty}^{\infty} G(\omega)e^{i\omega t} \, d\omega \qquad (12.88)$$

so that, inserting Eq. (12.88) into Eq. (12.87) and changing the integration order, we can write

$$\int_{-\infty}^{\infty} f(t)e^{-i\sigma t}\left[\frac{1}{2\pi} \int_{-\infty}^{\infty} G(\omega)e^{i\omega t} \, d\omega\right] dt = \frac{1}{2\pi} \int_{-\infty}^{\infty} G(\omega)\left[\int_{-\infty}^{\infty} f(t)e^{-i(\sigma-\omega)t} \, dt\right] d\omega$$

$$(12.89)$$

The expression enclosed by the brackets, however, can be identified as the Fourier transform

$$\int_{-\infty}^{\infty} f(t)e^{-i(\sigma-\omega)t} \, dt = F(\sigma - \omega) \qquad (12.90)$$

from which it follows that

$$2\pi \int_{-\infty}^{\infty} f(t)g(t)e^{-i\sigma t}\, dt = \int_{-\infty}^{\infty} G(\omega)F(\sigma - \omega)\, d\omega \tag{12.91}$$

Equation (12.91) represents the mathematical statement of the *frequency-domain convolution theorem*. It can be stated as follows: *The convolution of F($\omega$) and G($\omega$) and the product $2\pi f(t)g(t)$ represent a Fourier transform pair.*

The expressions for the Fourier transform pair, Eqs. (11.43) and (11.44), lack symmetry as the inverse Fourier transform contains the factor $1/2\pi$ and the Fourier transform does not. It is possible to symmetrize the expressions by assigning to both the factor $1/\sqrt{2\pi}$. A better way to achieve the same goal is to work with frequencies measured in hertz (cycles per second) instead of radians per second. Hence, introducing $\omega = 2\pi f$ into Eqs. (11.43) and (11.44), we obtain the Fourier transform pair

$$F(f) = \int_{-\infty}^{\infty} f(t)e^{-i2\pi ft}\, dt \tag{12.92}$$

$$f(t) = \int_{-\infty}^{\infty} F(f)e^{i2\pi ft}\, df \tag{12.93}$$

No confusion should arise from the fact that the same notation is used for the excitation and the frequency. Indeed, the first is a function of time and the second is not. Then, the time-domain convolution theorem can be given mathematically in the form

$$F(f)G(f) = \int_{-\infty}^{\infty} \left[ \int_{-\infty}^{\infty} f(\tau)g(t - \tau)\, d\tau \right] e^{-i2\pi ft}\, dt \tag{12.94}$$

$$\int_{-\infty}^{\infty} f(\tau)g(t - \tau)\, d\tau = \int_{-\infty}^{\infty} F(f)G(f)e^{i2\pi ft}\, df \tag{12.95}$$

and, letting $\sigma = 2\pi f$, the frequency-domain convolution theorem can be stated mathematically as follows:

$$\int_{-\infty}^{\infty} f(t)g(t)e^{-i2\pi ft}\, dt = \int_{-\infty}^{\infty} G(\gamma)F(f - \gamma)\, d\gamma \tag{12.96}$$

$$f(t)g(t) = \int_{-\infty}^{\infty} \left[ \int_{-\infty}^{\infty} G(\gamma)F(f - \gamma)\, d\gamma \right] e^{i2\pi ft}\, df \tag{12.97}$$

Throughout the remainder of this chapter, we will use the definition of the Fourier transform in terms of the frequency $f$.

## 12.9 FOURIER SERIES AS A SPECIAL CASE OF THE FOURIER INTEGRAL

In Sec. 12.8, we presented the time-domain convolution theorem in terms of the convolution between the excitation and impulse response functions. The theorem need not be so restricted and in general the time-domain convolution theorem can be stated as follows: *The convolution of $f(t)$ and $g(t)$ and the product $F(f)G(f)$ represent a Fourier transform pair,* where $f(t)$ and $g(t)$ are two arbitrary functions and $F(f)$ and $G(f)$ are their Fourier transforms, respectively. Of course, there is a restriction on $f(t)$ and $g(t)$ in that they possess Fourier transforms. No confusion should arise from the fact that in earlier discussions $g(t)$ denoted the impulse response and here it denotes an arbitrary function.

As an application of the time-domain convolution theorem, we will demonstrate that the Fourier series representation of a periodic function can be regarded as a special case of the Fourier integral. To this end, consider the periodic function $x(t)$ shown in Fig. 12.11. The function can be expanded in a Fourier series of the form

$$x(t) = \sum_{p=-\infty}^{\infty} C_p e^{i2\pi p f_0 t} \qquad f_0 = \frac{1}{T} \tag{12.98}$$

where the complex Fourier coefficients $C_p$ have the expressions

$$C_p = \frac{1}{T} \int_{-T/2}^{T/2} x(t) e^{-i2\pi p f_0 t} \, dt \qquad p = 0, \pm 1, \pm 2, \ldots \tag{12.99}$$

The same periodic function $x(t)$ can be expressed in the form of the convolution integral

$$x(t) = \int_{-\infty}^{\infty} f(\tau)g(t - \tau) \, d\tau \tag{12.100}$$

where the function $f(t)$ represents the single pulse shown in Fig. 12.12a and the function $g(t)$ represents the infinite set of equidistant unit impulses shown in Fig. 12.12b. The latter can be expressed mathematically as

$$g(t) = \sum_{p=-\infty}^{\infty} \delta(t - pT) \tag{12.101}$$

But, in general the Fourier transform of a function having the form of a convolution integral has the expression

$$X(f) = \int_{-\infty}^{\infty} x(t)e^{-i2\pi ft} \, dt = \int_{-\infty}^{\infty} \left[ \int_{-\infty}^{\infty} f(\tau)g(t - \tau) \, d\tau \right] e^{-i2\pi ft} \, dt \tag{12.102}$$

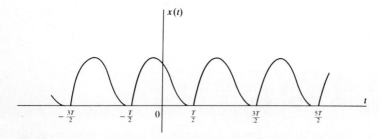

**Figure 12.11**

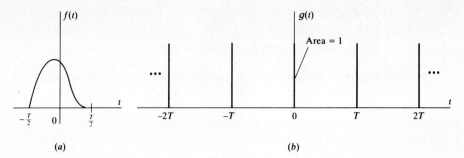

(a)                                                        (b)

**Figure 12.12**

so that, using the change of variables $t - \tau = \lambda, t = \lambda + \tau, dt = d\lambda$, Eq. (12.102) yields

$$X(f) = \int_{-\infty}^{\infty} \left[ \int_{-\infty}^{\infty} f(\tau)g(\lambda)\, d\tau \right] e^{-i2\pi f(\lambda + \tau)}\, d\lambda$$

$$= \int_{-\infty}^{\infty} f(\tau)e^{-i2\pi f\tau}\, d\tau \int_{-\infty}^{\infty} g(\lambda)e^{-i2\pi f\lambda}\, d\lambda = F(f)G(f) \quad (12.103)$$

where $F(f)$ and $G(f)$ are the Fourier transforms of $f(t)$ and $g(t)$, respectively. It can be shown,† however, that the Fourier transform of a sequence of unit impulses at equal distances $T$ is another sequence of impulses of magnitude $T^{-1}$ and at distances $T^{-1}$, so that the Fourier transform of $g(t)$, as given by Eq. (12.101), is

$$G(f) = \frac{1}{T} \sum_{p=-\infty}^{\infty} \delta\left(f - \frac{p}{T}\right) \quad (12.104)$$

Hence, inserting Eq. (12.104) into Eq. (12.103), we can write

$$X(f) = F(f)G(f) = F(f)\frac{1}{T} \sum_{p=-\infty}^{\infty} \delta\left(f - \frac{p}{T}\right)$$

$$= \frac{1}{T} \sum_{p=-\infty}^{\infty} F\left(\frac{p}{T}\right)\delta\left(f - \frac{p}{T}\right) \quad (12.105)$$

On the other hand, using Eq. (12.98), the Fourier transform of $x(t)$ is

$$X(f) = \int_{-\infty}^{\infty} x(t)e^{-i2\pi ft}\, dt = \int_{-\infty}^{\infty} \sum_{p=-\infty}^{\infty} C_p e^{i2\pi pf_0 t}e^{-i2\pi ft}\, dt$$

$$= \sum_{p=-\infty}^{\infty} C_p \int_{-\infty}^{\infty} e^{-i2\pi(f - pf_0)t}\, dt \quad (12.106)$$

But, it can also be shown‡ that

$$\int_{-\infty}^{\infty} e^{-i2\pi(f - pf_0)t}\, dt = \delta(f - pf_0) = \delta\left(f - \frac{p}{T}\right) \quad (12.107)$$

† A. Papoulis, *The Fourier Integral and Its Applications*, p. 44, McGraw-Hill Book Co., New York, 1962.
‡ A. Papoulis, op. cit., p. 281.

so that Eq. (12.106) reduces to

$$X(f) = \sum_{p=-\infty}^{\infty} C_p \delta \left( f - \frac{p}{T} \right) \tag{12.108}$$

where the coefficients $C_p$ are given by Eqs. (12.99). However, over the time interval $-T/2 < t < T/2$ the functions $x(t)$ and $f(t)$ are identical and $f(t)$ is zero everywhere else. Hence, the function $x(t)$ can be replaced by $f(t)$ in Eqs. (12.99), so that the expressions for the coefficients $C_p$ can be rewritten as

$$C_p = \frac{1}{T} \int_{-T/2}^{T/2} f(t)e^{-i2\pi p f_0 t}\, dt = \frac{1}{T} \int_{-\infty}^{\infty} f(t)e^{-i2\pi f_0 t}\, dt$$

$$= \frac{1}{T} F(p f_0) = \frac{1}{T} F\left( \frac{p}{T} \right) \tag{12.109}$$

where $F(p f_0) = F(p/T)$ is recognized as the Fourier transform of $f(t)$ in which the argument $f$ has been replaced by $p f_0 = p/T$. Inserting Eq. (12.109) into Eq. (12.106), we obtain the same Fourier transform as that given by Eqs. (12.105). It follows that, *for a periodic function, the coefficients of the ordinary Fourier series are the same as those derived by the Fourier integral divided by the period T.*

## 12.10 SAMPLED FUNCTIONS

Fourier transform pairs involve integrations both in the time and in the frequency domain and, except for some simple functions, their evaluation can cause difficulties. Hence, it appears desirable to develop a procedure for evaluating Fourier transforms on a digital computer. The discrete Fourier transform is such a procedure. Before discussing the discrete Fourier transform, however, it is necessary to introduce the concept of sampled functions. Sampled functions in the time domain were introduced in Sec. 12.6 and sampled functions in the time and frequency domains were used in Sec. 12.9. In this section we wish to formalize these concepts, as well as to discuss the subject of Fourier transforms of sampled functions.

Consider a function $f(t)$ that is continuous at $t = T$. Then, a sample of $f(t)$ at $t = T$ is defined as

$$\hat{f}(t) = f(t)\delta(t - T) = f(T)\delta(t - T) \tag{12.110}$$

The function $\hat{f}(t)$ can be interpreted simply as an impulse of magnitude equal to $f(T)$. If the function $f(t)$ is continuous at $t = nT$ ($n = 0, \pm 1, \pm 2, \ldots$), then

$$\hat{f}(t) = f(t) \sum_{n=-\infty}^{\infty} \delta(t - nT) \sum_{n=-\infty}^{\infty} f(nT)\delta(t - nT) \tag{12.111}$$

is called the *sampled function* $f(t)$ in which the sampling period is equal to $T$. Note that Eq. (12.105) represents a sampled function in the frequency domain. It will prove convenient to rewrite Eq. (12.111) in the form of the product

$$\hat{f}(t) = f(t)\Delta(t) \tag{12.112}$$

in which

$$\Delta(t) = \sum_{n=-\infty}^{\infty} \delta(t - nT) \tag{12.113}$$

is known as a *sampling function* and it consists of an infinite sequence of equidistant unit impulses, where the distance between any two adjacent impulses is $T$. The sampling process is shown in Fig. 12.13.

Next, let us consider the Fourier transform of a sampled function. According to the frequency-domain convolution theorem, Eq. (12.96), the Fourier transform of a product of two functions is equal to the convolution of the Fourier transforms of the two functions. But, the Fourier transform of $f(t)$ is $F(f)$, where $F(f)$ is shown in Fig. 12.14a. Moreover, the Fourier transform of $\Delta(t)$ is

$$\Delta(f) = \frac{1}{T} \sum_{n=-\infty}^{\infty} \delta\left(f - \frac{n}{T}\right) \tag{12.114}$$

and note that such a Fourier transform was encountered earlier in the form of Eq. (12.104). The Fourier transform $\Delta(f)$ is shown in Fig. 12.14b and we observe once again that $\Delta(f)$ is another infinite sequence of impulses of magnitude $T^{-1}$ and at distances $T^{-1}$. Hence, using the frequency-domain convolution theorem, we can write

$$\hat{F}(f) = \int_{-\infty}^{\infty} \hat{f}(t)e^{-i2\pi ft}\, dt = \int_{-\infty}^{\infty} f(t)\Delta(t)e^{-i2\pi ft}\, dt$$

$$= \int_{-\infty}^{\infty} F(\gamma)\Delta(f - \gamma)\, d\gamma \tag{12.115}$$

The transform $\hat{F}(f)$ is shown in Fig. 12.14c, from which we observe that the Fourier transform of a sampled function is a periodic function with period equal to $T^{-1}$ and with the function over one period resembling $F(f)$ except that the amplitudes are divided by $T$.

The above statement assumes that the sampling period $T$ is sufficiently small that the functions resembling $F(f)$ in Fig. 12.14c are separated. If the sampling period $T$ increases, then the impulses in $\Delta(f)$ draw closer together. If $T$ becomes too large, then a situation can arise in which the functions $F(f)$ are no longer separated but they overlap, as shown in Fig. 12.15. It follows that the Fourier transform of the sampled functions with low sampling rate undergoes a distortion relative to that with high sampling rate, a phenomenon known as *aliasing*. The question arises as to the sampling rate required to prevent aliasing. It is easy to verify from Fig. 12.14c that the minimum sampling period required to prevent overlapping is $T = 1/2f_c$, where $f_c$ is the highest frequency component in $F(f)$. Hence, to prevent aliasing the sampling period must satisfy

$$T < 1/2f_c \tag{12.116}$$

Note that absence of aliasing implies that the Fourier transform $F(f)$ of $f(t)$ is *band-limited*, i.e., that $F(f) = 0$ for $|f| > f_c$.

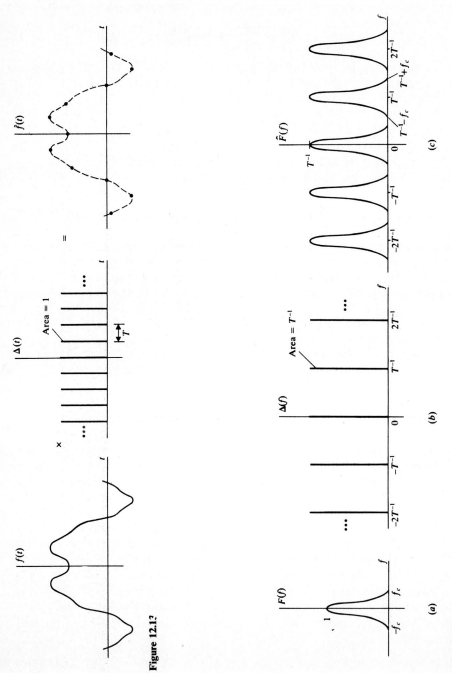

**Figure 12.13**

**Figure 12.14**

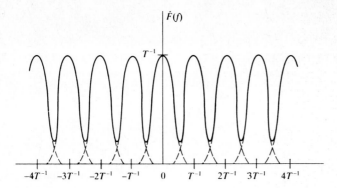

$|\hat{F}(f)|$

$T^{-1}$

$-4T^{-1} \quad -3T^{-1} \quad -2T^{-1} \quad -T^{-1} \quad 0 \quad T^{-1} \quad 2T^{-1} \quad 3T^{-1} \quad 4T^{-1}$

**Figure 12.15**

Choosing the sample interval $T$ so that aliasing does not occur is very important, as it permits a reconstruction of the continuous Fourier transform from samples of the continuous function. Moreover, if aliasing does not occur, these samples can be used to reconstruct the continuous function itself. Indeed, according to the sampling theorem, if the sampled values $f(nT)$ ($n = 0 \pm 1, \pm 2, \ldots$) are known and $T = 1/2f_c$, then the function $f(t)$ is given by

$$f(t) = T \sum_{n=-\infty}^{\infty} f(nT) \frac{\sin 2\pi f_c(t - nT)}{\pi(t - nT)} \tag{12.117}$$

Proof of the sampling theorem can be found in the book by E. O. Brigham.† As pointed out above, the sampling theorem requires that $f(t)$ be band-limited, which is seldom the case in practice. When $f(t)$ is not band-limited, sampling must be performed at a sufficiently fast rate so as to ensure that aliasing is negligible. Clearly, the sampling rate must be fast enough so as to reproduce with reasonable accuracy the highest harmonic component with significant participation.

## 12.11 THE DISCRETE FOURIER TRANSFORM

The discrete Fourier transform is simply a procedure for modifying Fourier transform pairs so as to permit their computation on a digital computer. Hence, the discrete Fourier transform is an approximation of the continuous Fourier transform. Of course, the object is that errors involved in the approximation be made as small as possible. The derivation of the discrete Fourier transform involves three steps: time-domain sampling, truncation, and frequency-domain sampling.

Consider the function $f(t)$ shown in Fig. 12.16a and the sampling function $\Delta_0(t)$ shown in Fig. 12.16b. Then, the sampled function has the expression

$$\hat{f}(t) = f(t)\Delta_0(t) = f(t) \sum_{k=-\infty}^{\infty} \delta(t - kT) = \sum_{k=-\infty}^{\infty} f(kT)\delta(t - kT) \tag{12.118}$$

The sampled function is plotted in Fig. 12.16c.

† *The Fast Fourier Transform*, p. 83, Prentice-Hall, Inc., Englewood Cliffs, New Jersey, 1974.

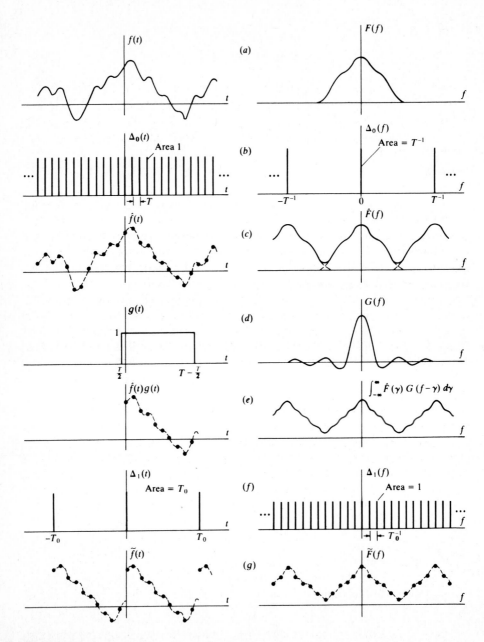

**Figure 12.16**

The sampled function $\hat{f}(t)$ involves an infinite number of samples. For practical reasons, $\hat{f}(t)$ must be truncated. To this end, we multiply $\hat{f}(t)$ by the rectangular function $g(t)$ given by

$$g(t) = u\left(t + \frac{T}{2}\right) - u\left(t - T_0 + \frac{T}{2}\right) \tag{12.119}$$

where $u(t - a)$ is the unit step function applied at $t = a$. Clearly, $g(t)$ has unit amplitude and its duration is $T_0$, as shown in Fig. 12.16d. Assuming that it is desired to retain $N$ samples, $T_0$ must satisfy $T_0 = NT$. The reason for starting $g(t)$ at $t = -T/2$ will be explained later in this section. The truncated sampled function can be written as

$$\hat{f}(t)g(t) = \left[\sum_{k=-\infty}^{\infty} f(kT)\delta(t - kT)\right]g(t) = \sum_{k=0}^{N-1} f(kT)\delta(t - kT) \tag{12.120}$$

The truncated sampled function is shown in Fig. 12.16e.

At this point, it is appropriate to pause and examine the Fourier transforms of the various functions discussed above. These Fourier transforms are shown on the right side in Figs. 12.16a–e. In particular, we notice the aliasing effect in Fig. 12.16c, where the figure was obtained via a frequency-domain convolution of the transforms of $f(t)$ and $\Delta_0(t)$. Clearly, faster sampling of $f(t)$ can reduce this aliasing. Figure 12.16e shows on the right side the Fourier transform of the truncated sampled function, which is obtained through a frequency-domain convolution of the Fourier transforms of $\hat{f}(t)$ and $g(t)$. We notice that the Fourier transform of the truncated sampled function contains the so-called *rippling effect*, which is typical of Fourier transforms of truncated sampled functions when compared to the Fourier transforms of the same functions before truncation. The rippling effect can be traced to the finite length $T_0$ of the truncation function $g(t)$. Indeed, the Fourier transform $G(f)$ of a rectangular function $g(t)$ of the type shown in Fig. 12.16d is proportional to $\sin f/f$. As $T_0$ approaches infinity, $\sin f/f$ approaches an impulse and the ripples disappear. In this regard, we must recognize that the convolution of a given function with the unit impulse reproduces the function. Hence, to reduce the rippling effect on the right side of Fig. 12.16e, the length $T_0$ of the rectangular truncation function should be made as large as possible. The task of deriving a Fourier transform pair that can be computed on a digital computer is not yet finished. Indeed, the Fourier transform of the truncated sampled function shown on the right side in Fig. 12.16e is continuous in the frequency $f$. Hence, to complete the task, we must sample it in the frequency domain, which amounts to multiplying it by a sampling function. But, according to the time-domain convolution theorem, multiplication in the frequency domain implies convolution in the time domain. Moreover, as established earlier, the inverse Fourier transform of a sampling function is another sampling function. The time-domain sampling function is

$$\Delta_1(t) = T_0 \sum_{j=-\infty}^{\infty} \delta(t - jT_0) \tag{12.121}$$

so that, by analogy with Eqs. (12.100) and (12.101), the approximation to the function $f(t)$ has the expression

$$
\begin{aligned}
\tilde{f}(t) &= \int_{-\infty}^{\infty} \hat{f}(\tau)g(\tau)\Delta_1(t-\tau)\,d\tau \\
&= \int_{-\infty}^{\infty} \left[ \sum_{k=0}^{N-1} f(kT)\delta(\tau - kT) \right]\left[ T_0 \sum_{j=-\infty}^{\infty} \delta(t - jT_0 - \tau) \right] d\tau \\
&= T_0 \sum_{k=0}^{N-1} \sum_{j=-\infty}^{\infty} f(kT)\delta(t - jT_0 - kT) \qquad (12.122)
\end{aligned}
$$

The functions $\Delta_1(t)$ and $\tilde{f}(t)$ are shown on the left side in Figs. 12.16f and g, respectively. We note once again that convolution of a function of duration $T_0$ with $\Delta_1(t)$ results in a periodic function with period $T_0$, which includes exactly $N$ samples. At this point we observe that the choice of the interval $-T/2 < t < T_0 - T/2$ for the rectangular function $g(t)$ given by Eq. (12.119) was to prevent aliasing in the time domain, which would have occurred if the interval were chosen $0 < t < T_0$, for example. Indeed, in this case the value of the $N$th sample in one period would have been added to the value of the first sample in the next period.

Now we are in the position to determine the Fourier transform of $\tilde{f}(t)$. In Sec. 12.9, we have shown that the Fourier transform of a periodic function is a sequence of equidistant impulses. In our case, by analogy with Eq. (12.108), the Fourier transform of $\tilde{f}(t)$ can be written in the form

$$
\tilde{F}(f) = \sum_{n=-\infty}^{\infty} C_n \delta(f - nf_0) \qquad f_0 = \frac{1}{T_0} \qquad (12.123)
$$

where

$$
C_n = \frac{1}{T_0} \int_{-T/2}^{T_0 - T/2} \tilde{f}(t)e^{-i2\pi nf_0 t}\,dt = \frac{1}{T_0} \int_{-T/2}^{T_0 - T/2} \tilde{f}(t)e^{-i2\pi nt/T_0}\,dt
$$
$$
n = 0, \pm 1, \pm 2, \ldots \quad (12.124)
$$

Inserting Eq. (12.122) into Eq. (12.124) and recognizing that integration is to be carried out over one period only, we obtain

$$
\begin{aligned}
C_n &= \frac{1}{T_0} \int_{-T/2}^{T_0 - T/2} T_0 \sum_{k=0}^{N-1} \sum_{j=-\infty}^{\infty} f(kT)\delta(t - jT_0 - kT)e^{-i2\pi nt/T_0}\,dt \\
&= \int_{-T/2}^{T_0 - T/2} \sum_{k=0}^{N-1} f(kT)\delta(t - kT)e^{-i2\pi nt/T_0}\,dt \\
&= \sum_{k=0}^{N-1} f(kT)e^{-i2\pi nkT/T_0} = \sum_{k=0}^{N-1} f(kT)e^{-i2\pi nk/N} \qquad n = 0, \pm 1, \pm 2, \ldots
\end{aligned}
$$
$$
(12.125)
$$

where it was recalled that $T/T_0 = 1/N$. Hence, Eq. (12.123) can be rewritten as

$$
\tilde{F}(f) = \sum_{n=-\infty}^{\infty} F(n)\delta(f - nf_0) \qquad (12.126)
$$

where $F(n) = C_n$.

It is easy to verify that the function $\tilde{F}(f)$ is periodic. Indeed, replacing $n$ by $n + N$ in the exponential in Eq. (12.125), we have

$$e^{-i2\pi(n+N)k/N} = e^{-i2\pi k}\, e^{-i2\pi nk/N} = e^{-i2\pi nk/N} \tag{12.127}$$

so that

$$F(n + N) = F(n) \tag{12.128}$$

Hence, there are only $N$ distinct values of $F(n)$, namely,

$$F(n) = \sum_{k=0}^{N-1} f(k)e^{-i2\pi nk/N} \qquad n = 0, 1, 2, \ldots, N - 1 \tag{12.129}$$

where the symbol $T$ has been omitted from $f(kT)$. The amplitudes $F(n)$ define the Fourier transform $\tilde{F}(f)$ completely. The function $\tilde{F}(f)$ is shown on the right side in Fig. 12.16$g$. For completeness, the Fourier transform $\Delta_1(f)$ of $\Delta_1(t)$ is shown on the right side in Fig. 12.16$f$.

The function $F(n)$ is the *discrete Fourier transform* sought. It can be identified as the discrete Fourier transform of $f(k)$. Indeed, it can be verified by substitution into Eq. (12.129) that $f(k)$ is the discrete inverse Fourier transform of $F(n)$, where $f(k)$ has the expression

$$f(k) = \frac{1}{N} \sum_{n=0}^{N-1} F(n)e^{i2\pi nk/N} \qquad k = 0, 1, 2, \ldots, N - 1 \tag{12.130}$$

Note that in substituting Eq. (12.130) into Eq. (12.129) the index $n$ must be replaced by a different one, say $p$. Then, from the orthogonality relation, it follows immediately that

$$\sum_{k=0}^{N-1} e^{i2\pi pk/N}e^{-i2\pi nk/N} = \sum_{k=0}^{N-1} e^{i2\pi(p-n)k/N} = N\delta_{np} \tag{12.131}$$

where $\delta_{np}$ is the Kronecker delta. Hence, $f(k)$ and $F(n)$ constitute a *discrete Fourier transform pair*.

It should perhaps be pointed out that Eqs. (12.129) and (12.130) can be derived through zero-order hold discretizations of the Fourier transform pair, Eqs. (12.92) and (12.93), both in the time and in the frequency domain and then truncating the resulting sums. The discretization involves the substitution $t = kT$ and $f = n/NT$, where $T$ is the sampling period.

## 12.12 THE FAST FOURIER TRANSFORM

The fast Fourier transform (FFT) is merely an algorithm for the efficient computation of the discrete Fourier transform. The FFT algorithm achieves its efficiency by taking advantage of the special form of the discrete Fourier transform, and in particular from the fact that the discrete Fourier transform involves the term $\exp(-i2\pi nk/N)$.

Let us introduce the notation

$$W = e^{-i2\pi/N} \tag{12.132}$$

so that the discrete Fourier transform, Eq. (12.129), can be written in the form

$$F(n) = \sum_{k=0}^{N-1} f(k)W^{nk} \qquad n = 0, 1, 2, \ldots, N-1 \tag{12.133}$$

The FFT algorithm can achieve exceptional efficiency when $N = 2^j$, where $j$ is an integer, $j = 2, 3, \ldots$. To present the basic ideas of the FFT algorithm, we consider the simplest case, $j = 2$, or $N = 4$, in which case Eqs. (12.133) can be written in the compact matrix form

$$\{F\} = [W]\{f\} \tag{12.134}$$

in which

$$\{F\} = \begin{Bmatrix} F(0) \\ F(1) \\ F(2) \\ F(3) \end{Bmatrix} \qquad \{f\} = \begin{Bmatrix} f(0) \\ f(1) \\ f(2) \\ f(3) \end{Bmatrix} \tag{12.135}$$

and

$$[W] = \begin{bmatrix} W^0 & W^0 & W^0 & W^0 \\ W^0 & W^1 & W^2 & W^3 \\ W^0 & W^2 & W^4 & W^6 \\ W^0 & W^3 & W^6 & W^9 \end{bmatrix} \tag{12.136}$$

Due to the nature of $W^{nk}$, certain simplifications can be made in the matrix $[W]$. In the first place, we note that $W^0 = 1$. Moreover, $W^{nk} = W^{nk \bmod N}$, where $nk \bmod N$ denotes the remainder after $nk$ has been divided by $N$. For example, in the case in which $n = 2$ and $k = 3$, we have

$$W^6 = e^{-i2\pi 6/4} = e^{-i2\pi}e^{-i2\pi 2/4} = e^{-i2\pi 2/4} = W^2 \tag{12.137}$$

because $\exp(-i2\pi) = \cos 2\pi - i \sin 2\pi = 1$. Hence, the matrix $[W]$ can be reduced to

$$[W] = \begin{bmatrix} 1 & 1 & 1 & 1 \\ 1 & W^1 & W^2 & W^3 \\ 1 & W^2 & 1 & W^2 \\ 1 & W^3 & W^2 & W^1 \end{bmatrix} \tag{12.138}$$

The number of multiplications and additions involved in Eq. (12.134) can be reduced appreciably by expressing the matrix $[W]$ as the product of three matrices, two having one-half of the entries equal to zero and the third being a permutation matrix. To demonstrate this, it is convenient to introduce a notation in terms of binary numbers.

Let us represent $n$ and $k$ as 2-bit binary numbers as follows:

$$n = (n_1, n_0) = 00, 01, 10, 11 \qquad k = (k_1, k_0) = 00, 01, 10, 11 \tag{12.139}$$

which correspond to $n = 0, 1, 2, 3$ and $k = 0, 1, 2, 3$, respectively. Hence, $n_1$ and $k_1$ correspond to $2^1 = 2$ and $n_0$ and $k_0$ correspond to $2^0 = 1$. These numbers can be written in the form

$$n = 2n_1 + n_0 \qquad k = 2k_1 + k_0 \qquad (12.140)$$

where $n_0$, $n_1$, $k_0$, and $k_1$ can assume the values of 0 and 1 only. Introducing Eq. (12.140) into Eq. (12.133), we obtain

$$F(n_1, n_0) = \sum_{k_0=0}^{1} \sum_{k_1=0}^{1} f(k_1, k_0) W^{(2n_1 + n_0)(2k_1 + k_0)} \qquad (12.141)$$

Next, let us consider

$$W^{(2n_1 + n_0)(2k_1 + k_0)} = W^{4n_1 k_1} W^{2n_0 k_1} W^{(2n_1 + n_0)k_0} \qquad (12.142)$$

But,

$$W^{4n_1 k_1} = (W^4)^{n_1 k_1} = (e^{-i2\pi})^{n_1 k_1} = 1^{n_1 k_1} = 1 \qquad (12.143)$$

Introducing Eqs. (12.142) and (12.143) into Eq. (12.141), we can write

$$F(n_1, n_0) = \sum_{k_0=1}^{1} \left[ \sum_{k_1=0}^{1} f(k_1, k_0) W^{2n_0 k_1} \right] W^{(2n_1 + n_0)k_0} \qquad (12.144)$$

Equation (12.144) forms the basis for the factorization of the matrix $[W]$ mentioned earlier.

The sum inside brackets in Eq. (12.144) can be rewritten as

$$f_1(n_0, k_0) = \sum_{k_1=0}^{1} f(k_1, k_0) W^{2n_0 k_1} \qquad (12.145)$$

which has the matrix form

$$\{f\}_1 = [W]_1 \{f\} \qquad (12.146)$$

where

$$\{f\}_1 = \begin{Bmatrix} f_1(0, 0) \\ f_1(0, 1) \\ f_1(1, 0) \\ f_1(1, 1) \end{Bmatrix} \qquad \{f\} = \begin{Bmatrix} f(0, 0) \\ f(0, 1) \\ f(1, 0) \\ f(1, 1) \end{Bmatrix} \qquad (12.147)$$

and

$$[W]_1 = \begin{bmatrix} 1 & 0 & 1 & 0 \\ 0 & 1 & 0 & 1 \\ 1 & 0 & W^2 & 0 \\ 0 & 1 & 0 & W^2 \end{bmatrix} \qquad (12.148)$$

In a similar manner, the second sum in Eq. (12.144) can be written as

$$f_2(n_0, n_1) = \sum_{k_0=0}^{1} f_1(n_0, k_0) W^{(2n_1 + n_0)k_0} \qquad (12.149)$$

which can be written in the matrix form

$$\{f\}_2 = [W]_2\{f\}_1 \tag{12.150}$$

in which

$$\{f\}_2 = \begin{Bmatrix} f_2(0, 0) \\ f_2(0, 1) \\ f_2(1, 0) \\ f_2(1, 1) \end{Bmatrix} \tag{12.151}$$

and

$$[W]_2 = \begin{bmatrix} 1 & 1 & 0 & 0 \\ 1 & W^2 & 0 & 0 \\ 0 & 0 & 1 & W^1 \\ 0 & 0 & 1 & W^3 \end{bmatrix} \tag{12.152}$$

Moreover, we note from Eqs. (12.144), (12.145), and (12.149) that

$$F(n_1, n_0) = f_2(n_0, n_1) \tag{12.153}$$

which implies that

$$\begin{Bmatrix} F(0) \\ F(2) \\ F(1) \\ F(3) \end{Bmatrix} = \begin{Bmatrix} f_2(0) \\ f_2(1) \\ f_2(2) \\ f_2(3) \end{Bmatrix} \tag{12.154}$$

so that the final output is in "scrambled" order. To unscramble the output, one can write simply

$$\{F\} = [P]\{f\}_2 \tag{12.155}$$

where

$$[P] = \begin{bmatrix} 1 & 0 & 0 & 0 \\ 0 & 0 & 1 & 0 \\ 0 & 1 & 0 & 0 \\ 0 & 0 & 0 & 1 \end{bmatrix} \tag{12.156}$$

is a permutation matrix. Equations (12.146), (12.152), and (12.155) can be combined into

$$\{F\} = [P][W]_2[W]_1\{f\} \tag{12.157}$$

so that, comparing with Eq. (12.134), we conclude that

$$[W] = [P][W]_2[W]_1 \tag{12.158}$$

which justifies the statement made earlier concerning the factorization of $[W]$.

Equations (12.146), (12.150), and (12.155) can be computed recursively. They represent the original formulation of the FFT algorithm by Cooley and Tukey for

$N = 4$. The formulation can be extended to $N = 8, 16, \ldots$, each time adding an extra matrix to the factorization of $[W]$. Another version of the Cooley-Tukey algorithm is that in which the input is scrambled and the output is in natural order.

The efficiency of the FFT algorithm compared with direct computation of the discrete Fourier transform increases as $j$ increases, where we recall that $N = 2^j$. It can be shown† that the ratio of the direct to the FFT computing time is $2N/j$, which increases exponentially with $j$.

A distinct form of the FFT algorithm, known as the Sande-Tukey algorithm, differs from the Cooley-Tukey algorithm in that the components of $n$ are separated instead of the components of $k$. In this case, Eqs. (12.142) and (12.143) are to be replaced by

$$W^{(2n_1 + n_0)(2k_1 + k_0)} = W^{4n_1 k_1} W^{2n_1 k_0} W^{(2k_1 + k_0)n_0}$$

$$= W^{2n_1 k_0} W^{(2k_1 + k_0)n_0} \tag{12.159}$$

so that Eq. (12.144) becomes

$$F(n_1, n_0) = \sum_{k_0 = 0}^{1} \left[ \sum_{k_1 = 0}^{1} f(k_1, k_0) W^{2n_0 k_1} W^{n_0 k_0} \right] W^{2n_1 k_0} \tag{12.160}$$

The factorization of $[W]$ follows the same pattern as that in the Cooley-Tukey algorithm, and once again there are two versions, one in which the output is scrambled and the other in which the input is scrambled.

The FFT algorithm need not be restricted to $N = 2^j$. Indeed, an algorithm can be developed for $N = j_1 j_2 \ldots j_m$, where $j_1, j_2, \ldots, j_m$ are integers. For details of this version of the FFT algorithm, as well as for a more in-depth discussion of the Cooley-Tukey and Sande-Tukey algorithms, the reader is urged to consult the book by E. O. Brigham.‡ The book also contains an FFT computation flow chart and listings for Fortran and Algol programs based on that flow chart.

## PROBLEMS

**12.1** Calculate the response of a mass-spring system to the excitation $F(t) = F_0 u(t)$, where $u(t)$ is the unit step function, by means of the approach based on the transition matrix.

**12.2** Verify by means of Eq. (12.21) that the transition matrix for a mass-spring system possesses the group property.

**12.3** Solve Prob. 12.1 for a mass-damper-spring system.

**12.4** Solve the problem of Example 4.11 by the approach of Sec. 12.4.

**12.5** Solve Prob. 12.1 by means of the convolution sum.

**12.6** Solve Prob. 12.3 by means of the convolution sum.

**12.7** Derive the discrete-time impulse response for a mass-spring system by means of the method based on the discrete-time transition matrix.

† See E. O. Brigham, op. cit., p. 152.
‡ Op. cit., chs. 10–13.

**12.8** Solve Prob. 12.1 by means of the method based on the discrete-time transition matrix, compare results with those obtained in Prob. 12.5 and draw conclusions as to the accuracy of the two approaches.

**12.9** Repeat Prob. 12.8 for the system of Prob. 12.6.

**12.10** Compute the solution of the nonlinear differential equation

$$\ddot{\theta}(t) + 4 \sin \theta(t) = 0$$

where $\theta(t)$ is subject to the initial conditions $\theta(0) = \pi/3$, $\dot{\theta}(0) = 0$. Use the fourth-order Runge-Kutta method.

# FOURIER SERIES

## A.1 INTRODUCTION

In many problems of engineering analysis it is necessary to work with periodic functions, i.e., with functions that repeat themselves every given interval, where the interval is known as the *period*. Periodic functions satisfy a relation of the type

$$f(t) = f(t + T) \tag{A.1}$$

where $T$ represents the period. Some of the simplest and most commonly encountered periodic functions are the trigonometric functions. Indeed, it is easy to verify that the functions $\sin nt$ and $\cos nt$ ($n = 1, 2, \ldots$) are periodic with period $2\pi$. Their period is actually $2\pi/n$, but any function with period $2\pi/n$ certainly repeats itself every $2\pi$. Clearly, trigonometric functions are special cases of periodic functions. Because trigonometric functions are relatively easy to manipulate, they are more desirable to work with than arbitrary periodic functions. Owing to this fact, it is well worth exploring the possibility of expressing any arbitrary periodic function $f(t)$ in a series of trigonometric functions. Such expansions are indeed possible and are known as *Fourier series*.

## A.2 ORTHOGONAL SETS OF FUNCTIONS

Let us consider a set of functions $\psi_r(t)$ ($r = 1, 2, \ldots$) defined over the interval $0 \leqslant t \leqslant T$. Then, if the functions are such that, for any two distinct functions $\psi_r(t)$ and $\psi_s(t)$,

$$\int_0^T \psi_r(t)\psi_s(t)\, dt = 0 \qquad r \neq s \tag{A.2}$$

the set $\psi_r(t)$ is said to be *orthogonal* in the interval $0 \leqslant t \leqslant T$, or more generally in any interval of length $T$. If the functions $\psi_r(t)$ are such that, in addition to satisfying

**519**

Eq. (A.2), they satisfy

$$\int_0^T \psi_r^2(t)\,dt = 1 \qquad r = 1, 2, \ldots \tag{A.3}$$

then the set is referred to as *orthonormal*. Hence, for an orthonormal set of functions we have

$$\int_0^T \psi_r(t)\psi_s(t)\,dt = \delta_{rs} \qquad r, s = 1, 2, \ldots \tag{A.4}$$

where $\delta_{rs}$ is the Kronecker delta, defined as being equal to unity for $r = s$ and equal to zero for $r \neq s$. It is easy to verify that the set of functions

$$\frac{1}{\sqrt{2\pi}}, \frac{\sin t}{\sqrt{\pi}}, \frac{\cos t}{\sqrt{\pi}}, \frac{\sin 2t}{\sqrt{\pi}}, \frac{\cos 2t}{\sqrt{\pi}}, \frac{\sin 3t}{\sqrt{\pi}}, \ldots \tag{A.5}$$

constitutes an orthonormal set. Indeed, we can write

$$\int_0^{2\pi} \frac{1}{\sqrt{2\pi}} \frac{\sin rt}{\sqrt{\pi}}\,dt = -\frac{1}{\sqrt{2\pi}} \left.\frac{\cos rt}{r}\right|_0^{2\pi} = 0$$
$$\int_0^{2\pi} \frac{1}{\sqrt{2\pi}} \frac{\cos rt}{\sqrt{\pi}}\,dt = \frac{1}{\sqrt{2\pi}} \left.\frac{\sin rt}{r}\right|_0^{2\pi} = 0 \qquad r = 1, 2, \ldots \tag{A.6}$$

Moreover, for $r \neq s$, we have

$$\int_0^{2\pi} \frac{\sin rt}{\sqrt{\pi}} \frac{\cos st}{\sqrt{\pi}}\,dt = \frac{1}{2\pi} \int_0^{2\pi} [\sin(r+s)t + \sin(r-s)t]\,dt$$

$$= -\frac{1}{2\pi}\left[\frac{\cos(r+s)t}{r+s} + \frac{\cos(r-s)t}{r-s}\right]_0^{2\pi} = 0$$
$$r, s = 1, 2, \ldots \tag{A.7}$$

and for $r = s$, we obtain

$$\int_0^{2\pi} \frac{\sin rt}{\sqrt{\pi}} \frac{\cos rt}{\sqrt{\pi}}\,dt = \frac{1}{2\pi} \int_0^{2\pi} \sin 2rt\,dt = -\frac{1}{4r\pi} \cos 2rt\Big|_0^{2\pi} = 0$$
$$r = 1, 2, \ldots \tag{A.8}$$

so that the set (A.5) satisfies Eq. (A.2); hence, it is orthogonal. On the other hand

$$\int_0^{2\pi} \left(\frac{1}{\sqrt{2\pi}}\right)^2 dt = 1$$

$$\int_0^{2\pi} \left(\frac{\sin rt}{\sqrt{\pi}}\right)^2 dt = \frac{1}{r\pi}\left[\frac{rt}{2} - \frac{\sin 2rt}{4}\right]_0^{2\pi} = 1 \qquad r = 1, 2, \ldots \tag{A.9}$$

$$\int_0^{2\pi} \left(\frac{\cos rt}{\sqrt{\pi}}\right)^2 dt = \frac{1}{r\pi}\left[\frac{rt}{2} + \frac{\sin 2rt}{4}\right]_0^{2\pi} = 1 \qquad r = 1, 2, \ldots$$

so that the set (A.5) is not only orthogonal but orthonormal.

If for a set of constants $c_r$ $(r = 1, 2, \ldots)$, not all equal to zero, there exists a homogeneous linear relation

$$\sum_{r=1}^{n} c_r \psi_r(t) = 0 \qquad (A.10)$$

for all $t$, then the set of functions $\psi_r(t)$ $(r = 1, 2, \ldots)$ is said to be *linearly dependent*. If no relation of the type (A.10) exists, then the set is said to be *linearly independent*. The set (A.5) can be shown to be a linearly independent set. Indeed, if we write the series

$$c_0 \frac{1}{\sqrt{2\pi}} + c_1 \frac{\sin t}{\sqrt{\pi}} + c_2 \frac{\cos t}{\sqrt{\pi}} + c_3 \frac{\sin 2t}{\sqrt{\pi}} + c_4 \frac{\cos 2t}{\sqrt{\pi}} + \cdots + c_{2p} \frac{\cos pt}{\sqrt{\pi}} = 0$$

$$(A.11)$$

multiply the series by any of the functions in (A.5), say $(\cos 2t)/\sqrt{\pi}$, and integrate with respect to $t$ over the interval $0 \leqslant t \leqslant 2\pi$, we obtain $c_4 = 0$. The procedure can be repeated for all constants, with the conclusion that $c_0 = c_1 = c_2 = \cdots = c_{2p} = 0$. Because this contradicts the stipulation that not all constants be zero, we must conclude that the set is linearly independent. Note that an orthogonal set is by definition linearly independent.

## A.3 TRIGONOMETRIC SERIES

An orthonormal set of functions $\psi_r(t)$ $(r = 1, 2, \ldots)$ is said to be *complete* if any piecewise continuous function $f(t)$ can be approximated in the mean to any desired degree of accuracy by the series $\sum_{r=1}^{n} c_r \psi_r(t)$ by choosing the integer $n$ large enough. In view of this, because the set (A.5) is complete in the interval $0 \leqslant t \leqslant 2\pi$, every function $f(t)$ which is continuous in that interval can be represented by the *Fourier series*

$$f(t) = \tfrac{1}{2}a_0 + \sum_{r=1}^{\infty} (a_r \cos rt + b_r \sin rt) \qquad (A.12)$$

where the constants $a_r$ $(r = 0, 1, 2, \ldots)$ and $b_r$ $(r = 1, 2, \ldots)$ are known as *Fourier coefficients*.

To establish the exact composition of the trigonometric representation of a given periodic function, it is necessary to calculate the Fourier coefficients. To this end, it will prove useful to summarize the following results derived in Sec. A.2:

$$\int_{0}^{2\pi} \cos rt \cos st \, dt = 0$$
$$\qquad\qquad r \neq s \qquad (A.13)$$
$$\int_{0}^{2\pi} \sin rt \sin st \, dt = 0$$

and

$$\int_0^{2\pi} \cos rt \sin st \, dt = \int_0^{2\pi} \sin rt \cos st \, dt = 0 \qquad \text{(A.14)}$$

where (A.14) is valid whether $r$ and $s$ are distinct or not. On the other hand, when $r = s$, the integrals in (A.13) are not zero but have the values

$$\int_0^{2\pi} \cos^2 rt \, dt = \begin{cases} \pi & \text{if } r \neq 0 \\ 2\pi & \text{if } r = 0 \end{cases} \qquad \text{(A.15)}$$

$$\int_0^{2\pi} \sin^2 rt \, dt = \pi \qquad \text{(A.16)}$$

Moreover, we can write

$$\int_0^{2\pi} \cos rt \, dt = \begin{cases} 0 & \text{if } r \neq 0 \\ 2\pi & \text{if } r = 0 \end{cases} \qquad \text{(A.17)}$$

$$\int_0^{2\pi} \sin rt \, dt = 0 \qquad \text{(A.18)}$$

At this time, let us multiply Eq. (A.12) by $\cos st$, integrate over the interval $0 \leqslant t \leqslant 2\pi$, interchange the order of integration and summation and obtain

$$\int_0^{2\pi} f(t) \cos st \, dt = \tfrac{1}{2} a_0 \int_0^{2\pi} \cos st \, dt$$

$$+ \sum_{r=1}^n a_r \int_0^{2\pi} \cos rt \cos st \, dt + b_r \int_0^{2\pi} \sin rt \cos st \, dt \quad \text{(A.19)}$$

For $s = 0$, (A.19) in conjunction with (A.17) and (A.18) yields

$$a_0 = \frac{1}{\pi} \int_0^{2\pi} f(t) \, dt \qquad \text{(A.20)}$$

so that $\tfrac{1}{2} a_0$ is identified as the *average value* of $f(t)$. If $s \neq 0$, we conclude that only one term survives from the series in (A.19), namely, that corresponding to the integral $\int_0^{2\pi} \cos rt \cos st \, dt$ in which $r = s$. Indeed, considering Eqs. (A.13) through (A.15), we can write

$$a_r = \frac{1}{\pi} \int_0^{2\pi} f(t) \cos rt \, dt \qquad r = 1, 2, \ldots \qquad \text{(A.21)}$$

Similarly, multiplying series (A.12) by $\sin st$, integrating over the interval $0 \leqslant t \leqslant 2\pi$, and considering Eqs. (A.13), (A.14), (A.16), and (A.18), we obtain

$$b_r = \frac{1}{\pi} \int_0^{2\pi} f(t) \sin rt \, dt \qquad r = 1, 2, \ldots \qquad \text{(A.22)}$$

thus determining the series (A.12) uniquely.

When $f(t)$ is an *even function*, i.e., when $f(t) = f(-t)$, the coefficients $b_r$

($r = 1, 2, \ldots$) vanish and the series is known as a *Fourier cosine series*. On the other hand, when $f(t)$ is an *odd function*, i.e., when $f(t) = -f(-t)$, the coefficients $a_r$ ($r = 0, 1, 2, \ldots$) vanish and the series is called a *Fourier sine series*. This can be more conveniently demonstrated by considering the interval $-\pi \leqslant t \leqslant \pi$ instead of $0 \leqslant t \leqslant 2\pi$.

If the function $f(t)$ is only piecewise continuous in a given interval, then a Fourier series representation using a finite number of terms approaches $f(t)$ in every interval that does not contain discontinuities. In the immediate neighborhood of a jump discontinuity, convergence is not uniform and, as the number of terms increases, the finite series approximation contains increasingly high-frequency oscillations which move closer to the discontinuity point. However, the total oscillation of the approximating curve does not approach the jump of $f(t)$, a fact known as the *Gibbs phenomenon*.

An an illustration, let us consider the periodic function $f(t)$ shown in Fig. A.1, where the function repeats itself every $2\pi$. The function is recognized as being an odd function of $t$, so that $f(t)$ can be represented by a Fourier sine series of the form

$$f(t) = \sum_{r=1}^{\infty} b_r \sin rt \tag{A.23}$$

The proof that $a_r = 0$ ($r = 0, 1, 2, \ldots$) is left as an exercise to the reader. The function $f(t)$ can be described mathematically by

$$f(t) = \frac{A}{\pi} t \qquad -\pi \leqslant t \leqslant \pi \tag{A.24}$$

so that the coefficients become

$$b_r = \frac{1}{\pi} \int_{-\pi}^{\pi} f(t) \sin rt \, dt = \frac{A}{\pi^2} \int_{-\pi}^{\pi} t \sin rt \, dt$$

$$= \frac{A}{\pi^2 r^2} (\sin rt - rt \cos rt) \Big|_{-\pi}^{\pi} = \frac{2A}{\pi r} (-1)^{r+1} \quad r = 1, 2, \ldots \tag{A.25}$$

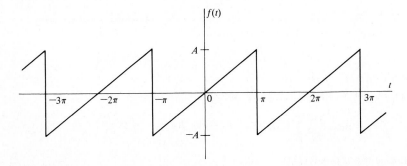

**Figure A.1**

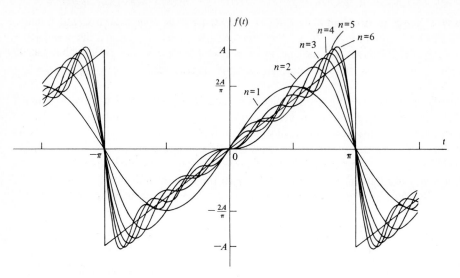

**Figure A.2**

Hence, the series becomes

$$f(t) = \frac{2A}{\pi} \sum_{r=1}^{\infty} \frac{(-1)^{r+1}}{r} \sin rt \qquad (A.26)$$

Fourier series are infinite series and on occasions they must be approximated by finite ones, as intimated earlier. This is done by replacing the upper limit in the series by a finite integer $n$, a process known as *truncation*. Figure A.2 shows the series representation for $n = 1, 2, \ldots, 6$. It is clear that the approximation improves with increasing $n$. Of course, if the accuracy of Fig. A.2 is not satisfactory, then additional terms should be included to bring the series representation to the desired level of accuracy. Note the Gibbs phenomenon at the discontinuity points $t = \pm\pi$.

## A.4 COMPLEX FORM OF FOURIER SERIES

The Fourier series can also be expressed in terms of exponential functions. Indeed trigonometric functions are related to exponential functions as follows:

$$\cos rt = \frac{e^{irt} + e^{-irt}}{2} \qquad \sin rt = \frac{e^{irt} - e^{-irt}}{2i} \qquad (A.27)$$

Inserting Eqs. (A.27) into Eq. (A.12), we obtain

$$f(t) = \tfrac{1}{2}a_0 + \frac{1}{2} \sum_{r=1}^{\infty} [a_r(e^{irt} + e^{-irt}) - ib_r(e^{irt} - e^{-irt})]$$

$$= \tfrac{1}{2}a_0 + \frac{1}{2} \sum_{r=1}^{\infty} [(a_r - ib_r)e^{irt} + (a_r + ib_r)e^{-irt}] \qquad (A.28)$$

Introducing the notation

$$C_0 = \tfrac{1}{2}a_0$$

$$C_r = \tfrac{1}{2}(a_r - ib_r) \qquad C_{-r} = C_r^* = \tfrac{1}{2}(a_r + ib_r) \qquad r = 1, 2, \ldots$$

(A.29)

where $C_r^*$ is the complex conjugate of $C_r$, Eq. (A.28) reduces to

$$f(t) = \sum_{r=-\infty}^{\infty} C_r e^{irt} \tag{A.30}$$

in which, using Eqs. (A.21) and (A.22), the coefficients $C_r$ have the form

$$C_r = \tfrac{1}{2}(a_r - ib_r) = \frac{1}{2\pi}\left[\int_0^{2\pi} f(t) \cos rt \, dt - i \int_0^{2\pi} f(t) \sin rt \, dt\right]$$

$$= \frac{1}{2\pi}\int_0^{2\pi} f(t)(\cos rt - i \sin rt)\, dt = \frac{1}{2\pi}\int_0^{2\pi} f(t) e^{-irt}\, dt$$

$$r = 0, 1, 2, \ldots \quad \text{(A.31)}$$

Equations (A.30) and (A.31) constitute the *complex form*, or *exponential form, of Fourier series.*

# B

# ELEMENTS OF LAPLACE TRANSFORMATION

## B.1 GENERAL DEFINITIONS

The Laplace transformation is an important tool in the study of linear systems with constant coefficients, particularly when the excitation is in the form of discontinuous functions. This introduction to the Laplace transformation method is very modest in scope, and its main purpose is to provide an elementary knowledge of the method and a certain degree of familiarity with the terminology.

The idea behind the Laplace transformation method is to transform a relatively complicated problem into a simpler one, solve the simpler problem and then perform an inverse transformation to obtain the solution to the original problem. The most common use of the method is to solve initial-value problems, namely, problems in which the system behavior is defined by ordinary differential equations, to be satisfied for all positive times, and by a given set of initial conditions. In such cases, the transformed problem involves algebraic expressions alone, with the initial conditions being taken into account automatically.

Let us consider a function $f(t)$ defined for all values of time larger than zero, $t > 0$, and define the (one-sided) Laplace transformation of $f(t)$ by the definite integral

$$\mathscr{L}f(t) = \bar{f}(s) = \int_0^\infty e^{-st} f(t)\, dt \tag{B.1}$$

where $e^{-st}$ is known as the kernel of the transformation and $s$ is referred to as a *subsidiary* variable. The variable $s$ is in general a complex quantity, and the associated complex plane $s = x + iy$ is called the $s$ plane, and at times the Laplace plane. Because transformation (B.1) is defined in terms of an integral, it is said to be an integral transformation, also commonly referred to as an integral transform.

The function $f(t)$ must be such that integral (B.1) exists, which places on $f(t)$ the restriction.

$$[e^{-st}f(t)] < Ce^{-(s-a)t} \qquad \text{Re } s > a \qquad (B.2)$$

where $C$ is a constant. Condition (B.2) implies that $f(t)$ must not increase with time more rapidly than the exponential function $Ce^{at}$. Another restriction on $f(t)$ is that it must be piecewise continuous. Most functions describing physical phenomena satisfy these conditions.

## B.2 TRANSFORMATION OF DERIVATIVES

Because our interest lies in the application of the method to differential equations, it becomes necessary to evaluate the transformation of derivatives of functions. Considering the transform of $df(t)/dt$, and integrating by parts, we obtain

$$\mathscr{L}\frac{df(t)}{dt} = \int_0^\infty e^{-st}\frac{df(t)}{dt}\,dt = e^{-st}f(t)\Big|_0^\infty - \int_0^\infty (-se^{-st})f(t)\,dt$$

$$= -f(0) + s\bar{f}(s) \qquad (B.3)$$

where $f(0)$ is the initial value of $f(t)$, namely, the value of $f(t)$ at $t = 0$.

Following the same pattern, the transform of $d^2f(t)/dt^2$ can be shown to be

$$\mathscr{L}\frac{d^2f(t)}{dt^2} = \int_0^\infty e^{-st}\frac{d^2f(t)}{dt^2}\,dt = -\dot{f}(0) - sf(0) + s^2\bar{f}(s) \qquad (B.4)$$

where $\dot{f}(0)$ is the value of $df(t)/dt$ at $t = 0$.

## B.3 TRANSFORMATION OF ORDINARY DIFFERENTIAL EQUATIONS

The differential equation of motion of a viscously damped single-degree-of-freedom system was shown in Sec. 1.3 to be

$$m\frac{d^2x(t)}{dt^2} + c\frac{dx(t)}{dt} + kx(t) = F(t) \qquad (B.5)$$

Introducing the notation $\mathscr{L}x(t) = \bar{x}(s)$, $\mathscr{L}F(t) = \bar{F}(s)$, transforming both sides of Eq. (B.5) and considering Eqs. (B.3) and (B.4), we obtain the algebraic equation

$$m[s^2\bar{x}(s) - sx(0) - \dot{x}(0)] + c[s\bar{x}(s) - x(0)] + k\bar{x}(s) = \bar{F}(s) \qquad (B.6)$$

where $x(0)$ and $\dot{x}(0)$ are the initial displacement and velocity, respectively. Recalling from Chap. 1 that $c/m = 2\zeta\omega_n$, $k/m = \omega_n^2$ and solving Eq. (B.6) for the transformed response $\bar{x}(s)$, we can write

$$\bar{x}(s) = \frac{1}{m(s^2 + 2\zeta\omega_n s + \omega_n^2)}\bar{F}(s) + \frac{s + 2\zeta\omega_n}{s^2 + 2\zeta\omega_n s + \omega_n^2}x(0)$$

$$+ \frac{1}{s^2 + 2\zeta\omega_n s + \omega_n^2}\dot{x}(0) \qquad (B.7)$$

which is called the *subsidiary equation* of the differential equation. To obtain the actual response $x(t)$, we must evaluate the inverse Laplace transformation of $\bar{x}(s)$. It is clear from Eq. (B.7) that the Laplace transformation method yields both the particular and the homogeneous solution simultaneously, with the implication that the method takes the initial conditions into account automatically.

## B.4 THE INVERSE LAPLACE TRANSFORMATION

As can be concluded from Eq. (B.7), the transformed response $\bar{x}(s)$ is a function of the subsidiary variable $s$. To obtain the time-dependent response $x(t)$, we must evaluate the inverse Laplace transform of $\bar{x}(s)$. The operation is denoted symbolically by

$$\mathscr{L}^{-1}\bar{x}(s) = x(t) \tag{B.8}$$

The rigorous definition (not given here) of the inverse transform (B.8) involves the evaluation of a line integral in the $s$ plane. In many cases, the integral can be replaced by a closed contour integral, which, in turn, can be evaluated by the residue theorem of complex algebra. By far the simplest way to evaluate inverse transformations is to decompose the function $\bar{x}(s)$ into a sum of simple functions with known inverse transformations. This is the essence of the method of partial fractions, to be described below. To expedite the inversion process, a table of commonly encountered Laplace transform pairs can be found at the end of this appendix.

Let us consider the case in which $\bar{x}(s)$ can be written as the ratio

$$\bar{x}(s) = \frac{A(s)}{B(s)} \tag{B.9}$$

where both $A(s)$ and $B(s)$ are polynomials in $s$. Generally $B(s)$ is a polynomial of higher degree than $A(s)$. Denoting by $s = a_k$ $(k = 1, 2, \ldots, n)$ the roots of $B(s)$, the polynomial can be written as the product.

$$B(s) = (s - a_1)(s - a_2) \cdots (s - a_k) \cdots (s - a_n) = \prod_{k=1}^{n} (s - a_k) \tag{B.10}$$

where $\Pi$ is the product symbol. The roots $s = a_k$ are known as *simple poles* of $\bar{x}(s)$. The partial fractions expansion of (B.9) has the form

$$\bar{x}(s) = \frac{c_1}{s - a_1} + \frac{c_2}{s - a_2} + \cdots + \frac{c_k}{s - a_k} + \cdots + \frac{c_n}{s - a_n} = \sum_{k=1}^{n} \frac{c_k}{s - a_k} \tag{B.11}$$

where the coefficients $c_k$ are given by the formula

$$c_k = \lim_{s \to a_k} [(s - a_k)\bar{x}(s)] = \frac{A(s)}{B'(s)}\Bigg|_{s=a_k} \tag{B.12}$$

in which $B'(s)$ is the derivative of $B$ with respect to $s$.

Noticing that

$$\mathscr{L} e^{a_k t} = \frac{1}{s - a_k} \tag{B.13}$$

it follows that

$$\mathscr{L}^{-1} \frac{1}{s - a_k} = e^{a_k t} \tag{B.14}$$

where (B.13) and (B.14) constitute a Laplace transform pair. In view of Eqs. (B.12) and (B.14), the inverse transform of $\bar{x}(s)$, Eq. (B.11), becomes

$$x(t) = \sum_{k=1}^{n} \frac{A(s)}{B'(s)} \bigg|_{s = a_k} e^{a_k t} = \sum_{k=1}^{n} \frac{A(s)}{B'(s)} e^{st} \bigg|_{s = a_k} \tag{B.15}$$

Quite often, however, it is simpler to consider Eq. (B.11) and write $A(s)$ in the form

$$A(s) = c_1 \prod_{\substack{i=2}}^{n} (s - a_i) + c_2 \prod_{\substack{i=1 \\ i \neq 2}}^{n} (s - a_i) + \cdots + c_n \prod_{i=1}^{n-1} (s - a_i)$$

$$= \sum_{k=1}^{n} c_k \prod_{\substack{i=1 \\ i \neq k}}^{n} (s - a_i) \tag{B.16}$$

Comparing the coefficients of $s^{j-1}$ ($j = 1, 2, \ldots, n$) on both sides of Eq. (B.16), we derive a set of algebraic equations that can be solved for the coefficients $c_k$ ($k = 1, 2, \ldots, n$).

As an illustration, let us consider the inverse of

$$\bar{x}(s) = \frac{s + 2\zeta\omega_n}{s^2 + 2\zeta\omega_n s + \omega_n^2} = \frac{A(s)}{B(s)} \tag{B.17}$$

Assuming that $\zeta < 1$, the roots of $B(s)$ are

$$\begin{aligned} a_1 \\ a_2 \end{aligned} = -\zeta\omega_n \pm i(1 - \zeta^2)^{1/2}\omega_n \tag{B.18}$$

so that

$$\begin{aligned} \bar{x}(s) &= \frac{c_1}{s - a_1} + \frac{c_2}{s - a_2} \\ &= \frac{c_1[s + \zeta\omega_n + i(1 - \zeta^2)^{1/2}\omega_n] + c_2[s + \zeta\omega_n - i(1 - \zeta^2)^{1/2}\omega_n]}{s^2 + 2\zeta\omega_n s + \omega_n^2} \end{aligned} \tag{B.19}$$

Comparing Eqs. (B.17) and (B.19), we conclude that

$$\begin{aligned} A(s) &= c_1[s + \zeta\omega_n + i(1 - \zeta^2)^{1/2}\omega_n] + c_2[s + \zeta\omega_n - i(1 - \zeta^2)^{1/2}\omega_n] \\ &= s + 2\zeta\omega_n \end{aligned} \tag{B.20}$$

from which it follows that

$$(c_1 + c_2)s + c_1[\zeta\omega_n + i(1 - \zeta^2)^{1/2}\omega_n]$$

$$+ c_2[\zeta\omega_n - i(1 - \zeta^2)^{1/2}\omega_n] = s + 2\zeta\omega_n \quad \text{(B.21)}$$

Equating the coefficients of $s^{j-1}$ $(j = 1, 2)$ on both sides of (B.21), we arrive at

$$c_1 + c_2 = 1$$

$$(c_1 + c_2)\zeta\omega_n + (c_1 - c_2)i(1 - \zeta^2)^{1/2}\omega_n = 2\zeta\omega_n \quad \text{(B.22)}$$

The solution of Eqs. (B.22) is simply

$$c_1 = \frac{1}{2}\left[1 + \frac{\zeta}{i(1 - \zeta^2)^{1/2}}\right] \qquad c_2 = \frac{1}{2}\left[1 - \frac{\zeta}{i(1 - \zeta^2)^{1/2}}\right] \quad \text{(B.23)}$$

Hence,

$$\bar{x}(s) = \frac{1}{2}\left[1 + \frac{\zeta}{i(1 - \zeta^2)^{1/2}}\right]\frac{1}{s + \zeta\omega_n - i(1 - \zeta^2)^{1/2}\omega_n}$$

$$+ \frac{1}{2}\left[1 - \frac{\zeta}{i(1 - \zeta^2)^{1/2}}\right]\frac{1}{s + \zeta\omega_n + i(1 - \zeta^2)^{1/2}\omega_n} \quad \text{(B.24)}$$

and, considering Eq. (B.14), we obtain the inverse transformation

$$x(t) = \frac{1}{2}\left[1 + \frac{\zeta}{i(1 - \zeta^2)^{1/2}}\right]e^{-[\zeta - i(1 - \zeta^2)^{1/2}]\omega_n t}$$

$$+ \frac{1}{2}\left[1 - \frac{\zeta}{i(1 - \zeta^2)^{1/2}}\right]e^{-[\zeta + i(1 - \zeta^2)^{1/2}]\omega_n t}$$

$$= e^{-\zeta\omega_n t}\left[\cos(1 - \zeta^2)^{1/2}\omega_n t + \frac{\zeta}{(1 - \zeta^2)^{1/2}}\sin(1 - \zeta^2)^{1/2}\omega_n t\right] \quad \text{(B.25)}$$

From Eqs. (B.7) and (B.17), we conclude that Eq. (B.25) represents the response of the damped single-degree-of-freedom system to an initial unit displacement, $x(0) = 1$.


## B.5 THE CONVOLUTION INTEGRAL. BOREL'S THEOREM

Let us consider two functions $f_1(t)$ and $f_2(t)$, both defined for $t > 0$. Moreover, let us assume that $f_1(t)$ and $f_2(t)$ possess Laplace transforms $\bar{f}_1(s)$ and $\bar{f}_2(s)$, respectively, and consider the integral

$$x(t) = \int_0^t f_1(\tau)f_2(t - \tau)\, d\tau = \int_0^\infty f_1(\tau)f_2(t - \tau)\, d\tau \quad \text{(B.26)}$$

The function $x(t)$, as defined by Eq. (B.26), sometimes denoted by $x(t) = f_1(t) * f_2(t)$, is called the *convolution* of the functions $f_1$ and $f_2$ over the interval

$0 < t < \infty$. The upper limits of the integrals in Eq. (B.26) are interchangeable because $f_2(t - \tau) = 0$ for $\tau > t$, which is the same as $t - \tau < 0$. Transforming both sides of Eq. (B.26), we obtain

$$\bar{x}(s) = \int_0^\infty e^{-st} \left[ \int_0^\infty f_1(\tau) f_2(t - \tau) \, d\tau \right] dt$$

$$= \int_0^\infty f_1(\tau) \left[ \int_0^\infty e^{-st} f_2(t - \tau) \, dt \right] d\tau$$

$$= \int_0^\infty f_1(\tau) \left[ \int_\tau^\infty e^{-st} f_2(t - \tau) \, dt \right] d\tau \qquad (B.27)$$

where the lower limit of the second integral was changed without affecting the result because $f_2(t - \tau) = 0$ for $t < \tau$. Next, let us introduce the transformation $t - \tau = \lambda$ in the last integral, observe that for $t = \tau$ we have $\lambda = 0$ and write

$$\bar{x}(s) = \int_0^\infty f_1(\tau) \left[ \int_0^\infty e^{-s(\tau + \lambda)} f_2(\lambda) \, d\lambda \right] d\tau$$

$$= \int_0^\infty e^{-s\tau} f_1(\tau) \, d\tau \int_0^\infty e^{-s\lambda} f_2(\lambda) \, d\lambda = \bar{f}_1(s) \bar{f}_2(s) \qquad (B.28)$$

From Eqs. (B.26) and (B.28), it follows that

$$x(t) = \mathscr{L}^{-1} \bar{x}(s) = \mathscr{L}^{-1} \bar{f}_1(s) \bar{f}_2(s)$$

$$= \int_0^t f_1(\tau) f_2(t - \tau) \, d\tau = \int_0^t f_1(t - \tau) f_2(\tau) \, d\tau \qquad (B.29)$$

The second integral in Eq. (B.29) is valid because it does not matter in which function the time is shifted. The integrals are called *convolution integrals*. This enables us to state the following theorem: *The inverse Laplace transformation of the product of two transforms is equal to the convolution of their inverse transforms.*

We recall that in Chap. 2 we derived a special case of the convolution integral, without reference to Laplace transforms, where one of the functions in the convolution was the impulse response.

## B.6 TABLE OF LAPLACE TRANSFORM PAIRS

| $f(t)$ | $\bar{f}(s)$ |
|---|---|
| $\delta(t)$ (Dirac delta function) | $1$ |
| $u(t)$ (unit step function) | $\dfrac{1}{s}$ |
| $t^n \qquad n = 1, 2, \ldots$ | $\dfrac{n!}{s^{n+1}}$ |
| $e^{-\omega t}$ | $\dfrac{1}{s + \omega}$ |
| $te^{-\omega t}$ | $\dfrac{1}{(s + \omega)^2}$ |
| $\cos \omega t$ | $\dfrac{s}{s^2 + \omega^2}$ |
| $\sin \omega t$ | $\dfrac{\omega}{s^2 + \omega^2}$ |
| $\cosh \omega t$ | $\dfrac{s}{s^2 - \omega^2}$ |
| $\sinh \omega t$ | $\dfrac{\omega}{s^2 - \omega^2}$ |
| $1 - e^{-at}$ | $\dfrac{\omega}{s(s + \omega)}$ |
| $1 - \cos \omega t$ | $\dfrac{\omega^2}{s(s^2 + \omega^2)}$ |
| $\omega t - \sin \omega t$ | $\dfrac{\omega^3}{s^2(s^2 + \omega^2)}$ |
| $\omega t \cos \omega t$ | $\dfrac{\omega(s^2 - \omega^2)}{(s^2 + \omega^2)^2}$ |
| $\omega t \sin \omega t$ | $\dfrac{2\omega^2 s}{(s^2 + \omega^2)^2}$ |
| $\dfrac{1}{(1 - \zeta^2)^{1/2}\omega} e^{-\zeta\omega t} \sin (1 - \zeta^2)^{1/2}\omega t$ | $\dfrac{1}{s^2 + 2\zeta\omega s + \omega^2}$ |
| $e^{-\zeta\omega t}\left[ \cos (1 - \zeta^2)^{1/2}\omega t + \dfrac{\zeta}{(1 - \zeta^2)^{1/2}} \sin (1 - \zeta^2)^{1/2}\omega t \right]$ | $\dfrac{s + 2\zeta\omega}{s^2 + 2\zeta\omega s + \omega^2}$ |

# ELEMENTS OF LINEAR ALGEBRA

## C.1 GENERAL CONSIDERATIONS

Linear algebra is concerned with three types of mathematical concepts, namely, matrices, vector spaces, and algebraic forms. Problems in mechanics, and particularly vibration problems, involve all three concepts. Vibration problems lead to algebraic forms. For computational purposes, however, the problems can be conveniently formulated in terms of matrices. The concept of vector spaces is quite helpful in providing a deeper understanding of linear transformations and their properties.

Our particular interest in linear algebra lies in the fact that it permits us to formulate problems associated with the vibration of discrete systems in a compact form, and it enables us to draw general conclusions concerning the dynamical characteristics of such systems. The discussion of linear algebra presented here is relatively modest in nature, and its main purpose is to familiarize us with some fundamental concepts of particular interest in vibrations.

## C.2 MATRICES

### a Definitions

Many problems in vibrations can be formulated in terms of rectangular arrays of scalars of the form

$$[a] = \begin{bmatrix} a_{11} & a_{12} & \cdots & a_{1n} \\ a_{21} & a_{22} & \cdots & a_{2n} \\ \cdots & \cdots & \cdots & \cdots \\ a_{m1} & a_{m2} & \cdots & a_{mn} \end{bmatrix} \tag{C.1}$$

where $[a]$ is called an $m \times n$ *matrix* because it contains $m$ rows and $n$ columns. It is also customary to say that the *dimensions* of $[a]$ are $m \times n$. Each element $a_{ij}$ $(i = 1, 2, \ldots, m; j = 1, 2, \ldots, n)$ of the matrix $[a]$ represents a scalar. For our purposes, the scalars will be regarded as real numbers. The position of the element $a_{ij}$ in the matrix $[a]$ is in the $i$th row and $j$th column, so that $i$ is referred to as the row index and $j$ as the column index.

In the special case in which $m = n$, matrix $[a]$ reduces to a *square matrix of order n*. The elements $a_{ii}$ in a square matrix $[a]$ are called the *main diagonal elements of* $[a]$. The remaining elements are referred to as the *off-diagonal elements of* $[a]$. If all the off-diagonal elements of $[a]$ are zero, then $[a]$ is said to be a *diagonal matrix*. If $[a]$ is a diagonal matrix and all its diagonal elements are equal to unity, $a_{ii} = 1$, then the matrix is called the *unit matrix*, or *identity matrix*, and denoted by $[1]$. Introducing the *Kronecker delta* symbol $\delta_{ij}$, defined as being equal to unity if $i = j$ and equal to zero if $i \neq j$, $[a]$ is diagonal if it can be written in the form $[a_{ij}\delta_{ij}]$. Similarly, the identity matrix can be written in terms of the Kronecker delta as $[\delta_{ij}]$.

A matrix with all its rows and columns interchanged is known as the *transpose* of $[a]$ and denoted by $[a]^T$, so that

$$[a]^T = \begin{bmatrix} a_{11} & a_{21} & \cdots & a_{m1} \\ a_{12} & a_{22} & \cdots & a_{m2} \\ \cdots\cdots\cdots\cdots\cdots\cdots \\ a_{1n} & a_{2n} & \cdots & a_{mn} \end{bmatrix} \tag{C.2}$$

Clearly, if $[a]$ is an $m \times n$ matrix, then $[a]^T$ is an $n \times m$ matrix.

When all the elements of a matrix $[a]$ are such that $a_{ij} = a_{ji}$, with the implication that the matrix is equal to its transpose, $[a] = [a]^T$, the matrix $[a]$ is said to be *symmetric*. When the elements of $[a]$ are such that $a_{ij} = -a_{ji}$ for $i \neq j$ and $a_{ii} = 0$, the matrix is said to be *skew symmetric*. Hence, $[a]$ is skew symmetric if $[a] = -[a]^T$. Clearly, symmetric and skew symmetric matrices must be square.

A matrix consisting of one column and $n$ rows is called a *column matrix* and denoted by

$$\{x\} = \begin{Bmatrix} x_1 \\ x_2 \\ \vdots \\ x_n \end{Bmatrix} \tag{C.3}$$

The transpose of the column matrix $\{x\}$ is the *row matrix* $\{x\}^T$. They are also known as a *column vector* and a *row vector*, respectively.

A matrix with all its elements equal to zero is called a *null matrix* and denoted by $[0]$, $\{0\}$, or $\{0\}^T$, depending on whether it is a rectangular, a column, or a row matrix, respectively.

## b Matrix Algebra

Having defined various types of matrices, we are now in a position to present some basic matrix operations.

Two matrices $[a]$ and $[b]$ are said to be equal if and only if they have the same number of rows and columns, and $a_{ij} = b_{ij}$ for all pairs of subscripts $i$ and $j$. Hence, considering two $m \times n$ matrices, the statement

$$[a] = [b] \tag{C.4}$$

implies that

$$a_{ij} = b_{ij} \qquad i = 1, 2, \ldots, m \qquad j = 1, 2, \ldots, n \tag{C.5}$$

*Addition and subtraction* of matrices can be performed if and only if the matrices have the same number of rows and columns. If $[a]$, $[b]$, and $[c]$ are three $m \times n$ matrices, then the statement

$$[c] = [a] \pm [b] \tag{C.6}$$

implies that, for every pair of subscripts $i$ and $j$,

$$c_{ij} = a_{ij} \pm b_{ij} \qquad i = 1, 2, \ldots, m \qquad j = 1, 2, \ldots, n \tag{C.7}$$

Matrix addition, or subtraction, is *commutative* and *associative*, namely,

$$[a] + [b] = [b] + [a] \tag{C.8}$$

and

$$([a] + [b]) + [c] = [a] + ([b] + [c]) \tag{C.9}$$

The *product of a matrix and a scalar* implies that every element of the matrix in question is multiplied by the same scalar. Hence, if $[a]$ is any arbitrary $m \times n$ matrix and $s$ an arbitrary scalar, then the statement

$$[c] = s[a] \tag{C.10}$$

implies that, for every pair of subscripts $i$ and $j$,

$$c_{ij} = sa_{ij} \qquad i = 1, 2, \ldots, m \qquad j = 1, 2, \ldots, n \tag{C.11}$$

The *product of two matrices* is generally *not a commutative process*. Hence, the relative position of the matrices is important, and indeed it must be specified. For example, the product $[a][b]$ can be described by the statement that $[a]$ is postmultiplied by $[b]$, or that $[b]$ is premultiplied by $[a]$. It is also customary to describe the product by the statement that $[a]$ is multiplied by $[b]$ on the right, or that $[b]$ is multiplied by $[a]$ on the left. For a product of two matrices to be possible the number of columns of the first must be equal to the number of rows of the second matrix. If $[a]$ is an $m \times n$ matrix and $[b]$ an $n \times p$ matrix, then the product of the two matrices is

$$[c] = [a][b] \tag{C.12}$$

where $[c]$ is an $m \times p$ matrix whose elements are given by

$$c_{ij} = a_{i1}b_{1j} + a_{i2}b_{2j} + \cdots + a_{in}b_{nj} = \sum_{k=1}^{n} a_{ik}b_{kj} \tag{C.13}$$

in which $k$ is a dummy index. We note that the element $c_{ij}$ is obtained by multiplying the elements in the $i$th row of $[a]$ by the corresponding elements in the $j$th column of $[b]$ and summing the products.

As an illustration, let us evaluate the following matrix product:

$$
\begin{bmatrix} 5 & 2 & 4 \\ 4 & -1 & 1 \\ 1 & 3 & -2 \end{bmatrix} \begin{bmatrix} 3 & 2 \\ 1 & 7 \\ -5 & 4 \end{bmatrix}
$$

$$
= \begin{bmatrix} 5 \times 3 + 2 \times 1 + 4(-5) & 5 \times 2 + 2 \times 7 + 4 \times 4 \\ 4 \times 3 + (-1) \times 1 + 1 \times (-5) & 4 \times 2 + (-1) \times 7 + 1 \times 4 \\ 1 \times 3 + 3 \times 1 + (-2) \times (-5) & 1 \times 2 + 3 \times 7 + (-2) \times 4 \end{bmatrix}
$$

$$
= \begin{bmatrix} -3 & 40 \\ 6 & 5 \\ 16 & 15 \end{bmatrix}
$$

In the above example, it is clear that the product is not commutative because the number of columns of the second matrix is 2, whereas the number of rows of the matrix is 3. Hence, when the position of the matrices is reversed the matrix product cannot be defined.

As an illustration of the case when both matrix products can be defined and the process is still not commutative, we consider the simple example

$$
\begin{bmatrix} 3 & 2 \\ 1 & -5 \end{bmatrix} \begin{bmatrix} 5 & 7 \\ 9 & 3 \end{bmatrix} = \begin{bmatrix} 3 \times 5 + 2 \times 9 & 3 \times 7 + 2 \times 3 \\ 1 \times 5 + (-5) \times 9 & 1 \times 7 + (-5) \times 3 \end{bmatrix}
$$

$$
= \begin{bmatrix} 33 & 27 \\ -40 & -8 \end{bmatrix}
$$

$$
\begin{bmatrix} 5 & 7 \\ 9 & 3 \end{bmatrix} \begin{bmatrix} 3 & 2 \\ 1 & -5 \end{bmatrix} = \begin{bmatrix} 5 \times 3 + 7 \times 1 & 5 \times 2 + 7 \times (-5) \\ 9 \times 3 + 3 \times 1 & 9 \times 2 + 3 \times (-5) \end{bmatrix}
$$

$$
= \begin{bmatrix} 22 & -25 \\ 30 & 3 \end{bmatrix}
$$

and it is clear that in general

$$
[a][b] \neq [b][a] \tag{C.14}
$$

Although there may be cases when a particular matrix product is commutative, these are exceptions and not the rule. One notable exception is the case in which one of the matrices in the product is the unit matrix, because it is easy to verify that

$$
[a][1] = [1][a] = [a] \tag{C.15}
$$

where $[a]$ must clearly be a square matrix of the same order as $[1]$.

The matrix product satisfies *associative laws*. Indeed, considering the $m \times n$ matrix $[a]$, the $n \times p$ matrix $[b]$, and the $p \times q$ matrix $[c]$, it can be shown that

$$
[d] = ([a][b])[c] = [a]([b][c]) \tag{C.16}
$$

where $[d]$ is an $m \times q$ matrix whose elements are given by

$$d_{ij} = \sum_{l=1}^{p} \sum_{k=1}^{n} a_{ik} b_{kl} c_{lj} = \sum_{k=1}^{n} \sum_{l=1}^{p} a_{ik} b_{kl} c_{lj} \qquad (C.17)$$

The matrix product satisfies *distributive laws*. If $[a]$ and $[b]$ are $m \times n$ matrices, $[c]$ is a $p \times m$ matrix, and $[d]$ is an $n \times q$ matrix, then it is easy to show that

$$[c]([a] + [b]) = [c][a] + [c][b] \qquad (C.18)$$

$$([a] + [b])[d] = [a][d] + [b][d] \qquad (C.19)$$

The matrix product

$$[a][b] = [0] \qquad (C.20)$$

*does not imply* that either $[a]$ or $[b]$, or both $[a]$ and $[b]$, are null matrices. The above statement can be easily verified by considering the example

$$\begin{bmatrix} 1 & 1 \\ 1 & 1 \end{bmatrix} \begin{bmatrix} 1 & -1 \\ -1 & 1 \end{bmatrix} = \begin{bmatrix} 0 & 0 \\ 0 & 0 \end{bmatrix} \qquad (C.21)$$

From the above discussion, we conclude that matrix algebra differs from ordinary algebra on two major counts: (1) matrix products are not commutative and (2) the fact that the product of two matrices is equal to a null matrix cannot be construed to mean that either multiplicand (or both) is a null matrix. Both these rules hold in ordinary algebra.

## c  Determinant of a Square Matrix

The determinant of the square matrix $[a]$, denoted by det $[a]$ or by $|a|$, is defined as

$$\det [a] = |a| = \begin{vmatrix} a_{11} & a_{12} & \cdots & a_{1n} \\ a_{21} & a_{22} & \cdots & a_{2n} \\ \multicolumn{4}{c}{\dotfill} \\ a_{n1} & a_{n2} & \cdots & a_{nn} \end{vmatrix} \qquad (C.22)$$

where $|a|$ is said to be of *order n*. Unlike the matrix $[a]$, representing a given array of numbers, the determinant $|a|$ represents a number with a unique value that can be evaluated by following certain rules for the expansion of the determinant. Although determinants have very interesting properties, we shall not study them in detail but confine ourselves to certain pertinent aspects.

We denote by $|M_{rs}|$ the *minor determinant* corresponding to the elements $a_{rs}$, where $|M_{rs}|$ is obtained by taking the determinant of $[a]$ with the $r$th row and $s$th column struck out. Hence, $|M_{rs}|$ is of the order $n - 1$. The signed minor determinant corresponding to the element $a_{rs}$ is called the *cofactor* of $a_{rs}$ and is given by

$$|A_{rs}| = (-1)^{r+s} |M_{rs}| \qquad (C.23)$$

With this definition in mind, the value of the determinant can be obtained by expanding the determinant in terms of cofactors by the $r$th row as follows:

$$|a| = \sum_{s=1}^{n} a_{rs}|A_{rs}| \tag{C.24}$$

or by the $s$th column in the form

$$|a| = \sum_{r=1}^{n} a_{rs}|A_{rs}| \tag{C.25}$$

where the value of $|a|$ is the same regardless of whether the determinant is expanded by a row or a column, any row or column. The expansions by cofactors are called Laplace expansions. The cofactors $|A_{rs}|$ are determinants of order $n - 1$, and if $n > 2$ they can be further expanded in terms of their own cofactors. The procedure can be continued until the minor determinants are of order 2, in which case their cofactors are simply scalars. As an illustration, we calculate the value of a determinant of order 3 by expanding by the first row, as follows:

$$\begin{vmatrix} a_{11} & a_{12} & a_{13} \\ a_{21} & a_{22} & a_{23} \\ a_{31} & a_{32} & a_{33} \end{vmatrix} = a_{11}|A_{11}| + a_{12}|A_{12}| + a_{13}|A_{13}|$$

$$= a_{11} \begin{vmatrix} a_{22} & a_{23} \\ a_{32} & a_{33} \end{vmatrix} - a_{12} \begin{vmatrix} a_{21} & a_{23} \\ a_{31} & a_{33} \end{vmatrix} + a_{13} \begin{vmatrix} a_{21} & a_{22} \\ a_{31} & a_{32} \end{vmatrix}$$

$$= a_{11}(a_{22}a_{33} - a_{23}a_{32}) - a_{12}(a_{21}a_{33} - a_{23}a_{31})$$

$$+ a_{13}(a_{21}a_{32} - a_{22}a_{31}) \tag{C.26}$$

From Eqs. (C.24) and (C.25) we conclude that

$$|a| = \det [a] = \det [a]^T \tag{C.27}$$

or the determinant of a matrix is equal to the determinant of the transposed matrix. It is easy to verify that the determinant of a diagonal matrix is equal to the product of the diagonal elements. In particular, the determinant of the identity matrix is equal to 1.

If the value of $\det [a]$ is equal to zero, then matrix $[a]$ is said to be *singular*; otherwise it is said to be *nonsingular*. Clearly, $\det [a] = 0$ if all the elements in one row or column are zero. It is easy to verify that the value of a determinant does not change if one row, or one column, is added to or subtracted from another. Hence, if a determinant possesses two identical rows, or two identical columns, its value is zero.

By definition, the *adjoint* $[A_{ji}]$ of the matrix $[a_{ij}]$ is the transposed matrix of the cofactors of $[a_{ij}]$, namely,

$$[A_{ji}] = [(-1)^{i+j}|M_{ij}|]^T \tag{C.28}$$

## d  Inverse of a Matrix

If $[a]$ and $[b]$ are $n \times m$ matrices such that

$$[a][b] = [b][a] = [1] \tag{C.29}$$

then $[b]$ is said to be the *inverse* of $[a]$ and is denoted by

$$[b] = [a]^{-1} \tag{C.30}$$

To obtain the inverse $[a]^{-1}$, provided the matrix $[a]$ is given, let us consider the product

$$[a_{ij}][A_{ji}] = \begin{bmatrix} a_{11} & a_{12} & \cdots & a_{1n} \\ a_{21} & a_{22} & \cdots & a_{2n} \\ \hdotsfor{4} \\ a_{n1} & a_{n2} & \cdots & a_{nn} \end{bmatrix}$$

$$\times \begin{bmatrix} |M_{11}| & -|M_{21}| & \cdots & (-1)^{1+n}|M_{n1}| \\ -|M_{12}| & |M_{22}| & \cdots & (-1)^{2+n}|M_{n2}| \\ \hdotsfor{4} \\ (-1)^{1+n}|M_{\cdot n}| & (-1)^{2+n}|M_{2n}| & \cdots & |M_{nn}| \end{bmatrix}$$

$$= \left[ \sum_{j=1}^{n} (-1)^{i+j} a_{kj} |M_{ij}| \right] \tag{C.31}$$

But a typical element of the matrix on the right side of (C.31) has the value

$$\sum_{j=1}^{n} (-1)^{i+j} a_{kj} |M_{ij}| = \begin{vmatrix} a_{11} & a_{12} & \cdots & a_{1n} \\ a_{21} & a_{22} & \cdots & a_{2n} \\ \hdotsfor{4} \\ a_{n1} & a_{n2} & \cdots & a_{nn} \end{vmatrix} = |a| \quad \text{if } i = k \tag{C.32}$$

On the other hand, if $i \neq k$ the determinant possesses two identical rows. This is because the determinant corresponding to $i \neq k$ is obtained from the matrix $[a]$ by replacing the $i$th row by the $k$th row and keeping the $k$th row intact. Hence, if $i \neq k$ the value of the element is zero.

Considering the above, matrix (C.31) can be written in the form

$$[a_{ij}][A_{ji}] = [a][A_{ji}] = |a|[1] \tag{C.33}$$

Premultiplying Eq. (C.33) throughout by $[a]^{-1}$ and dividing the result by $|a|$, we obtain

$$|a|^{-1} = \frac{[A_{ji}]}{|a|} \tag{C.34}$$

so that the inverse of a matrix $[a]$ is obtained by dividing its adjoint matrix $[A_{ji}]$ by its determinant $|a|$.

If det $[a]$ is equal to zero, then the elements of $[a]^{-1}$ approach infinity (or are indeterminate at best), in which case the inverse $[a]^{-1}$ is said *not to exist*, and the matrix $[a]$ is said to be *singular*. Hence, for the inverse of a matrix to exist its determinant must be different from zero, which is equivalent to the statement that the matrix must be *nonsingular*.

As the order of the matrix $[a]$ increases, formula (C.34) for the calculation of $[a]^{-1}$ ceases to be feasible, and other methods must be used. We shall present later a more efficient method of obtaining the inverse of a matrix, namely, the method based on Gaussian elimination in conjunction with back substitution.

### e Transpose, Inverse and Determinant of a Product of Matrices

If $[a]$ is an $m \times n$ matrix and $[b]$ an $n \times p$ matrix, then, according to Eq. (C.13), $[c] = [a][b]$ is an $m \times p$ matrix with its elements given by

$$c_{ij} = \sum_{k=1}^{n} a_{ik}b_{kj} \qquad (C.35)$$

Next consider the product $[b]^T[a]^T$. Because to any element $a_{ik}$ in $[a]$ corresponds the element $a_{ki}$ in $[a]^T$, and to any element $b_{kj}$ in $[b]$ corresponds the element $b_{jk}$ in $[b]^T$, we have

$$\sum_{k=1}^{n} b_{jk}a_{ki} = c_{ji} \qquad (C.36)$$

from which we conclude that

$$[c]^T = [b]^T[a]^T \qquad (C.37)$$

or the *transpose of a product of matrices is equal to the product of the transposed matrices in reversed order*. This statement can be generalized to a product of several matrices. Hence, if

$$[c] = [a]_1[a]_2 \cdots [a]_{s-1}[a]_s \qquad (C.38)$$

then

$$[c]^T = [a]_s^T[a]_{s-1}^T \cdots [a]_2^T[a]_1^T \qquad (C.39)$$

Let us consider again the product

$$[c] = [a][b] \qquad (C.40)$$

but this time $[a]$ and $[b]$ are square matrices of order $n$. Then, premultiplying Eq. (C.40) by $[b]^{-1}[a]^{-1}$, and postmultiplying the result by $[c]^{-1}$, we obtain simply

$$[c]^{-1} = [b]^{-1}[a]^{-1} \qquad (C.41)$$

or the *inverse of a product of matrices is equal to the product of the inverse matrices in reversed order*. Equation (C.41) can be generalized by considering the product (C.38) in which all matrices $[a]_i$ $(i = 1, 2, ..., s)$ are square matrices of order $n$. Following the same procedure as that used to obtain (C.41), it is easy to show that

$$[c]^{-1} = [a]_s^{-1}[a]_{s-1}^{-1} \cdots [a]_2^{-1}[a]_1^{-1} \qquad (C.42)$$

We state here without proof[†] that *the determinant of a product of matrices is equal to the product of the determinants of the matrices in question.* Hence, considering the product of matrices (C.38) in which $[a]_i$ $(i = 1, 2, ..., n)$ are all square matrices, we have

$$\det [c] = \det [a]_1 \det [a]_2 \cdots \det [a]_s \tag{C.43}$$

In view of Eqs. (C.29), (C.30) and (C.43), we conclude that *the value of* det $[a]^{-1}$ *is equal to the reciprocal of the value of* det $[a]$.

## f  Partitioned Matrices

At times it proves convenient to partition a matrix into submatrices and regard the submatrices as the elements of the matrix. As an example, a $3 \times 4$ matrix $[a]$ can be partitioned as follows:

$$[a] = \begin{bmatrix} a_{11} & a_{12} & a_{13} & a_{14} \\ a_{21} & a_{22} & a_{23} & a_{24} \\ a_{31} & a_{32} & a_{33} & a_{34} \end{bmatrix} = \begin{bmatrix} [A_{11}] & [A_{12}] \\ [A_{21}] & [A_{22}] \end{bmatrix} \tag{C.44}$$

where

$$[A_{11}] = \begin{bmatrix} a_{11} & a_{12} \\ a_{21} & a_{22} \end{bmatrix} \quad [A_{12}] = \begin{bmatrix} a_{13} & a_{14} \\ a_{23} & a_{24} \end{bmatrix} \tag{C.45}$$

$$[A_{21}] = [a_{31} \quad a_{32}] \quad [A_{22}] = [a_{33} \quad a_{34}]$$

are submatrices of $[a]$. Then if a second $4 \times 4$ matrix $[b]$ is partitioned in the form

$$[b] = \begin{bmatrix} b_{11} & b_{12} & b_{13} & b_{14} \\ b_{21} & b_{22} & b_{23} & b_{24} \\ b_{31} & b_{32} & b_{33} & b_{34} \\ b_{41} & b_{42} & b_{43} & b_{44} \end{bmatrix} = \begin{bmatrix} [B_{11}] & [B_{12}] \\ [B_{21}] & [B_{22}] \end{bmatrix} \tag{C.46}$$

where

$$[B_{11}] = \begin{bmatrix} b_{11} & b_{12} \\ b_{21} & b_{22} \end{bmatrix} \quad [B_{12}] = \begin{bmatrix} b_{13} & b_{14} \\ b_{23} & b_{24} \end{bmatrix}$$

$$[B_{21}] = \begin{bmatrix} b_{31} & b_{32} \\ b_{41} & b_{42} \end{bmatrix} \quad [B_{22}] = \begin{bmatrix} b_{33} & b_{34} \\ b_{43} & b_{44} \end{bmatrix} \tag{C.47}$$

the matrix product $[a][b]$ can be treated as if the submatrices were ordinary elements, namely,

$$[a][b] = \begin{bmatrix} [A_{11}] & [A_{12}] \\ [A_{21}] & [A_{22}] \end{bmatrix} \begin{bmatrix} [B_{11}] & [B_{12}] \\ [B_{21}] & [B_{22}] \end{bmatrix}$$

$$= \begin{bmatrix} [A_{11}][B_{11}] + [A_{12}][B_{21}] & [A_{11}][B_{12}] + [A_{12}][B_{22}] \\ [A_{21}][B_{11}] + [A_{22}][B_{21}] & [A_{21}][B_{12}] + [A_{22}][B_{22}] \end{bmatrix} \tag{C.48}$$

† For the proof, see D. C. Murdoch, *Linear Algebra*, p. 134, John Wiley & Sons, Inc., New York, 1970.

Note that $[A_{11}][B_{11}] + [A_{12}][B_{21}]$ and $[A_{11}][B_{12}] + [A_{12}][B_{22}]$ are $2 \times 2$ matrices, whereas $[A_{21}][B_{11}] + [A_{22}][B_{21}]$ and $[A_{21}][B_{12}] + [A_{22}][B_{22}]$ are $1 \times 2$ matrices, so that the product $[a][b]$ is a $3 \times 4$ matrix, as is to be expected.

If the off-diagonal submatrices of a square matrix are null matrices, then the matrix is said to be *block-diagonal*. In this case the determinant of the matrix is equal to the product of the determinants of the submatrices on the main diagonal. Considering the matrix (C.46), with $[B_{12}]$ and $[B_{21}]$ being identically equal to zero, we have

$$\det [b] = \det [B_{11}] \det [B_{22}] \qquad (C.49)$$

## C.3 VECTOR SPACES

### a Definitions

Let $\{V\}$ be a set of objects called *vectors* and $R$ any *field* with its elements consisting of a set of scalars possessing certain algebraic properties. Then, if $\{V\}$ and $R$ are such that two operations, namely, *vector addition* and *scalar multiplication*, are defined for $\{V\}$ and $R$, the set of vectors together with the two operations are called a *vector space* $\{V\}$ *over a field* $R$. A vector space is also referred to as a *linear space*.

We have considerable interest in *vector spaces of n-tuples*, that is to say, the vectors in the space possess $n$ elements of a field $R$. For two such vectors

$$\{u\} = \begin{Bmatrix} u_1 \\ u_2 \\ \vdots \\ u_n \end{Bmatrix} \qquad \{v\} = \begin{Bmatrix} v_1 \\ v_2 \\ \vdots \\ v_n \end{Bmatrix} \qquad (C.50)$$

and a scalar $c$ in $R$, the addition and multiplication are defined as follows:

$$\{u\} + \{v\} = \begin{Bmatrix} u_1 + v_1 \\ u_2 + v_2 \\ \vdots \\ u_n + v_n \end{Bmatrix} \qquad c\{u\} = \begin{Bmatrix} cu_1 \\ cu_2 \\ \vdots \\ cu_n \end{Bmatrix} \qquad (C.51)$$

The vector space of $n$-tuples over $R$ is denoted by $\{V_n(R)\}$.

### b Linear Dependence

Consider a vector space $\{V\}$ over $R$ and let $\{u\}_1, \{u\}_2, \ldots, \{u\}_k$ and $c_1, c_2, \ldots, c_k$ be $k$ vectors in $\{V\}$ and $k$ scalars in $R$, respectively. Then, the vector $\{u\}$ given by

$$\{u\} = c_1\{u\}_1 + c_2\{u\}_2 + \cdots + c_k\{u\}_k \qquad (C.52)$$

is called a *linear combination* of $\{u\}_1, \{u\}_2, \ldots, \{u\}_k$ with *coefficients* $c_1, c_2, \ldots, c_k$. The totality of linear combinations of $\{u\}_1, \{u\}_2, \ldots, \{u\}_k$ obtained by letting

$c_1, c_2, \ldots, c_k$ vary over $R$ is a vector space. The space of all linear combinations of $\{u\}_1, \{u\}_2, \ldots, \{u\}_k$ is said to be *spanned* by $\{u\}_1, \{u\}_2, \ldots, \{u\}_k$. If the relation

$$c_1\{u\}_1 + c_2\{u\}_2 + \cdots + c_k\{u\}_k = \{0\} \tag{C.53}$$

can be satisfied only for the *trivial case*, namely, when all the coefficients $c_1, c_2, \ldots, c_k$ are identically zero, then the vectors $\{u\}_1, \{u\}_2, \ldots, \{u\}_k$ are said to be *linearly independent*. If at least one of the coefficients $c_1, c_2, \ldots, c_k$ is different from zero, the vectors $\{u\}_1, \{u\}_2, \ldots, \{u\}_k$ are said to be *linearly dependent*, implying that one vector is a linear combination of the remaining vectors.

## c Bases and Dimension of a Vector Space

A vector space $\{V\}$ over $R$ is said to be *finite dimensional* if there exists a finite set of vectors $\{u\}_1, \{u\}_2, \ldots, \{u\}_n$ which span $\{V\}$, with the implication that every vector in $\{V\}$ is a linear combination of $\{u\}_1, \{u\}_2, \ldots, \{u\}_n$. For example, the space $\{V_n(R)\}$ is a finite dimensional because it can be spanned by a set of $n$ vectors, where $n$ is a finite integer.

Let $\{V\}$ be a vector space over $R$. A set of vectors $\{u\}_1, \{u\}_2, \ldots \{u\}_n$ which span $\{V\}$, is called a *generating system* of $\{V\}$. If $\{u\}_1, \{u\}_2, \ldots, \{u\}_n$ are linearly independent and span $\{V\}$, then the generating system is called a *basis of* $\{V\}$. If $\{V\}$ is a finite-dimensional vector space, any two bases of $\{V\}$ contain the same number of vectors.

If $\{V\}$ is a finite-dimensional vector space over $R$, then the *dimension* of $\{V\}$ is defined as the number of vectors in any basis of $\{V\}$. This integer is denoted by dim $\{V\}$. In particular, the vector space $\{V_n(R)\}$ has dimension $n$, because a basis of $\{V_n(R)\}$ contains $n$ linearly independent vectors.

Let $\{u\}$ be an arbitrary $n$-dimensional vector with components $u_1, u_2, \ldots, u_n$, where $\{u\}$ is in $\{V_n(R)\}$, and introduce a set of $n$-dimensional vectors given by

$$\{e\}_1 = \left\{\begin{array}{c} 1 \\ 0 \\ \vdots \\ 0 \end{array}\right\} \quad \{e\}_2 = \left\{\begin{array}{c} 0 \\ 1 \\ \vdots \\ 0 \end{array}\right\} \quad \cdots \quad \{e\}_n = \left\{\begin{array}{c} 0 \\ 0 \\ \vdots \\ 1 \end{array}\right\} \tag{C.54}$$

Then the vector $\{u\}$ can be written in terms of the vectors $\{e\}_i$ $(i = 1, 2, \ldots, n)$ as follows:

$$\{u\} = u_1\{e\}_1 + u_2\{e\}_2 + \cdots + u_n\{e\}_n = \sum_{i=1}^{n} u_i\{e\}_i \tag{C.55}$$

Hence, $\{V_n(R)\}$ is spanned by the set of vectors $\{e\}_i$ $(i = 1, 2, \ldots, n)$. Clearly, the set $\{e\}_i$ is a generating system of $\{V_n(R)\}$ and is generally referred to as *the standard basis of* $\{V_n(R)\}$.

## C.4 LINEAR TRANSFORMATIONS

### a The Concept of a Linear Transformation

Let us consider a vector $\{x\}$ in $\{V_n(R)\}$ and write it in the form

$$\{x\} = x_1\{e\}_1 + x_2\{e\}_2 + \cdots + x_n\{e\}_n = \sum_{i=1}^{n} x_i\{e\}_i \qquad (C.56)$$

where $x_i$ are scalars belonging to $R$ and $\{e\}_i$ are the standard unit vectors ($i = 1, 2, \ldots, n$). The scalars $x_i$ are called the *coordinates* of the vector $\{x\}$ with respect to the basis $\{e\}_1, \{e\}_2, \ldots, \{e\}_n$. Equation (C.56) is entirely analogous to the equation

$$\mathbf{x} = x_1\mathbf{i} + x_2\mathbf{j} + x_3\mathbf{k} \qquad (C.57)$$

expressing a three-dimensional vector $\mathbf{x}$ in terms of the cartesian components $x_1$, $x_2$, $x_3$, where $\mathbf{i}, \mathbf{j}, \mathbf{k}$ are unit vectors along rectangular axes. Next, consider an $n \times n$ matrix $[a]$ and write

$$\{x'\} = [a]\{x\} \qquad (C.58)$$

The resulting vector $\{x'\}$ is another vector in $\{V_n(R)\}$, so that Eq. (C.58) can be regarded as representing a *linear transformation* on the vector space $(V_n(R)\}$ which maps the vector $\{x\}$ into a vector $\{x'\}$.

Equation (C.56) expresses the vector $\{x\}$ in terms of the standard bassis. In many applications, the interest lies in expressing $\{x\}$ in terms of any arbitrary basis $\{p\}_1, \{p\}_2, \ldots, \{p\}_n$ for $\{V_n(R)\}$ as follows:

$$\{x\} = y_1\{p\}_1 + y_2\{p\}_2 + \cdots + y_n\{p\}_n = \sum_{i=1}^{n} y_i\{p\}_i = [P]\{y\} \qquad (C.59)$$

where

$$[P] = [\{p\}_1 \quad \{p\}_2 \quad \cdots \quad \{p\}_n] \qquad (C.60)$$

is an $n \times n$ matrix of the basis vectors and

$$\{y\} = \begin{Bmatrix} y_1 \\ y_2 \\ \vdots \\ y_n \end{Bmatrix} \qquad (C.61)$$

is an $n$-dimensional vector whose components $y_i$ are the coordinates of $\{x\}$ with respect to the basis $\{p\}_1, \{p\}_2, \ldots, \{p\}_n$. By the definition of a basis, the vectors $\{p\}_1, \{p\}_2, \ldots, \{p\}_n$ are linearly independent, so that the matrix $[p]$ is nonsingular. Similarly, denoting by $y'_1, y'_2, \ldots, y'_n$ the coordinates of $\{x'\}$ with respect to the basis $\{p\}_1, \{p\}_2, \ldots, \{p\}_n$, we can write

$$\{x'\} = [p]\{y'\} \qquad (C.62)$$

where

$$\{y'\} = \left\{ \begin{array}{c} y'_1 \\ y'_2 \\ \vdots \\ y'_n \end{array} \right\} \tag{C.63}$$

Inserting Eqs. (C.59) and (C.62) into Eq. (C.58), we can write

$$[p]\{y'\} = [a][p]\{y\} \tag{C.64}$$

so that, premultiplying both sides of Eq. (C.64) by $[p]^{-1}$, we obtain

$$\{y'\} = [b]\{y\} \tag{C.65}$$

where

$$[b] = [p]^{-1}[a][p] \tag{C.66}$$

Note that $[p]^{-1}$ exists by virtue of the fact that $[p]$ is nonsingular. The matrix $[b]$ represents the same linear transformation as $[a]$, but in a different coordinate system. Two matrices $[a]$ and $[b]$ related by an equation of the type (C.66) are said to be *similar* and the relationship (C.66) itself is known as a *similarity transformation*.

Next, let us consider the characteristic determinant associated with $[b]$, recall Eq. (C.66) and write

$$\begin{aligned} \det([b] - \lambda[1]) &= \det([p]^{-1}[a][p] - \lambda[p]^{-1}[1][p]) \\ &= \det([p]^{-1}([a] - \lambda[1])[p]) \\ &= \det[p]^{-1} \det([a] - \lambda[1]) \det[p] \end{aligned} \tag{C.67}$$

But

$$\det[p]^{-1} \det[p] = \det([p]^{-1}[p]) = \det[1] = 1 \tag{C.68}$$

so that

$$\det([b] - \lambda[1]) = \det([a] - \lambda[1]) \tag{C.69}$$

Equation (C.69) states that matrices $[a]$ and $[b]$ possess the same characteristic determinant, and hence the same characteristic equation. It follows that *similar matrices possess the same eigenvalues*.

One similarity transformation of particular interest is the orthonormal transformation. A matrix $[p]$ is said to be *orthonormal* if it satisfies

$$[p]^T[p] = [1] \tag{C.70}$$

from which it follows that an orthonormal matrix also satisfies

$$[p]^{-1} = [p]^T \tag{C.71}$$

Introducing Eq. (C.71) into Eq. (C.66), we obtain

$$[b] = [p]^T[a][p] \tag{C.72}$$

Equation (C.72) represents an *orthonormal transformation* and it implies that *eigenvalues are preserved under orthonormal transformation.*

An orthonormal transformation of special interest in vibrations is one for which *the matrix* $[b]$ *is diagonal,* because *then* $[b]$ *is the matrix of eigenvalues of* $[a]$. A number of computational algorithms for the solution of the eigenvalue problem consist of the diagonalization of $[b]$ by means of orthonormal transformations.† Note that the diagonalization of $[a]$ is carried out by means of a series of iterative steps, and the matrix $[p]$ is a continuous product of orthonormal matrices.

## b The Inverse of a Matrix by Elementary Operations

Equation (C.58) can be regarded as a set of simultaneous algebraic equations of the form

$$\sum_{j=1}^{n} a_{ij}x_j = x_i' \qquad i = 1, 2, \ldots, n \tag{C.73}$$

where $a_{ij}$ $(i, j = 1, 2, \ldots, n)$ are constant coefficients and $x_j$ $(j = 1, 2, \ldots, n)$ are the unknowns. The solution of the equation can be obtained by premultiplying both sides of Eq. (C.58) by $[a]^{-1}$, so that

$$\{x\} = [a]^{-1}\{x'\} \tag{C.74}$$

where $[a]^{-1}$ can be computed according to Eq. (C.34). When the dimension of $[a]$ is relatively large, the use of Eq. (C.34) is not very efficient computationally. A more efficient approach is to compute $[a]^{-1}$ by solving the set of algebraic equations. To this end, let us rewrite Eqs. (C.58) and (C.74) as follows:

$$[a]\{x\} = [1]\{x'\} \tag{C.75}$$

and

$$[1]\{x\} = [a]^{-1}\{x'\} \tag{C.76}$$

where the transition from the form (C.75) to the form (C.76) is carried out by means of a series of linear transformations designed to solve the set of algebraic equations. This can be done by means of the *Gaussian elimination method*, in conjunction with *back substitution*, according to which we use *elementary operations* on the rows of both sides of (C.75). Elementary operations consist of addition or subtraction of one row from another and multiplication or division of one row by a constant. The purpose is to reduce the square matrix on the left side of (C.75) to the identity matrix. When this is accomplished, the matrix on the right side will no longer be the identity matrix but the inverse $[a]^{-1}$. As an illustration, let us consider the inverse of a $3 \times 3$ matrix as follows:

$$\begin{bmatrix} 2 & -1 & 0 \\ -1 & 3 & -2 \\ 0 & -2 & 2 \end{bmatrix} \begin{bmatrix} 1 & 0 & 0 \\ 0 & 1 & 0 \\ 0 & 0 & 1 \end{bmatrix}$$

† See, for example, L. Meirovitch, *Computational Methods in Structural Dynamics*, chap. 5, Sijthoff & Noordhoff International Publishers. The Netherlands, 1980.

Dividing the first row by 2 and adding the result to the second row yields

$$\begin{bmatrix} 1 & -0.5 & 0 \\ 0 & 2.5 & -2 \\ 0 & -2 & 2 \end{bmatrix} \begin{bmatrix} 0.5 & 0 & 0 \\ 0.5 & 1 & 0 \\ 0 & 0 & 1 \end{bmatrix}$$

Next, divide the second row by 2.5, divide the third row by 2, and add the resulting second row to the third, so that

$$\begin{bmatrix} 1 & -0.5 & 0 \\ 0 & 1 & -0.8 \\ 0 & 0 & 0.2 \end{bmatrix} \begin{bmatrix} 0.5 & 0 & 0 \\ 0.2 & 0.4 & 0 \\ 0.2 & 0.4 & 0.5 \end{bmatrix}$$

Finally, multiply the third row by 5, add 0.8 of the result to the second, and then add 0.5 of the resulting second row to the first to obtain

$$\begin{bmatrix} 1 & 0 & 0 \\ 0 & 1 & 0 \\ 0 & 0 & 1 \end{bmatrix} \begin{bmatrix} 1 & 1 & 1 \\ 1 & 2 & 2 \\ 1 & 2 & 2.5 \end{bmatrix}$$

Clearly the matrix on the right side is the desired inverse, namely, $[a]^{-1}$. This can be easily verified by using the formula provided earlier, Eq. (C.34).

# BIBLIOGRAPHY

Bendat, J. S., and A. G. Piersol, *Random Data: Analysis and Measurement Procedures*, Wiley-Interscience, New York, 1971.

Brigham, E. O., *The Fast Fourier Transform*, Prentice-Hall, Inc., Englewood Cliffs, New Jersey, 1974.

Carnahan, B., H. A. Luther, and J. O. Wilkes, *Applied Numerical Methods*, John Wiley & Sons, Inc., New York, 1969.

Chetayev, N. G., *The Stability of Motion*, Pergamon Press, New York, 1961.

Courant, R., and D. Hilbert, *Methods of Mathematical Physics*, vol. 1, Wiley-Interscience, New York, 1961.

Crandall, S. H., and W. D. Mark, *Random Vibration in Mechanical Systems*, Academic Press, Inc., New York, 1963.

Huebner, K. H., and E. A. Thornton, *The Finite Element Method for Engineers*, 2d ed., John Wiley & Sons, Inc., New York, 1982.

Lasalle, J., and S. Lefschetz, *Stability by Liapunov's Direct Method*, Academic Press, Inc., New York, 1961.

Meirovitch, L., *Analytical Methods in Vibrations*, The Macmillan Co., New York, 1967.

Meirovitch, L., *Methods of Analytical Dynamics*, McGraw-Hill Book Co., New York, 1970.

Meirovitch, L., *Computational Methods in Structural Dynamics*, Sijthoff & Noordhoff International Publishers, The Netherlands, 1980.

Murdoch, D. C., *Linear Algebra*, John Wiley & Sons, Inc., New York, 1970.

Ralston, A., *A First Course in Numerical Analysis*, McGraw-Hill Book Co., New York, 1965.

Thompson, R. C., and A. Yaqub, *Introduction to Linear Algebra*, Scott, Foresman & Company, Glenview, Ill., 1970.

Zienkiewicz, O. C., *The Finite Element Method*, 3d ed., McGraw-Hill Publishing Company, Ltd., London, 1977.

# INDEX